OLDUVAI GORGE

VOLUME 2

The skull of *Australopithecus* (*Zinjanthropus*) *boisei*, reconstructed by R. J. Clarke under the author's supervision. Photograph by R. Campbell and A. R. Hughes (*see also* pls. 41 and 42).

OLDUVAI GORGE

EDITED BY DR L. S. B. LEAKEY

VOLUME 2

THE CRANIUM AND MAXILLARY DENTITION OF
AUSTRALOPITHECUS (ZINJANTHROPUS) BOISEI

BY
P. V. TOBIAS

WITH A FOREWORD BY
SIR W. E. LE GROS CLARK, F.R.S.

CAMBRIDGE
AT THE UNIVERSITY PRESS
1967

CAMBRIDGE UNIVERSITY PRESS
Cambridge, New York, Melbourne, Madrid, Cape Town, Singapore, São Paulo, Delhi

Cambridge University Press
The Edinburgh Building, Cambridge CB2 8RU, UK

Published in the United States of America by Cambridge University Press, New York

www.cambridge.org
Information on this title: www.cambridge.org/9780521105194

© Cambridge University Press 1967

This publication is in copyright. Subject to statutory exception
and to the provisions of relevant collective licensing agreements,
no reproduction of any part may take place without the written
permission of Cambridge University Press.

First published 1967
This digitally printed version (with additions) 2009

A catalogue record for this publication is available from the British Library

Library of Congress Catalogue Card Number: 66–21073

ISBN 978-0-521-06901-4 hardback
ISBN 978-0-521-10519-4 paperback

CONTENTS

List of text-figures	page	vii
List of plates		ix
List of tables		xi
Introductory Note		xiii
Foreword by Sir W. E. Le Gros Clark, F.R.S.		xv
Editor's Note		xvii
Acknowledgements		xviii

I	Introduction	1
	Anatomical terminology	3
	Classificatory nomenclature	4
	Plan of the study	5
II	Preservation and reconstruction of the cranium	6
III	The cranial vault	9
	A The curvature and components of the vault	9
	B The supra-orbital height index	16
	C The sagittal and nuchal crests	19
IV	The basis cranii externa	26
	A The occipital bone	26
	B The temporal bone	28
	C The sphenoid bone and related structures	42
V	Certain critical angles and indices of the cranium	43
	A The planum nuchale: tilt and height	43
	B The foramen magnum: position and plane	45
	C The porion position indices	48
	D The position of the occipital condyles and the poise of the head	49
VI	The interior of the calvaria	53
	A The endocranial surface of the frontal bone	53
	B The endocranial surface of the parietal bones	54
	C The basis cranii interna	54
	D The venous sinuses of the dura mater	63
VII	The thickness of the cranial bones	72
	A Robusticity owing to pneumatisation	72
	B Robusticity owing to ectocranial superstructures	page 73
	C The thinness of the parietal bones	73
VIII	The endocranial cast of *Zinjanthropus*	77
	A The cranial capacity of *Zinjanthropus*	77
	B The cranial capacities of the australopithecines	78
	C Australopithecine capacities compared with those of other hominoids	80
	D Cranial capacity in relation to body size	86
	E Morphological features of the endocranial cast	87
	F The pattern of meningeal vascular markings	93
IX	Metrical characters of the calvaria as a whole	95
	A Cranial length and the toro-occipital index	95
	B The position of euryon and the maximum cranial breadth	99
	C The postero-anterior tapering of the cranial vault	100
	D The height of the cranial vault	101
	E Some metrical features of the base of the calvaria	103
X	The structure of the face	104
	A The supra-orbital torus	104
	B The orbits and the interorbital area	106
	C The nose	109
	D The maxillary and zygomatic bones	113
	E Facial measurements and indices and calvariofacial indices	123
XI	The pneumatisation of the *Zinjanthropus* cranium	126
	A The maxillary sinus	126
	B The frontal sinus	127
	C Pneumatisation of the naso-orbital region	128
	D The sphenoidal sinus	129
	E Pneumatisation of the temporal bone	130
	F Developmental aspects of pneumatisation	130

CONTENTS

XII	The dental arcade and the palate		page 132
	A	The shape of the dental arcade, alveolar process and palate	132
	B	The front of the alveolar process	137
	C	The arrangement of teeth in the arcade: evidence of dental crowding	138
XIII	The pattern of dental attrition and occlusion, with comments on enamel hypoplasia		139
	A	Attrition of individual teeth	139
	B	The pattern of attrition and occlusion	140
	C	The state of the enamel	141
XIV	The size of individual teeth, absolute and relative		144
	A	Notes on methodology and terminology	144
	B	Dimensions of the incisors	147
	C	Dimensions of the canines	152
	D	Canine–premolar ratios	155
	E	Ratios of premolar dimensions to premolar–molar chords	157
	F	Dimensions of the premolars	158
	G	Dimensions of the molars	162
	H	Ratios of molar dimensions to premolar–molar chords	165
	I	Relative molar size	166
XV	The size of the dentition as a whole		170
	A	Flower's Dental Index	170
	B	'Tooth material'	170
XVI	The crown shape index of the teeth		172
XVII	The morphology of the teeth		175
	A	The maxillary incisors	175
	B	The maxillary canines	177
	C	The maxillary premolars	179
	D	The maxillary molars	182

XVIII	Summary of cranial and dental features of *Zinjanthropus*		page 193
	The cranial vault		193
	The basis cranii externa		195
	Certain critical angles and indices of the cranium		198
	The interior of the calvaria		200
	The thickness of the cranial bones		203
	The endocranial cast of *Zinjanthropus*		204
	Metrical characters of the calvaria as a whole		206
	The structure of the face		207
	The pneumatisation of the cranium		212
	The shape of the dental arcade and the palate		213
	The pattern of attrition and occlusion		213
	The state of the enamel		214
	The size of individual teeth		214
	Shape indices of the teeth		216
	Morphology of the teeth		217
XIX	The taxonomic status of *Zinjanthropus* and of the australopithecines in general		219
	A	General considerations	219
	B	*Zinjanthropus* a hominid and an australopithecine	223
	C	The generic and specific status of *Zinjanthropus*	224
	D	Formal definitions of *Australopithecus* and its species	233
XX	The cultural and phylogenetic status of *Australopithecus boisei* and of the australopithecines in general		236
	A	Cultural status	236
	B	The place of *Australopithecus boisei* and the other australopithecines in hominid phylogeny	240
References			245
Index of persons			253
Index of subjects			255

LIST OF TEXT-FIGURES

1 Craniograms of norma lateralis of *Paranthropus*, *Zinjanthropus* and *Australopithecus* — *page* 9

2 Craniograms of norma verticalis of *Paranthropus*, *Zinjanthropus* and *Australopithecus* — 10

3 Lateral craniograms of gorilla and *Zinjanthropus*, to show the landmarks from which are derived three cranial indices, CD/CE, AG/AB and FB/AB — 17

4 Diagrammatic representation of the values of three cranial indices in a variety of hominoid crania — 18

5 Dioptographic tracings of norma occipitalis of crania of chimpanzee, *Zinjanthropus* and gorilla — 20

6 Variations in the lateral margin of the tympanic plate in *Zinjanthropus*, *Australopithecus* (MLD 37/38), *Paranthropus* (SK 46, SK 48, SK 52 and SK 848) and *Homo erectus pekinensis* (107, 109 and 101) — 31

7 Diagrams to show the relationships of structures on the basis cranii, in *Zinjanthropus* and a variety of other hominoids — 39

8 Median sagittal craniograms of *Australopithecus* (Sts 5), *Zinjanthropus* and a female gorilla, showing the basicranial axis — 46

9 The interior of the occipital bone, to show the pattern of venous sinus grooves — 64

10 Scheme of venous sinus grooves in the posterior cranial fossa of *Zinjanthropus*, *Paranthropus* (SK 859) and three modern Bantu crania — 69

11 Cranial capacities of hominoids — 82

12 Range of estimates of 'extra' cortical neurones of hominoids — 88

13 Dioptographic tracings of norma ventralis (basalis) of endocranial casts of adult chimpanzee, adult gorilla and *Zinjanthropus* — 89

14 Dioptographic tracings of norma lateralis of endocranial casts of adult chimpanzee, adult gorilla and *Zinjanthropus* — 89

15 Dioptographic tracings of norma ventralis (basalis) of endocranial casts of juvenile chimpanzee, juvenile gorilla and *Zinjanthropus* — 90

16 Outline of norma dorsalis of endocranial cast of *Zinjanthropus* superimposed on that of *Australopithecus*—Sts II — *page* 90

17 Outline of norma dorsalis of endocranial cast of *Zinjanthropus* superimposed on that of *Australopithecus*—Sts 5 — 90

18 Outline of norma dorsalis of endocranial cast of *Zinjanthropus* superimposed on that of *Australopithecus* juvenile from Taung — 91

19 Dioptographic tracings of norma lateralis of endocranial casts of *Zinjanthropus* superimposed respectively on those of Sts II, the Taung juvenile and Sts 5 — 91

20 Dioptographic tracing of norma lateralis of endocranial cast of *Zinjanthropus*, to show the pattern of meningeal vessels on the right side of the brain — 93

21 Craniograms of norma facialis of *Paranthropus*, *Zinjanthropus* and *Australopithecus* — 112

22 The shape of the dental arcade in *Paranthropus*, *Zinjanthropus* and *Australopithecus* — 137

23 Mesiodistal diameters of maxillary teeth of *Zinjanthropus* compared with the ranges in *Paranthropus* and *Australopithecus* — 151

24 Buccolingual diameters of maxillary teeth of *Zinjanthropus* compared with the ranges in *Paranthropus* and *Australopithecus* — 151

25 Modules of maxillary teeth of *Zinjanthropus* compared with the ranges in *Paranthropus* and *Australopithecus* — 154

26 Crown areas of maxillary teeth of *Zinjanthropus* compared with the ranges in *Paranthropus* and *Australopithecus* — 155

27 Mesiodistal diameters of maxillary teeth of *Zinjanthropus*, compared with means for *Paranthropus*, *Australopithecus*, *Homo erectus pekinensis*, and the absolute values for *H. e. erectus* IV — 160

28 Buccolingual diameters of maxillary teeth of *Zinjanthropus*, compared with means for *Paranthropus*, *Australopithecus*, *Homo erectus pekinensis*, and the absolute values for *H. e. erectus* IV — 160

29 Mesiodistal diameters of maxillary teeth of *Zinjanthropus* compared with means for male and female Pongidae — 161

LIST OF TEXT-FIGURES

30 Buccolingual diameters of maxillary teeth of *Zinjanthropus* compared with means for male and female Pongidae — *page* 161

31 Modules of maxillary teeth of *Zinjanthropus*, compared with means for *Paranthropus*, *Australopithecus*, *Homo erectus pekinensis*, and with absolute values for *H. e. erectus* IV — 164

32 Crown areas of maxillary teeth of *Zinjanthropus*, compared with means for *Paranthropus*, *Australopithecus*, *Homo erectus pekinensis*, and with absolute values for *H. e. erectus* IV — 165

33 Modules of maxillary teeth of *Zinjanthropus* compared with mean modules in male and female Pongidae — 168

34 Crown areas of maxillary teeth of *Zinjanthropus*, compared with means for male and female Pongidae — 169

35 Crown shape indices of maxillary teeth of *Zinjanthropus* compared with the sample range of shape indices for *Paranthropus* — *page* 173

36 Crown shape indices of maxillary teeth of *Zinjanthropus* compared with the sample range of shape indices for *Australopithecus* — 174

37 Three earlier interpretations of the relationship between the Lower Pleistocene *Australopithecus africanus* and the Middle Pleistocene *A. robustus* — 241

38 Schema of Lower and Middle Pleistocene hominids, showing the position in space and time of the most important specimens discovered to date — 242

39 Provisional schema of hominid phylogeny from Upper Pliocene times to the Upper Pleistocene — 243

LIST OF PLATES

Frontispiece: *Zinjanthropus* with mandible (reconstructed by R. J. Clarke under the author's direction)

Between pp. 252 and 253

1. Norma facialis of the cranium of *Zinjanthropus*
2. Right norma lateralis of the cranium of *Zinjanthropus*
3. Left norma lateralis of the cranium of *Zinjanthropus*
4. Norma occipitalis of the cranium of *Zinjanthropus*
5. Norma verticalis of the cranium of *Zinjanthropus*
6. Norma basalis of the cranium of *Zinjanthropus*
7. The dental arcade and palate of *Zinjanthropus* compared with those of (A) two modern men (Bantu Negroids) and (B) a gorilla and chimpanzee
8. Norma basalis of the cranium of *Zinjanthropus* compared with that of the recently discovered cranium of the Makapansgat *Australopithecus* (MLD 37/38)
9. Norma occipitalis of the cranium of *Zinjanthropus* compared with that of MLD 37/38
10. The posterior part of the vault of the cranium as seen from above
11. The posterior part of the vault of the cranium as seen from the rear
12. The posterior part of the vault of the cranium as seen from (A) the left, (B) the right
13. The posterior part of the vault of the cranium seen from below, behind and to the right, to show the formation of the compound (temporal/nuchal) crest and the lateral divergence of the simple temporal and nuchal crests from it
14. The posterior part of the basis cranii externa (A) and interna (B)
15. The upper calvariofacial fragment, (A) from in front and (B) from above
16. Anterior view of the lower facial fragment, comprising most of the maxillae as well as the right zygomatic bone
17. The right maxilla seen from the right side
18. The maxilla of *Zinjanthropus*, (A) as seen from the left side, (B) as seen from behind, looking into the exposed antra
19. Left lateral view of the lower facial fragment: a straw has been passed through the bony canal for the inferior orbital nerve
20. Occlusal view of teeth and palate of *Zinjanthropus*
21. The parietal bones of *Zinjanthropus*, (A) from above and somewhat behind; (B) from above and to the left
22. The parietal bones seen from below and behind
23. The right temporal bone from the lateral (A) and medial (B) aspects
24. The left temporal bone seen (A) from above and (B) from below
25. Lateral view of the left temporal bone
26. The occipital bone seen from below and behind
27. The occipital bone from within
28. The endocranial cast of *Zinjanthropus* as seen in norma ventralis (basalis)
29. The dental arcade and palate of *Zinjanthropus*
30. The interior of the occipital bone of the juvenile *Paranthropus* from Swartkrans (SK 859)
31. The interior of the occipital bones of three Bantu crania with anomalous patterns of the venous sinuses
32. Labial, occlusal and lingual views of the front maxillary teeth
33. The tooth-row from lateral incisors to first molars
34. The maxillary canines and first premolars of *Zinjanthropus*
35. (A) Labial aspect of right canine and P^3; (B) labial aspect of right premolars and M^1; and (C) labial aspect of the left P^3 and P^4
36. The lingual aspect of the premolars and the first molar
37. Occlusal surfaces of the first and second molars
38. The lingual aspect of the molar teeth (M^2 and M^3 on right, and M^1, M^2 and M^3 on left)
39. Buccal views of the maxillary molars of *Zinjanthropus*
40. Maxillary third molars of *Zinjanthropus*, in occlusal and distal views
41. Norma facialis of reconstructed skull of *Zinjanthropus*
42. Norma lateralis dextralis of reconstructed skull of *Zinjanthropus*

LIST OF TABLES

1. Measurements and indices of the parietal bone in *Zinjanthropus* and other hominids — page 11
2. Ranges of measurements of the parietal bone in australopithecines (including *Zinjanthropus*), *Homo erectus* and modern man — 12
3. Ranges of indices of the parietal bone margins in australopithecines (including *Zinjanthropus*), *Homo erectus* and modern man — 13
4. Measurements and indices of the occipital bone in *Zinjanthropus* and other hominids — 13
5. The occipital index of *Zinjanthropus* and other hominids — 16
6. Differences between the supra-orbital height index of *Zinjanthropus* and other hominoids: (A) including the sagittal crest of *Zinjanthropus*; (B) excluding the sagittal crest of *Zinjanthropus* — 17
7. The petro-median angle in hominoid crania. (The angle between the axis of the petrous portion and the median sagittal plane, measured on the basis cranii externa) — 34
8. Dimensions and indices of the mandibular fossa in *Zinjanthropus* and other hominoids — 35
9. Differences between the nuchal area height index (AG/AB) of *Zinjanthropus* and other hominoids — 44
10. Differences between the condylar position index (CD/CE) of *Zinjanthropus* and other hominoids — 51
11. Dimensions and indices of the foramen magnum in *Zinjanthropus* and other hominoids — 61
12. Cranial capacity of *Zinjanthropus* and other australopithecines — 79
13. Ranges and means of cranial capacities of hominoids — 81
14. Cranial capacity of *H. erectus* — 81
15. The variability of hominoid cranial capacities — 85
16. Estimates of 'extra neurones' in hominoids — 87
17. Encephalic measurements and indices of *Zinjanthropus* and other australopithecines — 92
18. Measurements of the calvariae of *Zinjanthropus* and other hominids — 96
19. Toro-cristal length and index in hominids — 98
20. Indices of the calvariae of *Zinjanthropus* and other hominids — page 101
21. Facial measurements of *Zinjanthropus* and other hominids — 107
22. Indices of the face of *Zinjanthropus* and other hominoids — 107
23. Nasal measurements of *Zinjanthropus* and other hominoids — 110
24. Nasal indices of *Zinjanthropus* and other hominoids — 111
25. Calvariofacial indices of *Zinjanthropus* and other hominoids — 125
26. Palatal and arcadal dimensions and indices of *Zinjanthropus* — 134
27. Maxillo-alveolar index of *Zinjanthropus* and other hominoids — 135
28. Ranges of maxillo-alveolar indices in hominoid groups — 136
29. Crown dimensions and indices of individual teeth of *Zinjanthropus* — 144
30. Metrical characters of maxillary permanent teeth of *Australopithecus* (*A. africanus*) and *Paranthropus* (*A. robustus*) — 147
31. Metrical characters of maxillary permanent teeth of *H. erectus* — 148
32. Standard deviations of hominid teeth as computed by two different methods (S.D. 1—estimate by the method of deviations; S.D. 2—estimate from the sample range) — 149
33. Metrical characters of *Zinjanthropus* teeth as compared with those of *Paranthropus* (= *A. robustus*) — 150
34. Modules and crown areas of incisors and I^2/I^1 ratios in *Zinjanthropus* and other hominoids — 152
35. Crown dimensions, modules and crown areas of canines of *Zinjanthropus* and other hominoids — 154
36. Comparison of maxillary canine and premolar dimensions in Australopithecinae: (A) mesiodistal crown diameter, (B) buccolingual crown diameter, (C) module, (D) crown area — 156
37. Maxillary canine–premolar percentage ratios in *Zinjanthropus*, *Paranthropus* and *Australopithecus* — 157

LIST OF TABLES

#	Title	Page
38	Percentage ratios of premolar dimensions to P³–M³ chord	157
39	Percentage ratios of premolar dimensions to P⁴–M³ chord	158
40	Crown dimensions, modules and crown areas of premolars of *Zinjanthropus* and other hominoids	158
41	Crown dimensions, modules and crown areas of molars of *Zinjanthropus* and other hominoids	163
42	Percentage ratios of molar dimensions to P³–M³ chord	166
43	Percentage ratios of molar dimensions to P⁴–M³ chord	166
44	Maxillary 'tooth material' (sum of mesiodistal crown diameters of left and right I^1–M^1) of *Zinjanthropus* and other hominids	171
45	Shape indices (M.D./B.L.) of maxillary permanent teeth of *Zinjanthropus* and other hominoids	172
46	The frequency of Carabelli structures and protoconal cingulum in Australopithecinae	191
47	Coefficients of variation for buccolingual diameters of mandibular teeth	221
48	Coefficients of variation for M.D. and B.L. diameters of mandibular teeth	221
49	Mandibular canine–premolar ratios in *Paranthropus* and *Australopithecus*	227

Introductory Note to the 50th Anniversary of the Discovery of 'Zinjanthropus'

The Olduvai Gorge in the Republic of Tanzania came to the attention of the world shortly after my mother Mary discovered the 'Zinjanthropus boisei' skull on July 17th 1959. The field of African prehistory, and in particular the study of human evolution, has changed and developed dramatically over the past 50 years. I am particularly pleased that Cambridge University Press have decided to republish the 5 monographs that comprehensively cover the many scientific studies that have been undertaken on the Olduvai material collected by my parents, Louis and Mary, working with a number of colleagues. As the Golden Anniversary of the discovery approaches, it is timely to reflect on the importance of that find.

I was lucky to arrive at Olduvai two days after the discovery and I well recall the excitement of the occasion. My parents were operating on a very tight budget and the field season was short. Fortunately, on hand was world-renowned photographer Des Bartlett who, aided by his wife Jen, fully recorded on film the first few days of excavations and reassembly of bone fragments back in camp. As pieces were glued back together, and the shape of the skull and its morphology became clear, my parents showed uncharacteristic and unrestrained emotion! At the time, ages for fossils were wild guesses and radiometric dating had not been done anywhere in Africa. The best, guessed age for Zinj was a little more than 500,000 years. Some months later, a real Potassium/Argon date was obtained by Jack Evenden and Garniss Curtis, and the 1,750,000 age was announced. This ignited huge excitement worldwide and for the first time my father was able to raise financial support for extended field work at Olduvai. Everything changed. The unqualified enthusiasm and support of the National Geographic Society from 1960 onwards had a major impact on the later work at Olduvai, and indeed on the growing international interest of Africa as the cradle of humanity.

Since those first exciting years at Olduvai, the investigation of human origins has gone forward and extended to many other sites in Africa. The age of hominins has been taken back to beyond five million years and the collected fossils and lithic records are now numerous. International multi-disciplinary teams are working in many parts of the world and, with the exception of a few fundamentalist 'flat earth' types, the acceptance of the fossil record of our past is widely accepted. Much of this has come about because of the initial Olduvai finds.

The pioneering work at Olduvai was the launch of this fantastic 50-year period when we as a species have come to realize and appreciate our common evolutionary past. Olduvai, conserved and protected by the Republic of Tanzania, remains as a landmark in the epic story of humanity, and these monographs are a wonderful testimony to that landmark.

Richard Leakey, FRS

FOREWORD

By SIR W. E. LE GROS CLARK, F.R.S.

The discovery by Dr and Mrs Leakey of an australopithecine skull in Bed I of the deposits of the Olduvai Gorge is of major importance for two reasons. First, the skull was found at a stratigraphical level that has been dated with reasonable assurance by the potassium-argon method to well over a million years, indeed probably as much as one and three-quarters of a million. Second, the skull, though fragmented, was found to be practically complete except for the lower jaw; it was possible to piece together the broken fragments with fair accuracy and to demonstrate that it was almost free of distortion and deformation in spite of its prolonged period of fossilisation. In fact, apart from one specimen discovered by the late Dr Robert Broom at Sterkfontein in South Africa in 1947, no australopithecine skull is yet known that approaches in completeness the Olduvai skull.

It is particularly fortunate that the responsibility for the detailed study of this skull should have been given to Professor P. V. Tobias, for he has had many years of experience in dealing with fossil hominid material as well as a wide acquaintance with the skeletal structures of the higher Primates and their degree of variability. I do not suppose that any such meticulous and exhaustive description of a fossil hominid skull as is to be found in this monograph has ever before been made, even if account is taken of Boule's description of the Chapelle-aux-Saints skull, or of Weidenreich's account of the crania of Chinese representatives of *Homo erectus*. Not only does Professor Tobias describe and define with great clarity the various anatomical features of all parts of the skull, including its endocranial characters, he makes numerous statistical comparisons based on his extensive studies of large collections of skulls of modern apes and modern man, as well as comparisons with a number of fossil human skulls of which he himself has examined the original specimens.

When the Olduvai skull was discovered, Dr Leakey at first thought it to be generically distinct from any of the known South African australopithecines, and he gave it the name *Zinjanthropus boisei*. The careful studies of Professor Tobias have now convincingly shown that it is an East African representative of the genus *Australopithecus* and that even the use of *Zinjanthropus* as a sub-generic term is no longer justifiable. Following the recognition that the australopithecines should properly be included in one genus only, *Australopithecus*, possibly comprising not more than one or two species, several of the generic and specific terms that were coined during the early days of the australopithecine discoveries in South Africa, and which were the source of some misunderstanding in the initial controversies about these fossils, may now be discarded. Professor Tobias discusses at some length the details of the dental morphology in the African australopithecines (so far as these are at present known) and sees no reason on the basis of their variability to make any generic distinctions. The variability in the size and cusp pattern of the teeth is remarkable, to such an extent that some of the specimens of the dentition, for example those of smaller dimensions found at Sterkfontein, are not easy to distinguish from the more megadont individuals of the fossil species *Homo erectus*. It seems, indeed, that the dental characters by themselves do not have a high taxonomic relevance for making such a distinction; the latter must depend rather on contrasts in cranial characters, brain size and limb structure. Thus even the generic distinction of 'Telanthropus' may be open to question. This type has more recently been taken by some authorities to be a

South African representative of *Homo erectus*, but to the present writer the dental evidence seems equivocal, and the very fragmentary remains may with greater probability be assigned to small-sized individuals of the australopithecine group.

The cranial capacity of the Olduvai skull is small (530 c.c.) and thus comes within the existing australopithecine sample range. But it perhaps needs to be emphasised that this sample is still very limited, so that we do not yet know what the general population range may have been. If it was anything like that of the modern gorilla it may be assumed that the upper limits of the population range would have reached well over 600 c.c., perhaps nearer 700 c.c. In spite of its small cranial capacity, it is remarkable that in detail after detail described by Professor Tobias the Olduvai skull displays hominid and not pongid characters. It is thus no longer possible to controvert the allocation of the australopithecines to the family Hominidae rather than the Pongidae. The evidence for such a classification is now so complete that we can afford to forget the controversies that were aroused by the earlier discoveries in South Africa. For example, there is now no question of a sagittal crest (when it is present) being continuous with a high nuchal crest as it consistently is in the large apes. On the contrary, the nuchal area of the occipital bone is very restricted, and the external occipital protuberance is actually situated below the level of the Frankfurt plane. Other distinctively hominid characters include a large pyramidal mastoid process of typical hominid form even in an immature individual, a parabolic dental arcade with no diastema, spatulate canine teeth wearing down flat from the tip only, the degree of flexion of the cranial base, the anterior position of the occipital condyles and the nearly horizontal plane of the foramen magnum, the detailed construction of the mandibular fossa of the temporal bone, the moderate subnasal prognathism, and so forth. Of course, these hominid features of the australopithecine skull were already known from the South African fossil material but, as much of the latter consisted of fragmentary specimens, it is highly satisfactory to find such an emphatic confirmation of them in the almost complete and undistorted Olduvai skull.

No doubt one of the most important implications of the Leakeys' discovery is the apparently long time gap between the estimated date of the Olduvai skull and that of the almost identically similar skulls of the robust variety of *Australopithecus* in South Africa, a gap of a million years or more. Clearly the genus *Australopithecus* was long-lived (though not more so than some fossil genera in other mammalian groups). With such a wide temporal and geographical dispersion it would be surprising if the East African populations of *Australopithecus* did not show some degree of morphological difference from the South African populations, though some may doubt whether such differences are adequate to postulate generic, or even specific, distinctions—it may even be that they warrant no more than sub-specific distinctions.

Remains of australopithecines have been reported from other parts of the world outside Africa, but such reports are based on very friable evidence. It may be, therefore, that *Australopithecus* evolved somewhere in Central Africa, and that in the course of many centuries members of the group trekked down to South Africa where they persisted unchanged into the Middle Pleistocene long after the genus *Homo* had arisen elsewhere. Since the discovery of the 'Zinjanthropus' skull was announced, further relics of fossil hominids have come to light in the deposits of the Olduvai Gorge as the result of the energetic field-work of Dr and Mrs Leakey. The interpretation of these later finds is still obscure because full reports on them have not yet been completed and published. The correct assessment of their real significance will doubtless depend on a recognition of the wide range of variability in the different groups of the genus *Australopithecus* already known, with due attention to the differential taxonomic relevance of various characters of the skull, dentition and limb structure. In this monograph, Professor Tobias has shown that he is well aware of the need to take such factors into account.

EDITOR'S NOTE

This is the second volume of the new series of publications dealing with the discoveries made at Olduvai Gorge since 1951. The title of the first volume included the dates '1951–1961'. These have been dropped from the present volume, and will also not occur in the remainder of the series, since, inevitably, data obtained since 1961 will need to be included. The next two volumes in this series will deal with the cultural sequence and excavations of the 'living floors' in Beds I and II, at Olduvai, by Mary D. Leakey, and the detailed geology of the deposits exposed in the Gorge by Richard L. Hay. It is not yet certain which of the two will be ready for publication first.

A fifth volume is in preparation by Phillip Tobias, the author of this volume on '*Zinjanthropus*'. In it he will describe, in detail, the fossil hominid remains from Olduvai that are attributed to *Homo habilis*, and to *Homo erectus*, as well as the new mandible of an australopithecine of '*Zinjanthropus*' type from Peninj, north of Olduvai. The fifth volume will also include chapters dealing with the postcranial hominid material from Olduvai by Dr John Napier, Dr Peter Davis and Dr Michael Day.

In the present volume, my friend and colleague, Phillip Tobias, has produced a truly magnificent piece of research—a major contribution to the study of the anatomy of the australopithecines, or 'near men' as I call them. This volume is without doubt the most detailed study of any one single fossil hominid skull that has ever been made, and the facts that have emerged from his studies are presented with admirable clarity. I need not enlarge upon this, since Professor Sir Wilfrid Le Gros Clark has done so in his Foreword.

Naturally, Tobias concludes his study of the facts about *Zinjanthropus* with an expression of his views as to the taxonomic status of the members of the sub-family Australopithecinae. He believes that the ancestral stock which gave rise to both *Homo habilis* and to *A. africanus* was an Australopithecine. I would prefer to refrain from suggesting that the common ancestor was closer in morphology to one or the other. We do not yet know. Similarly, it would seem better to leave open for the present the question of whether *Homo erectus* was a derivative of *Homo habilis*.

Finally, it remains only for me to congratulate Phillip Tobias upon a magnificent piece of anatomical study. This book will stand as a model for a very long time as to how a study of a fossil skull should be conducted.

L. S. B. LEAKEY

ACKNOWLEDGEMENTS

Various parts of this study have been generously assisted by the British Council, the Wenner-Gren Foundation for Anthropological Research, the South African Council for Scientific and Industrial Research, the Boise Fund, the National Geographic Society and the University of the Witwatersrand, Johannesburg.

I am grateful to both Professor R. A. Dart, who guided me into physical anthropology, and Dr J. T. Robinson, for allowing me freely to study specimens in their collections and for helpful discussions. Doctor Robinson also kindly assisted me with the loan of an unpublished median sagittal craniogram of *Australopithecus* (Sterkfontein 5). Professor Sir W. E. Le Gros Clark, F.R.S., kindly contributed the Foreword and made available a median sagittal craniogram of a female gorilla. In addition, I wish to thank the Keeper of Palaeontology and the Keeper of Zoology at the British Museum (Natural History), Dr K. P. Oakley, Dr J. C. Trevor, Dr D. R. Hughes, Mr D. Brothwell, Miss Rosemary Powers, Professor H. V. Vallois, Dr C. Arambourg, Professor S. Sergi, Dr D. Hooijer, Professor G. H. R. von Koenigswald, Dr V. FitzSimons, Professor L. H. Wells and the Uganda Gorilla Research Unit of the University of the Witwatersrand.

The extremely difficult and laborious task of casting each of the delicate and frequently undercut fragments of the cranium was successfully undertaken by Mr T. W. Kaufman; the casts have been skilfully painted to simulate the originals by Mrs C. R. Esson. The reconstruction of the cranium and the plaster and endocranial cast have been made by Mr A. R. Hughes in conjunction with the author, and it is a pleasure to acknowledge my indebtedness to him for freely imparting the fruits of his long experience at handling, preparing and reconstructing fossil material, as well as in the accurate preparation of dioptographic tracings and drawings. The X-rays were made by Miss J. Dreyer and Mrs D. Coetzee, through the co-operation of Professor J. Kaye, Head of the Department of Radiology at the Johannesburg General Hospital. Miss C. Orkin and Miss J. Soussi sacrificed their personal vacation arrangements to execute painstakingly many of the drawings, graphs, diagrams and dioptographic tracings, while Miss Soussi was largely responsible for the punctilious tabulation and statistical reduction of data, the tests for significance of differences, and for the careful checking and collation of the manuscript. Miss A. Wright helped with such tedious details as the cutting out and mounting of the photographs. Mr T. E. Badenhuizen assisted with the calculation of indices, as well as with the statistical reduction of comparative data in the literature. Miss C. Orkin and Mr D. Gillmer helped me check the proofs. The entire task of photographing the skull—its several components, individual teeth and the assembled reconstruction—was undertaken with finesse, artistry and devotion by Mr R. Klomfass. In all some 200 photographs were taken. A proportion of these is included in this work and eloquently testifies to his skill. Mrs L. V. Hitchings, as well as Mrs R. W. Kaplan and Mrs B. E. Wilson, graciously lavished many weeks of their time on the meticulous and thoughtful typing of the manuscript, the tables and the legends. To all of this fine team, I express my warmest gratitude.

The advice, inspiration and encouragement of Dr and Mrs Leakey have been forthcoming at all times: I am deeply in their debt for giving me the privilege of working on so important and anatomically ideal a specimen.

Finally, I sincerely appreciate the patient, helpful and friendly co-operation of those many people at the Cambridge University Press—in the Pitt Building, at the Printing House, and in Bentley House, London—who have participated in producing this book. It has been a joy to work with them.

P.V.T.

CHAPTER I

INTRODUCTION

On 15 August 1959 the first announcement of the discovery of a new fossil cranium from Olduvai appeared in *Nature* (Leakey, 1959a). A week later, on Saturday 22 August 1959, at the formal opening ceremony of the fourth Pan-African Congress on Prehistory in Léopoldville, Dr L. S. B. Leakey publicly announced to the astonished and enthusiastic delegates more details of the important new specimen (Tobias, 1960a). He reiterated his view, expressed already in *Nature*, that the big-toothed cranium was that of a young male, representing a new genus of the subfamily Australopithecinae. To this genus he gave the name *Zinjanthropus*, *Zinj* being 'the ancient name for East Africa as a whole'. He enumerated twenty features in which he claimed the new specimen differed from members of either of two previously-recognised australopithecine genera, *Australopithecus* and *Paranthropus*. He went on to designate the species as *Zinjanthropus boisei* sp.nov., the specific name being in honour of Mr Charles Boise, whose generous financial help had made the discovery possible. Leakey's (1959a) preliminary diagnosis of its specific features reads as follows: 'A species of *Zinjanthropus* in which the males are far more massive than the most massive male *Paranthropus*. The face is also excessively long. Males have a sagittal crest, at least posteriorly. Upper third molars smaller than the second.'

Leakey stressed the tentative nature of his generic and specific diagnosis, recognising that, if and when further material were found, 'the diagnosis will need both enlarging and possibly modifying'.

Less than one month after Mary Leakey's discovery of the cranium on 17 July 1959, the Leakeys brought it to South Africa to compare it with the known australopithecine specimens in the Anatomy Department of the University of the Witwatersrand, Johannesburg, and in the Transvaal Museum, Pretoria. During this brief visit on the eve of the Léopoldville Congress, an opportunity to examine the new cranium was provided for Professor Raymond A. Dart, Dr John T. Robinson and the author. Our first impressions were that the new cranium undoubtedly represented a big-toothed, heavy-muscled member of the Australopithecinae and that its nearest affinities were with the more robust and large-toothed of the existing australopithecine forms, namely *Paranthropus* of Swartkrans and Kromdraai in the Krugersdorp District. From the outset, the wisdom of creating a new genus was questioned, since there seemed to be a *prima facie* case for regarding the new form as an East African member of the genus *Paranthropus*.

In passing, it may be mentioned that the generic distinctness of *Paranthropus* from *Australopithecus*, the criteria for which had been formulated by Robinson (1954a, 1956, 1961–3), is not accepted by most recent writers on the subject. Among those who still accept the generic distinctness of *Paranthropus* and *Australopithecus* are Heberer (1960a, 1962), Napier (1961) and Coppens (1964). On the other hand, ever since 1948, Dart has maintained that all the australopithecines, gracile and robust, belong to a single genus (1948a, b, 1955a). Likewise, Le Gros Clark (1955, 1964) recognises only a single genus, *Australopithecus*, within the subfamily, Australopithecinae, and relegates the distinctions between the more robust and the more gracile forms to specific differences. This unigeneric classification of the australopithecines is accepted as well by von Koenigswald (1960), Campbell (1962, 1963), Coon (1963), Simpson (1963), Hulse (1963), Mayr (1963a), Harrison and Weiner (1964) and others, and it was apparently the consensus at an international symposium on classification and human evolution held by the Wenner–Gren Foundation at its European conference centre of Burg Wartenstein in the summer of 1962 (Washburn, 1963). Furthermore, the unigeneric view was the consensus at a symposium on the taxonomy of fossil man, in which the author participated, during the holding of the Seventh International Congress of Anthropological and Ethnological Sciences in Moscow, 1964. Robinson, on the other hand, has found that he can explain the morphological differences between the two South African forms only by attributing major ecological differences to them (1954d, 1956, 1961–3). On the basis of such inferred ecological variations, he feels justified in maintaining two genera.

The several merits of these classifications will be discussed later in this work (chapter XIX). A recent compromise viewpoint may be mentioned here: Leakey, Tobias and Napier (1964) have suggested the recognition of a single australopithecine genus (*Australopithecus*), with, for the moment, three subgenera, *Australopithecus*, *Paranthropus* and *Zinjanthropus*. Since a major part of my description of the Olduvai cranium involves comparison of it with each of the South African forms, it will be convenient in this account to refer to the Olduvai Bed I australopithecine as *Zinjanthropus*, and to use the other two subgeneric labels, *Australopithecus* and *Paranthropus*, to designate the two forms of australopithecine described from the South African sites.

A further opportunity to examine the new cranium and discuss its affinities was provided during the course of the Léopoldville Congress. All physical and palaeo-anthropologists at the Congress were given a private viewing and an intensive discussion followed. Much of the interchange centred around the systematic status of the specimen. Following his examination of the South African specimens in Pretoria and Johannesburg, Leakey averred that, while the Tanganyikan cranium had a number of features in common with each of *Paranthropus* and *Australopithecus*, there was a still bigger list of respects in which it was distinct from both; if, therefore, both of these warranted generic distinction, *Zinjanthropus*, too, should be regarded as a distinct genus. Some were opposed to this view, while others felt that a separate name might be justified as a distinctive label for the time being.[1] The view was represented to Leakey that, if he indeed persisted in regarding the Olduvai specimen as representing a new genus, the new generic name would likely be sunk very soon—probably with the discovery of a second specimen at Olduvai! Leakey was prepared to take this risk; he felt that the name should remain, at least *for the time being*, and until more was known of the range of variation of the Tanganyikan australopithecines.

Thus, the name entered into the literature, despite the misgivings expressed beforehand by many scientists, while others questioned it subsequently (e.g. Sergi, 1959; Heberer, 1960a; Kurth, 1960).

Robinson has challenged the validity of attributing separate generic status to the Olduvai australopithecine. In a paper in *Nature* in May 1960 he analysed many of the twenty characteristics which Leakey had earlier enumerated: his analysis led him to conclude that 'separate generic status seems unwarranted and biologically unmeaningful'. He proposed that the Olduvai form be regarded as a distinct species within the genus *Paranthropus* and that its name be *Paranthropus boisei* (Leakey).

In reply Leakey (1960b) commented, 'Inevitably, different scientific workers have different ideas of what characters justify specific, generic, and even superfamilial rank.' He summarised the position as follows: 'Dr Robinson and I agree that *Zinjanthropus boisei* is closely related to the Australopithecinae; we agree that it has certain resemblances to *Paranthropus*, and we disagree mainly in that he believes the differences to be insufficient to justify separate generic rank, while I think they do.'

As only a very brief and preliminary report of the morphology of *Zinjanthropus* has so far appeared, and as most of the points at issue are

---

[1] The same view was later expressed by Heberer (1960b): 'Diesen Namen (*Zinjanthropus*) darf man wohl nur als individuelle Fundbezeichnung, nicht aber als systematisch zu wertende Benennung versteken' (p. 317).

INTRODUCTION

based on differing interpretations of the morphology, the disputation was clearly premature. A detailed description of the specimen is the first requirement; from this, the assessment of its systematic affinities should become more readily apparent. Dr Leakey has entrusted the specimen to me for detailed morphological treatment. In the present work, a fairly comprehensive description is presented and comparisons are made with the crania of other australopithecines, hominines and pongids.

In preparing the present report, I have been handicapped by the fact that although a bibliography on the australopithecines included over 380 titles already in 1954 (Musiker, 1954), not a single cranium of this group has thus far been fully described. True, Dart had prepared a 300-page monograph on the Taung child in the 'twenties, but when Keith's lengthy account of *Australopithecus* appeared in *New Discoveries Relating to the Antiquity of Man* (1931), Dart's monograph remained unpublished. The only aspect of australopithecine morphology which has been exhaustively described is the dentition—in Robinson's masterly and meticulous monograph (1956). In the description of the dentition of *Zinjanthropus*, I have leaned heavily upon this work. Inevitably, the teeth of *Zinjanthropus* have received more detailed treatment in the present work than other aspects of the fossil's anatomy, since more published comparative material is available. At the same time, I have enjoyed free access to the original specimens in the Transvaal Museum, Pretoria, as well as to Professor Dart's specimens, the repository of which is the Anatomy Department of the University of the Witwatersrand, Johannesburg. It has thus been possible to compare *Zinjanthropus* with the entire collection of African australopithecines, with the exception of one maxillary fragment with two teeth from Garusi in Tanzania, a specimen which Robinson (1953*a*, 1955) has shown is australopithecine in character.

In making comparisons between *Zinjanthropus* and non-australopithecine fossil hominids, I have had the opportunity of studying the originals of fossil human crania in the British Museum (Natural History) at South Kensington, London; the Musée de l'Homme and the Institut de Paléontologie Humaine in Paris; the Rijksmuseum in Leiden and the Rijksuniversiteit in Utrecht; Professor S. Sergi's apartment in Rome; the National Museum (formerly the Coryndon Museum), Nairobi; the Department of Anatomy at the University of Cape Town; the Port Elizabeth Museum; and the Department of Anatomy of the University of the Witwatersrand, Johannesburg. In addition, in the latter department, a most comprehensive collection of plaster casts of fossil crania is available, as well as a large collection of endocranial casts of modern man, fossil man, australopithecines and other Primates. For adequate comparison between *Zinjanthropus* and the Pongidae, a study has been made of pongid skulls in the British Museum (Natural History), the Powell–Cotton Museum at Birchington in Kent, the National Museum of Kenya and the Department of Anatomy of the University of the Witwatersrand.

Anatomical terminology

In this description, the author has adhered to the recently adopted Paris Nomina Anatomica, as amended by the Sixth International Congress of Anatomists at New York in April 1960. For the most part, the English equivalents of the international Latin forms have been employed.

In the text, the term 'skull' has been used where the mandible is included. 'Cranium' is used for the skull without the mandible. The cranium, in turn, comprises a facial skeleton and 'calvaria' (brain-case). The calvaria consists of two parts, the 'calotte' or top of the brain-case, and the 'basis cranii' or base of the brain-case (Trevor, 1950).

In the description of the teeth, the terms *length* and *breadth* have been abandoned in favour of *mesiodistal diameter* and *buccolingual* or *labiolingual diameter*. In this account, for simplicity, the inner aspect of the teeth is spoken of as *lingual*; the outer aspect as *labial* in respect of incisors and canines and as *buccal* in respect of premolars and molars.

Classificatory nomenclature

The Hominoidea or, simply, hominoids, refers to the superfamily comprising both apes and man. The Pongidae or the pongids refers to the members of the family of anthropoid apes. The Hominidae or hominids refers to the members of the family of man.

The Homininae or hominines refers to the members of a subfamily of Hominidae, comprising members of the genus *Homo* and including non-australopithecine fossil hominids such as *Pithecanthropus*, *Atlanthropus* and *Hemanthropus*, which are commonly today lumped in the genus *Homo*.

The Australopithecinae or australopithecines, occasionally dubbed prehominids, near-men, half-men or ape-men, refers to the members of another subfamily, comprising those fossils which have been designated *Australopithecus* (including the former 'Plesianthropus'), *Paranthropus* and *Zinjanthropus*, and possibly also *Meganthropus*.[1] This australopithecine subfamily is classified by most students as a subfamily of the Hominidae and, as such, it is regarded in this account; although a few workers still regard it as a subfamily of the Pongidae.

The terms *Sinanthropus* and *Pithecanthropus* are not used in this study. Instead, the specific term *Homo erectus* is used to designate the taxon represented by the group of fossils formerly called by these terms.[2] On this usage, *H. erectus* includes, too, *Atlanthropus* and the cranium of the individual formerly known as 'Chellean Man' (Leakey, 1961*a*) from the upper part of Bed II, Olduvai Gorge. Possibly 'Telanthropus capensis', the hominine from Swartkrans in the Transvaal, should be included as well in this group, as was suggested by Simonetta (1957) and Robinson (1961), and supported by Tobias and von Koenigswald (1964).

Within the species *H. erectus*, the Choukoutien fossils are regarded here as falling into the *pekinensis* subspecies and the Javanese into the *erectus* subspecies, as in Campbell's scheme (1963). The north-west African forms, it is suggested, could well be classified as a third subspecies which should be called *mauritanicus* after the specific name of *Atlanthropus* (Arambourg, 1954). The decision whether or not the Olduvai representative of *H. erectus* ('Chellean Man') falls into any of these three subspecies or constitutes a fourth minor taxon within the species must wait upon the detailed description of the cranium (in preparation).

The Javanese cranium called by Weidenreich *Pithecanthropus robustus* and by von Koenigswald *P. modjokertensis* is generally called *Homo erectus erectus* IV or, simply, *H. e. erectus* IV in the present study.

For ease of reference, hominid remains are sometimes referred to in this text by their sites of origin. Thus, the forms lumped as *Paranthropus* come from Kromdraai and Swartkrans, whilst those included in the subgenus *Australopithecus* have been found at Taung (formerly spelt Taungs), Sterkfontein and Makapansgat. Following the classificatory system formerly employed by Robinson at the Transvaal Museum, Pretoria, I have sometimes designated individual crania from Sterkfontein by the 'official' abbreviation *Sts*, e.g. Sts 5 for cranium number 5 from that site;[1] in the same way, remains from Swartkrans are designated *SK*. Individual specimens from Makapansgat are identified by the prefix MLD, standing for Makapansgat Limeworks Deposit (Boné and Dart, 1955), e.g. MLD 1 for the first-discovered australopithecine specimen, the occiput, or MLD 37/38 for the most recently discovered and most complete cranium from that site.

Other special terminological usages are explained in the text.

---

[1] In a recent re-examination of the data and some of the material, Tobias and von Koenigswald (1964) concluded that, whereas the African species of *Meganthropus* (*M. africanus*) seemed indisputably to be australopithecine, the Asian species (*M. palaeojavanicus*) showed more advanced features than the australopithecines and might therefore have to be classified as a member of the Homininae.

[2] I at first hesitated to substitute the term *H. erectus* for *Pithecanthropus*, for reasons which I have discussed elsewhere (Tobias, 1962). Subsequently, however, definitions of the augmented genus, *Homo*, have been published by Robinson (1962) and by Leakey et al. (1964). As a result, my objections against the use of the term *H. erectus* have fallen away.

[1] Where Roman numerals are used with Sterkfontein or Sts, e.g. Sts VII, Sts VIII, the reference is to the 'skull numbers' employed by Broom, Robinson and Schepers (1950), and not to the official Transvaal Museum Catalogue number.

INTRODUCTION

Plan of the study

Following an account of the state and degree of preservation and of the reconstruction of the cranium (chapter II), the ensuing description will comprise the following parts:

(1) The calvaria from its outer surface, including the curvature of the vault, the supra-orbital height index, the sagittal and nuchal crests, the basis cranii externa, the mandibular fossa, the position and the plane of the foramen magnum, the indices of porion position, the position of the occipital condyle and the nuchal height index.

(2) The calvaria from its inner surface, including the endocranial surface of the frontal and of the parietal bones, the pattern of the venous sinuses, the basis cranii interna—posterior and middle cranial fossae, the endocranial cast and cranial capacity.

(3) The facial skeleton, including the supra-orbital torus, the orbits and the interorbital area, the nose, the maxilla with special emphasis on prognathism, the zygomatic bone, facial measurements and indices, and calvariofacial indices.

(4) The teeth, including the shape of the dental arcade and the palate, the arrangement of the teeth in the arcade, the pattern of occlusion and attrition, the state of the enamel, the dimensions and indices of individual teeth, tooth size as a whole, and the morphology of individual teeth.

(5) The cranium as a whole, including measurements and indices.

Thereafter will follow a summary of the cranial features of *Zinjanthropus* and a general discussion, with special reference to the affinities, taxonomy and definitions of the australopithecines in general and *Zinjanthropus* in particular; and the cultural status and phylogenetic position of *Zinjanthropus*.

CHAPTER II

PRESERVATION AND RECONSTRUCTION OF THE CRANIUM

The fossil remains, to which the name '*Zinjanthropus boisei*' has been given, comprise an almost complete cranium with the entire maxillary dentition. In the same level of Bed I were found some fragments of a second cranium and parts of the shafts of a gracile hominid tibia and fibula, while from the slope just below were recovered an upper lateral incisor, an upper molar and a lower premolar of a very gracile dentition.[1] According to Leakey, 'the hominid skull was found as a single unit within the space of approximately one square foot by about six inches deep...The expansion and contraction of the bentonitic clay, upon which the skull rested and in which it was partly embedded, had resulted, over the years, in its breaking up into small fragments...' (1959a, p. 491). However, despite its fragmentary condition, the individual pieces of bone are in a beautiful state of preservation: there is no sign of warping or squashing or other pressure effects, nor are the surfaces defaced by weathering. As a result, a wealth of anatomical detail is preserved in the cranium, which compares most favourably with the very best of the australopithecine remains hitherto described from South Africa. In completeness and state of preservation, the *Zinjanthropus* cranium must take its place alongside the child skull from Taung and the adult cranium (Sts 5) from Sterkfontein as the most perfect of the early hominid crania thus far found in Africa.

Reconstruction of the cranium was greatly facilitated because of this freedom from distortion and defacement. Nevertheless, it was a long and tedious process to assemble the hundreds of fragments, many of which were minute and lacking in manifest anatomical character. In this task, Dr Leakey enjoyed the patient and invariably fortunate co-operation of Mary Leakey. To these co-discoverers of the fossil is owing, in the main, the reconstruction of the cranium. The final assembling of the parts was made by Mr A. R. Hughes under the author's direction. The marked degree of pneumatisation of the cranium and the numerous and intricate septa subdividing the air-sinuses provided an additional source of irritatingly undistinctive fragments and, even at the stage of writing this report, scores of minute particles remain unaffixed to the reconstruction.

The portions of bone have been assembled in the first instance into six parts of the cranium and it will facilitate the subsequent description to enumerate these parts:

Part I. *The maxillofacial fragment* (pls. 16–18)

This comprises most of the lower part of the face: the nasal and subnasal parts of the maxillae including the nasal crest; the alveolar processes and the contained permanent teeth; the maxillary and palatine components of the hard palate; the maxillary antra; much of the body of the maxilla on each side, including the anterior surface and the opening of the infra-orbital foramen, the infra-temporal surface and the zygomatic process; most of the right zygomatic bone, including the infero-lateral margin of the right orbit, the zygomatic arch as far back as the serrations for the zygomaticotemporal articulation, and a small part of the temporal border of the zygomatic (the frontal processes and much of the orbital surfaces of the maxillae and of the right zygomatic, as well as most of the left zygomatic bone, form part of the upper or calvariofacial element). Attached to the

---

[1] The teeth and fragments of the second cranial vault have been recorded as a paratype of *Homo habilis* (Leakey et al. 1964). It is impossible to determine to which of the two hominids represented on this living-floor the tibia and fibula belong (Davis, Day and Napier, 1964).

back of the maxillary tuberosities are much of the left and right pterygoid processes of the sphenoid bone.

Part II. The calvariofacial fragment (pl. 15)

In this fragment are the nasal bones; the frontal processes and parts of the orbital and nasal surfaces of the maxillae, including parts of the margins of the orbit and of the piriform aperture; a small remnant of the lacrimal bone; much of the body of the left zygomatic bone including its malar and temporal surfaces; the frontal processes and most of the orbital surfaces of both zygomatic bones, including the related portion of the orbital margin; much of the frontal bone, including the frontal squame with its supra-orbital torus; the nasal part, orbitomarginal part and orbital surface, zygomatic process, temporal surface and internal surface of the frontal bone, as far back as a short distance in front of the coronal suture. It is in this position, just behind the postorbital constriction, that most bone is missing from the posterior border of the frontal and the anterior border of the parietals.

Part III. The biparietal fragment (pls. 21 and 22)

The third fragment comprises most of the two parietal bones. They are articulated together along the sagittal suture, which is overlaid for part of its anteroposterior extent by the left and right temporal crests coming into apposition as a sagittal crest. The posterior (lambdoid) margin is practically intact; a small broken-off part of the occipital remains articulated to the right lambdoid suture close to lambda. The lateral (squamosal) margins are likewise intact; but the anterior (coronal) margin is defective, no part of the coronal suture being apparent.

Part IV. Right temporal bone (pl. 23)

The fourth part of the cranium comprises the virtually complete right temporal bone. Slight defects are present along the upper edge of the squame, over the surface of the pars mastoidea and near the apex of the petrous part. The zygomatic process is broken across at the level of the articular eminence, that is, some distance behind the zygomaticotemporal suture. In front, part of the greater wing of the sphenoid is attached.

Part V. Left temporal bone (pls. 24 and 25)

The fifth part of the cranium comprises most of the left temporal bone. There are somewhat more defects than on the right, including much loss of the pars mastoidea and of the posterior half of the squame. On the other hand, most of the petrous process is preserved and the zygomatic process extends forward as far as the zygomaticotemporal suture. In front, part of the greater wing of the sphenoid is attached: this includes the base of the pterygoid process and part of the lateral pterygoid lamina.

Part VI. Occipitosphenoid fragment (pls. 26 and 27)

Finally, the sixth fragment comprises virtually the entire occipital bone, save for the detached fragment in the region of lambda mentioned above and slight defects in the lateral (exoccipital) part. Fused to the front of the basilar part of the occipital is part of the body of the sphenoid, completing the clivus and including the dorsum sellae and posterior part of the hypophyseal fossa. On the inferior aspect of the sphenoid, its rostrum bears part of the superior border and the alae of the vomer.

Apart from the defects enumerated above, the posterior parts of the walls of the orbits, the posterior parts of the nasal cavity, the posterior and medial parts of the anterior cranial fossa, and the anterior part of the middle cranial fossa are missing. All these defects follow from the lack of the ethmoid bone and of much of the sphenoid bone, especially the anterior part of its body, its lesser wings and much of its greater wings.

These defects are, however, relatively minor and it is not difficult to associate the six parts so as to assemble an almost entire cranium. Parts III–VI—the temporals, parietals, occipital and body of the sphenoid—articulate perfectly with one another, so as to constitute most of the calvaria, from its posterior limit almost as far forward as the plane of the coronal suture (pls. 10–13).

Parts I and II—the maxillofacial and calvariofacial fragments—make good contact with each other in the inferolateral part of the right orbital margin; a further approximation below the left orbit permits faithful alignment of the upper and lower parts of the face. The orientation of the upper fragment on the lower is further controlled by the co-planarity of the internasal suture above with the intermaxillary suture below, as well as by the relationship between the temporal surfaces of the frontal bone above, and those of the zygomatic and maxillary bones below (pls. 1–3).

The cranium is thus assembled into anterior faciocalvarial and posterior calvarial moieties. The joining of these major portions depends upon the approximation between two parts of the left lateral pterygoid lamina attached respectively to the maxillary tuberosity below and in front, and the greater wing of the sphenoid above and behind. A further precise guide is provided by the happy circumstance that the temporal process of the right zygomatic bone is intact as far back as the zygomaticotemporal suture, while the zygomatic process of the left temporal bone is intact as far forward as the same suture. It is thus a relatively simple matter to align the anterior and posterior moieties, so that the centres of the sloping sutural surfaces of these two processes lie in the same transverse plane. Such alignment satisfies the contours of anterior and posterior portions of the calotte and cranial base. The hafting of the facial on to the calvarial part has been facilitated by modelling the missing parts of the zygomatic arch on each side and by making an endocranial cast. In this way, the entire cranium has been assembled on an endocranial cast (pl. 5).

Our final reconstruction differs only slightly from that made by Dr and Mrs Leakey and published in *Nature* (1959a) and the *Illustrated London News* (19 September 1959b, and 4 March 1961a).[1] For instance, our final estimate of the maximum cranial length (from glabella to opisthocranion, which in this instance practically coincides with inion) is 173·0 mm., as compared with Leakey's (1959a) figure of 174·0 mm., obtained on the provisional assembly and reconstruction of the skull. Our estimate of the length is, however, an absolute minimum, as shown by the approximation of the parts of the lateral pterygoid plate and the alignment in the same coronal plane of the left and right zygomaticotemporal sutures.

---

[1] The reconstruction published by the Leakeys in 1959 is less complete than the later reconstruction, because additional parts were recovered subsequently.

CHAPTER III

THE CRANIAL VAULT

A. The curvature and components of the vault

1. *The vault as a whole*

The posterior part of the calvaria of *Zinjanthropus* is well filled and well rounded (pls. 9–12). From the external occipital protuberance, the posterior parieto-occipital plane rises steeply for a considerable distance before turning forwards over the summit of the vault. This striking feature, brought out in Fig. 1, was stressed by Leakey as the third of his twenty diagnostic criteria of *Zinjanthropus* (1959a, p. 492). However, Robinson (1960) claimed that this feature applied also to *Paranthropus*. In Fig. 1, Robinson's (1961) reconstruction of *Paranthropus*, based on the crushed specimen SK 48, has a remarkably similar parieto-occipital contour to that of *Zinjanthropus*, whereas the earlier reconstruction of SK 48 by Broom and Robinson (1952, p. 11) had a very different parieto-occipital contour from that of *Zinjanthropus*. As none of the specimens of *Paranthropus* is sufficiently undistorted to permit the contour in this region to be reconstructed accurately, it seems very likely that the *Zinjanthropus*-like contour in Robinson's later (1961) reconstruction has been influenced, at least subconsciously, by the intact parieto-occipital contour of *Zinjanthropus*. Robinson's (1960) claim that the steep parieto-occipital plane occurs as well in *Paranthropus* may therefore be discounted, at least until more intact cranial material is discovered. *Australopithecus* (Sts 5) has a more evenly-curved parieto-occipital surface.

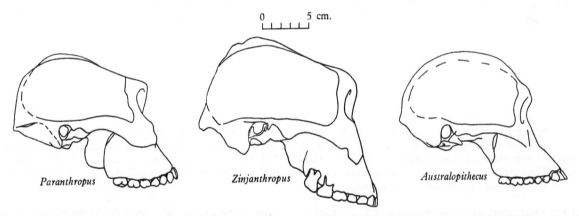

Fig. 1. Craniograms of norma lateralis of *Paranthropus*, *Zinjanthropus* and *Australopithecus*. The dioptographic tracing of *Zinjanthropus* was made with the cranium in the F.H.; the missing parts have been restored. The other two craniograms, also aligned on the F.H., are from those published by Robinson (1961) and are based on the reconstruction and restoration of SK 48 (*Paranthropus*) and on Sts 5 (*Australopithecus*).

Part of the full rounded contour of the vault is contributed by the high, steep, parietotemporal walls (pls. 9 and 11). They rise almost vertically before bending over to become a dome, as Leakey pointed out in the eleventh of his twenty criteria of *Zinjanthropus*. When this contour is compared with that of the *Australopithecus* cranium, MLD 37/38 from Makapansgat (Dart, 1962b), it is seen that the latter, too, is characterised by steeply rising side-walls of the vault; if anything those of MLD 37/38 are somewhat steeper than those of *Zinjanthropus* (pl. 9).

By and large, the vault of *Zinjanthropus* is spheroidal in shape as seen in norma verticalis, tapering from a greatly expanded posterior portion to a narrow postorbital constriction (Fig. 2). In this respect, it more closely resembles *Paranthropus*, as reconstructed by Robinson (1961), than *Australopithecus*, as represented by Sterkfontein 5 (Fig. 2). In comparison with *Paranthropus*, however, *Zinjanthropus* is clearly a 'long-spheroid' cranium, while that of Swartkrans has

sagittal readings in *Zinjanthropus* have been taken just to the left and just to the right of the sagittal crest; so these values may be taken as reflecting the dimensions of the vault itself, excluding its superstructures.

Most of the dimensions (mm.) of the parietals in *Zinjanthropus* fall within the range of variation of four *Australopithecus* crania from Sterkfontein and Makapansgat. The only exceptions are the coronal margin chord (69·4), which somewhat exceeds the

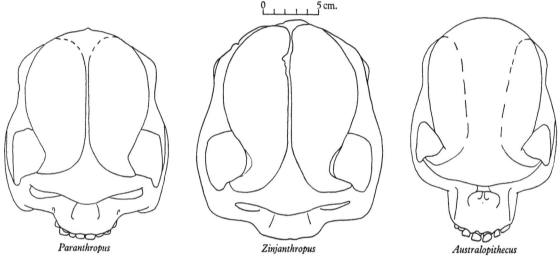

Fig. 2. Craniograms of norma verticalis of *Paranthropus*, *Zinjanthropus* and *Australopithecus*. The dioptographic tracing of *Zinjanthropus* was made with the cranium orientated in the F.H.; the missing parts have been restored. The other two craniograms were enlarged from those published by Robinson (1961) and are based on the reconstruction and restoration of SK 48 (*Paranthropus*) and on Sts 5 (*Australopithecus*). Note that in *Zinjanthropus*, a small part of the mastoid process protrudes laterally beyond the supramastoid crest.

been reconstructed as a 'short-spheroid' or more truly spheroid cranium. In this respect, *Zinjanthropus* approaches slightly the condition in *Australopithecus* (Sts 5), which has a long, ovoid cranium, with a lesser degree of postorbital constriction. *Zinjanthropus* thus appears to be intermediate in calvarial shape between the two South African australopithecines, though obviously resembling more closely that of *Paranthropus*, as reconstructed by Robinson in 1961 and 1962.

The dimensions and curvature of individual vault-bones may next be considered.

2. The parietal bones

In Table 1, the chord and arc along each of the four borders of the parietal bones of *Zinjanthropus* are compared with those of other hominids. The

top of the range (60·5 to ±66·5) for three *Australopithecus* crania; the temporal margin chord (81·8) which falls slightly above the value of 78·0, measured on two *Australopithecus* crania; and the temporal margin arc (89·3) which lies above the range (83–85) for the same two *Australopithecus* crania.

In Table 2, values measured on *Zinjanthropus* have been added to those of *Australopithecus* to make a total australopithecine sample of five crania. The values for Sts 58 have, however, been omitted from the range, since they could be measured only on the endocranial surface and are thus too small. The australopithecine parietal dimensions are compared with those of samples of *Homo erectus* from Java, China and north-west Africa (Ternifine) and of modern man. The values

Table 1. Measurements and indices of the parietal bone in Zinjanthropus and other hominids (mm.)

	Zinjanthropus		Sts 5		Sts 25		Sts 58	MLD 37/38		H. e. pekinensis* ($n = 4$–5 individuals)	H. e. erectus				H. e. mauritanicus	Modern man†
	Left	Right						L	R		I	II	III	IV		
Parietal sagittal chord	73·6‡	76·2‡	84·5		±73·5		70·3§	77		86–106 ($\bar{x} = 95·2$)	87·5	91	88·5	89	92	106·7–109·4
Parietal sagittal arc	80·0‡	81·0‡	91·5		±78·0		74·5§	85		92–113 ($\bar{x} = 100·8$)	90–91	94	94	90	95	117·3–133·5
Parietal sagittal index	92·0‡	94·1‡	92·3		94·2		94·4§	90·6		93·5–95·8 ($\bar{x} = 94·5$)	96·1–97·1	96·8	94·1	96·8	96·8	88·7–91·9
	L	R	L	R	L	R		L	R							
Coronal margin chord	60·5	69·4‡	—	—	—	±66·5	—	?	66·3	79–93 ($\bar{x} = 86·2$)	—	—	77	—	90	83·7–91·1
Coronal margin arc	67·0	74·0‡	—	—	—	±82·0	—	?	79·0	?97–111 ($\bar{x} = 103·5$)	—	—	88	—	105	104·1–116·7
Coronal margin index	90·3	93·8‡	—	—	—	81·1	—	?	83·9	71·2–88·6 ($\bar{x} = 82·6$)	—	—	87·5	—	85·7	83·8–86·6
Lambdoid margin chord	56·5	57·8	59·5	59·0	±54·0	—	—	54·4	53·8	80–90 ($\bar{x} = 83·5$)	—	—	78	—	90	76·0–83·7
Lambdoid margin arc	59·5	59·0	62·5	62·0	±55·5	—	—	58	58	90–103 ($\bar{x} = 95·8$)	—	—	84	—	95	88·0–103·1
Lambdoid margin index	94·9	98·0	95·2	95·2	±97·3	—	—	93·8	92·8	77·7–92·5 ($\bar{x} = 87·3$)	—	—	92·9	—	94·7	88·2–96·9
Temporal margin chord	—	81·8‡	—	78·0	—	—	—	?	78·0	87–99 ($\bar{x} = 93·3$)	—	—	91	—	102	88·7–94·0
Temporal margin arc	—	89·3‡	—	±83·0	—	—	—	?	85	92–106 ($\bar{x} = 97·6$)	—	—	95	—	105	94·0–104·9
Temporal margin index	—	91·6‡	—	94·0	—	—	—	?	91·8	90·6–98·9 ($\bar{x} = 95·6$)	—	—	95·8	—	97·1	91·9–94·3

* Means and ranges of several fossil crania.
† Ranges of population means, based on Martin (1928) and Pittard and Kaufmann (1936).
‡ These measurements and indices are based on incomplete edges of the parietal bones.
§ Measured on endocranial surface.

Table 2. *Ranges of measurements of the parietal bone in australopithecines* (*including* Zinjanthropus), Homo erectus *and modern man* (*mm.*)

(Sample size in brackets refers to number of individuals, not of parietal bones)

	Australopithecines	*Homo erectus*	Modern man*
Parietal sagittal chord	73·5–84·5 (4)	86–106 (10)	106·7–109·4
Parietal sagittal arc	78·0–91·5 (4)	90–113 (10)	117·3–133·5
Coronal margin chord	60·5–69·4 (4)	77– 93 (6)	83·7– 91·1
Coronal margin arc	67·0–82·0 (4)	88–111 (6)	104·1–116·7
Lambdoid margin chord	53·8–59·5 (4)	78– 90 (7)	76·0– 83·7
Lambdoid margin arc	±55·5–62·5 (4)	84–103 (7)	88·0–103·1
Temporal margin chord	78·0–81·8 (3)	87–102 (6)	88·7– 94·0
Temporal margin arc	±83·0–89·3 (3)	92–106 (6)	94·0–104·9

* Ranges of population means, as in Table 1.

of the sagittal chord and arc of the australopithecines lie closest to those of *H. erectus*, the only actual overlap being of the sagittal arc. Thus, the parietal sagittal arc of Sts 5 (which far exceeds the arcs of the other australopithecines) is 91·5 mm., while in *H. e. erectus* I it is 90 or 91 mm., in *H. e. erectus* IV 90 mm., and in the *H. e. pekinensis* cranium with the smallest parietal 92 mm. Similarly, the australopithecine temporal margin is not far short of the lowest values in *H. erectus* (Table 2), though the two ranges are separated by gaps of 5·2 and 2·7 mm. respectively.

On the other hand, the coronal and lambdoid marginal measurements are appreciably smaller than those of *H. erectus*, the sample ranges being separated by gaps of 7·6, 6·0, 18·5 and 21·5 mm. respectively. In contrast, the ranges for these features overlap appreciably when *H. erectus* is compared with modern man. Clearly, in anteroposterior extension, whether measured along the sagittal or the temporal margins, the australopithecine parietal is slightly smaller than that of *H. erectus*; whereas in mediolateral extension, whether measured along the coronal or the lambdoid suture, the australopithecine parietal is far smaller than that of *H. erectus*.

The degree of curvature of the parietal bone may be roughly assessed by expressing the chord as a percentage of the arc along each of the four margins. The higher the index, the lower the curvature, and vice versa.

From the figures in Table 1, the parietal bone is slightly less curved along the coronal margin in *Zinjanthropus* than in *Australopithecus*; whereas the curvatures of the other margins of the parietal bones are comparable in the two forms. Table 3, which has been prepared in the same manner as Table 2, shows that, in general, there is a tendency for the australopithecine crania to be more curved anteroposteriorly and less curved mediolaterally than the crania of *H. erectus*. However, all the ranges overlap.

The mean anteroposterior extension of the parietal bone in pongids is far smaller than in the Australopithecinae. A few figures for pongids, culled from Martin (1928), are:

	Mean parietal sagittal chord (mm.)	Mean parietal sagittal arc (mm.)
Gorilla ♀[1]	68·5	71·5
Orang-utan ♂	59·0	—
Orang-utan ♀	57·7	59·8
Chimpanzee ♂	62·2	65·3
Chimpanzee ♀	64·2	67·6

3. *The parieto-occipital ratio*

In Table 4, the sagittal and transverse dimensions of the occipital bone of *Zinjanthropus* are compared with those of other australopithecines and hominines. The first point to note is the relative lengths of the sagittal arc dimensions of parietal and occipital: these cannot be determined for *Zinjanthropus*, since the front parts of the parietals

[1] The values could not be determined on adult male gorillas because of the cranial superstructures.

Table 3. *Ranges of indices of the parietal bone margins in australopithecines (including* Zinjanthropus), Homo erectus *and modern man (mm.)*

(Sample size in brackets refers to number of individuals)

	Australopithecines	*Homo erectus*	Modern man*
Parietal sagittal index	90·6–94·2 (4)	93·5–97·1 (10)	88·7–91·9
Coronal margin index	81·1–93·8 (4)	71·2–88·6 (6)	83·8–86·6
Lambdoid margin index	92·8–98·0 (4)	77·7–94·7 (7)	88·2–96·9
Temporal margin index	91·6–94·0 (3)	90·6–98·9 (6)	91·9–94·3

* Ranges of population means, as in Table 1.

Table 4. *Measurements and indices of the occipital bone in* Zinjanthropus *and other hominids (mm.)*

	Zinj-anthropus	*Australopithecus*			*H. e. pekinensis**	*H. e. erectus*			Modern man†
		MLD 1	MLD 37/38	Sts 5		I	II	IV	
Lambda–inion chord	36·2	44·1	36·0	c. 38·2	47 – 52·5‡	—	45	44	54·5– 70·0
Lambda–inion arc	36·5	46·5	37·0	c. 39·0	49 – 55	—	47	—	58·5– 75·7
Chord–arc index ('cerebral index')	99·2	94·8	97·3	c. 97·9	95·5– 96·1	—	95·8	—	88·6– 96·3
Inion–opisthion chord	43·7	37·8	c. 33·5	c. 31·6	57 – 63‡	—	48	64	36·3– 49·0
Inion–opisthion arc	46·0	38·5	c. 34·0	c. 32	60 – 67	—	52	—	38·5– 50·8
Chord–arc index ('cerebellar index')	95·0	98·2	c. 98·5	c. 98·7	94·0–?96·7	—	92·3	—	94·5–100·0
Occipital sagittal chord (S_3')	57·8	68·0	59	c. 58·5	84 – 86	78	75	79·5	85·2–102·0
Occipital sagittal arc (S_3)	82·0	85·0	71	c. 72	108 –118	103	100	110	103·5–123·0
Chord–arc index (S_3'/S_3)	70·5	80·0	83·1	c. 81·3	72·9– 77·8	75·7	75·0	72·3	80·8– 87·9
Chord index of upper and lower scales $\left(\frac{\text{inion-opisthion chord}}{\text{lambda-inion chord}}\right)$	120·7	85·7	94·1	c. 84·9	100 –144	—	106·7	145·2	52·4– 86
Chord breadth of occipital (biasterionic br.)	89·2	85	79·5	c. 76	?113 –117	—	?120	130	103·5–114·4
Arc breadth of occipital	112·0	115	90	c. 90	145 –155	—	—	—	—
L/B index of occipital squama (chords)	64·8	80·0	74·2	c. 77	71·7– 76·1	—	?62·5	61·2	80·9– 98·0
L/B index of occipital squama (arcs)	73·2	73·9	78·9	80·0	73·2– 76·6	—	—	—	—

* Ranges of individual values.
† Ranges of population means, from Martin (1928) and Sauter (1941–6).
‡ These values from Table XIX of Weidenreich (1943) differ from those in Table VI of the same work.

are lacking, but incomplete as they are, the parietal sagittal arcs (80+ mm., 81+ mm.) are almost equal to that of the occipital (82 mm.). It seems certain that, in the complete cranium, the parietal would have been greater anteroposteriorly than the occipital (and this despite the large nuchal crest which serves to enhance the occipital sagittal arc in *Zinjanthropus*). In two other australopithecine crania, Sts 5 and MLD 37/38, a similar relationship holds, the parietal and occipital sagittal arcs being 91·5 and 72 mm. in the former and 85 and 71 mm. in the latter. The parietal preponderance of these crania of *Australopithecus* is probably greater because of the absence of large nuchal crests.

Parietal predominance over the occipital is essentially a feature of modern man (Sauter, 1941–6); *H. e. erectus* and *H. e. pekinensis* show the opposite condition, the parietal being appreciably smaller in this measurement than the occipital. An occipitoparietal arc index (per cent) in *H. e. pekinensis* has a mean of 119·7, in *H. e. erectus* of Java a mean of 114, and in modern man a mean of 87·6 (Weidenreich, 1943, p. 130); the values in australopithecines are 83·5 (MLD 37/38) and 78·3 (Sts 5). In this respect, *Australopithecus* is nearer to modern man than is *H. erectus*!

For the pongids Martin (1928) quotes mean values which show a slight occipital preponderance over the parietal in every series except the chimpanzee: in female orang, the occipital and parietal sagittal arcs contribute respectively 33 and 30 per cent of the total sagittal arc; in male gorilla ?38 and ?26 per cent; and in female gorilla 32 and 31 per cent. In chimpanzee there is occipitoparietal parity, each bone contributing 30 per cent of the total sagittal arc in males and 31 per cent in females. If we omit the dubious values of male gorilla, the pongids in general seem very close to equality of contribution by parietal and occipital to the total sagittal arc. Weidenreich quoted the total percentage contributions of frontal, parietal and occipital bones to the total sagittal arc in 'the great apes' as frontal bone 34·9, parietal bone 29·0 and occipital bone 27·0. While at first glance this suggests a parietal preponderance over the occipital—at variance with the figures of Martin (1928)—the three values quoted by Weidenreich (1943, p. 130) add up to 90·9 instead of 100 per cent, and therefore cannot be correct.

To summarise the trends revealed in this analysis, it seems that pongids are in a neutral position of parieto-occipital equality, if anything veering slightly towards occipital preponderance; *H. erectus* has clear-cut occipital precedence in common with Solo man; while modern man, fossil sapient and Neandertal man of Europe and the Australopithecinae have an equally clear parietal pre-eminence.

4. *The occipital bone*

In absolute dimensions (mm.), the occipital bone of *Zinjanthropus* compares fairly well with those of the three crania of *Australopithecus*. Its sagittal chord is small (57·8), the range in *Australopithecus* being 58·5–68·0, but, enhanced by the large nuchal crest, its sagittal arc (82·0) is well within the range of *Australopithecus* (71–85). On the other hand, the *Zinjanthropus* occipital is broader than those of *Australopithecus*, so that the L/B index is appreciably smaller (64·8 as against 74·2–80 per cent). The australopithecine occipital measurements are all much smaller than those of the Homininae.

If inion be taken as the dividing point in the median plane between planum nuchale and planum occipitale of the occipital squama, the relative sizes of the two parts may be gauged by comparing the chord and arc from lambda to inion with those from inion to opisthion. The values for these components of the squama are recorded in Table 4. In *Zinjanthropus* the upper (or cerebral) arc and chord are smaller than the lower (or cerebellar) arc and chord, giving a ratio of lower to upper chords of 120·7 per cent, whereas the converse is true of the three crania of *Australopithecus* (chord ratios 84·9–94·1 per cent). When these values are compared with those of the Homininae, it is seen that *Zinjanthropus* agrees with *H. erectus* (chord ratios 100–145·2 per cent), whereas *Australopithecus* is close to modern man (mean chord ratios 52·4–86 per cent). The feature which distinguishes both *Zinjanthropus* and *H. erectus* from the other Hominidae is the presence of large occipital superstructures, an occipital torus in *H. erectus* and a nuchal crest in *Zinjanthropus*.

In an elegant analysis, Delattre and Fenart (1960) have shown how, in the ontogeny and growth of an anthropoid cranium with little cerebral expansion and with heavy ectocranial embellishments, the inion is carried progressively farther from its foetal or neonatal starting-point. The angle subtended by the cartilaginous occipital (opisthion to inion) at the intersection of the vestibular axes increases from birth to adulthood by 13° in the gorilla and only 10° in the chimpanzee. During the same period, the angle subtended by the membranous occipital (lambda to inion) decreases by 26° in the gorilla and 18° in the chimpanzee. In this way, the opisthion–inion segment increases at the expense of the lambda–inion component. In modern man, with a much greater cerebral expansion and no marked migration up the calvaria of the nuchal muscles, an exactly opposite process occurs: the angle subtended by the cartilaginous occipital (opisthion–inion) *decreases* by 6° between birth and adulthood, while that subtended by the membranous occipital (lambda–inion) *increases* by 16°. Hence, the upper (lambda–inion) segment of the squama increases in modern man at the expense of the lower (opisthion–inion) component.

If we apply Delattre's elucidation to *Zinjanthropus*, we see that with its restricted cerebral expansion and gorilloid extension of the nuchal muscles, resulting in the formation of a crest, the growth conditions may be expected closely to resemble those obtaining in the gorilla. We may thus expect an increase of the cartilaginous part of the occipital (opisthion–inion) at the expense of the membranous part (lambda–inion), resulting in a preponderance of the lower part of the squama. The same sequence would be expected to apply to *H. erectus* with its heavy occipital torus. It would seem therefore that the lower scale preponderance of *Zinjanthropus* and *H. erectus* and the upper scale preponderance of *Australopithecus* and modern man do not have any special taxonomic significance; rather do the former two depart from the hominid pattern by their possession in common of certain developmental conditions, which the latter two groups lack.

The degree of curvature of the occipital bone may be expressed by the percentage relationship of the sagittal chord and arc of the entire bone, as well as of its constituent upper and lower scales. This curvature was formerly expressed by Karl Pearson's Occipital Index (Oc.I.), which expressed the ratio of the radius of curvature of the occipital bone (from lambda to opisthion in the median sagittal plane) to the lambda–opisthion chord. The radius of curvature was calculated by assuming that the lambda–opisthion arc was an even curve, part of the circumference of a single circle. However, in a series of studies begun at the instance of Dr J. C. Trevor, Director of the Duckworth Laboratory, Cambridge, I showed that the simple chord–arc index, besides being shorter and easier to calculate, had a number of distinct advantages over Pearson's Oc.I. (Tobias, 1959*b*, 1960*b*). Accordingly, I recommended that in future craniological studies, Oc.I. be abandoned and replaced by the simple chord–arc index. The higher the value of this index, the lower the curvature, and vice versa.

The chord–arc indices for the upper and lower scales and for the whole occipital squama of *Zinjanthropus* are given in Table 4. Comparing, first, the 'cerebral' or upper curvature with the 'cerebellar', we see that the values are higher for the cerebellar index than for the cerebral index in both *Australopithecus* and modern man; that is, the opisthion–inion curve is flatter than the lambda–inion curve. One exception in the table is *Zinjanthropus*, in which the upper component of the occipital squama is much flatter than the lower. Again this is clearly the result of nuchal crest formation; as the nuchal muscles have moved up the occipital, they have carried the inion upwards and backwards, thus tending to straighten out the planum occipitale between lambda and the edge of the crest. In fact, they have gone further and produced a hollow between lambda and the crest. Thus, the upper arc is slightly greater than the upper chord, because the tape-measure follows a concavity (not a convexity) between lambda and inion! This hollow corresponds closely to the *sulcus supratoralis* of *H. e. pekinensis* (Weidenreich, 1943, p. 39). In *H. erectus*, too, the occipital cerebral indices are higher than the occipital cerebellar indices, in three out of four crania, only *H. e. pekinensis* III having a slightly higher cerebellar index (?96·7) than cerebral (95·9). In general, these results agree with those of *Zinjanthropus* in showing stronger curvature of the *lower* segment.

Finally, we may consider the occipital chord–arc index for the entire bone. The values for *Zinjanthropus* and other australopithecines are compared with those of other hominids in Table 5. In three crania of *Australopithecus*, the indices range from 80·0 to 83·1 per cent, values which lie well within the range for modern man. *Zinjanthropus* is again rendered peculiar in having an index ten points lower (70·5 per cent), less than the smallest

Table 5. *The occipital index of* Zinjanthropus *and other hominids* (*per cent*)

	Occipital index ($S'_3/S_3 \times 100$)		Reference
Australopithecinae (individual values)			
Zinjanthropus	70·5		Present study
Australopithecus MLD 1	80·0		Tobias (1959*d*)
MLD 37/38	83·1		Present study
Sts 5	*c.* 81·3		Present study
Homininae	Mean	Range	
Homo erectus (6)	74·5	72·3–77·8	After Weidenreich (1943, 1945, 1951)
Solo (6)	72·0	?67·3–74·6	After Weidenreich (1951)
Neandertaloids of Africa (2)	75·7	?75·2–76·2	Tobias (1959*d*)
Neandertalers of Europe (2)	77·7	76·9–78·4	After Morant (1927); McCown and Keith (1939)
Tabūn–Skhūl (5)	77·9	?70·5–?83·3	Tobias (1959*d*), after McCown and Keith (1939)
Upper Palaeolithic (Europe)			
♂ (13)	80·8	76·6–??83·7	Morant (1930)
♀ (6)	80·6	76·2–81·5	Morant (1930)
Modern European series	—	80·8–83·9*	Martin (1928); Sauter (1941–6)
Modern African series	—	81·2–87·9*	Tobias (1958, 1959*a*)

* Range of population means.

values yet recorded for any hominines save for one or two exceptional crania. If the queried value for Solo X is correct, 67·3 per cent would seem to be the lowest chord–arc index on record for an adult human cranium (Tobias, 1959*a*). Thereapart, Tildesley (1921) reported two modern Burmese crania with curvatures which are the equivalent of indices of 70·0 per cent. Thus, *Zinjanthropus* falls at the lower extreme of the range of values for the Homininae. Its low index is the result of its formidable nuchal crest (see pp. 21–3), which greatly increases the disparity between arc and chord (as, for example, in the Broken Hill cranium, Tobias, 1959*c*).

B. The supra-orbital height index

In 1950, Le Gros Clark proposed a supra-orbital height index, to give metrical expression to certain relations between the brain-case and the upper part of the face, to which he had drawn attention in his 1947 paper. This index was the percentage ratio between the height of the calotte above the upper margin of the orbit (FB) and the total calvarial height (AB) above the Frankfurt Horizontal (Fig. 3).

With its full rounded vault, cranium 5 from Sterkfontein gave an index of 68 per cent in Le Gros Clark's (1950*b*) original study. This figure was amended to 74 by Ashton and Zuckerman (1951) for reasons of 'photographic distortion'. Subsequently, however, Robinson (1962) gave the value as 61. At the same time, Robinson determined the indices of several Choukoutien crania from illustrations: they ranged from 'about 63 to 67' (1962, p. 134). Robinson's value for Sts 5 (61) lies near the lower end of the modern human range determined by Ashton and Zuckerman (1951) on three cranial series: the indices of their modern crania range from about 63 to about 77 per cent and have sample means of 71·3 per cent (Spitalfields English), 70·4 per cent (West African) and 70·6 per cent (Australian) (*see* Table 6). On the other hand, percentage values for the great apes are lower, the means ranging from 49·2 in orang to 54·0 in one series of gorilla (Ashton and Zuckerman, 1951). Thus, the value of 61 in Sts 5 lies between the indices for pongids and those for hominines. (It may be pointed out, in passing, that the values estimated for Sts 5 by Le Gros Clark and by Ashton and Zuckerman would remove this cranium much further from the upper limits of the pongid ranges and place it well and truly within the hominine range.)

No South African cranium of *Paranthropus* is sufficiently undistorted to permit this index to be

CRANIAL VAULT

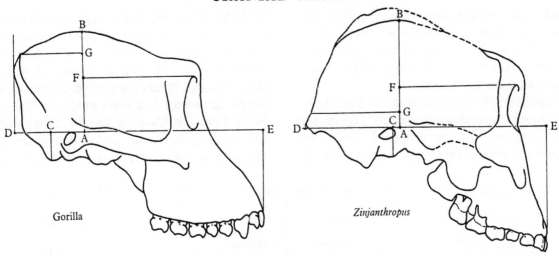

Fig. 3. Lateral craniograms of gorilla and *Zinjanthropus*, to show the landmarks from which are derived three cranial indices, CD/CE, AG/AB and FB/AB. DE is the F.H. (based on the two poria and the left orbitale). AB is the perpendicular from the highest point of the calvaria (B) to the F.H. F is the projection on to this perpendicular of the highest point of the superior orbital margin; and G is the projection on to AB of the highest point reached by the nuchal muscles on either side of the external occipital protuberance. C is the projection on to the F.H. of the centre of the curvature of the occipital condyle. (After Le Gros Clark, 1950*b*.)

Table 6. *Differences between the supra-orbital height index (FB/AB per cent) of* Zinjanthropus *and of other hominoids**

(A) *Including the sagittal crest of* Zinjanthropus

Zinjanthropus compared with:	Index in Zinjanthropus	Mean index of other hominoid group	S.D.	Diff.	Diff./S.D.
English (Spitalfields)	58·61	71·30	2·00	−12·69	−6·35
West African	58·61	70·41	2·03	−11·80	−5·81
Australian	58·61	70·57	2·69	−11·96	−4·45
Gorilla (1st series)	58·61	51·79	6·75	+6·82	+1·01
Gorilla (2nd series)	58·61	54·03	8·18	+4·48	+0·56
Chimpanzee (1st series)	58·61	49·77	4·68	+8·84	+1·89
Chimpanzee (2nd series)	58·61	50·80	3·43	+7·81	+2·28
Orang-utan	58·61	49·24	5·60	+9·37	+1·67

(B) *Excluding the sagittal crest of* Zinjanthropus

Zinjanthropus compared with:	Index in Zinjanthropus	Mean index of other hominoid group	S.D.	Diff.	Diff./S.D.
English (Spitalfields)	52·33	71·30	2·00	−18·97	−9·49
West African	52·33	70·41	2·03	−18·08	−8·91
Australian	52·33	70·57	2·69	−18·24	−6·78
Gorilla (1st series)	52·33	51·79	6·75	+0·54	+0·08
Gorilla (2nd series)	52·33	54·03	8·18	−1·70	−0·21
Chimpanzee (1st series)	52·33	49·77	4·68	+2·56	+0·55
Chimpanzee (2nd series)	52·33	50·80	3·43	+1·53	+0·45
Orang-utan	52·33	49·24	5·60	+3·09	+0·55

* Pongid means in both sections A and B of the table include crested individuals and the values, especially for gorillas, are thus too high.

assessed with any accuracy. However, from his later reconstruction of *Paranthropus* from Swartkrans, Robinson (1962) obtained a figure of 50 per cent, well within the pongid ranges and appreciably below the hominine range.

The *Zinjanthropus* cranium is the first specimen of the robust australopithecines to permit the supra-orbital index to be measured accurately. *Zinjanthropus* gives a lower value than *Australo-*

FB′/AB′, where B′ is the highest point on the surface of the vault, on either side of the base of the crest. This gives a value of 52·33 for *Zinjanthropus* (Table 6B), which must be compared with *non-crested* pongid crania (or crested crania on which similar allowance has been made for the crest). In Le Gros Clark's pongids, the highest indices, after making allowance for the crests, are 50–55. *Zinjanthropus* thus falls at the top of the pongid

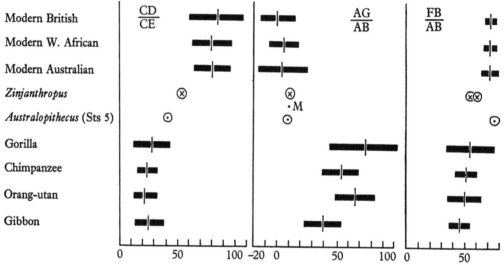

Fig. 4. Diagrammatic representation of the values of three cranial indices in a variety of hominoid crania. CD/CE represents the occipital condyle position index; the higher the index the further forward the condyle. AG/AB represents the nuchal area height index; the higher the index, the higher the nuchal muscle area on the occipital bone. FB/AB is the supra-orbital height index of the calvaria; the higher the index, the more the calvaria is elevated above the level of the upper margin of the orbits. The vertical line on each solid block in the diagram represents the mean; the limits of the solid blocks represent the greatest and least values at 98 per cent confidence limits. The dot-in-a-circle is the value for each index in Sts skull 5; the X-in-a-circle represents the values in *Zinjanthropus* (the two values shown under FB/AB are based on measurements with and without the crest, from left to right respectively); the dot marked M represents the value in a third australopithecine cranium, namely the new MLD 37/38, from Makapansgat. (After Ashton and Zuckerman, 1951.)

pithecus: when both FB and AB are measured *to the top of the crest*, the mean index for the height above the left and right orbits is 58·61 per cent. This value is well below the hominine means, differing from them by 4·45 to 6·35 times the standard deviation. On the other hand, it lies well within the ranges for gorilla, exceeding the means of two series by only 1·01 and 0·56 s.d.'s. Likewise, its value exceeds the means of two chimpanzee series by 1·89 and 2·28 s.d.'s and of an orang series by 1·67 s.d. (Table 6A).

The crest has the effect of raising the index, as Le Gros Clark (1950b) and Ashton and Zuckerman (1951) have stressed. I have therefore calculated

range (Fig. 4). It differs appreciably from *Australopithecus* and the Homininae, in possessing a calvaria which rises but little above the upper margin of the orbits. This is a feature which *Zinjanthropus* shares with *Paranthropus*.

Yet, the failure of the brain-case of *Zinjanthropus* to rise appreciably above the orbits does not betoken a failure of the brain to expand. Nor does it reflect a necessary difference in brain morphology, as Robinson seems to suggest (1962, p. 134). In fact, as we show in the section on the endocranial cast, the endocranial form and volume of *Zinjanthropus* are very similar to those of previously studied australopithecines. Rather, it indicates that the whole calvaria is hafted on to the facial skeleton at a lower level than in *Australopithecus*: this is clearly seen in a comparison of the lateral craniograms of the three creatures (Fig. 1). That for *Paranthropus* is based on Robinson's (1961) restoration of SK 48 from Swartkrans; that for *Australopithecus* is based on Sts 5 from Sterkfontein. This lower hafting of brain-case to facial skeleton is a pongid-like feature of *Zinjanthropus* and *Paranthropus*.

C. The sagittal and nuchal crests

1. *The sagittal crest*

The frontal bone of *Zinjanthropus* is characterised by very prominent and keeled temporal lines on the posterior surface of the supra-orbital torus (pl. 15B): they are more strongly developed than in any existing specimen of the Australopithecinae. They converge posteriorly over the frontal squame, but through damage in this area, their exact point of meeting is not preserved. However, extrapolation of the two converging curves has made it possible to estimate their point of meeting as some 60 mm. from glabella. The most anterior part of the sagittal crest is thus not available, but an appreciable portion is well preserved on the parietal bones. Although the lateral parts of the parietals are preserved for some 27 mm. in front of the surviving anterior limit of the crest, the medial parts of the parietals bearing the portion of the crest which extends forwards to the frontal bone have not been found. The exact anteroposterior extent of the sagittal crest can thus not be determined. From our restoration of the missing parts of the calotte, it would seem that it may have had a total chord length of some 97·5 mm.; the available part of the crest has a chord length of 52·0 mm.

About 3–4 mm. behind the anterior edge of its surviving portion, the crest reaches a height of 12·0 mm. Anterior to this point, it drops away to 10·2 mm. at the broken edge. Descriptively, we may recognise three parts of the crest as seen in side view (Fig. 1), an anterior segment, extending from the point of first contact between left and right temporal crests, back to the 12·0 mm. peak; an intermediate segment, which extends from the 12·0 mm. peak as a slightly hollowed plateau to another peak, 9·0 mm. high, and 22·5 mm. behind the anterior peak; and a posterior segment tailing down from the posterior peak, as a series of low foot-hills, until a point is reached posteriorly where the temporal crests begin to diverge.

Most of the anterior segment is missing, save for the 3 or 4 mm. immediately in front of the anterior peak; small as this area is, it shows an important feature. The two juxtaposed laminae or temporal crests, which comprise the sagittal crest, are here most intimately fused with each other (Fig. 5). This fusion can be seen both in norma verticalis (pl. 21) and in the broken front edge of the crest. Since fusion of the laminae must represent an advanced stage in the development of the crest, it might be concluded that the anterior segment is the most advanced, that is to say, that the temporal lines or crests first met in this anterior area, as Robinson (1954c, 1958) maintained for *Paranthropus*.

The intermediate segment lies a little forward of what must have been the middle third of the parietal bregma-lambda arc. From the anterior peak backwards, this segment is clearly bilaminar, fusion being apparent only deep within the sulcus between the laminae and in one position just in front of the posterior peak, reaching the upper edge of the smaller left lamina (pl. 21). Throughout this segment, the right lamina is both thicker and higher than the left (pl. 21B). At the posterior peak, the right lamina reaches a maximum mediolateral thickness of 6·1 mm., whereas the left lamina is nowhere thicker than 4·4 mm. and is generally 2–2½ mm. thick. The difference in height between the left and right laminae is minimal in the anterior segment and maximal in the intermediate segment, reaching close to 4·0 mm. in favour of the right lamina at the posterior peak. These differences, which are apparent also in the posterior segment of the crest, bespeak an imbalance between the left and right temporalis muscles: there is a clear right-sided temporalis preponderance, suggesting asymmetrical functioning of the jaw. Later we shall see further evidence of such asymmetrical functioning in the striking inequality of attrition patterns on the right and left maxillary teeth.

In the posterior segment, immediately behind

the posterior peak, there is a sharp decline in the height of the crest, affecting both laminae. They drop to about 3·0 mm. in height, in a sudden descent which is more obvious on the right lamina, then gradually tail down to end as a clearly elevated, bilaminar crest some 30 mm. behind the summit of the posterior peak. Although the relative heights of left and right laminae are somewhat variable in the posterior segment, there being one or two points where the left lamina rises

uncrested. The smooth area behind the crest forms a triangle, the sides of which are formed by the diverging temporal lines and the base of which is made up of the central part of the nuchal crest; perhaps quadrilateral would be a more appropriate term, since the base is itself V-shaped, with its powerfully down-curved external occipital protuberance (Fig. 5). This smooth area is comparable with those of other australopithecines, e.g. the Makapansgat calvarial fragment, MLD 1

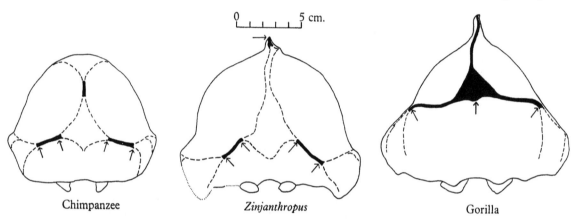

Fig. 5. Dioptographic tracings of norma occipitalis of crania of chimpanzee, *Zinjanthropus* and gorilla. The interrupted lines indicate the simple temporal and nuchal crests; the solid areas represent compound crests, formed by the fusion of two apposed simple crests. The solid areas indicated by arrows in the nuchal region indicate the mediolateral extent of compound (temporal/nuchal) crests; the arrow above the *Zinjanthropus* cranium indicates the point, situated well forward, at which the two temporal crests fuse intimately. The region of intimate fusion of the two temporal crests is much farther back in the pongids. The triangular or rhomboidal area above the external occipital protuberance is a 'bare area' in chimpanzee and *Zinjanthropus*; in the gorilla it represents the highest part of the compound (temporal/nuchal) crest, where it is intersected by the sagittal crest.

slightly higher than the right, the right-sided preponderance is evident in the greater thickness of the right lamina. The sulcus between the two laminae is wider posteriorly than in the middle and anterior segments; no fusion whatsoever is evident, the two laminae being separated by a gap of 1·2–2·2 mm. In parts, the sagittal suture (which is still open on the inner aspect of the calvaria) can be seen in the floor of the interlaminar sulcus.

2. The 'bare area of the skull' (Dart)

Behind the posterior limit of the sagittal crest, there is a further distance of 22 mm. of parietal bone as far as lambda, completely devoid of a sagittal crest; and a distance of 58·5 mm. in the median plane between the caudal end of the sagittal crest and the inion. All this distance is smooth and

(Dart, 1948*a*). As there is, too, a powerful nuchal crest in *Zinjanthropus* and, as the whole of the parieto-occipital area is beautifully preserved with much fine surface detail, it is possible to say forthrightly that the sagittal and nuchal crests are not continuous with each other, the former being developed far forwards on the calotte and fading out a great distance anterior to the inion and the nuchal crest.

In his description of the first Makapansgat australopithecine fragment, Dart, too, commented on the smooth central area uncovered by muscles (except for the occipitofrontalis muscle sheet and its *galea aponeurotica*). He proposed that this highly significant though hitherto unnamed area might happily be dubbed the 'bare area of the skull', by analogy with the bare areas of viscera left uncovered by peritoneum or other serous membranes. That part of the bare area of the skull occupying the squama occipitalis and lying between the superior nuchal lines and the lambdoid suture, he proposed to call the 'bare area of the occiput'. This bare area of the occiput is non-existent in the adult male gorilla (Fig. 5), but for a female chimpanzee cranium in the Transvaal Museum, Dart quoted a size of

$$\tfrac{1}{2}(L \times B) = \tfrac{1}{2}(20 \times 30) = 300 \text{ mm.}^2.$$

In *Australopithecus* (MLD 1), the area is

$$\tfrac{1}{2}(55 \times 76) = 2,140 \text{ mm.}^2,$$

whilst in the heavily muscled *Zinjanthropus* the area is nevertheless as large as

$$\tfrac{1}{2}(36 \cdot 5 \times 40 \cdot 3) = 735 \cdot 48 \text{ mm.}^2.$$

Despite the excessively developed muscles of *Zinjanthropus*, his occipital bare area is $2\tfrac{1}{2}$ times bigger than that of the female chimpanzee with its relatively light muscles: this apparently paradoxical situation is simply another reflection of the different pattern of sagittal crest formation in the australopithecines from that in the pongids. In the former, the crest forms much more anteriorly, thus leaving a large bare area posteriorly, even when the musculature is well developed. In the pongids, on the other hand, the sagittal crest is first formed far back on the calvaria, thus making inroads into the bare area even when the muscles are not excessively developed.

Holloway (1962) has reported finding a single gibbon skull, in which there was a well-developed sagittal crest. Unlike the position in most of the crested pongids, the crest on this gibbon cranium is situated relatively far forwards, leaving a bare area no less than 25 mm. long in the median plane between the posterior end of the crest and the inion. This specimen, like Dart's chimpanzee, indicates that 'even among the Pongidae there is no common morphogenetic process which can be said to be common to all members of this family' (*op. cit.* p. 529).

3. *The nuchal crest*

Immediately behind the posterior segment of the sagittal crest, both temporal lines diverge slightly lateral to the medial or sagittal margin of the respective parietal bones, then continue to run posteriorly parallel to the sutural edge but 2–3 mm. from it. At a point some 13 mm. behind the sagittal crest, the temporal lines start curving away laterally, so that at the level of lambda, the right temporal line is almost 10 mm. from the mid-line. Just to the left of lambda, the left temporal line leaves the parietal bone in a somewhat more medial position; it must have passed across a small sutural (Wormian) ossicle which is missing from a position just left of lambda. (The presence of such an ossicle is inferred because the sutural edges above and below a small gap are intact.) Where the lines cross on to the occipital squame, they are apparent as slightly elevated linear roughnesses almost 21 mm. apart. On the occipital squame, the temporal lines curve downwards and lateralwards, helping to delimit the quadrilateral smooth area previously mentioned (Fig. 5). Thereafter, the temporal line or crest is inseparably fused with the nuchal crest as far laterally as asterion.

This fused crest corresponds with what Robinson (1958) has called a compound temporal–nuchal crest or compound (T/N) crest, a form of crest which he has not been able to identify in any of the known australopithecine specimens, either of *Paranthropus* or of *Australopithecus*. In *Zinjanthropus*, the length of nuchal crest which is of this compound character is some 25 mm. on the right, beginning at a point 17 mm. from the median plane (shortest distance) or 23 mm. from inion along the margin of the nuchal crest. On the left, the length of compound (T/N) crest is about 28 mm., beginning at a point 17·5 mm. from the median plane (shortest distance) or 23 mm. from inion along the margin of the nuchal crest (Fig. 5).

Towards the lateral end of the nuchal crest, the inferior temporal line leaves the crest, starting to diverge forwards close to asterion, while the superior nuchal line passes inferiorly, away from the temporal line (pl. 13). The temporal line, now elevated as a sharp temporal crest, runs forwards over the asterionic corner of the occipital squame. It crosses the occipitotemporal suture a few millimetres below the asterionic angle of the parietal bone, which here imbricates over the occipital and mastoid temporal bones at the occipitoparietal (lambdoid) and parietomastoid sutures. Gaining the mastoid part of the temporal bone, the temporal crest continues as the robust, rounded supramastoid crest, coursing over the upper root of the mastoid process. The crest is separated from the mastoid process proper by a deep sulcus below. The supramastoid crest runs over the well-pneumatised pars mastoidea and this fact, coupled with the prominence of the crest itself, throws the crest to one of the most lateral positions on the hinder part of the skull: thus, the maximum breadth across the supramastoid crests (139·5 mm.) is exceeded only by the bimastoid breadth (142·0 mm.) in this part of the calvaria. Anteriorly, the supramastoid crest continues directly into the posterior root of the zygoma: from asterion to this position, the crest forms a wide shelf. The continuity of the crest and root of zygoma contrasts with the apparent arrangement of SK 46, in which the supramastoid crest seemingly does not continue directly to the root of the zygoma, but lies a little below the root (Robinson, 1958, pp. 409–10).

The nuchal crest itself begins in the middle line as a strong, prominent external occipital protuberance, the character of which will be described in the section on the Height of the Nuchal Area and of the Inion. Lateral to inion, the crest slopes steeply upwards to a sharp angular summit, which corresponds with the highest extension of the nuchal musculature and with the usually curved summit of the superior nuchal line. The height of this summit is used in the computation of Le Gros Clark's nuchal area height index (see below). The part of the nuchal crest from inion to the summit on each side is a simple nuchal crest, the temporalis muscle *not* participating in its formation. The elevation of the crest from the surface of the planum nuchale is about 9·0 mm. on the left and 8·5 mm. on the right, close to inion. The elevation diminishes laterally and, at the angular summit, is about 5·5 mm. on the left and 4·1 mm. on the right. The edge of this more medial part of the crest is fairly sharp.

At or close to the summit, the superior nuchal line is joined by the inferior temporal line to form the aforementioned compound (T/N) crest. The lateral downward slope from the summit is rugged but not as well developed as the more medial simple crest. Its elevation varies from 3·5 to 8·8 mm. on the left, there being a rugged high peak immediately lateral to the summit. On the right, the elevation of this part of the crest varies from 2·8 to 10·1 mm., the more lateral part being more strongly developed in height and thickness than on the left. About 23·0 mm. from the occipitomastoid suture on the right, the temporal and nuchal lines begin to diverge; the actual point of divergence on the left is not present, there being a small piece of bone missing in the crucial position. However, it is possible to estimate the position to within a millimetre or so: it is only about 6 mm. from the occipitomastoid suture.

After the departure of the inferior temporal line, the lateral part of the superior nuchal line continues once more as a simple nuchal crest. This turns sharply downwards alongside the occipitomastoid suture forming, at least on the left, a high, down-turned flange, the occipitomastoid crest. The corresponding position on the right is occupied by a much smaller flange, a corner of which is missing. Both flanges articulate on their lateral face with the mastoid part of the temporal bone, thus contributing greatly to the thickness of the occipitomastoid suture in this area. The right superior nuchal line continues across the suture on to the inferior surface of the mastoid process: here it seems to curve downwards towards the tip of the process, but some local damage near the tip makes its exact course difficult to determine. The relevant area of the left mastoid process is largely missing, only a small part of the superior nuchal line on pars mastoidea adjacent to the occipitomastoid suture being preserved.

The nuchal crest is thus a powerful structure, but it cannot be described as shelf-like: the shelving, if we are to use the term, is on the *undersurface* of the crest; there is no true upper surface of the crest, but simply the flattish or slightly hollow, nearly vertical face of the planum occipitale. In the central area above the simple nuchal crest, the planum occipitale is slightly concave from above downwards: when the calvaria is seen in norma lateralis, it is clear that this gentle concavity is insufficient to create a shelf-like upper surface of the nuchal crest. In the region of the compound (T/N) crest further laterally, the planum occipitale is somewhat more concave from above downwards, and a slight tendency towards shelving is apparent here. The concavity continues further laterally above the simple temporal crest, but only on the temporal bone are the degree of concavity above and the prominence of the supramastoid crest sufficient to create the appearance of a broad shelf. With maturation and ageing, the shelving might well have become further accentuated.

4. *The pattern of cresting*

It is not proposed here to enter into an account and evaluation of the lengthy controversy over the patterns of cresting in hominoids (Zuckerman, 1954; Robinson, 1954 b, c; Ashton and Zuckerman, 1956; Robinson, 1958). However, through the extraordinarily good state of preservation of the *Zinjanthropus* cranium, this study sheds light on several of the points which have been at issue.

First, *Zinjanthropus* provides an exception to the generalisation arrived at by Robinson (1958) that 'the crests of the australopithecines...differ from those in the pongids in that...the small partial nuchal crest (of the australopithecines) is not a compound (T/N) nuchal crest' (p. 427). Robinson attributed these differences in the crests to the fact that, in the australopithecines, the nuchal plane is more nearly horizontal than it is in the pongids, 'thereby moving the nuchal musculature to a position which normally precludes the possibility of their reaching the temporalis muscle' (p. 426). On the other hand, it might be argued that changes in the position of the nuchal muscles in no way limit the ability of the temporalis muscles to reach *them*! Indeed, although *Zinjanthropus* shares with other members of the Australopithecinae a nearly horizontal planum nuchale, yet it has an unmistakable compound (T/N) crest. It seems that crest formation is related not to the orientation of the nuchal plane, but to the degree of development of the musculature.

Perhaps this conclusion should have been foreseen: for in some of the australopithecine crania, Robinson found that the inferior temporal line approached very closely to the superior nuchal line, the smallest separation being 9 mm. in SK 49, 7·6 mm. in SK 46 and 1 mm. in MLD 1 (Makapansgat). A somewhat greater degree of development of the musculature might surely have been expected to superimpose the temporal crest on the already partly formed nuchal crest. This is precisely what has occurred in *Zinjanthropus*, the temporal crest coalescing with the nuchal crest over rather more than the intermediate one-third of the nuchal crest on each side. The more medial and more lateral parts of the crest remain as a simple nuchal crest, to the formation of which the temporalis muscles have contributed nothing. On the basis of this finding, it seems likely that further discoveries of *Paranthropus* and *Australopithecus*, especially of well-preserved males and older specimens, will show that in these forms, a compound (T/N) nuchal crest could arise, when muscular development reached a sufficient degree, despite the nearly horizontal planum nuchale.

Secondly, the position in *Zinjanthropus* convincingly disproves the *a priori* argument of Ashton and Zuckerman (1956, p. 605), namely: 'The significant point is that a nuchal crest, of the kind defined in accounts of the actual fossils, and as can be seen on casts, would not have been present in *Paranthropus* if the posterior fibres of the animal's temporalis muscle had not already reached the superior nuchal line.' Clearly such a crest has formed in *Zinjanthropus*, at least medially and laterally, without the participation of the temporalis; only in the intermediate area is there an indication that the posterior fibres of temporalis have assisted in the formation of the nuchal crest. It is relevant to comment, too, that the most strongly developed part of the nuchal crest is the *medial third*, which is a simple nuchal crest to which temporalis fibres have made no contribution. When the evidence of *Zinjanthropus* is considered along with that assembled by Robinson for the other Australopithecinae, it is reasonable to conclude that in these forms the sequence of events in nuchal crest formation was as follows: first, a simple nuchal crest was formed by the nuchal muscles alone; secondly, with the growth backwards of the posterior fibres of the temporalis muscle, the inferior temporal line eventually came into tangential contact with the pre-existing nuchal crest, converting part of the simple nuchal crest into a compound (T/N) crest along the line of contact. This is the stage that we find in *Zinjanthropus*. As the M³'s of *Zinjanthropus* were still to come into occlusion, it is likely that further development of the musculature would have occurred. This, in turn, would have increased the extent of contact between temporal and nuchal lines, thus converting a large proportion of the originally simple crest into a compound (T/N) crest. This view of cresting patterns suggests that the pongid and hominid sequences of nuchal crest formation are not sharply defined, fundamental, family distinctions, but that they are essentially the same processes which attain different degrees of development in the pongids and hominids. With its excessive musculature, *Zinjanthropus* nicely bridges the gap between the two hominoid groups.

The foregoing argument applies to the nuchal crest. In respect of the sagittal crest, this study provides confirmation in *Zinjanthropus* of Robinson's conclusion that the sagittal crest, when present in the australopithecines, does not extend back to the superior nuchal line, despite the great development of the muscles in the Olduvai specimen. It is possible that further development of its muscles might have occurred, had the creature lived longer; but from the position of the crest in the specimen, it seems unlikely that the crest would have extended as much as a further 60 mm., down the parieto-occipital plane to meet the nuchal crest. The sagittal crest in *Zinjanthropus* clearly took origin near the front of the calotte, as our detailed study of the surviving parts of the crest has led us to conclude, and did not start near the back of the calotte. The generalisation arrived at by Zuckerman (1954) from his study of the crania of living Primates, namely that, during individual development, 'the primary meeting-point of the sagittal lines (*sic*) of the two sides is usually just in front of the external occipital protuberance' (pp. 322–3) does not apply to *Zinjanthropus*. From my study of the crested specimens from Swartkrans, I can corroborate Robinson's claim that this generalisation does not apply to those australopithecines either; nor is it valid for *Australopithecus* from Makapansgat.

Apparently, Zuckerman's generalisation does not apply to the Homininae either, in those modern human crania in which the temporal lines approach most closely to the mid-line. In a study and review of such skulls, Riesenfeld (1955) shows a photograph of Hrdlička's (1910) remarkable Eskimo cranium, in which the temporal crests approach to within 7 mm. of each other: the point of closest approximation of the two crests lies far forward, a short distance behind bregma. Riesenfeld also quotes da Costa Ferreira's (1920) case of a female microcephalic skull in which coalescence of the temporal lines occurs *near bregma*. It seems that what happens in the australopithecines is simply the extreme development of a process that may still occur in modern man, especially in Eskimos, Melanesians and Australians. As Riesenfeld puts it, '...if in spite of his lighter jaw and the larger insertion area for the temporal muscles, the temporal lines in (modern) man can move up so high as almost to meet in the median line, the appearance of a sagittal crest in *Australopithecus crassidens* loses a great deal of its specific quality and would seem to be only the extreme expression of this principle of variability...' (Riesenfeld, 1955, pp. 609–10).

This study therefore supports Robinson's claim that sagittal crests in the australopithecines differ from those in the pongids, in that the sagittal crest, when present, does not start from nor extend back to the external occipital protuberance. It does not support his claim that the australopithecines do not have a compound (T/N) nuchal crest, for one is clearly present in *Zinjanthropus*. Furthermore, from the close approach of the inferior temporal and superior nuchal lines to each other in several young adult and female australopithecine calvariae, it seems to this worker that compound (T/N) crests should be expected in australopithecines generally, especially adult males, with big teeth and heavy musculature. On the other hand, while confirming Zuckerman's claim that the temporal and nuchal lines did meet in some australopithecines, the above analysis does not support his view that the meeting of these lines *initiates* nuchal crest formation and that, without such meeting, no nuchal crest can form. There is good evidence that nuchal muscles alone may form a nuchal crest, to which the inferior temporal line (or temporal crest) may later be added in some australopithecines, thereby converting part of it into a compound (T/N) crest. This sequence of events for the nuchal crest in the Australopithecinae is at variance with those sequences obtaining in most Pongidae and in non-hominoid Primates.

In the same way, the sequence of events in the formation of sagittal crests differs in the australopithecines from that in pongids and non-hominoid Primates. Both of these australopithecine distinctions point in the same direction, namely, that the anterior and middle parts of the temporalis muscle develop more rapidly than the posterior parts in the australopithecines. Thus, the temporal lines first reach the mid-line *anteriorly*, while they reach the superior nuchal line (or nuchal crest) later.

CHAPTER IV

THE BASIS CRANII EXTERNA

The external surface of the basis cranii of *Zinjanthropus* is beautifully preserved (pls. 6 and 14). With the exception of a few parts of the left temporal bone and the apex of the right petrous pyramid, the entire basis is preserved, down to the finest anatomical detail, as far forward as the body and greater wings of the sphenoid bone. Furthermore, the parts which are missing or damaged on one side are intact on the opposite side.

A. The occipital bone

1. *The planum nuchale*

The character of the nuchal crest has been dealt with in the preceding chapter. Immediately below the nuchal crest, on either side of the external occipital protuberance and of the posterosuperior part of the external occipital crest, is a clear oval impression, about 27 by 16 mm., in the area between the superior and inferior nuchal lines. In modern man, the *semispinalis capitis* muscle is attached to a corresponding impression. Anterolateral to the last impression is another somewhat irregular impression, corresponding to the attachment of *m. obliquus capitis superior*. Posterolaterally, the lower lateral limb of the nuchal crest is rugged, as it flows towards the adjacent pars mastoidea and mastoid process: this part probably provided attachment to the *splenius capitis* muscle.

The inferior nuchal line is not very clearly defined. It runs below and in front of the impression for *semispinalis capitis*; then continues laterally for a short distance, turning sharply forwards opposite the impression for *m. obliquus capitis superior*, finally flowing over the lateral part of the exoccipital and terminating in the jugular process. This forwardly directed portion of the inferior nuchal line is thrown into a prominent spur of bone on the left, some 4 mm. high. The spur, represented on the right only by a slightly elevated linear roughness, bounds a broad, smooth, anteroposteriorly directed area, between the spur and the occipitomastoid suture: the latter area is presumably for the occipital artery.

Antero-inferior to the inferior nuchal line are two oval facets, one on either side of the external occipital crest: they correspond to the attachment areas of the left and right *rectus capitis posterior minor* muscles. Lateral to this impression on each side is a long comma-shaped impression, filling the gap between the oval facet and the forwardly-directed part of the inferior nuchal line, and extending anteriorly, somewhat nearer to the margin of the foramen magnum: this large comma-shaped area must be the impression for the *rectus capitis posterior major* muscle.

The fine anatomy of the planum nuchale is preserved so much better in *Zinjanthropus* than in any other australopithecine that it has been deemed worthwhile to place the foregoing detailed description on record. In the description of the type specimen from Sterkfontein (Sts I), only the following statement occurs: 'The supraoccipital fragment is much thicker than the parietals or frontal but there are no ridges such as (are) seen in the gorilla, and often in the male chimpanzee' (Broom and Schepers, 1946, p. 53). Again, in respect of Sts 5, Broom and Robinson (1950, p. 22) state: 'As the whole occipital region is much less developed than in man, the part of the skull behind the foramen (magnum) is relatively short, and it slopes much more upward than in man, but it does not slope as much upward as in the chimpanzee or other anthropoids, and is much flatter.' A more detailed description is given of Sts VIII (Broom and Robinson, 1950, p. 28): according to this, 'the portion of the occiput behind and to the sides of the foramen magnum is remarkably like that of man, and though the foramen and the condyles are smaller than in man the lower part of the occiput is as wide as in some Bushmen skulls'. Sts VIII has no postcondylar foramen (in contrast to *Zinjanthropus*), but 'there is a distinct concavity (condylar fossa) behind the outer side of each condyle'. The comparable region in *Paranthropus* is well preserved only in juvenile crania, of those published.

2. *The lateral (condylar) part of the occipital bone*

The *occipital condyles* in *Zinjanthropus* are extraordinarily small for so massive a cranium (pl. 6). That on the left is intact and measures 20·0 mm. (chord length) by 10·3 mm. (maximum breadth). That on the right is broken across near its anterior extremity, the chord length of the surviving part being 17·7 mm. and the maximum breadth 10·9 mm. No previous measurements of australopithecine condyles could be found in the literature, but of Sterkfontein cranium VIII Broom and Robinson (1950, p. 28) stated, 'The condyles are relatively small and about twice as long as broad.' No condylar parts of the Pekin crania were available at the time Weidenreich prepared his classical monographs.

The condyles converge anteriorly, so that their minimum distance apart is 12·0 mm.; near their posterior poles, they are 27·5 mm. apart. Their surfaces are somewhat everted, as in modern man, and they encroach very slightly on the anterolateral margin of the foramen magnum. They are highly convex from before backwards. Each condyle is kidney-shaped, there being a distinct fissure in the region of the hilus, presumably testifying to the dual developmental origin of the condyle. The forward position of the condyles in relation to the cranial base and to the biporionic axis is discussed in the section on the position of the occipital condyles (chapter v, D). All in all, the condyles of *Zinjanthropus* are very much more like those of man than those of the pongids, as is true of other australopithecines in which the condyles are preserved. The hypoglossal canal is single on the right, but double on the left.

A deep oval hollow on the inferior surface of the exoccipital marks the attachment area usually occupied by the *rectus capitis lateralis* muscle. Medial to this fossa, and immediately posterolateral to the occipital condyle, is a small groove in the position of the condylar fossa: it is not as well developed as the condylar fossa which, in modern man, receives the posterior edge of the superior articular facet of the atlas when the head is extended. The floor of this condylar groove in *Zinjanthropus* is perforated by a condylar canal, such as usually transmits an emissary vein from the sigmoid sinus.

The *jugular process* of the occipital has an inferiorly directed paramastoid process on its lower surface: part of it, preserved only on the right, seems to have met the transverse process of the atlas, which has imprinted a small facet on the inferior aspect of the process. Such a paramastoid process is extremely rare in modern man (De Villiers, 1963).

3. *The basilar part of the occipital bone*

The inferior surface of the *basi-occipital* shows a strongly marked, downturned process where the median plane intersects the anterior margin of the foramen magnum; ectobasion is the lowest point on this process. The process tallies with the attachment point of the ligament of the apex of the dens. Anterolateral to the ectobasion are two roughened areas, corresponding with the impressions for the *rectus capitis anterior*. A pair of strong elevations protrudes downwards and lateralwards, just in front of the last: they are probably for the attachment of *longus capitis*. Similar elevations occur in Sts 5, Sts VIII and MLD 37/38, although in the Sterkfontein crania they were interpreted as the attachments of rectus capitis anterior (Broom and Robinson, 1950, p. 28). Between the two elevations in *Zinjanthropus* is a moderate median tubercle for the pharyngeal raphe. The basi-occipital is well inflated and has a generally puffed appearance.

The site of the now closed spheno-occipital synchondrosis cannot be detected on the external surface.

B. The temporal bone

1. *The mastoid process*

The mastoid process of *Zinjanthropus* is very large and prominent, protruding both laterally to provide the most lateral point on the cranium, apart from the zygomatic arch, and inferiorly as a well-defined process. It protrudes only slightly medialwards and, in this respect, the process is much more like that of modern man than is the mastoid process of *Homo erectus erectus* IV. In the latter, the mastoid process is very large and projects far downwards, but it does not descend vertically, as in modern man, but turns so steeply inward that, when viewed from below, its tip lies halfway between a plane passing through the most lateral projection of the supramastoid crest and one through the medial border of the condyle (Weidenreich, 1945, p. 24). In *Zinjanthropus*, seen in similar view (pl. 14 A), the tip of the mastoid process extends inwards for only one-third of the way between the same two planes. Thus, Leakey (1959a, p. 492) could claim as the sixth of his twenty criteria for the recognition of *Zinjanthropus*, '...the mastoids are more similar to those seen in present-day man, both in size and shape'. However, Schultz (1950 a, b) has found that, in large numbers of adult anthropoid ape crania examined by him, the mastoid region is extraordinarily variable. He states:

In chimpanzees and, especially, in gorillas there can appear late in growth true mastoid processes which are fully comparable in size and shape with those of man. In the latter, however, these processes develop in all cases, whereas in the apes they remain small, except in a minority of the specimens. In man the mastoid processes begin to form early in post-natal life, but in apes they never become noticeable before the approach of adulthood and then only in occasional specimens. The largest mastoid processes have been encountered by the writer in old gorillas and chimpanzees. From this can be concluded that the formation of mastoid processes is the result of an ontogenetic innovation which has led to the late development of these structures in only a part of the population of apes and to their comparatively early and constant development in man. Only in this restricted sense can the mastoid process be regarded as a distinction of man (Schultz, 1950a, p. 447).

Seen in this developmental light, the presence of very large mastoid processes in the still immature *Zinjanthropus* aligns him with the hominids rather than with the pongids. The conclusion is supported by the fact that a large mastoid process of modern human shape is present in all australopithecine crania in which the region is preserved. Even stronger support for the hominine affinities of the australopithecine mastoid comes from the fact that younger crania even than that of *Zinjanthropus*, namely the two juvenile crania of *Paranthropus* from Swartkrans, show well-developed mastoid processes at perhaps 7 and 11 years of age. It seems that a large mastoid process in the australopithecines is, first, a constant feature, and, secondly, develops early in life. In both of these respects, the Australopithecinae differ from the Pongidae and resemble modern Homininae.

The mastoid crest is strongly developed, while a deeply-indented supramastoid sulcus separates the upper part of the mastoid process and crest from the powerful supramastoid crest, as in *H. e. pekinensis*. The presence of a foramen mastoideum cannot be detected in *Zinjanthropus*; nor is a suprameatal spine present on either side.

2. *The supramastoid crest*

The supramastoid crest is the strongly ridged and pneumatised continuation of the temporal crest (or inferior temporal line), after the latter turns forwards from the compound (temporal/nuchal) crest and crosses from the occipital on to the temporal bone. The supramastoid crest bulges maximally above the mastoid process, protruding laterally almost as much as the mastoid process itself. Passing forwards, it retreats medialwards slightly, above the external acoustic meatus. Then, as the posterior root of the zygoma, it again juts laterally, this time to surpass its maximum protrusion further posteriorly and it becomes here the most lateral point on the cranium. In this undulating contour of the supramastoid crest, *Zinjanthropus* is closely paralleled by other australopithecine crania, especially the type specimen from Kromdraai, in which the relevant area is very well preserved, and Sts VIII.

3. *The mastoid notch*

The exact nature of the digastric fossa cannot be determined in *Zinjanthropus*, because the area is missing on the left, whilst on the right a small area of surface damage has laid bare the air cells on the medial aspect of the mastoid process. However, the position of the digastric fossa is clear on the right, and it is in a straight line with the *foramen processus styloidei* and the *foramen stylomastoideum*; this line runs obliquely from anteromedially to posterolaterally, exactly as in modern man. In Pekin Man, while the foramen processus styloidei and the digastric fossa are in the same straight line, the foramen stylomastoideum lies outside this line. In *Zinjanthropus*, as in *H. e. pekinensis*, the medial lip of the mastoid notch is formed by a high, pneumatised ridge which runs along the occipitomastoid suture. This mastoid ridge abuts against a thin, non-pneumatised ridge on the lateral margin of the occipital bone: the two ridges articulate as part of the occipitomastoid sutural joint, the thickness of which is thereby greatly increased (pl. 4). The combined ridge corresponds with the *crista occipitomastoidea* of Weidenreich (1943, p. 64). The occipital contribution to the crest in *Zinjanthropus* is moderately strong on the left and weaker on the right, so that the lowest 2–3 mm. of the medial surface of the mastoid contribution to the crest on the right protrudes below the occipital component of the crest and is thus non-articular. A similar occipitomastoid crest is found in two Swartkrans crania, SK 846 and the adolescent SK 47, in which P^4 is erupting and M^3 beginning to erupt, and in the Makapansgat cranium, MLD 37/38. In all four australopithecine crania, the main component of the crest is temporal and the occipital contribution is less than in *H. e. pekinensis*. In fact, in MLD 37/38, the large, pneumatised crest seems to be exclusively mastoid, the occipitomastoid suture itself apparently lying in a slight depression. A prominent occipitomastoid crest is characteristic not only of Pekin Man, but of Neandertal crania, including Swanscombe (Stewart, 1964). In fact, Stewart states:

...ancient skulls with big brow ridges and coarse facial features generally have occipitomastoid crests that rival in size, or dwarf, their mastoid processes, whereas modern skulls with small brow ridges and delicate facial features either lack occipitomastoid crests or have such small ones that they are scarcely noticeable alongside of the usually large mastoid process. Stated in another way, the occipitomastoid crest provides a seemingly reliable way of telling whether an ancient occipital or temporal bone belongs to a human being of primitive or modern form (Stewart, 1964, pp. 157–8).

A distinction should perhaps be drawn between the type of crest in the australopithecines and that in later hominines. In *Zinjanthropus*, for instance, the occipital component of the crest takes the form of an ectocranial superstructure, a crest properly speaking, which greatly thickens the occipitomastoid suture. In Swanscombe, however, the effect of a crest is achieved by a down-turning of the whole occipital bone near the occipitomastoid suture. Thus, in Swanscombe, the recurvation of this part of the occipital bone is apparent even on the endocranial aspect and there is no marked thickening of the occipitomastoid suture. Further study is necessary on the distribution and morphological relations of these two patterns of cresting.

Stewart (1964) has focused attention on another process which, in modern crania, is commonly present in the mastoid notch. It delimits the digastric sulcus laterally from the sulcus for the occipital artery medially. 'This eminence', writes Stewart (p. 157), 'is ignored in modern text-books of anatomy'—which is certainly true of textual descriptions; but the process is, nevertheless, *depicted* separating the digastric and arterial sulci.[1] The only name which Stewart could trace for this process was 'apophyse-mastoïde surnuméraire', quoted by Le Double (1903) and attributed by the latter to Zoja (1864). According to Stewart, 'An eminence in this position has not been reported for ancient man'. It is therefore worth recording here that such a process is present in the well-preserved mastoid notches of two *Paranthropus* crania, SK 846 and the adolescent SK 47. In SK 846, it clearly separates a digastric groove laterally from a more medial *sulcus arteriae occipitalis*, which in turn is demarcated medially by a blunt occipitomastoid

---

[1] See, for instance, fig. 119 of *Buchanan's Manual of Anatomy* (8th ed.); fig. 141 of *Morris' Human Anatomy* (11th ed.); fig. 333 of *Gray's Anatomy* (33rd ed.); and figs. 67–14 of *Anatomy* by Gardner, Gray and O'Rahilly (1st ed.).

crest. In SK 47, the intermediate process becomes confluent posteriorly with the occipitomastoid crest, thus closing off the more medial sulcus behind. The exact course of the occipital artery in this instance is difficult to determine, but a linear smooth area *medial* to the occipitomastoid crest and apparently distinct from the area for m. *obliquus capitis superior* suggests that the artery may have run over the surface of the occipital bone medial to the occipitomastoid crest. The imperfectly preserved corresponding area of MLD 37/38 suggests that in this australopithecine, too, the artery might have pursued a similar course.

4. *The tympanic plate*

The tympanic plate is perfectly preserved on the right, but part of the lateral margin is broken off on the left.

The lateral margin of the tympanic plate, limiting the external acoustic porus, is overhung somewhat by a strong shelf comprising the supramastoid crest (or posterior root of the zygoma). This feature constituted the fifth and seventh of Leakey's twenty diagnostic criteria of *Zinjanthropus* (1959a, p. 492). The porus itself thus lies some 5–6 mm. medial to the sagittal plane through the auriculare point. This distance is much less than in *H. e. pekinensis*, in which it varies from 10 to 15 mm.; according to Weidenreich, in modern man it hardly ever exceeds 10 mm. In this respect, the chimpanzee is similar to *Zinjanthropus* and modern man, while the orang-utan is closer to Pekin Man. The gorilla tympanic seems to be highly variable, the porus sometimes protruding further laterally than the supramastoid crest just above.

In *Zinjanthropus*, as in modern man, the tympanic plate is orientated nearly vertically. It extends from an upper margin, which meets the squamous portion of the temporal bone along the tympanosquamosal fissure laterally and the petrotympanic (Glaserian) fissure medially, downwards and somewhat backwards to an inferior margin. This inferior margin is a free petrous crest in its medial half, while its lateral half comes into direct contact with the mastoid process. The anteriorly inclined face of the tympanic plate forms the posterior wall of the mandibular fossa. In anthropoid apes, the inclination of the plate is much more nearly horizontal, there being an anterior and a posterior margin, instead of a superior and an inferior margin (Weidenreich, 1943, p. 53). *Zinjanthropus* resembles *H. e. pekinensis* in being intermediate between modern man and the apes in this respect, though it approaches more closely to modern man.

Although the tympanic plate is most variable in the hominoids, in the Homininae it seldom reaches the great mediolateral elongation and anteroposterior restriction it often shows in anthropoid apes. In *Zinjanthropus*, the tympanic plate is large both mediolaterally and supero-inferiorly: thus, its proportions resemble more closely those of *H. e. pekinensis* and of modern man than those of the apes. It may be noted here that, as the eighth feature in his list of twenty diagnostic criteria of *Zinjanthropus*, Leakey (1959a, p. 492) listed 'the shape and form of the tympanic plate, whether seen in *norma lateralis* or in *norma basalis*'. He added, 'In this character the new skull has similarities with the Far Eastern genus *Pithecanthropus*.'

The surface of the tympanic plate in *Zinjanthropus* is concave from side-to-side and from above downwards, although the supero-inferior concavity is somewhat interrupted by a slight transverse crest. This crest serves to demarcate the superior part of the tympanic plate which participated in the glenoid articulation, from the inferior part which apparently did not. However, the transverse crest does not alter the fundamentally concave surface of the tympanic, which thus differs from the plane or even convex surface in *H. e. pekinensis* and anthropoid apes.

The thickness of the tympanic plate is most apparent at its lateral edge where it partially circumscribes the external acoustic porus. In *Zinjanthropus*, this border has a striking arrangement (Fig. 6): it is of variable thickness around the periphery of the meatus, being thinnest in the floor of the meatus where, in modern man, it is often thickest. The thickest part in the Olduvai cranium is the postero-inferior portion which attains a thickness practically as great as the anteroposterior diameter of the porus itself (about 8·5 mm.)! The anterior part is also thickened and shows a curious doubling as though the anterosuperior margin of the plate had folded over on itself immediately behind the postglenoid process (pl. 23A and Fig. 6). The inner lamina of this doubled portion is of the same thickness as the thin floor of the meatus (2·5 mm.), but, if one takes the combined thickness of the bilaminar structure, it is 5·5 mm., making this area the second thickest part of the plate. Thus, in *Zinjanthropus*, the tympanic plate thickens anteriorly and posteriorly to the floor, tapering again towards the posterosuperior limit of the part-ring. The margin is damaged on the left bone.

The arrangement in *Zinjanthropus* is compared with that of various other hominids in Fig. 6. The pattern in MLD 37/38 (*Australopithecus* from Makapansgat) is similar to that in *Zinjanthropus*, save for the absence of the anterior doubling, while in Kromdraai (*Paranthropus*), the floor is very thin and the anterosuperior and posteroinferior thickenings are not detectable. The Pekin crania are very variable in this respect; one has the tympanic plate cleft into anterior and posterior lips; another has a slit, while yet a third has a notch. The curious cleavage into two lips in skull III was initially described in detail by Davidson Black (1931); later, however, Weidenreich doubted the constancy of the cleft and regarded it as an 'abnormality' or individual variation. However, as shown in Fig. 6, at least two other crania of *H. e. pekinensis* show a tendency towards notching or indenting of the lateral margin, though in neither does this tendency go as far as in skull III. It is interesting to note this tendency towards thinning or cleavage in the floor of the meatus in three out of six Pekin crania, in a similar position to that in which *Zinjanthropus* and other australopithecines have the thinnest portion of the tympanic rim. The inferior tympanic thinning in the Australopithecinae and the tendency towards inferior tympanic indentation or cleavage in *H. e. pekinensis* may be different expressions of the same phenomenon—namely, a poor or incomplete degree of fusion between the two developmental moieties of the tympanic bone. Thus, the two thickest parts of the tympanic margin in these early hominids would correspond to the two areas in which the ossific centres of the tympanic ring make their first appearance.

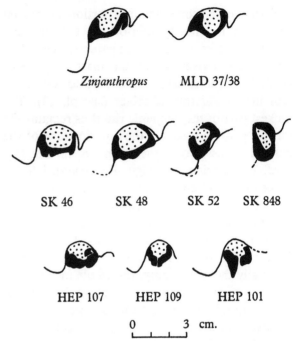

Fig. 6. Variations in the lateral margin of the tympanic plate in *Zinjanthropus*, *Australopithecus* (MLD 37/38), *Paranthropus* (SK 46, SK 48, SK 52 and SK 848) and *Homo erectus pekinensis* (HEP 107, HEP 109 and HEP 101). The tympanic plate is shown in black, the external acoustic meatus is stippled, while the postglenoid process, the mastoid process and the supramastoid crest are shown in outline. (The diagrams for SK 48, SK 52 and SK 848 have been taken from Robinson, 1960; those for the Pekin crania from Weidenreich, 1943; while those for *Zinjanthropus*, MLD 37/38 and SK 46 have been drawn from the originals by Mr A. R. Hughes. The diagrams of HEP 109, SK 46 and MLD 37/38 have been reversed from drawings of the left ear.)

If this simple explanation suffices to explain the features of the tympanic rim in the two fossil groups, it may pertinently be enquired how it comes about that, from similar ontogenetic origins, the floor of the meatal aperture in modern man may be said to be the *thickest* part of the ring (cf. Weidenreich, 1943, p. 54). In fact, the part of the tympanic periphery which constitutes the floor (or inferior wall) of the meatal aperture, with the skull in the Frankfurt Horizontal, seems to have altered in the development of modern man. A study of a large number of modern crania shows that, whereas in the australopithecines the thickest part of the rim is postero-inferior, in modern man the corresponding portion has moved forward to lie directly inferiorly to the aperture. This forward movement is apparently related to the forward movement of the mastoid process, which in modern man points further anteriorly than in the australopithecines (*see* pl. 23). The thickest part of the tympanic rim thus remains the portion plastered against the anterior face of the mastoid process; only the latter, in moving forward, has carried the originally postero-inferior thickening into an inferior position, whilst the thinner intermediate part which, in *Zinjanthropus*, lies in the floor, has in modern man been 'pushed' into the anterior wall of the meatus. The relationships of the thin and thick parts thus remain essentially the same; only, the whole meatal wall has been rotated into a more vertical disposition. The rotation is clockwise on the left side and anticlockwise on the right. A study of Weidenreich's illustrations of temporal bones in Pekin Man, as well as of casts, shows that the amount of rotation is approximately intermediate between that in *Zinjanthropus* and that in modern man; the thickest part of the rim still lies just behind the midst of the meatal floor.

Aside from the rotation of the meatal wall, there has, of course, been a general thinning of the tympanic plate in modern man, as Weidenreich (1943, p. 54) has stressed. This thinning seems to have affected the rim less than it has the body of the plate: in modern races, the rim often becomes much thickened. The thinning or lesser development of the body of the plate may result in a deficiency, the foramen of Huschke, persisting in the tympanic plate to adulthood. On one tympanic of the adult Sterkfontein cranium, Sts 5, there is a breach of continuity strongly suggestive of a foramen of Huschke; the same is true on both left and right sides in the young adult, Sts 19, in which the M^3 is just erupted and in occlusion. These gaps in the tympanic (which I have observed on the original specimens) can just be detected in Text-figure 9, Plate 2 (fig. 7), and Plate 3 (Fig. 10) of Broom and Robinson (1950). In *Australopithecus* of Makapansgat, MLD 37/38, each tympanic has a few tiny perforations instead of a single foramen of Huschke. In a juvenile Swartkrans specimen, SK 47, with P^4 and M^3 erupting, there are again suggestions of Huschke foramina. The last specimen would have been a little younger in individual age than *Zinjanthropus*, which has no foramen of Huschke. The foramen of Huschke occurs with variable incidence in modern cranial series. Thus De Villiers (1963) found it in 1·2 per cent of 740 South African Bantu negroid crania; Wood Jones (1931*a*) in 3 per cent of 100 Hawaiian crania; Krogman (1932) in 5·5 per cent of 146 Australian aboriginal crania, and Akabori (1933) in 12 per cent of North Chinese crania. The values may rise as high as 30·7 per cent in 41 Bushman crania (De Villiers, 1963), 32 per cent in 66 prehistoric crania from Guam (Wood Jones, 1931*b*; Akabori, 1933) and 33·3 per cent among certain tribes of Amerindians (Oetteking, 1930). This foramen, which represents the persistence into adulthood of a normal infantile and juvenile gap, occurs in the area where the two ontogenetic components of the tympanic plate normally meet: thus, it too is a reflection of poor or incomplete fusion between these two elements, as manifested in a cranium with an overall *thin* tympanic plate. A thin meatal floor and a notched or cleft meatal rim, as in some specimens of *Homo erectus*, reflect a similar process in crania with a *thick* tympanic plate. A developmental connection between the two forms of defect —the cleft or notched meatal rim and the foramen of Huschke—was recognised by Weidenreich (1943, pp. 54–5). In support of such a link, he cited cases in which the free border was indented and the fissure terminated in a circular foramen; he illustrated an example published by Bürkner in 1878. These various manifestations—thinning of the tympanic floor, marginal indentation and a foramen of Huschke—represent in varying degrees the persistence of a juvenile condition into adulthood, a phenomenon to which the term pedomorphism has been applied.

5. *The external acoustic porus (meatal aperture)*

The form of the external acoustic porus in *Zinjanthropus* is elliptical. Its long axis is nearly vertical, sloping slightly backwards, though not as far backwards as is the *wall* of the meatus (pls. 23 and 25). Thus its form may be described as obliquely to vertically elliptical. The height of the aperture on the right is 11·6 mm. and its anteroposterior diameter 9·7 mm. The vertically elliptical form seems to prevail in modern man, while in *H. e. pekinensis*, cranium V was described as having a 'vertically elliptic' meatal aperture; cranium XI a 'round' aperture; and crania III and XII a 'horizontally elliptic' form, the anteroposterior diameter in the latter two crania exceeding the vertical diameter (Weidenreich, 1943, p. 55). The form of the meatal aperture is variable in anthropoid apes: Weidenreich (*loc. cit.*) comments that there are chimpanzee skulls with round, horizontally and vertically elliptical apertures. Nevertheless, he comments, 'It is of great interest that Broom's *Paranthropus robustus* plainly deviates from the anthropoids and approaches the hominids in this regard. The aperture is round with a diameter of 11 mm. in each direction.'

Zinjanthropus and the other australopithecines agree with the Homininae in having relatively larger meatal apertures than do the apes. In the latter, the diameters hardly ever exceed 10 or 6 mm. in one or the other direction (Weidenreich, 1943, p. 56); in *Zinjanthropus* the respective values are nearly 12 and nearly 10 mm.; in other australopithecines, too, the diameters are large, while in *H. e. pekinensis*, the average for the greater diameter is 11·5 and 8·9 mm. for the lesser. In size of meatal aperture, the australopithecines are clearly aligned with fossil hominines. The dimensions in modern races of man are very variable, those of Europeans being greater than those of Eskimos and Amerindians (Stewart, 1933).

6. *The vaginal process and the styloid process*

The more medial part of the inferior margin of the tympanic plate is thrown into a very prominent vaginal process of the styloid process; here, as in *H. e. pekinensis* (Davidson Black, 1931), no trace of a styloid process can, however, be detected. Weidenreich (1932, 1943) drew a distinction between the *vagina processus styloidei* and his spine of the crista petrosa: the latter in *H. e. pekinensis*, he said, rises a good deal more medially than the vagina normally would. In their relationships, the spine of the crista in *H. e. pekinensis* and the process in *Zinjanthropus*, which I have above called the vagina processus styloidei, correspond closely with each other and with the vaginal process of modern man; immediately medial to each is the jugular fossa and lateral to each the front of the mastoid process; posteromedial to each is the stylomastoid foramen. Thus, they occupy corresponding positions. In the one group, it seems, the styloid process is missing, in the other it is present. Clearly, the two formations are related to each other. *Zinjanthropus* agrees with *Paranthropus* (Broom and Schepers, 1946, p. 92), *Australopithecus* (Sts VIII, MLD 37/38, etc.), *H. e. pekinensis* and pongids in having no ossified styloid process.

7. *The petrous part of the temporal bone*

The petrous portion as seen from the external surface lies in almost the same axial line as does the tympanic plate, the angle between the two axes being 30° on the left and 35° on the right. The intersection between the two axial lines lies close to the carotid foramen. This approach to a co-axial orientation of the two parts is encountered as well in Sts VIII, MLD 37/38 and some of the Swartkrans *Paranthropus* skulls, and closely approaches the position in *H. e. pekinensis*, in which a similar angle exists between the axes of the two parts. In modern man, the angle of axial deviation is very slight indeed, whereas in anthropoid apes the deviation is much greater. Thus, the angle between the tympanic and petrous axes varies from 60° to 73° in a small series of gorilla crania, from 53° to 75° in a small series of chimpanzee, and is 55–58° in orang-utan. The *Zinjanthropus* values of 30–35° fall well outside the pongid ranges. *Australopithecus* and *Paranthropus* are intermediate between pongids and modern man in this feature, but lie closer to modern man. Weidenreich (1943, p. 57) has pointed out that the reason for the difference among hominoids is an alteration not in the direction of the axis of the tympanic plate, but in that of the petrous pyramid, which turns from a more anteroposterior direction in anthropoids to a more transverse one in modern man. This change may be expressed by the angle which the petrous axis makes with the median sagittal plane, an angle which might conveniently be called the *petro-median angle*. In Table 7, values are quoted for *Zinjanthropus* and other australopithecines, in comparison with those of other hominoids for which Weidenreich (1943) has quoted values or for which I have measured the angles.

Table 7. *The petro-median angle in hominoid crania*

(The angle between the axis of the petrous portion and the median sagittal plane, measured on the basis cranii externa)

Cranium	Angle (deg.)	Reference
Modern hominines		
Bantu negroid ♂	38	Present study
Bantu negroid ♂	44	Present study
Bantu negroid ♂	46·5	Present study
Bantu negroid ♀	39	Present study
Bantu negroid ♀	46	Present study
European ♂	63	Weidenreich (1943)
Fossil hominines		
Homo erectus pekinensis III	40	Weidenreich (1943)
Australopithecines		
Zinjanthropus	47·5	Present study
Australopithecus Sts 5	32·0	Present study
Sts VIII	32·0	Present study
MLD 37/38	33·5	Present study
Pongids		
Orang-utan ♂	30	Weidenreich (1943)
Orang-utan ♀	17·5	Present study
Gorilla ♂	10	Present study
Gorilla ♂	15	Weidenreich (1943)
Gorilla ♂	23	Present study
Chimpanzee ♀	16·5	Present study
Chimpanzee ♀	20	Present study

The individual values in three specimens of *Australopithecus* cluster closely (32–33·5°), just outside the maximum value of the angle recorded for a pongid (30°), and a little short of the lowest value recorded for a modern man (38°). *Zinjanthropus*, however, with a value of 47·5°, falls well outside the pongid range and well inside the range for modern man. It is of interest to note that in a small series of prognathous negroid crania, the values range from 38 to 46·5°, whereas in a single orthognathous Caucasoid cranium measured by Weidenreich the value is given as 63°. Similarly, in the more prognathous *Australopithecus*, the angle measures from 32 to 33·5°, whereas in the more orthognathous *Zinjanthropus*, the value is 47·5°. It is suggested that these two features are seemingly associated in this way, because both gnathism and the angle between the petrous and the median plane are related to a third, common factor, the shortening of the cranial base. It would seem that both the hyperprognathism and the low petro-median angles of apes reflect a state of maximum expansion (or lack of constriction) of the cranial base. In modern man, with lesser grades of prognathism and wider petro-median angles, there is evidence of marked contraction or rolling up of the cranial base. The australopithecines and Pekin Man lie more or less between these two extremes, the more prognathous *Australopithecus*, with its lower petro-median angle, being closer to the upper limits of the pongid range, the less prognathous *Homo erectus* and *Zinjanthropus* having wider petro-median angles and lying well within the modern human range.

Thus, the change in orientation of the pyramid axis, together with a certain shortening of the pyramid, are already present in the Australopithecinae, as part of a general shortening of the basis cranii, especially the spheno-occipital part. This shortening may be expressed metrically by a *spheno-occipital index* which relates the distance from basion to hormion (the point where the posterior border of the vomer underlying the sphenoid intersects the median plane) to the maximum cranial length. In *Zinjanthropus*, the values are 25 and 173 mm., giving an index of 14·5 per cent. Weidenreich's values for recent man range from 11·8 to 18·5 per cent; for adult gorilla from 23·9 to 29·0 per cent; for adult chimpanzee from 22·5 to 32·0 per cent and for adult orang-utan from 30·8 to 39·9 per cent. *Zinjanthropus* thus lies well within the range for modern man. The value in *Australopithecus* (Sts 5) is 21·5 per cent, between the pongid and hominine ranges. The australopithecine values provide an interesting confirmation of the view that the shortening of the base is not simply a relative shortening caused by enlargement of the brain-case: for the basion-hormion measurements and indices of the australopithecines are small, despite the relative non-expansion of the brain-case. The absolute distance from basion to hormion in three australopithecines (pl. 8) is 25–31·5 mm., and in modern man 22–31 mm.; while in adult pongids it is 30–41 mm. (chimpanzee), 39–50 mm. (gorilla) and 37–48 mm. (orang-utan). As in *H. erectus*, there are clear indications that the shortening of the spheno-occipital part of the cranial base has begun in australopithecines, though it has not yet reached the degree attained in modern man.

8. *The mandibular fossa of* Zinjanthropus

The mandibular fossa is perfectly preserved on the right (pl. 12B), every detail of the fossa, the post- and entoglenoid processes and the tympanic bone being clearly preserved. On the left (pl. 14A), a triangular area of the anterior wall and articular tubercle is missing from the specimen, likewise a triangular wedge of the anterior and inferior parts of the tympanic bone. There is also some slight loss of bone in the deepest part of the floor of the fossa. Despite these deficiencies, the main features of the anatomy of the left fossa are clear.

The first important feature is that the fossa is very large, especially in mediolateral breadth. It is difficult to express its dimensions metrically, because of the absence of clearly defined landmarks. The measurements in Table 8 are based on those employed by Weidenreich (1943, p. 46). They are compared with measurements I have made on the fossae of the latest Makapansgat cranium, MLD 37/38 (Dart, 1962a), the original Kromdraai specimen, and a Swartkrans cranium, SK 48, through the kindness of Professor R. A. Dart and Dr J. T. Robinson. In the table, I have included data for Homininae and Pongidae culled from Weidenreich (1943, p. 46).

The anteroposterior length of the fossa in *Zinjanthropus* (27·6, 27·9 mm.) is practically as great as in the australopithecine with the longest fossa, namely *Paranthropus* from Kromdraai (28·1 mm.) and as in two Javanese specimens of

Table 8. *Dimensions and indices of the mandibular fossa in* Zinjanthropus *and other hominoids* (mm.)

	Length	Breadth	Depth	L/B index	D/L index	D/B index
Zinjanthropus (L)	c. 27·6	35·9	8·5	c. 76·9	c. 30·8	23·7
(R)	27·9	32·9	8·8	c. 84·8	31·5	26·7
Paranthropus*						
SK 48 (L)	c. 26·4	c. 31·4	?9·5	c. 84·1	c. 36·0	c. 30·3
(R)	c. 25·2	c. 32·4	9·0	c. 77·8	c. 35·7	c. 27·8
Kromdraai† (L)	28·1	31·5	10·0	89·2	35·6	31·7
Australopithecus*						
MLD 37/38 (L)	23·1	31·6	7·5	73·1	32·5	23·7
(R)	22·3	29·3	8·0	76·1	35·9	27·3
Homo erectus						
H. e. pekinensis (mean)	18·8	25·0	13·9	75·2	74·0	55·6
H. e. erectus II	28	?23	13	?123	46·4	56·5
H. e. erectus IV	28	?28	18	?100	64·3	64·3
Homo sapiens						
European ♂ (mean)	23·5	21·5	12·5	109·5	53·2	58·2
New Caledonian ♂ (mean)	27·0	26·0	16·5	104·0	61·1	63·4
Amerindian ♀ (mean)	23·0	26·0	16·0	88·4	69·5	61·5
Pongidae						
Gorilla ♂ (mean)	27	46	10	58·7	37·1	22·1
Chimpanzee ♂ (mean)	25	29	7	86·3	27·9	24·1
Orang-utan ♂ (mean)	18	40	9	45·0	50·0	22·5

* These measurements were made by myself on the original specimens, through the courtesy of Dr John T. Robinson and of Professor R. A. Dart.

† The Kromdraai dimensions, presumably based upon a cast, were given by Weidenreich (1943) as 27, 29 and 11·5 mm. respectively (Table X, p. 46).

H. erectus (28 mm.). It is slightly bigger than the mean for male New Caledonians (27 mm.) and for male gorilla (27 mm.). The *range of means* in the hominines is 18·8–27 mm., that in the pongids almost exactly the same (18–27 mm.), and the *range of individual values* in the australopithecines 22·3–28·1 mm.

The mediolateral breadth of the fossa sorts the hominoids somewhat more effectively. The values in *Zinjanthropus* (35·9, 32·9 mm.) are absolutely greater than in three other australopithecines, *H. erectus*, *H. sapiens* and male chimpanzee; on the other hand, they are smaller than in male orang-utan and male gorilla. The *range of means* in the hominines is 21·5–26·0 mm., that in the pongids 29–46 mm., while the *range of individual values* in the australopithecines is 29·3–35·9 mm. The large breadth of the mandibular fossa in the australopithecines is in keeping with their large mandibular condyles, and is a feature which tends to align this group with the pongids rather than with the hominines.

The shallow depth of the mandibular fossa is another pongid-like feature of the australopithecines. The depth of 8·5–8·8 mm. in *Zinjanthropus* falls within the australopithecine range of 7·5–10·0 mm., and tallies well with the range of pongid means (7–10 mm.). On the other hand, a striking feature of the hominine mandibular fossa, as Ashton and Zuckerman (1954) have shown, is that it is usually anteroposteriorly 'compressed', i.e. deeper and shorter than in apes. This is reflected in the hominine range of mean depths, namely 12·5–16·5 mm.

The relative dimensions of the mandibular fossa may be expressed by various indices. Thus, the depth–length and depth–breadth indices of *Zinjanthropus* and other australopithecines are closer to the pongid than to the hominine values (Table 8). On the other hand, the length–breadth indices (per cent) of the australopithecines (73·1–89·2) overlap the bottom of the range of hominine values (75·2–±123), and, especially, are near the mean value for three specimens of *H. e. pekinensis* (range 72·0–78·3), while they overlap the top of the range of pongid means (45·0–86·3). Only the chimpanzee has a mean L/B index (86·3) well within the hominid range; whereas the means for gorilla (58·7) and orang (45·0) fall well below the values in Australopithecinae and Homininae.

The significance of the differences for these dimensions and indices could not be tested, as Weidenreich did not provide standard deviations for his hominoid means, nor the individual data from which they could be computed. It seems clear, however, that the anteroposterior 'compression' of the mandibular fossa which is so marked a feature of the hominines is not yet apparent in the Australopithecinae. Ashton and Zuckerman (1954) have shown that the mandibular fossa in the young human skull is no more compressed than it is in the ape skull of corresponding age, thus confirming an earlier observation by Petrovits (1930). The differences between the fossae of adult pongids and hominines, they conclude, are 'essentially due to the fact that in the ape growth appears to take place in all directions, whereas in man relatively less growth occurs in the anteroposterior axis of the fossa than in other directions' (p. 46). While, of course, the fossa itself does not grow, its walls and related structures do grow and so does the condyle of the mandible lodged in the fossa. The shape of the fossa would seem to depend upon an interaction between the rate and direction of growth of the condyle of the mandible and those of the surrounding structures which form the walls of the mandibular fossa. It may well be that the limiting factor in the growth of the fossa is not so much uni- or multi-axial growth of the walls of the fossa, as the growth potential of the head of the mandible, since its size and shape are reflected in the contours and dimensions of the mandibular fossa. The size and shape of the head of the mandible, in turn, most probably bear some relationship to dental size, as well as mode of occlusion and mastication. In that event, the breadth and depth of the australopithecine fossa might be a function *inter alia* of the large mandibular tooth size which is characteristic of this group, as it is of the pongids.

The morphological characteristics of the mandibular fossa in *Zinjanthropus* are very similar to those in other australopithecines.

The anterior wall of the glenoid fossa is well-curved mediolaterally. From without inwards, this wall slopes posteromedially from the anterior root of the zygoma to the entoglenoid process. A third slope to this surface extends from antero-inferior (the 'summit' of the articular tubercle as seen from below) to posterosuperior (the deepest part of the floor of the mandibular fossa), this face sloping at an angle of about 45° to the F.H. The face is smooth as far as its lower edge, and thereafter we find a linear roughness marking the lowest part of the articular tubercle. This combination of characteristics of the anterior wall—which it would be an oversimplification to dismiss simply as the height (or depth) of the articular tubercle—has not been seen in any one of a great number of pongid crania examined in South and East Africa and in the United Kingdom. Generally, the anterior wall of the fossa in pongid crania is situated more transversely, instead of sloping from anterolateral to posteromedial; and its anteroposterior slope is at a much smaller angle to the F.H. than 45°—sometimes indeed the anterior wall is in a plane parallel to the F.H. On the other hand, the pattern of the anterior wall of the mandibular fossa in *Zinjanthropus* finds an almost exact parallel, although of course on a smaller scale, in the Broken Hill cranium of Zambia, and in the Neandertal cranium of La Chapelle-aux-Saints, as well as in a number of crania of modern Bantu-speaking negroids of Southern Africa. It is also the typical pattern of the australopithecines, such as SK 46, SK 48, Kromdraai type, MLD 37/38, Sts 5 and Sts VIII.

Apart from the angle of the anterior articular surface, which we have discussed in the preceding paragraph, Weidenreich (1943, p. 47), following Lubosch (1906), has stressed that two other factors influence the relative development or accentuation of the articular tubercle. The first is the form and size of the fossa itself: the wider and flatter the fossa anteroposteriorly, the less pronounced the tubercle. This factor, in turn, is influenced, in part at least, by the conformation of the tympanic plate: where the latter is nearly horizontal, as in many pongid crania, the fossa is represented by a mere furrow; where it is nearly vertical, as in modern man, the tympanic plate forms the posterior wall of a deep fossa. The second factor is the orientation of the basal surface of the alisphenoid (the roof of the infratemporal fossa or *preglenoidal plane* of Weidenreich in front of the articular tubercle). Sometimes this preglenoidal plane is more or less horizontal; sometimes, it slopes upwards from the articular tubercle as far as the superior orbital fissure. If, as in *Zinjanthropus*, the surface slopes upwards, the articular tubercle is rendered more prominent, since the angle between the preglenoid surface and the anterior wall of the mandibular fossa is more acute. If the *absolute* slope of the anterior articular wall is likewise more vertical, as in *Zinjanthropus*, this will further increase the acuteness of the angle and yet further enhance the salience of the tubercle.

In their study of the mandibular fossa in man and apes, Ashton and Zuckerman (1954) expressed the degree of development of the articular tubercle solely in terms of the projection depth of 'the point of maximum convexity of the articular eminence' below the deepest point of the mandibular fossa. They apparently omitted to take into consideration either the absolute angle of slope of the front wall of the articular fossa or the plane of the roof of the infratemporal fossa (the preglenoidal plane of Weidenreich). Until these other aspects are fully investigated, there is no adequate basis upon which to accept their conclusions with regard to 'the prominence of the articular eminence' in modern men and apes, nor in regard to the australopithecines, of whom they concluded, 'An examination of plaster casts provides little reason for supposing that the condition of the australopithecine articular fossa is materially different from that found in the great apes' (Ashton and Zuckerman, 1954, p. 49). On the contrary, the properties of the anterior wall of the mandibular fossa of the australopithecines sharply distinguish the crania of this group from all pongid crania examined for these features.

The *entoglenoid process* is large and robust, being much thicker on the right than on the left. It is formed entirely by the squamous part of the temporal and not at all by the alisphenoid, the sphenosquamosal suture lying wholly medial to the process. This tallies with the arrangement in

Paranthropus, but differs from that in *Australopithecus* as shown in MLD 37/38 (Fig. 7) and Sts 19. In the latter, the sphenosquamosal suture passes across the entoglenoid process, virtually dividing it into squamosal and alisphenoid moieties. The two parts are subequal in size. This arrangement in MLD 37/38 and in Sts 19 is very much closer to the modern human pattern, in which the major part of the process occurs on the alisphenoid (and is known as the sphenoid spine), while the entoglenoid lip is a small, slightly downturned tongue of the squamosal, abutting on the lateral face of the sphenoid spine, with the sphenosquamosal suture intervening. In some modern human crania, indeed, there is no trace of the entoglenoid lip; then the lateral face of the sphenoid spine forms the inner wall of the mandibular fossa.

In *H. e. pekinensis*, on the other hand, the arrangement closely resembles that in *Zinjanthropus* and *Paranthropus*, in that the entoglenoid process is wholly squamosal, the suture lying immediately medial to the process. However, whereas in *Zinjanthropus* even foramen spinosum is in the squamous bone, in the Pekin crania this foramen is bounded mainly by the alisphenoid (as is foramen ovale), but is completed laterally by the squamosal, i.e. the suture intersects the lateral part of foramen spinosum: an arrangement which is a little closer to that of modern man. Thus, in MLD 37/38 and in Sts 19, the relative positions of foramen ovale, foramen spinosum, the sphenosquamous suture and the entoglenoid process are intermediate between those of Pekin Man and of modern man, i.e. well within the range of variation in the Homininae.

Figure 7 shows the relative arrangements in the four groups discussed: it is seen that there is a morphological series with *Zinjanthropus*/*Paranthropus* at one extreme and modern man at the other.

When we compare the arrangements in these hominids with those of pongids, we see that, in general, there is a tendency in gorilla for the structures to be crowded medialwards on to the alisphenoid, whereas in both orang-utan and chimpanzee there is a lateral tendency towards the squamosal. The following are some detailed observations:

	Entoglenoid process	Foramen spinosum	Foramen ovale
Gorilla 1	Overrides the sphenosquamous suture	On alisphenoid	On alisphenoid
Gorilla 2 (L)	Overrides the sphenosquamous suture	On process of squamosal extending into alisphenoid	On alisphenoid
Gorilla 2 (R)	Overrides the sphenosquamous suture	No separate foramen	On alisphenoid
Gorilla 3	Overrides the sphenosquamous suture	On alisphenoid	On alisphenoid
Gorilla 4	Overrides the sphenosquamous suture	No separate foramen	On alisphenoid
Orang-utan 1	On squamosal	No separate foramen	Overrides the sphenosquamous suture
Orang-utan 2	On squamosal	No separate foramen	Overrides the sphenosquamous suture
Chimpanzee 1	On squamosal	Overrides the sphenosquamous suture	On alisphenoid
Chimpanzee 2	On squamosal	Overrides the sphenosquamous suture	On alisphenoid
Chimpanzee 3	On squamosal	Overrides the sphenosquamous suture	On alisphenoid
Chimpanzee 4 (pygmy)	On squamosal	Overrides the sphenosquamous suture	On alisphenoid

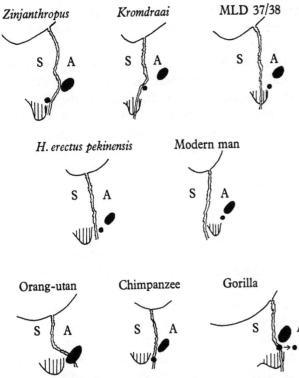

Fig. 7. Diagrams to show the relationships of structures on the basis cranii, in *Zinjanthropus* and a variety of other hominoids. S = squama of temporal bone; A = alisphenoid (greater wing of sphenoid bone). The double serrated line represents the sphenosquamosal suture; the dark oval the foramen ovale; the dark circle the foramen spinosum; and the hatched protuberance the entoglenoid process (or sphenoid spine, in the case of modern man). All the diagrams were based on a study of original specimens, save that for *H. e. pekinensis*, which is based on the illustrations and textual descriptions of Weidenreich (1943).

The chimpanzee is intermediate in these relations between *Zinjanthropus* and *H. e. pekinensis*; the orang would seem to be at the squamosal extreme of the range; while the gorilla, like MLD 37/38, is between *H. e. pekinensis* and modern man.

In the pongids with medium-sized teeth and condylar process of the mandible (orang and chimpanzee), the foramina and entoglenoid process tend towards the squamosal. This trend is carried so far in the orang that even the foramen ovale, which in other pongids lies wholly within the alisphenoid, lies astride the sphenosquamous suture and is completed laterally by the squama! However, in the pongid with large teeth and a large mandibular condyle (gorilla), the foramina and entoglenoid process are displaced, as it were, on to the alisphenoid, so that even the most laterally-situated of the structures being considered, namely the entoglenoid process, straddles the sphenosquamous suture.

These features in the big-toothed, big-jawed gorilla differ interestingly from those of the big-toothed, big-jawed *Zinjanthropus*. In the latter, despite the marked size of the mandibular fossa, both foramen spinosum and the entoglenoid process are wholly within the squamosal, while foramen ovale, although within the alisphenoid, is bounded laterally only by the thinnest tongue of alisphenoid bone separating it from the squamosal. It seems as though *Zinjanthropus* has achieved its very big mandibular fossa, not by displacing the entoglenoid process inwards, but by throwing the outer margin of the fossa outwards. The gorilla, on the other hand, seems to have gained the same end by displacing the entoglenoid process medialwards on to the alisphenoid.

Corroboration of the inference that the glenoid fossa of the gorilla has expanded inwards and that of *Zinjanthropus* outwards is afforded by a study of the relationship of the fossa to the base and sidewalls of the crania, in both creatures. This may be done in at least two different ways.

In the first place, when we examine the relationship of the articular tubercle to the infratemporal fossa in modern man, it is seen that the medial two-thirds or even three-quarters of the mediolateral extent of the tubercle is fronted by infratemporal fossa, before the calvarial wall turns upwards in the line of the postorbital constriction. In *Gorilla* with a relatively less expanded frontal part of the brain and brain-case, about the medial half of the tubercle is fronted by infratemporal surface. In *Zinjanthropus*, with a comparable poor expansion of the frontal part of the brain and brain-case, only about one-quarter to one-third of the mediolateral extent of the tubercle is fronted by infratemporal fossa; the lateral two-thirds to three-quarters protrudes beyond the plane of the side-wall of the calvaria. The striking degree of protrusion, well seen on the norma basalis of *Zinjanthropus* (pl. 6), supports the inference that the enlargement of the mandibular fossa in the Olduvai australopithecine occurred by lateral expansion, in contrast with enlargement by medial expansion in the gorilla.

Secondly, if we measure the distance apart of the left and right entoglenoid processes, they are seen to be further apart absolutely and relatively in *Zinjanthropus* than in the gorilla. An index may be devised relating the distance between the outer faces of the left and right entoglenoid processes to the biporial distance or to the '*biglenoid distance*' (the distance between the outermost points of the glenoid articular surface). The former index (the *interglenoid-biporial index*) gives values (per cent) of 64·5 for *Zinjanthropus* and ?62·3 for *Paranthropus* (SK 48), but only 48·8 and 52·6 for two gorilla crania. The latter index (*interglenoid-biglenoid index*) is 56·2 in *Zinjanthropus* and ?55·0 in SK 48, but 44·2 and 48·4 in two gorilla. These *ad hoc* indices confirm the view that the glenoid fossa approaches nearer the mid-line in the gorilla than in *Zinjanthropus*.

In comparing *Gorilla* and *Zinjanthropus*, we are comparing two forms with large teeth, a large mandibular fossa, and a poorly expanded frontal part of the brain-case. When such hominoids are compared with modern man, with his well-expanded frontal part of the brain and brain-case, it is seen that the very expansion of the brain converts the mandibular fossa from being mainly outside the sagittal plane of the lateral cranial wall, to being mainly or even wholly within this lateral plane—a point which Weidenreich was at pains to emphasise (1943, p. 50).

The possible significance of the *lateral* expansion of the mandibular fossa in *Zinjanthropus*, in contrast with its *medial* expansion in *Gorilla*, lies in the fact that the interglenoid and biglenoid distances reflect the intercondylar and bicondylar distances of the mandible. That is, the tendency of the entoglenoid processes of *Zinjanthropus* to move outwards relative to the cranial base is correlated with a wide intercondylar distance. It seems that *Zinjanthropus* has thus achieved an increase in dental, mandibular and condylar size, without narrowing the space between the two halves of the mandible.

A further thought on entoglenoid topography is that the precise relationships of the entoglenoid process, the foramina and the sphenosquamous suture seemingly depend on two distinct factors. The first is the tendency of the mandibular fossa with its contained condylar process to expand in certain hominoids and to become smaller in others; associated with this factor is the direction of expansion—medialwards or lateralwards. The second is a tendency towards lateral expansion of the alisphenoid with hominisation, the sphenosquamous suture developing progressively more laterally and successively engulfing the foramen ovale, foramen spinosum and, ultimately, part or all of the entoglenoid process. The former tendency —towards expansion of the fossa—is evident in *Zinjanthropus* and other Australopithecinae, but it is an expansion laterally: the second process, the hominising tendency towards lateral expansion of the alisphenoid, is just evident in the australopithecines, the foramen ovale alone having been engulfed by the alisphenoid. The exact relations of the structures and the suture in any hominoid will depend on the resultant between these two developmental trends.

The *postglenoid process* of *Zinjanthropus* is intact on the right but its tip is broken off on the left. It is relatively small and is closely applied to the anterior surface of the tympanic plate, so that its plane continues medially directly into the plane of the anterior part of the tympanic plate. It is not extensive mediolaterally (10·2 mm.) and its maximum height of 7·1 mm. extends downwards for just over half the height of the external acoustic meatus. The process is thus a little shorter mediolaterally and a little higher inferosuperiorly than the process in the Kromdraai *Paranthropus* and a little smaller than that in SK 48. By general hominoid standards it is small and nearer to the hominine condition though, as Ashton and Zuckerman (1954) have shown, an occasional pongid may have so small a postglenoid process. Apart from the abbreviation of its length and height, its thickness in *Zinjanthropus* is strikingly reduced anteroposteriorly as compared with most pongid crania. From a study of pongid crania with damage in this area, it is clear that the general pneumatic inflation of the temporal bone in anthropoid apes extends, too, into the postglenoid process, and that this factor is largely responsible for the common full and rounded appearance of the process in pongid crania. Through the fracture across the tip of the left postglenoid process in *Zinjanthropus*, the internal structure of the process can be clearly visualised: the process is not pneumatised at all in its lower part, the structure being that of a nearly flat sheet of compact bone. This is in spite of the heavy pneumatisation which is so striking a feature of the Olduvai australopithecine cranium.

The limited extent of the postglenoid process leaves uncovered the medial half to two-thirds of the anterior surface of the tympanic plate, which thus provides the medial part of the posterior articular surface. The articulating part of the tympanic is smooth and clearly demarcated, in contrast with the rough surface of the non-articular parts. This face of the tympanic plate is practically vertical to a point just below the tip of the postglenoid process; then it slopes backwards and downwards, to reach its lowest point near the posterior edge of the plate. The arrangement in *Zinjanthropus* tallies with that in *Paranthropus* of Kromdraai (Broom and Schepers, 1946) and of Swartkrans (Broom and Robinson, 1952). It contrasts with the pongid arrangement, in which the tympanic is virtually excluded from the articular fossa by a much larger postglenoid process. At the most, in pongid crania, a small, commonly triangular part of the anterior surface of the tympanic contributes to the joint surface in a posterior recess of the glenoid fossa, between the postglenoid and the entoglenoid processes.

An interesting variation in the postglenoid tympanic area is present on the right temporal of *Zinjanthropus* (pl. 23A). Immediately behind the postglenoid process is a second process of about the same height. However, unlike the true postglenoid process, the second process is not part of the squamosal element; but is attached to the tympanic plate and is similar in texture to the tympanic. It was noted earlier, in the description of the external acoustic porus (p. 31), where it was described as a seemingly folded-over part of the anterior wall of the tympanic, the two layers being continuous above and laterally at the antero-superior angle of the external acoustic meatus. Medially, the two layers blend with each other in the plane of the medial extension of the postglenoid process. Through breakage and loss it cannot be determined whether this feature was present on the left. Although this variation seems to be an individual one, it is relevant to note that Broom and Robinson (1950, p. 19) reported of Sterkfontein 5 that the rather large postglenoid was 'peculiar in having a small secondary process behind it, and it is to this that the tympanic bone is attached'. They added, 'The condition is unlike that in any human skull we have seen, and even more unlike that of any chimpanzee we have been able to examine.' From their description, and from my re-examination of the original, it is clear that the secondary process in Sts 5 is a squamosal rather than a tympanic derivative, so it would apparently not be strictly comparable with the variation in *Zinjanthropus*.

Closer re-study of Sts 5 reveals that the 'small secondary process' is simply a part of the right postglenoid process partially separated from the rest of the postglenoid by a distinct sulcus. This sulcus runs mediolaterally below, near the lowest point of the postglenoid; then it curves smoothly upwards over the posterior surface of the postglenoid process, towards the zygomatic process of the temporal bone, and it disappears just before it reaches the lateral surface of the zygoma. Below and medially, it is deep and narrow (2·0 mm. in width); as it curves upwards, it widens and becomes shallower. A trace of a similar groove is present on the left side in Sts 5 as well. It is this groove which, by cutting deeply into the posterior surface of the postglenoid, partially subdivides the process and conveys the impression that a small secondary process exists behind it. Partial subdivision of the postglenoid tubercle in modern man was reported by Cabibbe (1902). The groove in Sts 5 resembles a vascular sulcus; traced medially, it approaches the squamotympanic (or Glaserian) fissure. It is possible that the groove transmitted a somewhat unusual anterior tympanic artery through the fissure, or through its medial continuation, the petrotympanic fissure, to the middle ear. The anterior tympanic artery is usually an *ascending* branch of the first part of the maxillary artery; it traverses the petrotympanic fissure to enter the tympanic cavity. In Sts 5, we have a groove *descending* across the zygoma and the postglenoid process: one possible explanation is that in this individual from Sterkfontein, the anterior tympanic artery was a branch of the superficial temporal artery, rather than of its usual parent-trunk, the maxillary artery. Alternatively, the groove may have been owing to the presence in Sts 5 of a *petrosquamous sinus* of Luschka. This inconstant venous sinus is considered to represent the continuation of the embryonic transverse sinus to the primitive jugular system. When present in the adult, it enters posteriorly into the transverse sinus close to the junction of the latter with the sigmoid sinus. Anteriorly, it may pass through the cranial wall as an emissary vein, traversing one or other of the so-called 'spurious jugular foramina' (Waltner, 1944), most commonly either a squamosal foramen or a postglenoid foramen (which in the adult may appear as a persistent passage in the line of the squamotympanic fissure), to enter the deep temporal vein or some other indirect tributary of the external jugular vein. The postglenoid sulcus in Sts 5 is well placed to have lodged such a petrosquamous sinus.

C. The sphenoid bone and related structures

The special feature of the alisphenoid in relation to the articular tubercle is discussed in the preceding section (The Mandibular Fossa).

1. *The body of the sphenoid and the vomer*

The fused spheno-occipital part of the basis cranii extends forwards beyond the putative position of the synchondrosis as far as the sphenoid rostrum (the triangular spine on the inferior surface of the body of the sphenoid for articulation with the vomer). Part of the vomer is present, straddling the rostrum (pl. 14A). The sphenoid sinus projects into and inflates the rostrum, which deeply and widely grooves the sphenoid furrow on the upper surface of the vomer. The invading air-sinus has, as it were, forced the vomerine plate and the adjacent wall of the sphenoidal rostrum together on each side and in one area the two juxtaposed layers have fused.

2. *The pterygoid process of the sphenoid bone*

The break between the anterior and posterior portions of the cranium of *Zinjanthropus* has sundered the pterygoid process into two parts (pls. 14 and 20). From behind, a part of the greater wing of the sphenoid containing the foramen ovale protrudes forwards and downwards as a strong elevation, the root or attachment of the lateral pterygoid lamina, with part of the sphenoid sinus invading the base. Nearby, from the anterior calvariofacial part, the lateral pterygoid lamina protrudes backwards: the two parts come very close to articulating, but do not quite meet. In the lateral craniogram of *Zinjanthropus* (Fig. 1, p. 9), the missing parts of the lateral pterygoid lamina have been restored by joining up the tracings of the adjacent parts. Enough is preserved of the laminae to show that both were powerfully developed, broad, high structures, for the attachment of presumably stout pterygoid muscles. Although the sphenoid sinus extends downwards into the base of the pterygoid process, the air cells do not extend far into the laminae themselves (*see* chapter XI, D).

3. *The pyramidal process of the palatine bone*

The palatine bone sends a substantial pyramidal process upwards behind the maxillary tuberosity, to form a large part of the pterygoid fossa between the medial and lateral pterygoid laminae. The same is true of Sts VIII (Broom and Robinson, 1950, p. 31), in which the palatine bone is said to form 'a considerable part' of the pterygoid fossa. The pyramidal process in all pongid crania examined is tiny, whereas in modern man it is commonly a long thin process. Unfortunately few comparative data are available from which to deduce the possible significance of this feature in *Zinjanthropus* and *Australopithecus*.

CHAPTER V

CERTAIN CRITICAL ANGLES AND INDICES OF THE CRANIUM

A. The planum nuchale: tilt and height

1. *The tilt of the planum nuchale*

The planum nuchale of the occipital squama is tilted at only a slight angle to the plane of the foramen magnum which, in turn, is tilted upwards at an angle of 7° to the F.H. (*see* section on 'The Plane of the Foramen Magnum'). Thus the planum nuchale of *Zinjanthropus* is much more horizontal than is that of the Pongidae, in which the nuchal surface rises steeply upwards to the nuchal crest. This low, nearly horizontal planum nuchale is characteristic, too, of *Australopithecus* (e.g. MLD 37/38 and Sts 5), though in the very unusual and apparently deformed cranium VII from Sterkfontein, the planum is much steeper and more pongid in appearance. Broom and Robinson (1950, p. 25) suggested that this unique feature in Sts VII, which they likened to the effects of artificial deformation on Amerindian crania, was due to 'slow postmortem crushing without very manifest breaking of the bones'. In Sts VIII, as well, the planum nuchale 'makes an angle with the main part of the base of the basioccipital of about 49°' (Broom and Robinson, 1950, p. 28). The latter plane is thus not nearly as horizontal as in *Zinjanthropus*. Of the *Paranthropus* crania published, the only ones in which this area is satisfactorily preserved are those of two juveniles from Swartkrans, in which the plane of the nuchal surface is very similar to that in *Zinjanthropus* (Broom and Robinson, 1952, pp. 26–30). In the orientation of the planum nuchale of *Zinjanthropus* and other australopithecines (except Sts VII and to a lesser extent VIII), we have a clear hominid resemblance, differing strongly from the corresponding orientation in pongids.

2. *The height of the nuchal area and of inion*

In 1947, Le Gros Clark stressed that the nuchal musculature of the Australopithecinae was less powerfully developed than in the modern apes. He inferred this from the lower position and poorer development of the nuchal crest in the adult australopithecine fossils then known—from two sites in the Sterkfontein valley. Subsequently, with the greatly extended cranial series following the 1947–8 excavations by Broom and Robinson, Le Gros Clark (1950*b*) was able to determine a 'nuchal area height index': he compared the distance, AG, of the uppermost limit of the nuchal area above the F.H., with the maximum height of the calvaria, AB, above this base-line (Fig. 3, p. 17). It should be noted that the highest point of the nuchal area is commonly not at inion but at the summit of the arched superior nuchal lines lateral to inion. It is this summit which was used by both Le Gros Clark (1950*b*) and Ashton and Zuckerman (1951) in studies on this index; in *Zinjanthropus*, there is a difference of 15·6 mm. in the projection height between inion and the summit of the superior nuchal lines.

The value of the index in Sts 5 is 8 per cent, i.e. the nuchal musculature rises only a short distance above F.H. The low position of the nuchal area is confirmed by observations on two other Sterkfontein crania (Sts I and Sts VII) and on the Makapansgat occipital (MLD 1) (Dart, 1948*a*). The value of the index in *Zinjanthropus* is 8·2 per cent, the nuchal muscles rising to just 7·1 mm. above the F.H. In arriving at this value of the index, I have followed the procedure used by both Le Gros Clark (1950*b*) and Ashton and Zuckerman (1951) on crested gorilla crania, and have included the *Zinjanthropus* crest in the measurement of the maximum height, AB (86·0 mm.). However, as the former has pointed out, the inclusion of the crest has the effect of *lowering* the index. Hence the value 8·2 per cent is slightly lower than it would be if the crest were excluded. When AG is related to maximum calvarial height, measured on both sides of the base of the crest (75·0 mm. in *Zinjanthropus*), the value of the index becomes 9·5 per cent.

The australopithecine values are very different from those of the great apes (Table 9), the means for which range from 50·81 to 76·83 per cent, and they fall well outside the 98 per cent fiducial limits on either side of these means (Ashton and Zucker-

diagnosis of the new genus, *Zinjanthropus*, Leakey (1959a) listed the following as the second item in his list of 20 points: 'The inion, despite the great evidence of muscularity, is set lower (when the skull is in the Frankfurt plane) than in the other two genera.'

This was one of the points queried by Robinson (1960). He wrote: 'In both *Australopithecus* and *Paranthropus* the base of the external occipital protuberance is almost exactly in the Frankfurt plane, as seems to be the case with the Olduvai specimen.' (It is presumed that Robinson was referring to the apex or the lowest point, not the base, of the external occipital protuberance.) To

Table 9. *Differences between the nuchal area height index (AG/AB) of* Zinjanthropus *and other hominoids (per cent)*

Zinjanthropus compared with:	Index in *Zinjanthropus*	Mean indices of hominoids	S.D.	Diff.	Diff./S.D.
English (Spitalfields)	+9·5	− 0·57	6·22	+10·07	+1·62
West African	+9·5	+ 4·52	5·00	+ 4·98	+1·00
Australian	+9·5	+ 3·39	8·83	+ 6·11	+0·69
Gorilla (1st series)	+9·5	+76·83	11·38	−67·33	−5·92
Gorilla (2nd series)	+9·5	+71·00	11·52	−61·50	−5·34
Chimpanzee (1st series)	+9·5	+50·81	6·96	−41·31	−5·93
Chimpanzee (2nd series)	+9·5	+53·30	6·19	−43·80	−7·08
Orang-utan	+9·5	+64·72	7·07	−55·22	−7·81

man, 1951; Zuckerman, 1954). Only in a pygmy chimpanzee did Le Gros Clark find an index as low as 21 per cent; the individual minima for gorilla and orang-utan were 44 and 35 per cent respectively. On the other hand, the two australopithecine values fall comfortably within the ranges for modern man; in fact they lie not far from the means (Fig. 4, p. 18).

Data for *Homo erectus* do not seem to be available, Weidenreich (1943, 1945) having used other indices to demonstrate the relative position of inion and opisthocranion, but none to demonstrate the maximum height of the nuchal area.

The nuchal area height indices of *Zinjanthropus* and of Sts 5 align these australopithecines with the Homininae.

The position of the inion, as distinct from that of the maximum height of the nuchal area, is of importance in *Zinjanthropus*. In his original

this, Leakey replied that, 'whereas in *Paranthropus* and *Australopithecus* (as Dr Robinson says) the external occipital protuberance lies more or less on the Frankfurt plane, in *Zinjanthropus*, it lies below it'.

Leakey's statement is borne out by my study of the Olduvai cranium. When the skull is carefully aligned in the Frankfurt Horizontal, the lowest point on the external occipital protuberance lies 8·5 mm. below the F.H. In this feature, *Zinjanthropus* does not appear to have a parallel in the known specimens of *Australopithecus*, while none of the *Paranthropus* crania is sufficiently undistorted to permit a direct reading to be taken. Robinson's (1961) restoration of *Paranthropus*, possibly influenced by *Zinjanthropus*, shows the inion at or close to the level of the F.H. It seems fairly clear that the strong development of the external occipital protuberance in *Zinjanthropus* is part of the general tendency towards extreme muscularity in this specimen: as such, it is rather doubtful whether the level of the lowest tip of the process could be regarded as of generic value. A closer look at its morphology may throw light on its muscular relations.

The protuberance is robust and down-curved (pl. 26) and at its apex has a nearly round, hollowed area, looking downwards and slightly backwards (pl. 13); to this rounded area would doubtless have been attached the somewhat thickened, dorsal margin of the *ligamentum nuchae*, which, in turn, gives origin to the *trapezius* muscle. The powerful, down-curved protuberance and the hollow attachment area near its apex connote strong development of the trapezius and of the dorsal margin of the ligamentum nuchae. The deeper, anterior portion of the ligament gives origin to the *splenius* muscles which, like trapezius, are strong extensors of the head. It is of interest to mention that the attachment line of this deeper, more ventral part of ligamentum nuchae—the external occipital crest, extending from the protuberance to opisthion in the median plane of the posterior margin of foramen magnum—is not strongly developed at all; it is a thin, median, linear impression, which is not elevated more than 2 mm. from the surface of planum nuchale. This suggests that the part of ligamentum nuchae which was best developed was that associated with trapezius. The relatively rudimentary state of the ligamentum nuchae in man is generally correlated with his erect position. In many pronograde mammals on the other hand, the ligament is well developed and assists the extensor muscles in supporting the head and neck. The evidence in *Zinjanthropus*, of strong development both of the nuchal muscles and of the ligamentum nuchae, suggests a postural function of these structures greater than that subserved in the Homininae. In other words, this evidence points to the probability that the head of the Olduvai creature was not as well balanced on its occipital condyles as in hominines, but tended to incline forwards somewhat. This conclusion, we shall see, is supported by the placement of the occipital condyles, the position index of which is intermediate between those of the Homininae and those of the Pongidae.

The nature of the nuchal crest itself has already been discussed (chapter III).

B. The foramen magnum: position and plane

1. *The foramen magnum position index*

In his first, now historical, paper on the Taung australopithecine published over 40 years ago, Dart (1925) stressed the relatively forward position of the foramen magnum as one of the humanoid characters of *Australopithecus africanus*. He devised a 'head-balancing index', by expressing the basion–inion distance as a percentage of the basion–prosthion length. Estimates of these two measurements in the Taung child gave an index (per cent) of 60·7, as compared with 50·7 in an adult chimpanzee, 83·7 in the fossil cranium of Broken Hill, 90·9 in a dolichocephalic European and 105·8 in a brachycephalic European. Dart commented, 'It is significant that this index, which indicates in a measure the poise of the skull upon the vertebral column, points to the assumption by this fossil group of an attitude appreciably more erect than that of modern anthropoids' (Dart, 1925, p. 197).

Dart's head-balancing index would, of course, alter with other variables than the position of the foramen magnum. Because he used prosthion as the anterior terminal, varying degrees of prognathism would introduce additional variability; furthermore, his use of the absolute distance from basion to inion would likewise bring a further source of variability according as to whether inion was low (as in hominines) or high (as in pongids). Other workers have therefore sought to avoid such sources of error, by using nasion or glabella as the anterior terminal and *projecting* the position of opisthocranion on to a base-line, either the F.H. or the nasion–opisthion base-line.

One method is to project opisthion and opisthocranion on to the F.H. and to measure the horizontal distance between them: this is the *horizontal occipital length*. This value may then be expressed as a percentage of the maximum cranial length to give an *occipital length index I* (Weidenreich, 1943, p. 131). The lower the index the farther back the position of the opisthion; the higher the index the more forward its placement. In *Zinjanthropus*, the horizontal occipital length is 37·7 mm. and the index 21·8 per cent.

This value falls short of the range of indices in *Homo erectus pekinensis* (25·2–26·1 per cent) as well as the range cited by Weidenreich for modern man (25·1–36·9 per cent). On the other hand, it falls within the sample range for the Neandertalians (20·4–29·1 per cent) and outside the range quoted for pongids (10·0–14·1 per cent). That is to say, in *Zinjanthropus* opisthion is placed relatively forward on the base of the cranium, just within the hominine range and well outside the pongid range. The value of 21·8 per cent is undoubtedly somewhat exaggerated by the large nuchal crest which has carried opisthocranion posteriorly; nevertheless, even if allowance is made for this, it remains true that opisthion is farther forward than in pongid crania which have a low value for this index (Fig. 8).

Weidenreich preferred to project opisthion and opisthocranion on to an extension of the nasion–opisthion line. This projective occipital length was expressed as a percentage of the nasion–opisthion length to give an *occipital length index II*. Because the projection of opisthocranion falls behind opisthion, the index was expressed with a minus sign. In *Zinjanthropus*, the projective distance of opisthocranion on this base-line is 36·5 mm. and the occipital length index II is −26·8 per cent. The mean index (per cent) in *H. e. pekinensis* is −24·0, the value in *H. e. erectus* II is −22·5 and in modern man it is −25·5 (if the utmost posterior point is used when the cranium is viewed on the nasion–opisthion base-line), or −21·8 (if the usually defined opisthocranion is determined with the F.H. as base). In the anthropoid apes, the index amounts to only −6·4 per cent, with, according to Weidenreich, a very small range of variation. In the reconstruction of *H. e. erectus* IV, the foramen is 'surprisingly central', the value of the index being over −25, even when the enormous thickness of the occipital torus is deducted. If we make a similar deduction in *Zinjanthropus*, the index drops to −18·7. This is close to the value of −17·5 determined for *Australopithecus* (Sts 5) which has no powerful crest or occipital torus. These

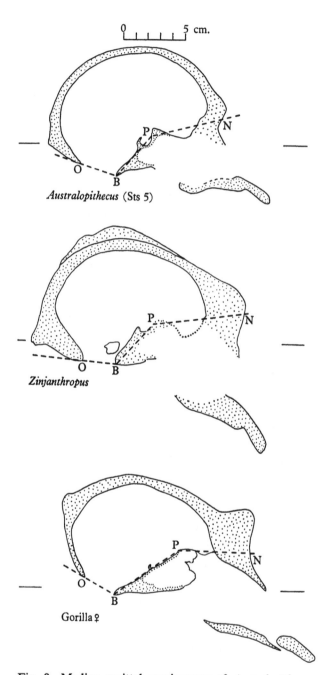

Fig. 8. Median sagittal craniograms of *Australopithecus* (Sts 5), *Zinjanthropus* and a female gorilla, showing the basicranial axis. N = nasion; P = prosphenion; B = basion (ectobasion); O = opisthion. The craniogram of Sts 5 was very kindly prepared and made available by Dr J. T. Robinson, while that of the gorilla was prepared by the Department of Human Anatomy, Oxford, and kindly made available by Professor Sir W. E. Le Gros Clark.

australopithecine values still remain well within the hominine range (the lowest value in a modern human cranium was −11·8 per cent) and well beyond the values for apes, in which opisthion lies close to the utmost posterior point of the skull.

From his studies of *H. erectus*, Weidenreich concluded that, in respect of the position of opisthion, 'the different stages of hominids, represented by fossil finds at hand, show no recognisable tendency to approach the anthropoids more and modern man less. I, therefore, consider the more central position of the occipital foramen, in obvious contrast to the conditions in anthropoids, as a specific character' (1943, p. 132).

It is clear from data on *Zinjanthropus* and Sts 5, that Weidenreich's general statement applies to them as well: the foramen magnum position indices are one further important feature aligning these australopithecines with the Hominidae. This forward position of the foramen magnum reflects itself as well in the forward position of the brain-stem, as seen on the endocranial cast of *Zinjanthropus*, in striking contrast with the posterior placement of the medulla oblongata in the pongid brain (*see* Fig. 13, p. 89, and chapter VIII).

2. *The plane of the foramen magnum*

The plane of the foramen magnum is almost directly horizontal, as a glance at Fig. 8 will show. In this regard, *Zinjanthropus* is even more hominine-like than *Australopithecus* (Sts 5), and contrasts markedly with the pongids in which the foramen faces backwards and downwards. The shape and plane of the foramen magnum constituted the fourth of Leakey's criteria for the genus *Zinjanthropus* (1959a, p. 492). The exact angle made by the foramen with the F.H. is 7·0°; in Sts 5 it is 19·5°; while in a female gorilla, the median sagittal craniogram of which was kindly supplied by Professor Sir Wilfrid Le Gros Clark, the angle is 28·5°. The measurements confirm that the plane of the foramen magnum is almost horizontal in *Zinjanthropus*.

While the angle between the plane of the foramen magnum and the F.H. conveys an impression of the orientation of the plane when the cranium is held in a standard position, it does not provide information on the relationship between the foramen magnum plane and the plane of the rest of the basis cranii. Several methods have been devised to express this relationship.

The first is to measure the opisthion–basion–prosphenion angle on a median sagittal craniogram. In *Zinjanthropus*, unfortunately, the point prosphenion is missing, though the posterior part of the body of the sphenoid is present. However, the position of prosphenion can be estimated with reasonable assurance. In Fig. 8, the median sagittal craniogram of *Zinjanthropus* is compared with those of Sts 5 and of a gorilla. The angle opisthion–basion–prosphenion (OBP) is 126° in *Zinjanthropus*, 113° in *Australopithecus* (Sts 5) and 118° in the gorilla. This angle depends on two different sets of factors operating in opposite directions. On the one hand, the hominid tendency to change the plane of the foramen magnum, from facing somewhat backwards to facing somewhat forwards, will tend to open out the angle OBP. On the other hand, the hominid tendency towards inbending of the basicranial axis will, by raising the point P, have the effect of closing the angle OBP. It might be expected then that little change would come about in this angle during hominisation, since the two trends would presumably cancel out their respective effects on the angle OBP. This inference would seem to be borne out by the fact that the gorilla reading is intermediate between the two australopithecine readings.

The same general comment applies to the use of klition (the point where the median plane intersects the highest part of the dorsum sellae) as an anterior terminus, since klition, like prosphenion, may be expected to be thrust into the interior of the brain-case during the bending of the basicranial axis.

Hence the bending of the cranial base will vitiate any conclusions which may be drawn about either of the two preceding angular measurements of the foramen magnum plane. It is clear that some other anterior terminus should be selected, which would not be manifestly affected by the bending of the basicranial axis. Weidenreich used nasion for this purpose, measuring basion–nasion, nasion–

opisthion and opisthion–basion, as the three sides of a triangle, in order to arrive at the angle between the foramen magnum plane and the nasion–basion axis. The three distances in *Zinjanthropus* are (mm.):

endobasion–nasion	112·5
nasion–opisthion	136·4
opisthion–endobasion	25·6

These three sides of the triangle give an angle at endobasion of 157°. This figure compares very well with the angles in a series of Homininae (*H. e. pekinensis* 156–159°, Ngandong 154°, Broken Hill 153°, La Chapelle-aux-Saints 150°, modern man mean 156°, range 145–171°), but departs considerably from the pongid mean of 127° and range of 121–134°.

In *Zinjanthropus*, therefore, not only does the plane of the foramen magnum face almost directly downwards, thus relating the specimen to fossil hominine crania, but the angle which this plane makes with the nasion–basion line is completely hominine and wider than in anthropoid apes.

The angle at basion will be smaller, the longer the nasion–basion line is in relation to the nasion–opisthion line, and vice versa. Weidenreich was able to deduce that 'in all hominids, including *Sinanthropus*, the length of the nasion–basion line should amount to three-quarters of the length of the nasion–opisthion line, whereas in anthropoids the ratio is much greater, the nasion–basion line totalling 85 or more per cent of the nasion–opisthion line' (Weidenreich, 1943, p. 136). In this passage, by 'hominids', Weidenreich clearly referred only to the subfamily Homininae, since he quoted no data for the other hominid subfamily, Australopithecinae. The value of the ratio in *Zinjanthropus* is 82·5 and in *Australopithecus* (Sts 5) 80·0 per cent. In other words, these two australopithecine crania lie midway between the Pongidae and the Homininae in this respect, the larger faced Olduvai specimen inclining more towards the pongids.

The plane of the foramen magnum in *Zinjanthropus*, both absolute (when the cranium is in the F.H.) and relative (to the nasion–basion plane), thus provides further evidence relating the Australopithecinae—to which *Zinjanthropus* undoubtedly belongs—to the Homininae. The significance of this hominid basicranial feature will be considered later, in conjunction with other such features.

C. The porion position indices

Weidenreich (1943) demonstrated that (1) the anteroposterior position of porion in relation to the nasion–opisthion line differs relatively slightly among anthropoid apes and hominines and, also, among the different evolutionary stages of the Homininae; and (2) there is a clear tendency for the porion to rise from a lower to a higher level during human evolution. It will be interesting to see what the porion of *Zinjanthropus* reveals.

Following Weidenreich, we have made a median sagittal craniogram and projected the exact position of porion on to the mid-sagittal plane. The nasion–opisthion line has been taken as a basis and a perpendicular has been dropped from porion on to this line. The distance between the foot-point of the perpendicular and nasion has been determined; this distance has been expressed as a percentage of the total nasion–opisthion line, to give the *porion position length index*. The value in *Zinjanthropus* is 78·3 per cent; this value is somewhat higher than any of those recorded by Weidenreich (anthropoid apes 64·2–73·3 per cent; *H. e. pekinensis* 66–71·6 per cent; Ngandong 68·5–72·7 per cent; Neandertalians 67–69·3 per cent; modern man 64·8–74·5 per cent). This suggests that in *Zinjanthropus* porion occupies a somewhat more posterior position than in other hominoids. However, the tremendous thickness of the supra-orbital torus has carried nasion far forwards (*see* Fig. 8), thus giving an exaggeratedly high value to the porion–nasion distance and thus to the index. When allowance is made for this, the position of porion probably lies within both the pongid and hominine ranges.

If the height of the perpendicular from porion to the base-line be measured and expressed as a percentage of the total nasion–opisthion line, the *porion position height index* is obtained. The value in *Zinjanthropus* is +5·3 per cent: that is to say, porion is a little distance above the nasion–opisthion base-line. This value lies well outside the range for anthropoid apes, in most of which porion lies below the base-line (mean −3·3, range +1·6 to −8·6 per cent). On the other hand, the value in *Zinjanthropus* is well within the range of the Homininae, being in fact somewhat nearer to those of modern man than are the values for *H. e. pekinensis* and the Neandertalians. The comparative values (per cent) are: *H. e. pekinensis* mean +3·0, range +0·7 to +4·1; *H. s. soloensis* mean +1·2, range −2·0 to +5·5; Neandertalians mean +2·4, range −0·9 to +5·0; modern man mean +6·6, range +2·8 to +10·9.

Zinjanthropus thus clearly shows the characteristic hominine tendency for porion to rise above the nasion–opisthion base-line. This markedly elevated position of porion in the Olduvai australopithecine contrasts sharply with its depressed position in the Pongidae. Since the rise of porion has been shown by Weidenreich to be part of a general hominising transformation of the skull base, it may be inferred that *Zinjanthropus*, too, shows such a hominising pattern in its cranial base, for which inference other evidence has already been considered.

D. The position of the occipital condyles and the poise of the head

Dart's (1925) suggestion of a head-balancing index was re-emphasised by several investigators, including Sollas (1926), Broom and Schepers (1946) and Le Gros Clark (1947). The latter affirmed this forward position not only for *Australopithecus* from Taung, but for *Australopithecus* from Sterkfontein and for *Paranthropus* from Kromdraai. In Kromdraai, Le Gros Clark (1947, p. 308) pointed out that 'the posterior margin of the occipital condyle is on a transverse level with the posterior margin of the tympanic bone, while in the gorilla and chimpanzee it is considerably behind this level'. This was subsequently confirmed for Sts 5 and Sts VII discovered by Broom at Sterkfontein in 1947–8: in cranium 5, the point of maximum convexity of the condyle coincides in transverse level with the posterior margin of the external acoustic aperture, a position which Le Gros Clark (1950b) found in only two out of over 400 ape crania.

To express this hominid feature of Sts 5 metrically, Le Gros Clark devised an index relating the central point on the convexity of the condyle to the maximum length of the cranium in the F.H. In Fig. 3, the relevant measurements are shown and the *condylar position index* is represented by CD/CE × 100. The value of this index in Sts 5 was found to be 40 per cent, or, as re-calculated by Ashton and Zuckerman (1951), 39 per cent. In a series of pongid crania on which this index was determined by Le Gros Clark, the value for Sts 5 was found to lie at or just beyond the highest value (39 per cent in a pygmy chimpanzee skull) for the sample of apes, save for one gorilla skull (B.M. 23.11.29.6), in which the postcondylar segment of the cranium was grossly exaggerated by an extreme posterior prolongation of the sagittal crest, imparting a value for this index of 45 per cent. The mean values (per cent) for the index were found to be 23·9 in 22 adult gorilla crania; 23·0 in 27 crania of the ordinary chimpanzee; 34·8 in 5 pygmy chimpanzee crania; and 20·6 in 34 orang crania. The maxima for these four groups were respectively 32 (excluding the unusual gorilla cranium mentioned above), 34, 39 and 29. Clearly, the position in the *Australopithecus* cranium from Sterkfontein is at or close to the extreme upper limit for Pongidae: the condyles are therefore well forward as compared with the means and ranges in apes. At the same time, Le Gros Clark made clear that the condyles are by no means so far forward relative to the total cranial length as they are in the modern human cranium.

Ashton and Zuckerman (1951) computed a condylar position index in further samples of hominoids, as well as in cercopithecoids and ceboids. Unfortunately, their procedure was not identical with that of Le Gros Clark, as pointed out by him (1952a): they measured the pre- and postcondylar segments of the cranial length from the *lowermost* point of the condyle. In gorilla skulls, especially old males, according to Le Gros Clark (1952a), the condyles frequently slope markedly downwards and forwards, so that the lowermost point would be at the front of the condyle. This would tend somewhat to increase the relative length of the postcondylar segment of the gorilla skull; whereas in Sts 5 and *H. sapiens*, the condyles are more nearly horizontal, so that the lowermost point is near the centre of the condyle. Le Gros Clark had recommended using the central point of the convexity of the condyle. Ashton and Zuckerman also disregarded the influence on the index of excessive development of sagittal and nuchal crests. Nevertheless, their results largely confirmed those of Le Gros Clark; at the same time, although they verified that the cranium of Sts 5 had an index close to the extreme upper limit of apes, they were able to show that its value was not significantly different from the mean for gorillas. It was significantly greater than those of the other pongid crania and significantly smaller than the values for three series of modern human crania. It may be commented that the failure to find Sts 5 differing significantly from the gorilla crania was owing at least in part to the very large variability of this index in the gorillas, a variability which is correlated no doubt with the variable development of the posterior prolongations of cranial crests. Thus, the S.D.'s for the two gorilla samples were 7·45 and 6·28, in contrast with those for the other pongids (4·03 and 3·40 in chimpanzee, 3·87 in orang-utan).

The value for Le Gros Clark's condylar position index in *Zinjanthropus* is 55·1: this was based on measurements made directly on the specimen whilst in the F.H., so there is no need to introduce the correction factor of Ashton and Zuckerman (1951) for photographic distortion. The value is much closer to the means and ranges depicted by these workers for modern human samples. Fig. 4 based on fig. 2 of Ashton and Zuckerman (1951) shows the value in *Zinjanthropus* as compared with those in other hominoids. It would, of course, be far more appropriate to compare the australopithecine values with those of *H. erectus* rather than with those of modern races of *H. sapiens*. Unfortunately, not one of the Choukoutien fossils (*H. e. pekinensis*) has the condylar area preserved; but it would be possible to determine the index in the reconstruction of *H. e. erectus* IV (Weidenreich, 1945). However, Weidenreich has used two occipital indices which express the relation of opisthion (the median point of the posterior margin of foramen magnum) to opisthocranion (the point on the hinder part of the calvaria in the median plane, which is furthest from glabella), in order to show the forward position of the foramen in *H. e. pekinensis*.

We may express the 'distance' of the *Zinjanthropus* value from the means for hominines and pongids in terms of standard deviations, since the latter have been provided for their samples by Ashton and Zuckerman (1951). Table 10 conveys the absolute and standardised values of the differences between *Zinjanthropus* on the one hand and each other group. From the table, it is seen that the value of the index in *Zinjanthropus* occupies an intermediate position between the modern hominine means and the pongid means; it is somewhat nearer to *H. sapiens*, both in absolute and standardised terms, differing from the means by absolute values of 21·87–26·17 per cent and by standardised values of 2·81–3·54 S.D.'s. On the other hand, the value in the Olduvai fossil differs from those of the pongid series by absolute values of 27·60–33·46 and by standardised values of 3·85–9·26 S.D.'s. In respect of this index, *Zinjanthropus* thus lies much closer to the hominines than does *Australopithecus* (Sts 5) with its index of 39·40 per cent. We have already mentioned how, among gorilla crania, the posterior expansion of the sagittal and nuchal crests will raise this index, although of course the position of the condyles relative to the biporial plane will not be altered. Precisely the same effect is seen here: the non-crested australopithecine from Sterkfontein has a relatively low index, while the crested australopithecine from Olduvai has a raised index, although in both, the condyles lie in the coronal plane through the biporial axis. There is clearly, however, a real difference in the average position of the condyles in the gorillas and australopithecines: cresting raises the index of a gorilla to about the same value as in the *non-crested* australopithecine, while cresting in *Zinjanthropus* raises its indicial value to about the bottom of the range for non-crested hominines—and this despite the fact that the degree of cresting in *Zinjanthropus* is not nearly as great as in many gorilla crania.

Accepting such a difference in the average position of the condyles between australopithecine and

CRITICAL ANGLES AND INDICES

Table 10. *Differences between the condylar position index (CD/CE) of* Zinjanthropus *and other hominoids (per cent)*

Zinjanthropus compared with:	Index in Zinjanthropus	Mean indices of other hominoids	S.D.	Diff.	Diff./S.D.
English (Spitalfields)	55·1	81·27	9·32	−26·17	−2·81
West African	55·1	76·97	6·98	−21·87	−3·13
Australian	55·1	78·25	6·54	−23·15	−3·54
Gorilla (1st sample)	55·1	26·07	7·54	+29·03	+3·85
Gorilla (2nd sample)	55·1	27·50	6·28	+27·60	+4·39
Chimpanzee (1st sample)	55·1	24·96	4·03	+30·14	+7·48
Chimpanzee (2nd sample)	55·1	23·62	3·40	+31·48	+9·26
Orang-utan	55·1	21·64	3·87	+33·46	+8·65

gorilla, what may we legitimately infer from the difference? Ashton and Zuckerman (1951) have pointed out that the condylar position index may act as a head-balancing index if the cranium or head is regarded as dead-weight, when supported on the occipital condyles and when orientated in the F.H. 'It also provides a relative measure of the force which must be supplied by the nuchal musculature in order to maintain the poise of the head in the ear–eye horizontal—the heads of presumably all primates, as Schultz has shown, being heavier in front of the occipital condyles than behind, when orientated in this plane' (1951, p. 281).

Ashton and Zuckerman (1952) have demonstrated striking age differences in the condylar position index of chimpanzee and gorilla: from the time of the deciduous dentition up to adulthood, indices dropped from 50–52 to 25–30 per cent. Yet, 'it is our impression that there are no striking, or indeed, obvious differences in the gait of young and adult great apes, nor in the way they carry their heads' (1952, p. 282). To prove the latter point would require far more detailed longitudinal studies of the posture of living great apes than are at present available. In any event, however, it seems to be a *non sequitur* to claim that, since there are no obvious changes in the carriage of the head with changes in the index, 'the precise numerical relation of the occipital condyle–prosthion and occipital condyle–inion dimensions would hardly seem, therefore, to be of much value in showing how the head is poised on the vertebral column...' (1952, pp. 282–3). The age changes in the index may be assumed to reflect the fact that, from birth until all the permanent teeth have erupted, the face of apes grows relatively more than the cranium. It would be a logical inference that, under such circumstances, the load placed upon the nuchal musculature would be progressively increased. Such an increase in load would not automatically lead to an alteration in the poise of the head on the vertebral column; only if the nuchal muscles were *not* equal to the additional force now required of them would one expect a change in the carriage of the head to become manifest. The very fact that no striking change in the position of the head has been observed, despite the increasing weight of the precondylar part of the head, suggests that the nuchal musculature possesses sufficient powers of response to overcome the tendency for the head to sag forwards. One could test this inference that the ontogenetic decrease in the index is accompanied by an increase in activity and in growth of the nuchal muscles in two ways.

In the first place, electromyographic studies of the muscles—if feasible—would throw light on the degree and extent of their activity. In the second place, increasing activity of the nuchal muscles would be expected to have an effect on the nuchal area of the skull and in particular on the nuchal lines or crests. The development of the nuchal crest during the time when the condylar position index is changing is now well established (Zuckerman, 1954; Ashton and Zuckerman, 1956; Robinson, 1958). We may take this as confirmation that, with a preponderance of precondylar growth, the nuchal musculature increases its activity and its modelling influence on the cranium.

The argument thus far relates to differences in the condylar position index between different age groups of the same species. What of differences in index between adults of different groups? With no less assurance, we may see in them a reflection of different degrees of pre- and postcondylar imbalance, hence of different loads upon the nuchal muscles in correcting, counteracting or yielding to the precondylar heaviness. It is a simple physical principle that in a lever, the nearer the fulcrum to the effort force, the greater will that force have to be in order to overcome the resistance of the load force. If the condyles were relatively far back, that is, if the load arm were longer and the effort arm shorter, the tendency to imbalance would be greater and the requisite nuchal muscular force (the effort force) would be larger: it would be less likely that, in the living, the nuchal musculature would be capable of maintaining the head in anything like the F.H. (even supposing the vertebral column were relatively erect). Conversely, if the condyles were relatively further forward, and the lever arms approached equality, the tendency to imbalance would be much smaller, and the nuchal muscular effort required would be less: it would be more likely that, in the living, the nuchal musculature would be capable of maintaining the head in a position approaching the ear–eye horizontal.

Therefore, the fact that the index in *Zinjanthropus* is intermediate between those of erect hominines and of obliquely quadrupedal or semi-erect pongids, suggests that his head was not as well balanced as in man but was better balanced than in the pongids. This conclusion is supported by the bony evidence of his nuchal crest, which is much more strongly developed than is usual in hominines, but not as marked as in many pongids. It might be inferred that the nuchal muscles of *Zinjanthropus* were perforce more active in maintaining any sort of head balance, against the twin challenges constituted by the situation of the condyles, which are not as far forward relatively as in modern man, and by the excessive precondylar heaviness of the greatly elongated face and massive teeth. Confronted by these functional demands, it is likely that the head of *Zinjanthropus* was not maintained as horizontally as in the Homininae, but sagged forward slightly, though not nearly to the same degree as in most of the Pongidae.

CHAPTER VI

THE INTERIOR OF THE CALVARIA

A. The endocranial surface of the frontal bone

Only a small part of the endocranial surface of the frontal bone is present. The angle between the squama and the floor of the anterior cranial fossa is difficult to determine, because only the rostral end of the floor is present; it seems, however, to be much nearer the angle in *Homo erectus pekinensis* (50°) than that of modern man (about a right angle—Weidenreich, 1943, p. 32). Laterally, the posterior turn or recurvation of the squama at the postorbital constriction is clear, giving an extremely narrow frontal region. From the mid-line, the chord to the line of recurvation is only 31·0 mm. The endocranial width between the left and right lines of recurvation is 59·4 mm.

There is a thin though strong *frontal crest*, as in Pekin Man and other hominines. In pongids, on the other hand, it is missing or, at most, is represented 'by a low, insignificant ridge' (Weidenreich, 1943, p. 32). The relationship of the frontal crest to the sulcus for the superior sagittal sinus is described in Section D of this chapter (p. 63). Weidenreich regarded the internal frontal crest as part of the sagittal reinforcing system of the cranial vault of Pekin Man; he drew attention to the fact that the frontal crest fades out just where the sagittal thickening (which he unfortunately designated 'sagittal crest') originates on the outer surface. Hence he considered that the sagittal reinforcing system starts internally at the foramen caecum, extending forwards and upwards as the frontal crest, emerging externally as his 'sagittal crest' (which we might better name *sagittal torus*) and, finally, again manifesting itself internally as the sagittal limb of the cruciate eminence leading down to the foramen magnum (Weidenreich, 1943, pp. 160–1). However, in view of the massiveness of the sagittal torus and of the sagittal limb of the cruciate eminence in Pekin Man, it is difficult to see how the slender frontal crest could be regarded as a part of such a 'reinforcement system'. The frontal crest marks the attachment of the anterior part of the powerful dura mater ligament, the *falx cerebri*: therefore, we should perhaps look to possible variations in the degree and pattern of development of the falx, as between hominids and pongids, to explain the presence in the former and the virtual absence in the latter of a well-developed frontal crest.

On either side of the frontal crest is a marked fossa, deeper on the right than on the left, and more or less parallel to the frontal crest. It extends vertically or almost vertically downwards to become continuous with a caving in of the floor of the most rostral part of the anterior cranial fossa. I have been unable to detect whether or not a foramen caecum is present: it is absent in pongids and apparently in *H. e. pekinensis* (Weidenreich, 1943, p. 32). From lateral to medial, the floor of the anterior cranial fossa slopes gently at first, then descends abruptly and steeply towards the root of the frontal crest. The site where foramen caecum might be expected is depressed about 16 mm. inferior to the lateral part of the floor of the fossa. The corresponding position in *H. e. pekinensis* is 'more than 15 mm.' below the floor of the fossa, whereas in recent man 'it is at the same level or only slightly below' (Weidenreich, 1943, p. 32). In a number of pongid crania (or as reflected on a number of endocranial casts of pongids), the site of the foramen caecum is much more than 16 mm. below the lateral part of the floor. In the modelling of this rostral part of the frontal bone, and especially in the deep olfactory recess, *Zinjanthropus* resembles *H. e. pekinensis*: and both are here closer to the pongid pattern than to that of modern man.

The *impressiones gyrorum* are well marked, not only on the orbital surface of the frontal, but on the internal surface of the squame, as far as it is preserved. Thus, on the surviving part of the frontal squame, there is no trace of a *limen coronale* (the rather abrupt transition between the lower part of the frontal squame marked by strong endocranial relief and the upper part from which it is absent). The impressions of the superior, middle and inferior frontal gyri, separated by the superior and middle frontal sulci, are extraordinarily clear.

The frontal sinus will be described in the chapter on the pneumatisation of the cranium (chapter XI).

B. The endocranial surface of the parietal bones

Almost the entire parietal bone is present on the right, stopping short anteriorly behind the coronal suture. On the left, rather more of the anterior margin is missing. In mediolateral extent, both bones are complete from sagittal to squamosal sutures. A few areas of endocranial surface (inner table) are missing, though the outer table is present in these places.

Parallel to the sagittal suture is a distinct ridge on each bone, the two ridges and the shallow sulcus between them forming the groove for the superior sagittal sinus (*see* account of the Venous Sinuses of the Dura Mater, pp. 63–71).

The sphenoid angle of the parietal bone is not present on either side in *Zinjanthropus*, the parietal on the right stopping short of the anterior edge of the temporal squame by about 6·0 mm. Nevertheless, there is evidence of a moderately elevated and broad crest extending in the direction of the centre of the parietal bone; such a low *crista Sylvii* resembles more that described by Schwalbe (1902) in modern man than the well-developed cristae of *H. e. pekinensis*. The lower slope of the crest is not confined to the parietal bone, but is seen on the pterionic angle of the temporal squame, as well as on the small adjacent portion of the alisphenoid. Such a broad, low crest extending over the three adjacent bones is encountered in some gorilla, chimpanzee and orang-utan crania. It is not the same as the distinctive entity which in Homininae may flow from the lesser wing of the sphenoid on to the parietal. On the left, it is not possible to detect whether a crista existed; if it did, it certainly did not extend as far backwards as the anterior limit of the presently surviving part of the left parietal. The Kromdraai *Paranthropus* shows such a crest of moderate proportions.

The mastoid angle of the parietal is wide and obtuse, the adjacent lambdoid and squamosal margins being but little 'indented' or notched, in association with the relatively poor expansion of the occipital and temporal squamae respectively. Since both these adjacent margins are not as hollowed as is usual in modern man, the mastoid angle projects downwards only slightly on both sides. On the inner aspect of the mastoid angle, a groove is present in the position commonly occupied by the transverse sinus as it passes over the inferolateral corner of the parietal prior to descending as the sigmoid sinus. However, this groove in *Zinjanthropus* is narrow and, when the adjacent bones are articulated, is seen to be a continuation of the groove for the posterior ramus of the middle meningeal artery. Below and behind this groove, the mastoid angle of the parietal bulges inwards to form a *torus angularis* of modest dimensions. It is less developed on the left than on the right, and on both sides slight endocranial defects make the precise definition of the torus difficult. The slightness of the torus is reflected in thickness measurements: at the mastoid angle the thickness rises to 6·5–7·0 mm., whereas just above it is 5·5–6·0 mm. There is thus a thickening of only 1 mm. in the region of the torus angularis of *Zinjanthropus*. A similar poorly developed torus angularis occurs in three specimens of *Australopithecus*. In contrast, the thickness at the mastoid angle may reach 17·4 mm. in *H. e. pekinensis*.

C. The basis cranii interna

The best-preserved part of the basis cranii is the posterior cranial fossa (pl. 14B), which is complete save for the apex of the petrous portion. The middle cranial fossa is well represented laterally by the anterior surface of the petrous temporal, as well as by a part of the floor of the fossa, especially on the right; on both sides it extends as far forwards as the sphenosquamosal suture and a part of the greater wing of the sphenoid. In the median part of the fossa, the *dorsum sellae* and most of the hypophyseal fossa are intact. However, whereas, on the external surface, the body of the sphenoid extends as far forwards as the articulation with the vomer, on the internal aspect the fossa does not reach the *sphenoidale* (a point in the median plane immediately in front of the *tuberculum sellae*), still less the *prosphenion* (where the median plane intersects the spheno-ethmoid suture). The anterior part of the middle cranial fossa has been lost, likewise the posterior part of the anterior cranial fossa, including all trace of the lesser wings of the sphenoid, the optic foramina and all but the most anterior part of the orbital surface of the frontal bone. Thus, the anterior cranial fossa is the least preserved of the three fossae.

1. *The posterior cranial fossa*

The greatest breadth of the posterior cranial fossa, measured with an internal caliper, is 89·8 mm. The limits of this measurement lie on the temporal bones just behind the base of the petrous pyramid. The corresponding measurement in Sts 19 (cranium VIII from Sterkfontein) is 84·2 mm. The venous sinus system in the posterior cranial fossa is discussed later in this chapter (pp. 63–71).

The cruciate eminence and the related cerebral and cerebellar fossae

The cruciate eminence is a striking formation on the internal aspect of the occipital bone. The inferior part of the sagittal limb of the cruciate eminence in *Zinjanthropus* is nearer to the position in anthropoid apes, being a short prominence, instead of a sharp internal occipital crest. The formation in the Makapansgat *Australopithecus* (MLD 1) is, however, much nearer to a crest and therefore much nearer to that of the hominines than is that of *Zinjanthropus*. The inferior limb is single above; then it is crossed by a communication between the left and right occipital sinus grooves; finally, just behind the foramen magnum, the limb bifurcates to enclose a small, slightly hollowed *vermian* or *vermiform fossa*. This arrangement is commonly found in hominines and pongids, but in several of the gorilla crania examined, the inferior limb is double from the beginning, the two components diverging to straddle or flank the foramen magnum. By so doing, they leave a large 'vermian' fossa between them, extending from the foramen magnum, practically back to the internal occipital protuberance. The arrangement in Sts 5, of which the posterior cranial fossa is perfectly preserved, is not mentioned in the description of the cranium (Broom *et al*. 1950), but in the illustration of Sterkfontein VIII, the arrangement as depicted (*op. cit*. p. 29) seems to be the same as in *Zinjanthropus*, except that there is no trace of the enlarged occipital and marginal sinus-grooves which characterise the type specimen of *Zinjanthropus*.

The sagittal limb of the cruciate eminence separates the left and right cerebral and cerebellar fossae from each other. The part of the occipital squame which rises above the internal occipital protuberance to the lambda is small, lambda being slightly nearer to the protuberance than is opisthion. At first sight, this gives the impression that the cerebral fossae are not as extensive as the cerebellar fossae; if one were to judge by the occipital bone alone, this would certainly be true. However, a careful study of the endocranial surface, when the parietals are articulated with the occipital, shows that only the deepest part of the cerebral fossae lies on the occipital bone, immediately above the transverse limb of the cruciate eminence: this part lodges the occipital pole itself. But the cerebral fossa as a whole, lodging not just the occipital poles but the occipital lobes, clearly extends beyond the occipital squame on to the posterior parts of the parietal bones. The fossa, as a whole, has the form of a relatively large and somewhat hollow area extending on either side of the mid-line right across the lambdoid suture. The same is true of the *Australopithecus* calvarial fragment from Makapansgat (MLD 1). A similar type of cerebral fossa is seen in a number of modern African negroid crania, in which the fossa is not confined to the occipital squame. If we take into account the total parieto-occipital fossa in *Zinjanthropus*, then in its vertical and transverse extent, the cerebral fossa exceeds the cerebellar fossa in the ratio of about 4 to 3. In *H. e. pekinensis*, on the other hand, the area of the cerebellar fossa is much smaller than that of the cerebral fossa, 'almost the half of the latter' (Weidenreich, 1943). In modern man, in contrast, Weidenreich points out, the cerebellar fossae are distinctly larger than the cerebral ones in both longitudinal and transverse directions.

In *Zinjanthropus* and *Australopithecus* the cerebellar fossae have a somewhat flattened floor, more so on the right than on the left, whereas the cerebral fossae are more evenly hollowed, being limited medially by a well-elevated superior limb of the cruciate eminence. The inferior limb of the eminence is about as prominent as the upper limb, but its elevation is somewhat offset by the large grooves and their bounding lips, for the occipital and marginal sinuses. These grooves deeply invade the cerebellar fossae on each side, thus further diminishing the volume available for the cerebellum.

Lambda in the mid-line internally is only 26·5 mm. above the central point of the internal occipital protuberance, which in turn is 27·0 mm. from opisthion (chord distances).

The position in *Australopithecus* (MLD 1) presents an interesting contrast. First, the inferior sagittal limb is much more elevated than in *Zinjanthropus* and there are no occipital and marginal sinus grooves to offset this prominence. This high inferior limb thus throws into strong relief the cerebellar fossae, which are appreciably deeper than the cerebral fossae. The cerebral fossae are very slight in the Makapansgat specimen, and the superior limb of the cruciate eminence is elevated barely at all above the plane of the cerebral fossae flanking it. The distances from the internal occipital protuberance to lambda and opisthion respectively are 41·5 and 18·5 mm. It should be mentioned that there is a very much clearer summit of the protuberance in the Makapansgat *Australopithecus* than in *Zinjanthropus*. On closer examination, one sees that, in the Makapansgat specimen, the 'very slight cerebral fossae' are only the lowest, deepest part of the fossa, housing the occipital pole alone and lying over the inner aspect of the occipital squame.

In the F.H., the inion of *Zinjanthropus* is appreciably lower than the internal occipital protuberance; in marked contrast with the position in *Australopithecus* (MLD 1), where the internal occipital protuberance is lower and lies near the opisthion. In the latter form, the small distance between opisthion and the internal occipital protuberance (18·5 mm.) is reminiscent of the position in *H. e. pekinensis* in which 'the posterior extremity of the occipital foramen comes very close to the internal protuberance' (Weidenreich, 1943, p. 41).

In *Zinjanthropus*, in contrast with *Australopithecus* and *H. e. pekinensis*, the internal occipital protuberance is situated halfway up the occipital bone. At the same time, the inion has been carried inferiorly on the marked nuchal crest: thus the distance between the inion and the internal protuberance is 26·8 mm., the protuberance being about 20 mm. *higher* than the inion. In all examples of *H. e. pekinensis*, the distance between protuberance and inion is greater (27·5–38·0 mm.), but in these instances the protuberance is *lower* than the inion. From a comparison of the absolute measurements, it is clear that the low level of the inion relative to the internal protuberance in *Zinjanthropus* is not simply the result of the descent of inion with the nuchal crest, but also results from the ascent of the internal protuberance, marking the upper limit of the cerebellum. It would seem reasonable to infer that there is a relatively more expanded cerebellum in *Zinjanthropus* than in *Australopithecus* (MLD 1). A comparison of the relevant parts of the endocasts confirms this deduction. In this regard, *Zinjanthropus* seems to have moved in a hominine direction, since, in modern man, the cerebellar fossae are more extensive than the cerebral.

Zinjanthropus has very slight impressions in the cerebellar fossae, mainly taking the form of two transverse hollow impressions, roughly parallel to the transverse limb of the cruciate eminence. A faint, horizontal, elongate eminence separates these two features: it corresponds in position with the *fissura horizontalis cerebelli*. If this is correct, the impressiones gyrorum above and below it are due to the *lobuli semilunares superior et inferior*. Further anteriorly, in the angle between the grooves for the sigmoid and marginal sinuses, is a faint oval fossa which could have lodged the *lobulus biventer*. At the lateral ends of the two upper transverse impressions is a full rounded elevation at right angles to them and parallel to the sigmoid sinus groove: it probably corresponds with the deep sulcus between the superior and inferior surfaces of the cerebellar hemisphere.

The petrous pyramids: superior margin and posterior surface

The two petrous pyramids point forwards, their upper borders being at about 45° to the horizontal. When the lines of the superior margins are extended to meet, they do so at an angle of just less than a right angle (85°). This is similar to the angle in Sts 5 and in modern man. Unfortunately, it is difficult to assess the range of variation of this feature, since the overwhelming majority of pongid crania which have been available to me have not been sawn open. It should be noted that this is not the angle referred to by Weidenreich (1943, p. 58), in his Table XIII, as the angle the pyramid forms with the mid-sagittal plane: that angle is the one measured on the basal surface of the petrous, as seen in norma basalis.

As in modern man, the posterior surface of the petrous part of the temporal is practically vertical, when the cranium of *Zinjanthropus* is held in the F.H. The height of the surface, measured near the base of the petrous from the anterior lip of the sigmoid sulcus to the superior petrosal sulcus, is 21·1 mm. According to Weidenreich (1943, p. 67), this height does not exceed 18 mm. in any cranium of *H. e. pekinensis*, while it may reach 23 mm. in modern man. To add to its modern human appearance, in the lateral part of this surface, a depression undermines the superior margin. In these respects, this surface in *Zinjanthropus* looks indistinguishable from that in modern man; in fact, to a certain extent, it is more modern-looking than in *H. e. pekinensis*. The superior margin is thus, for the most part, a sharp edge which partly overhangs the posterior surface, whereas in *H. e. pekinensis* it is a blunt, rounded margin with the adjacent anterior and posterior surfaces pressed down, as it were, towards the floor of the fossa, as in anthropoid apes.

The left internal acoustic meatus is round, with bevelled anterior, superior and inferior margins; the posterior margin takes the form of a nearly vertical thin edge of bone. On the right, the petrous pyramid is broken short at the anterior margin of the meatus, but the other margins conform in descriptive details with that of the left, except that the sharp posterior margin is somewhat thicker and more rounded.

A short distance below, and slightly behind, the meatus is an inferiorly directed triangular process of bone, under cover of the anterior margin of which is the small opening for the aqueduct of the cochlea. Eight millimetres behind the posterior margin of the meatus is a small cleft or indentation, the *subarcuate fossa*. Below and just behind that is a downwards directed cleft for the aqueduct of the vestibule. In all these respects the appearance is indistinguishable from that in modern man.

In none of the pongid crania examined is this combination of features encountered. Generally, the posterior surface is not nearly vertical, but at an angle of 20–30° to the vertical; the sharp posterior margin of the internal auditory meatus may be crescentic instead of vertical and the ledge of bone may continue upwards beyond the limits of the meatus. An additional feature is present in some pongid crania, an overgrowing shelf of bone which bridges the petro-occipital gap: when present, this feature gives the apex of the petrous a squarish or truncated appearance, rather than the usual acuminate apex.

The clivus

The clivus in *Zinjanthropus* is relatively long and narrow, extending forwards and upwards from a flat, nearly straight, transverse lower edge, the anterior margin of the foramen magnum (pl. 14B). Below, the clivus is overlapped from both sides by a strongly-developed boss of bone, the *tuberculum jugulare*, which is more marked on the right. This boss overhangs the inner opening of the hypoglossal canal. It is altogether absent or slightly developed in the gorilla crania examined, the inner openings of the hypoglossal canal being thus very obvious from above. The tubercles are slightly more strongly developed in a small group of chimpanzee and orang-utan crania, though some lack it altogether, while in none is it as marked as in *Zinjanthropus*. Where it is somewhat better developed in the pongids, the lower part of the clivus is rendered distinctly concave from side to side. In a series of modern human crania, the process is slight in Caucasoid crania and moderately strong in a series of African negroid crania, but in none does it approach the degree of development in *Zinjanthropus*.

In the articulated skull of *Zinjanthropus*, a distinct *petro-occipital fissure* intervenes between the apical part of the petrous and the basilar part of the occipital. The clear-cut separation of the two bony components in the Olduvai cranium contrasts sharply with the position in a series of chimpanzee crania: in the latter, to a greater or lesser extent, the medial margin of the petrous overflows as a big, bony excrescence which roofs over the petro-occipital cleft and often meets and even articulates with the basi-occipital. For instance, in a chimpanzee cranium in the Kenya National Museum (M.T. 82, VI. 1942), the petrous shelf makes a sutural joint with the jugular tubercle, while in another (C. 2), the process almost meets the occipital bone in front of the jugular tubercle. From the illustration of Sts 5, it appears as though such a shelf is present, at least on the right (Broom and Robinson, 1950, p. 22), but this cannot be confirmed today as the calotte has been sealed on to the basis cranii. It is not present in Sts 19 (cranium VIII of Sterkfontein). The petrous shelf is lacking in several gorilla crania, but is present in a series of modern human crania.

Deeply concave in its lower part, the clivus flattens as it approaches the dorsum sellae. Just posterior to the dorsum sellae, roughly in the position of the now completely fused spheno-occipital synchondrosis, are two distinct bony elevations on the upper face of the clivus, opposite the tip of the apex of the petrous part. Each is a small hemispherical eminence and their summits are separated by 7·5 mm. It is feasible to suppose that they result from the ossificatory events which closed the synchondrosis. In an orang cranium in which the synchondrosis is still open, two knob-like processes are seen growing backwards and downwards from the basisphenoid towards the basi-occipital. Nothing comparable has been seen in a series of gorilla crania, though slight excrescences about the line of closure occur in several chimpanzee crania. A single pygmy chimpanzee and a single gibbon cranium have bilateral eminences reminiscent of those in *Zinjanthropus*, as do several crania of *Colobus* and *Cercopithecus*. Slight linear eminences *crossing* the line of the synchondrosis appear in the cranium of a Frenchman, while in a Bushman-like skull in the Kenya National Museum, there is a slight, oval, *median* eminence at the level of the synchondrosis.

It appears that there is a fairly widespread tendency in the cercopithecoid and hominoid Primates for regular or irregular bony prominences to mark the upper surface of the clivus at the level of the spheno-occipital synchondrosis, and the feature in *Zinjanthropus* should perhaps be viewed in this light.

The dorsum sellae

The dorsum sellae is narrow, being 12 mm. broad, and consists of solid bone throughout. There is slight damage to the posterolateral aspects of the dorsum on each side and the points of the posterior clinoid processes are missing; however, a small basal knob protrudes forwards from the base of each process. The processes are not manifestly connected by ossified dural ligaments with either the apex of the petrous or the anterior clinoid process. Just posterolateral to the dorsum sellae (on the left) is a *petrosal process* articulating with the apex of the petrous part of the temporal bone. The *foramen lacerum* lies immediately in front of this point of contact, the *carotid canal* being inferolateral. On the right the apex of the petrous as well as the petrosal process of the sphenoid have been broken off and are missing.

Thus, the dorsum sellae of *Zinjanthropus* is typically hominid in form, and totally unlike the poorly developed dorsum of the hypophyseal fossa presented by pongids. In all of the ape crania which I have examined for this point, the dorsum sellae is either low and squat, consisting largely of a posterior 'overflowing' of bone over the clivus; or completely absent, being represented merely by two elevations in the position of the posterior clinoid processes; or present though very thin and perforate, as in a number of chimpanzee crania. The anatomy of this region in *Zinjanthropus*, as well as in *Australopithecus* (Sts 5), is extraordinarily reminiscent of that of modern man. Unfortunately, the basisphenoid, basi-occipital and exoccipital areas are not preserved in any of the specimens of *H. erectus*, either those from Choukoutien or those from Indonesia.

2. *The foramen magnum*

In chapter v, B, an account has been given of the anteroposterior position of the foramen magnum on the basis cranii and of the plane of the foramen.

The shape of the foramen magnum in *Zinjanthropus* is unusual: it is more or less heart-shaped, with a blunt point posteriorly (pl. 6). The bone in the vicinity is massive and the border of the foramen is thick in parts. Anteriorly, in the region of the basion, the border comprises a thin shelf of bone superiorly, underlaid in the mid-line by a downturned process of bone (ectobasion), flanked by two small fossae. Lateral to the downturned *ectobasionic process*, the rim of the foramen as far laterally as the condyles is only as thick as the superior shelf of bone, i.e. 2·5 mm.; whereas, in the mid-line anteriorly, it reaches 5·0 mm. in thickness. The lateral rim of the foramen is curved in J-fashion, reaching its greatest distance from the mid-line far anteriorly, just in front of the posterior extremity of the occipital condyles. It is therefore in this position—towards the 'base' of the 'heart'—that the greatest width of the foramen is read. Posteriorly, a slight *vermian (or vermiform) fossa* abuts on the margin of the foramen, between the two diverging branches of the inferior sagittal limb of the cruciate eminence. It is a more clearcut and definite fossa in *Zinjanthropus* than in any of the pongid crania available for comparison. Generally, in the ape crania examined, the area between the diverging branches of the inferior sagittal limb is strongly developed and flush with the two branches, rather than recessed or hollowed as a fossa: in a pongid cranium, therefore, one should speak of a vermian plane (*planum vermiforme*) rather than a vermian fossa.

It was at first thought that the curious truncation of the anterior part of the foramen magnum in *Zinjanthropus* was owing to pneumatisation of the basi-occipital, by extension backwards of the sphenoidal air-sinus. However, on probing the exposed cells of the sphenoidal sinus posteriorly, one was able to detect their extension just posterior to the presumed site of the spheno-occipital synchondrosis, to the level of a faint transverse crest on the inferior surface joining two tubercles for the *longus capitis* muscles. From the line of this crest forwards the thickness of the basi-occipital/basisphenoid suddenly jumps from 6–9 mm. to 12–15 mm., such is the effect of the pneumatic inflation. But behind this line, there is no evidence —either from the bony thickness or from probing —that the pneumatisation had extended as far back as the posterior part of the basi-occipital. This factor, therefore, cannot be adduced as an explanation for the strange cardioid shape of the foramen.

The extension of the sphenoidal sinus backwards into the basilar part of the occipital bone, such as *Zinjanthropus* manifests, does occur in anthropoid apes, according to L. Hofman (quoted by Weidenreich, 1943, p. 169), and my own observations on the fine collections of pongid crania in the Powell-Cotton Museum and in the British Museum (Natural History), many of which have damage in the relevant area, permitting one to explore the posterior extent of the sphenoidal air-sinus. A recent study of eleven 'ape crania' has likewise confirmed the backward extension of the sphenoidal sinus into the basi-occipital, it being noted that the most posterior extent of the sinus approaches to within a mean distance of 26·9 mm. of basion (Weiner and Campbell, 1964).

In hominine crania, Le Gros Clark (1938a) drew attention to the extension in the Swanscombe occipital of the sphenoidal sinus into the basi-occipital. He commented, 'Such a backward extension is very uncommon in modern human skulls, though it does occasionally occur... possibly it indicates a strong development of the facial part of the skull' (p. 59). Weiner and Campbell (1964) have recently confirmed that backward enlargement of the sinus is 'a feature very exceptional in modern man' (p. 179). Of thirty-four mid-sectioned or disarticulated modern human skulls, only three were found to approach Swanscombe in the degree of backward extension of the sinus. Apart from Swanscombe, skiagrams show that the sinus 'does appear to extend well back' in the Gibraltar I and Broken Hill crania, while Washburn and Howell (1952) reported that the sphenoidal sinus extends well back into the basi-occipital in some of the Solo crania. Thus, backward extension is a feature in this group of Neandertaloid crania. Unfortunately, the basi-occipital is not preserved in any of the *H. erectus* crania from Indonesia, Choukoutien or Olduvai, save for a small, dislocated fragment in *H. e. erectus* IV (illustrated in text-figure 1*e* and Plate 2*a* of Weidenreich, 1945): although the latter skull is characterised by a large sphenoidal sinus 'which occupies the base of the pterygoid process' (p. 26), no indication is given in either the text or the illustration whether it extended into the surviving part of the basi-occipital.

The length of the foramen magnum, as measured from ectobasion, is 26·4 mm. This markedly abbreviated length is owing to the above-mentioned anterior truncation. When the length is considered in relation to the breadth of 26·1 mm., the foramen magnum of *Zinjanthropus* has the unusually high index of 98·6 per cent. These values are compared with those of several other australopithecine crania, some culled from the literature (e.g. Broom and Robinson, 1952, p. 29) and some measured personally, as well as of several pongid crania and the ranges of means quoted by Martin (1928) for modern man (Table 11). The ?11-year-old Swartkrans child, SK 47, has been omitted from the australopithecine ranges. The length of 26·4 mm. is smaller than those of other australopithecines, except Sterkfontein VIII, which has a generally small foramen magnum (25 × 21·3 mm.), although its cranial capacity of 530 c.c. is identical with that of *Zinjanthropus*. The length is much smaller than the lowest mean in modern man and the lowest values in gorilla and orang; only a single pygmy chimpanzee with a foramen magnum length of 25·0 mm. is smaller in this respect. The range for six chimpanzee foramina excluding the pygmy chimpanzee is 27·0–37·5 mm. The length of the foramen magnum in *Zinjanthropus* is very short by general hominoid standards.

The breadth of 26·1 mm. is just smaller than the lowest mean among modern men and the minimum value among half-a-dozen gorilla. It falls within the sample range for other australopithecines and for orang, but is broader than the maximum width in seven chimpanzee crania.

The foramen magnum index of 98·6 per cent is beyond the range for all other hominoids reflected in Table 11, although a single gorilla foramen reaches a value of 98·2 per cent. The values in the other australopithecines fall into the dolichotrematous (x–81·9) or mesotrematous (82·0–85·9) category; only *Zinjanthropus* is brachytrematous (86·0–x).

The significance of the abbreviation of the foramen magnum in *Zinjanthropus* is not clear: it is unlikely that this is a part of the process of shortening of the base of the cranium, for the latter process usually involves the basis cranii anterior to the basion.

3. *The middle cranial fossa*

The middle cranial fossa is represented by its posterior wall (the anterior surface of the petrous pyramid), parts of the floor, both central and lateral portions, and parts of the lateral walls.

The petrous pyramid: anterior surface

The modelling of the anterior surface of the petrous pyramid is very clear. The *arcuate eminence* is particularly prominent and protrudes above the summit of the pyramid. Anterior to the arcuate eminence and close to the apex is a shallow *trigeminal impression*. Just below this the carotid canal is apparent, its anterior wall being deficient; part of the canal is present on the right just behind the break across the apex of the petrous.

On the right, a narrow cleft, the *petrosquamous fissure* marks the separation between the squamous and petrous portions of the temporal; it pursues a wavy course across the middle cranial fossa very closely behind, and for the most part parallel to, the sulcus for the parietal or posterior branch of the middle meningeal artery. This area of bone is missing on the left.

INTERIOR OF THE CALVARIA

Table 11. *Dimensions and indices of the foramen magnum in* Zinjanthropus *and other hominoids*

	Length (ectobasion–opisthion) (mm.)	Breadth (maximum) (mm.)	B/L index (%)
Zinjanthropus	26·4	26·1	98·6
Paranthropus			
SK 48	c. 28·0	c. 21·5	76·8
SK 47 (juv.)	27·5	19·0	69·1
Kromdraai	?	c. 30·0	—
Australopithecus			
Sts I	?	c. 24·0	—
Sts 5	c. 31·4	c. 24·3	77·4
Sts VIII	25·0	21·3	85·2
MLD 37/38	c. 30	c. 25	83·3
Australopithecinae (excluding *Zinjanthropus*)	25·0–31·4 (n = 4)	21·3–30·0 (n = 6)	76·8–85·2 (n = 4)
Homo erectus*	?	?	?
Modern man†	32·6–37·1	26·5–34·3	72·6–89·1
Gorilla (n = 58) (Schultz, 1962)	27 –47	23 –35	—
Gorilla (n = 8) (present study)	27·5–40·9	26·2–31·7	77·5–98·2
Orang-utan (n = 4) (present study)	32·0–36·0	21·0–28·0	65·6–84·8
Chimpanzee (n = 7) (present study)	25·0–37·5	20·0–25·5	68·0–85·2

* The anterior margin of the foramen magnum is not intact in a single cranium of either the Indonesian, Chinese or African *H. erectus*.
† The figures quoted for modern man are ranges of *means* culled from Martin (1928).

The tegmen tympani and the middle ear

Further laterally the tegmen tympani is apparent (where it is present on the right) and bulges slightly upwards. On the left, it is missing and the middle ear cavity is exposed. Through the external acoustic meatus on the left, one is able to see the oval window (*fenestra vestibuli*) above and to the front, the *promontory* below that, and the round window (*fenestra cochleae*) below and behind. The posteromedial wall of the middle ear slopes upwards to the *aditus ad antrum*; the *epitympanic recess* and the *antrum mastoideum* are visible, giving way almost immediately to a great cluster of air cells which completely fills the pars mastoidea.

The sella turcica and carotid sulcus

The central part of the floor of the middle cranial fossa is represented by the dorsum sellae, which has been described with the posterior cranial fossa, and by the posterior half or two-thirds of the *sella turcica*. Such part of the sella as is present is deep and well defined and like that of modern man. There is a slit-like foramen in the floor of the fossa. On either side posteriorly, a large *carotid sulcus* is present, running upwards towards the side of the body of the sphenoid and leading to a very shallow sulcus flanking the sella turcica. On the right, a distinct *lingula* projects backwards from the posterior margin of the greater wing, limiting the carotid sulcus laterally. On the left, it is broken off.

The foramina of the middle cranial fossa

At the posterolateral angle of the alisphenoid on the right is the inner opening of the *foramen ovale*: it is round rather than oval. On the anterolateral part of its periphery is a bay or inlet off the main foramen, from which a groove runs laterally, some 10 mm. in front of the foramen spinosum and the groove for the stem of the middle meningeal artery. The bay of the foramen ovale might have been thought to represent an incompletely separated *foramen of Vesalius*; however, since the latter inconstant foramen transmits an emissary vein from the cavernous sinus which lies *medial* to foramen ovale, it is unlikely that the antero*lateral* bay and its *lateral* groove would represent the same thing. Nor is it likely to represent the *canaliculus innominatus* which generally lies behind and medial to the foramen ovale. It seems probable that the bay and the lateral groove must be owing to the accessory (small) meningeal artery, which generally traverses the foramen ovale in company with the mandibular branch of the trigeminal nerve. The inner opening of the foramen ovale on the right is encircled by the greater wing of the sphenoid (alisphenoid), two delicate processes extending around the foramen and apparently completing it laterally. On the left it is not possible to detect whether the foramen is completely encircled by alisphenoid, since the lateral rim of the foramen is broken off and missing. A deep sulcus leads upwards and medialwards from the medial margin of the foramen ovale: this probably lodged the mandibular division as it ran down from the semilunar (Gasserian) ganglion. The preserved portions of the alisphenoid do not extend far enough forwards to include even the posterior margin of the *foramen rotundum*.

The *foramen spinosum* is preserved. On the right, its inner opening lies between the anterior margin of the petrous temporal and the back of the temporal squame in its inferior or undercurved part, close to where the squame ends medially by articulating with the alisphenoid (pl. 24). However, the foramen is entirely surrounded by temporal bone; the alisphenoid and its petrosal process play no part in completing the periphery of the foramen. The sphenosquamous suture is clearly preserved medial to the foramen. On the left, the posteromedial part of the wall of the foramen is lacking through damage, but the opening receives no manifest contribution from the more medially situated alisphenoid. This contrasts with the common arrangement in modern man, in whom the foramen lies totally within the alisphenoid, although it is sometimes incompletely walled when it lies near the edge of the alisphenoid.

In *H. e. pekinensis*, the position of foramen spinosum is somewhat intermediate, approximating a little more closely to the modern human condition: it lies in Choukoutien cranium III 'on a plane very close to the sphenosquamous suture' (Weidenreich, 1943, p. 47), though it seems from the illustration to be excavating the alisphenoid rather than the squamous temporal. On the base, it is seen that, associated with the foramen spinosum, the entoglenoid process of *Zinjanthropus* likewise stems wholly from the squamous part and not at all from the alisphenoid: this interesting relationship is discussed in the chapter on the 'Basis Cranii Externa' (pp. 37–9).

The grooves for the middle meningeal artery

From the foramen spinosum situated in the squamous part of the floor of the middle cranial fossa, a short wide groove runs laterally and slightly anteriorly for 10·5 mm. (right): this is the sulcus for the common trunk of the *middle meningeal artery*. Then, almost at right angles to itself, the trunk-groove gives off a large posterior branch-groove, which must have lodged the parietal (or posterior) branch of the middle meningeal artery. The groove for the parietal branch courses posterolaterally and slightly superiorly, over the cerebral surface of the squama, just in front of and for the most part parallel to the petrosquamous fissure (pl. 23). The trunk-groove continues forwards, veering slightly more anteriorly and climbing fairly steeply up the inner face of the vertical part of the squame for another 7 mm. Then it bifurcates again: one branch-groove runs forwards, rising slightly, to the anterior edge of the squame; the other turns steeply upwards, curving slightly backwards, before its preserved course is interrupted by a defect on the endocranial aspect of the right squame.

The short length of the common stem-groove for the middle meningeal artery is reminiscent of the arrangement in the Taung specimen (*Australopithecus africanus*), as well as in Sts VII and possibly VIII: in the latter three specimens, the bifurcation of the middle meningeal artery occurs on the incurved part of the squame (i.e. on the *inferior* surface of the temporal lobe), so that when the artery appears on the lateral surface of the temporal lobe, the two main branches are apparent. In contrast, the artery bifurcates over the *lateral* surface of the temporal lobe in Sts I and II (*Australopithecus*) and Kromdraai (*Paranthropus*) (Schepers, 1946, 1950). With its basal bifurcation, the middle meningeal artery of *Zinjanthropus* approaches more closely to the pattern in the former group. As far as can be ascertained, the pattern in the Swartkrans *Paranthropus* has not been assessed. It seems, however, that variability is great and that no real taxonomic significance attaches to the site of bifurcation of the artery.

The two divisions of the frontal (or anterior) branch seemingly correspond to the two major branches of the anterior trunk described by Schepers in Sts I, II, 5 and VII, Taung and Kromdraai. However, in their further course, which is described with the endocranial cast, the branches in *Zinjanthropus* differ somewhat from those of other australopithecines.

On the left side, the groove for the common trunk of the middle meningeal artery cannot be followed, because a segment of the endocranial surface of the undercurved part of the squame is missing. Beyond the break, the bifurcation of the presumed anterior trunk can just be detected, one branch curving steeply backwards, the other running slightly forwards and upwards parallel to the anterior margin of the squame. Just before the superolateral angle of the squame, this anterior branch curves somewhat backwards along the superior border of the squame.

The squamosal suture

The side wall of the middle cranial fossa, inasmuch as it is composed of the temporal squame, is long and low, as in *H. e. pekinensis*. When viewed from the inner aspect, the line of the squamosal suture (with the cranium in F.H.) rises from the junction between the squamosal and parietomastoid sutures posteriorly, at an angle of about 25–30° to the horizontal, to reach a summit in the biporial axis. It then declines at a gentler angle down to pterion, the anterior limb of the suture (endocranially) being about twice as long as the posterior limb.

This endocranial contour of the squamosal suture is very similar to that encountered in modern human crania. In a number of these examined for the purpose, there is a posterior steepish rise, followed by a gentler rise to a summit in or close to the biporial plane, the suture then declining anteriorly towards pterion. In a number of pongid crania, however, the *Zinjanthropus* pattern could not be duplicated in a single instance. In a group of gorilla, chimpanzee and orang-utan crania especially examined for the purpose, the highest point of the squamosal suture is invariably well behind the biporial axis, at or just behind the point where the margin of the squame crosses the root of the superior margin of the petrous pyramid. From the petrous forwards, the suture endocranially pursues a long, sometimes undulating, gently downhill course to pterion; in none of the pongid crania showing undulations of the suture does any part rise to a higher level than the posterior extremity. In the arrangement of its squamosal suture, *Zinjanthropus* differs strikingly from the Pongidae and closely resembles the Homininae.

D. The venous sinuses of the dura mater

The sulci for the venous sinuses are for the most part clearly impressed on the endocranial surface of the calvaria of *Zinjanthropus*.

The *sulcus for the superior sagittal sinus* begins on the calvariofacial portion. The slender *frontal crest* divides to form the two lips of the sagittal sulcus about 6 mm. posterosuperior to the most salient point of the crest and about 27 mm. (straight or chord distance) from the beginning of the frontal crest, which is preserved close to the broken lower edge of the endocranial surface of the frontal bone. The two lips diverge at a low angle until, at the point where the frontal squame is broken posterosuperiorly, they are about 3·5 mm. apart, and, instead of continuing as elevated twin crests, they are beginning to flatten into the general contour of the endocranial surface.

Behind the frontoparietal gap, the sulcus may be detected again, probably not far behind bregma and some 25 mm. in front of the biporial transverse plane. At this point it is 7 mm. broad, shallow and rather poorly defined. It is represented for about 15 mm., followed by a gap of a further 15 mm. from which the inner table is missing, and then another area where the sulcus is poorly represented through damage to the inner table. Posterior to this area, it once again appears a little more clearly defined, being 6–7 mm. broad. It appears to attain its maximum width in the middle reaches rather than posteriorly, exactly as Schepers found on his Sterkfontein endocranial cast 'type 2' (Schepers, 1946, p. 235).

The groove for the superior sagittal sinus reflects itself on the endocranial cast as a median ridge. This ridge is wholly comparable with those described by Schepers (1946, p. 236) in the Sterkfontein endocasts. Referring to the Kromdraai type cranium, Schepers had lamented that the vault of this *Paranthropus* specimen was defective, so that it was not possible to determine if it, too, possessed such a median ridge. *Zinjanthropus* provides us with the first evidence that one of the robuster australopithecines, too, possessed this feature. Schepers drew attention to the fact that Keith (1916), who had made an extensive search for similar structures in a wide variety of casts, had shown 'that comparable anterior median vascular ridges are only to be seen in Modern Man, and, in a modified form, in Neanderthal Man' (p. 236).

The exact course of the superior sagittal sulcus, as it approaches the cruciate eminence, is not as clearly marked. There is no obvious groove, sweeping around either to the right or to the left; instead, the main continuation of the sulcus is directly downwards, just to the right of the median plane (Fig. 9). After surmounting the slight elevation caused by the right half of the transverse limb of the cruciate eminence, it expands into an oval hollow in the angle between that transverse limb and the inferior part of the sagittal limb.

From the oval hollow, a clearly defined groove for a *right occipital sinus* runs along the right side of the inferior part of the sagittal limb of the cruciate eminence. About half-way between the internal occipital protuberance and the opisthion, this sulcus diverges to the right as a *right marginal sinus* groove. It runs anterolaterally close to the margin of the foramen magnum. Ascending slightly over the jugular process of the pars lateralis (exoccipital), the sulcus then flows over the summit, turning sharply downwards as the *jugular notch* forming the posterior boundary of the jugular foramen. Just as this marginal sulcus approaches the summit of the jugular process, it is joined from the right by a much narrower sulcus,

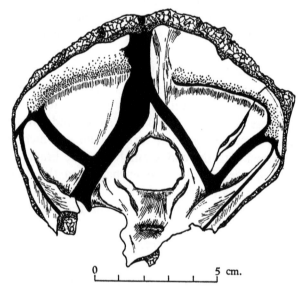

Fig. 9. The interior of the occipital bone, to show the pattern of venous sinus grooves.

for the right *sigmoid* sinus. Whereas the right marginal sulcus varies from 6 to 8 mm. in breadth, the right sigmoid sulcus on the temporal bone is only 4–5 mm. broad.

The sulcus for the *sigmoid sinus* does not arise in the usual way, as the continuation of the transverse sinus sulcus; instead it is formed close to the base of the petrous pyramid, by the junction of the sulcus for the small *petrosquamous sinus* with the clearly defined sulcus for the *superior petrosal sinus*. After converging posteriorly, these two sulci turn sharply downwards and medialwards, then pursuing the usual course of the sigmoid sinus.

It is impossible to detect any clear-cut right transverse sinus sulcus, either flowing from the area of the internal occipital eminence, or entering into the formation of the sigmoid sinus. One cannot entirely exclude the possibility that a small amount of blood may have followed this more usual route; but if so, it was so small in bulk as to have left no clear sinus impression on the lower part of the posterior cerebral fossa. The possibility has been left open in the diagram (Fig. 9) by means of a faint stippled impression.

The left transverse sinus, too, seems to have been missing, there being no clear sulcus detectable on the bone nor elevation on the plaster endocast of this area. Instead, there is a clearly defined depression about 4 mm. wide on the left aspect of the inferior sagittal limb of the cruciate eminence (Fig. 9). Above, this is continuous with the oval hollow on the right by means of a shallow sulcus which crosses the inferior part of the sagittal limb. Below, after an extremely short sagittal course (as the sulcus of the *left occipital sinus*), it diverges leftwards as the sulcus of the *left marginal sinus*, running parallel to the foramen magnum. The left sulcus is a little deeper, though narrower than the right. It ascends the jugular process of the left exoccipital, where it is joined from the left by the sulcus for the *left sigmoid sinus*. Finally, as on the right, it dips steeply inferiorly as the jugular notch forming the posterior boundary of the jugular foramen.

The groove for the left sigmoid sinus, like that for the right, does not seem to receive a definite contribution from the groove for the left transverse sinus; however, there is rather more likelihood of a passageway for blood through the lower part of the left posterior cerebral fossa (i.e. the position of the left transverse sinus) than on the right, but the impression is so uncertain as to leave the matter indeterminate (Fig. 9 and pl. 27). As on the right, the left sigmoid sinus sulcus is formed mainly by the junction of the short *left petrosquamous sinus* sulcus and the rather poorly defined *left superior petrosal sinus* sulcus.

The pattern of the venous sinuses in *Zinjanthropus*, as reflected in the grooves left on the interior of the cranium, is distinctive. It seems that most if not all of the blood from the superior sagittal and straight sinuses ran into the occipital and marginal sinuses, thus by-passing the transverse-sigmoid sinus system. The blood in the marginal sinuses and in the occipital sinus—when these are present in modern man—usually passes *towards* the internal occipital protuberance. This direction of flow must have been reversed in *Zinjanthropus*, the blood following a seemingly shorter and more direct route to the jugular bulb.

The absence of clear markings for the transverse sinus contrasts sharply with the position in the *Australopithecus* 'type 1' endocast from Sterkfontein: in the latter, the impression testifies to a very distinct and deeply engraved sulcus of an unusual degree of fullness (Schepers, 1946). However, as Le Gros Clark (1947) has pointed out, the diameter of the impression in the latter specimen is only 5·5 mm., corresponding well with that in modern apes. The absence (or relative absence) of transverse sinuses finds no parallel at all in five *Australopithecus* endocasts from Sterkfontein and one from Taung, nor in the markings on *Australopithecus* from Makapansgat (MLD 1). In all seven *Australopithecus* individuals from three sites, blood from the sagittal and straight sinuses seems to have followed the commonly encountered route— through the transverse and sigmoid sinuses to the jugular bulb. This is true as well of the specimens of *H. e. pekinensis* in which the area is preserved.

Only one hitherto-published fossil, the type specimen of *Paranthropus robustus* from Kromdraai, has come near to shedding any light on the pattern of venous sinuses in the posterior cranial fossa of the robust australopithecines. According to Schepers (1946, p. 235), 'The preservation of the sigmoid sinus is relatively poor in the *Paranthropus robustus* cast'. Nevertheless, from his plaster endocast, he claimed that the lower sigmoid portion of the transverse sinus is 'distinctly thicker, suggesting an increased vascular capacity for the brain'. He added what may be a significant statement: 'It is somewhat less prominent over the cerebellar hemispheres.' Does this suggest—what his reconstruction drawing does not—a similar occipital-marginal sinus drainage pattern to that in *Zinjanthropus*? Unfortunately, the endocranial surface of the specimen is appallingly preserved and the specimen is broken short and damaged just behind the base of the petrous pyramid. It is thus impossible to tell whether the sigmoid sinus groove received any contribution from a transverse sinus groove. It would be hazardous in the extreme to attempt to draw any conclusions on this point from the specimen or its plaster endocast.

However, one juvenile occipital bone from Swartkrans, SK 859, provides definite and compelling evidence. This specimen is recorded in the Catalogue of the Transvaal Museum as having been discovered by Dr J. T. Robinson in the 'lower breccia' at Swartkrans in 1952, and I am indebted to Dr V. FitzSimons, then Director of the Museum, Dr C. K. Brain, Professional Officer in charge of Vertebrate Palaeontology and Physical Anthropology, and Dr J. T. Robinson, his predecessor, for permission to examine this and other specimens and to use such observations in the present comparative study of *Zinjanthropus*. SK 859 comprises most of a juvenile occipital, with part of the right parietal articulated, including the sagittal suture and lambda, and with a small, supra-asterionic part of the left parietal articulated. The joint between the supra-occipital and ex-occipital on each side has not yet fused and parts of both left and right exoccipitals are present. The left exoccipital portion, indeed, extends as far forwards as the joint-surface for articulation with the basi-occipital, and this, too, shows no sign of fusion. The fusion of the two intra-occipital joints is usually said to occur in about the third year and at about six years respectively in a modern human child (Breathnach, 1965). If conditions were at all comparable in *Paranthropus*, the child to which SK 859 belonged is unlikely to have been much more than 2 years of age.

The endocranial surface of the SK 859 occipital shows the sulcus for the superior sagittal sinus flowing into an unmistakable large, right-sided occipital sinus groove on the supra-occipital. As a marginal sinus groove 5 mm. in diameter, it flows on to the right lateral occipital. A smaller left marginal sinus groove, 4 mm. in diameter, continues on to the left lateral occipital. A left transverse sinus groove is probably present. In other words, venous sinus drainage in the Swartkrans child was similar to that in the *Zinjanthropus* youth (Fig. 10 and pl. 30).

In SK 46 from Swartkrans, the left sigmoid sinus groove does *not* appear to receive any contribution from a transverse sinus groove. A small part of the area where the transverse sinus should turn downwards into the sigmoid is preserved and, though slightly crushed, there is no impression at all of the lateral end of the transverse sinus.

SK 847, a further Swartkrans specimen, which I do not believe has been published, has a large left sigmoid groove on one of the fragments of this broken up and probably adult cranium. From this incomplete evidence, it is possible that this specimen had a well-developed lateral sinus system comparable to that in *Australopithecus*, Pekin Man and modern man.

It is at least striking that both *Zinjanthropus* and the only *Paranthropus* specimen having all of the area in question should both show a similar departure from the modal pattern in hominids, while conditions on the one preserved side of another specimen of *Paranthropus* (SK 46) are compatible with a similar pattern having existed. It is pertinent to enquire whether such a departure is known among other hominoids.

This enquiry involved a search through the literature, as well as personal examination of some 300 pongid and hominine crania. Although the reading of the marks on dry bone cannot yield as precise results as the study of preserved dissection specimens, they provide a fair indication of the main drainage pattern. Moreover, Woodhall (1936, 1939) demonstrated that the markings on the cranium corresponded closely to the differences in the volumes of the left and right lateral sinuses and followed exactly the distribution of these sinuses. This correlation was, however, questioned by Browning (1953). In another study, Woodhall and Seeds (1936) showed a good correlation between the roentgenologic appearances of sinuses in the living subject and the incidence of various patterns of cranial marking.

In a survey of the sinus grooves among a sample of pongid crania in the Powell-Cotton collection at Birchington in Kent, in the British Museum (Natural History), in the National Museum of Kenya, and in the Anatomy Department of the University of the Witwatersrand, Johannesburg, I could find no single cranium of about 100 examined which duplicated the conditions in *Zinjanthropus*. Only in one pygmy chimpanzee cranium in the Kenya National Museum were there distinct occipital and right marginal sinus grooves which reached the jugular foramen; the marginal sinus groove differed from those in *Zinjanthropus* in that it ran so near to the margin of the foramen magnum that, instead of passing over the superior surface of the jugular process, it dropped below the medial rim of the anterior condylar canal, then turned laterally to traverse the canal and so gain the jugular foramen. Even in this pygmy chimpanzee cranium, however, the marginal sinus was supplementary to the main drainage of the superior sagittal sinus into the right transverse sinus. In general, grooves for the occipital or marginal sinuses were commoner in chimpanzee (18 out of 31 crania) than in gorilla (9 out of 23). It may be of interest to record that the superior sagittal sinus passed predominantly into the *right* transverse sinus in 11 out of 23 gorilla (48%) and 14 out of 31 chimpanzee crania (45%); into the *left* transverse sinus in 1 out of 23 gorilla (4%) and 4 out of 31 chimpanzee crania (13%); whilst drainage occurred nearly equally into the left and right transverse sinuses in 11 out of 23 gorilla (48%) and 13 out of 31 chimpanzee crania (42%).

The only references I could trace to variant patterns of sinuses in fossil hominines refer to the Swanscombe and Předmostí remains. In his description of the general features of the Swanscombe bones, Le Gros Clark (1938a) stated that 'The groove for the right lateral sinus is considerably narrower than that for the left, *and its direction indicates that the sinus turned downwards medially into the occipital sinus*—a not uncommon variation in modern skulls' (italics mine). This interpretation is open to question: first, to judge by a cast of good quality, as well as excellent published photographs, there is no suggestion of grooves for either occipital or marginal sinuses. The endocranial cast of Swanscombe (Clark, 1938b) confirms the absence of impressions for occipital and marginal sinus grooves. Secondly, in several among 211 crania of Bantu-speaking negroid Africans, the medial end of that transverse sinus groove which manifestly did *not* receive the superior sagittal sinus groove was curved downwards near the cruciate eminence; yet, in all such instances, the down-curved sinus groove did *not* lead inferiorly to an occipital or marginal sinus groove, but laterally to contribute in large measure to the formation of the sigmoid sinus sulcus. The downward curve of the medial end of one transverse sinus groove was noted only in those crania where the superior sagittal sinus groove clearly led directly to the *opposite* transverse sinus (exactly as occurs in the occipital from Swanscombe). It would seem that the down-curved transverse sinus had received blood from the straight sinus, and that the straight sinus had flowed into the lower part of the confluence area. The median part of the tentorium cerebelli must have attached to a rather low position on the cruciate eminence, since the straight sinus runs in the margin of the falx cerebri attaching along the upper surface of the tentorium. To attain its position on the upward slope of the ridge dividing the cerebral from the cerebellar fossae, the transverse sinus would then have had to ascend somewhat, thus producing the down-curved type of sulcus present in the Swanscombe occipital. Keith (1916) apparently first referred to the general tendency of the transverse sinus to sink to a lower level when traced lateromedially and he attributed this to an expansion of the occipital pole. He observed this feature in the Piltdown cranium and the Australian aboriginal, but not in *Gorilla*. The point was commented on again by Schepers (1946).

I am indebted to Professor L. Borovansky, Head of the Department of Anatomy of Charles University, Prague, for bringing to my attention a study by Matiegka (1923) of the venous sinuses in the Předmostí crania from Moravia. A number of variations of the venous sinuses in these fossil crania are described and figured by Matiegka; they include cases in which 'La plus grande partie du sang n'est pas conduite par les sinus horizontaux, mais par le sinus inférieur à la partie inférieure de l'écaille occipitale' (extract from French résumé, *op. cit.* p. 37). From the accompanying illustrations, it is clear that what Matiegka calls the 'inferior sinus' or 'inferior longitudinal sinus' corresponds with the enlarged occipital sinus with right or left marginal sinus. He comments, too, on the occasional absence of a left or right transverse sinus; in one of his crania (his figure 1, p. 32), the pattern is virtually identical with that in SK 859 (*see* Fig. 10). It is of interest to read Matiegka's final observation:

La présence en masse de certaines anomalies (déviation prononcée vers le côté droit, l'existence de la gouttière longitudinale inférieure) chez les crânes de Předmostí n'est pas un critérium spécial d'une race diluvienne, mais simplement celui du lien de parenté entre les membres du groupe de Předmostí et causé par l'hérédité familiale (*op. cit.* p. 38).

Matiegka refers as well to variations in other European fossil crania. Thus, absence of the groove for the transverse sinus is encountered among the Krapina and Obercassel crania, while marked rightward deviation of the sulcus for the superior sagittal sinus is encountered not only among the Předmostí crania, but in the Neandertal calotte.

A search through the literature on modern man, as well as a personal survey of 211 Bantu negroid crania, revealed that the condition present in *Zinjanthropus* is known in modern man, but is an extreme rarity.

Knott (1882) reported that in two out of forty-four modern human heads examined by him, the right lateral sinus was almost completely absent, 'only a small venous canal of 1½ mm. diameter following its course as far as the mastoid foramen, through which it disappeared' (p. 31). Knott quoted Lieutaud (*Essais Anat.* p. 332) as recording a case of complete absence of the left lateral sinus, as far as the usual point of entry of the superior petrosal sinus. In two of Knott's specimens, the left occipital sinus continued as a marginal sinus to join the sigmoid sinus at the jugular foramen. Finally, the full picture similar to that represented by *Zinjanthropus* occurred in one of his series: both lateral sinuses were small and the greater part of the blood was transmitted along the occipital and marginal sinuses to join the internal jugular vein at the jugular foramen. He commented that Henlé had alluded to 'this abnormality'. I am indebted to Professor Sir W. E. Le Gros Clark for assisting me to find the reference to Knott's work.

Several more recent reviews have summarised and classified the diversity of patterns of the venous sinuses in the posterior cranial fossa (e.g. Woodhall, 1936, 1939; Waltner, 1944; Browning, 1953 and Petříková, 1963). Most of them cite a case of Streit (1903), the original publication of which I have been unable to see: the superior sagittal sinus apparently continued directly into the jugular bulb to the right of the foramen magnum. There was in addition an attenuated right lateral sinus and a full-sized left lateral sinus, while apart from the large channel which continued from the superior sagittal sinus, a pair of narrow occipital sinuses led down through marginal sinuses to the respective jugular bulbs (see figures and descriptions in Woodhall, 1939 and Waltner, 1944). This case differs from the position in *Zinjanthropus* in the possession of a large left lateral sinus, while it is not possible in the fossil to detect narrow occipital sinus grooves over and above the main channel. A somewhat similar case was reported by Vernieuwe (1921), quoted by Woodhall (1939).

Woodhall (1936, 1939) found two out of 100 autopsy specimens in which there occurred what he described as 'a preservation of the fetal pattern in the adult sinus system'. In each, there was a single occipital sinus, dividing into two marginal sinuses which emptied into the jugular bulb. Each marginal sinus was 6–8 mm. in diameter and was as large as the lateral sinuses. In these cases, it seems, the occipital-marginal sinus system provided a collateral drainage, apart from that afforded by the lateral sinus system. In addition, there were four cases with 'an exaggeration of the occipital sinus and one marginal sinus, connecting either with the torcular or with either major sinus and aiding materially in the distribution of the volume of the blood' (1936, p. 307). Two of these four were similar to the cases of Streit and Vernieuwe. Thus, in six out of 100 subjects, Woodhall found that the occipital-marginal system contributed appreciably to the drainage of venous blood.

According to Petříková (1963), Köppl (1947) found that the occipital sinus system played a major role in three out of 216 specimens.

In his series of 100 dissections, Browning (1953) found two cases in which the occipital sinus connected the confluence with the left and right jugular bulbs respectively, and he stated, 'Presumably the occipital sinuses served as collateral channels for the smaller of the transverse sinuses' (p. 316).

In her own series of 125, Petříková (1963) reported the 'occipital type' of drainage pattern in '3%' of cases. I am grateful to Dr R. Čihak of the Charles University, Prague, for drawing my notice to the last study and for making available a copy of Miss Petříková's article.

trace of a groove for the right lateral sinus, and the right sigmoid sinus groove was correspondingly puny, entering the last part of the right marginal sinus as a small feeder groove (Fig. 10 and pl. 31). The left transverse sinus groove led to a well-developed left sigmoid sinus groove which, just before its termination, received the left marginal sinus groove. Presumably the straight sinus drained into the left transverse sinus. Conditions were similar in A 2125, except that the superior sagittal sinus seems to have drained partly into a right

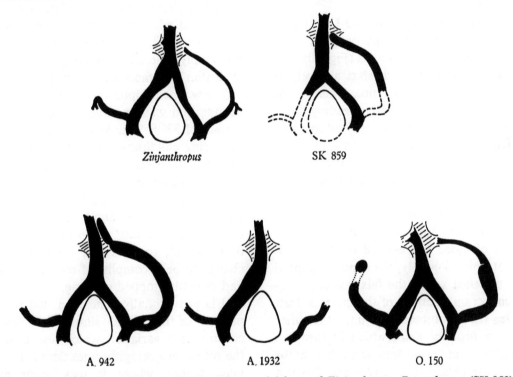

Fig. 10. Scheme of venous sinus grooves in the posterior cranial fossa of *Zinjanthropus*, *Paranthropus* (SK 859) and three modern Bantu crania. All five specimens show enlarged occipital and marginal sinus grooves. The hatched area represents the internal occipital protuberance; the dotted line represents a faintly or uncertainly marked groove; while the interrupted lines represent the probable sinus grooves on missing areas of bone in specimen SK 859.

Among 211 dried Bantu crania in the collection of the Department of Anatomy, University of the Witwatersrand, Mr D. B. Kaye and I found thirteen with marked grooves for the occipital and/or marginal sinuses. Four of these showed indications that the occipital-marginal system was responsible for an appreciable part of the drainage. In cranium A 942, the superior sagittal sinus evidently flowed straight into a right-sided occipital and left and right marginal sinuses. There was no

transverse sinus and partly into an occipital sinus leading to two enlarged marginal sinuses. In both of these crania, the occipital-marginal system shared the drainage with the lateral sinus system. The most striking example, however, and one which came closest to the apparent position in *Zinjanthropus*, was cranium A 1932 (Fig. 10 and pl. 31). The superior sagittal sinus groove flowed straight across the cruciate eminence into a prominent occipital sinus sulcus situated just to the right of the internal occipital crest, thence via a large right marginal sinus groove to the jugular foramen. The straight sinus seems to have drained into the same pathway, as the transverse sinus grooves are not in evidence and the sigmoid sinus grooves are very small. As in *Zinjanthropus*, however, we cannot rule out the possibility of a small left transverse sinus contribution to the left sigmoid. In this cranium, the markings as a whole suggested a similar pattern to that of *Zinjanthropus*, namely, one in which virtually the entire drainage passed through the occipital-marginal system.

An additional Bantu cranium, O 150, shows extremely clear markings for the occipital sinus and, especially, the marginal sinuses (Fig. 10). The occipital sinus groove seems to start abruptly at the internal occipital protuberance. There is *no* groove for the superior sagittal sinus leading down to the protuberance and the pathway adopted by the blood in this sinus is not at all clear. One possibility is suggested by the character of the left transverse sinus groove: from the protuberance laterally, it is attenuated. Near the junction of the transverse with the sigmoid sinus grooves, it suddenly becomes both wider and deeper, as though an accession of blood had entered the transverse sinus at that point. Possibly, the superior sagittal sinus had travelled *within* the falx cerebri, away from the endocranial surface of the bone, at least just above the protuberance; had thereafter flowed into a *tentorial sinus*, which in turn ran lateralwards to enter the left transverse sinus at the site of the bulge (Fig. 10). In any event, this is a further example of a modern Bantu cranium—the fourth out of 211—in which the bony markings indicate that an appreciable part of the drainage was through an enlarged occipital-marginal sinus system.

Streeter (1915, 1918) has described how the transverse sinus changes position during development: 'In embryos between 35 and 50 mm. long, we can recognise a main channel of the tentorial plexus that is to become the transverse sinus. If we disregard the sigmoid portion of it, it forms a fairly straight line with the internal jugular vein. In the interval between the 50 mm. embryo and the adult, the transverse sinus bends backwards until it comes to lie at an angle of 90° with the internal jugular.' Browning (1953) reported that, in a small series of late foetal or neonatal sinus systems, the same basic patterns were found as in adults. However, the occipital sinuses were always very large, each having a capacity similar to that of a transverse sinus. Woodhall (1939) described the persistence of a large single or paired occipital sinus with large marginal sinuses as 'comparable to the pattern that is so common in the fetus or during the first few years of life' (p. 991).

In the light of these ontogenetic aspects, it might be queried whether the pattern of sinuses in the *Paranthropus* child from Swartkrans (SK 859) does not simply reflect the young age of the individual at death. Two facts would seem to speak against this view: the occipital-marginal sinus grooves are definite and well engraved upon the endocranial surface of SK 859; unless such grooves come and go during childhood more readily than might be expected, they give one the impression that they are permanent sulci. Secondly, *Zinjanthropus* is already subadult and shows a similar pattern of grooves; if his childhood pattern were to have changed to an 'adult' pattern during his postnatal development, one would have expected this change to be accomplished by the time the brain had virtually stopped growing.

It is provisionally concluded that the drainage pattern of the dural sinuses of *Zinjanthropus* constitutes a variation which was commoner in the robust australopithecines than it is in modern man, among whom it may occur as a rare individual variation. In the development of this variation, it would seem that an infantile arrangement has been retained with minimal modification and is utilised as the major drainage route in the adult, while the more usual major drainage channel, around the attached margin of the tentorium cerebelli, has not developed. One possible factor influencing the selection of this route in place of the lateral sinus route is the enlargement of the cerebellum of *Zinjanthropus*, to which reference has already been made. Perhaps, if the cerebellum had been subject to rapid enlargement, the development of the customary trans-cerebellar route to the jugular bulb might have been precluded during ontogeny in a proportion of cases. However, not nearly enough is known about the factors influencing the development of the normal pattern of venous sinuses in man. Until such knowledge is available, and until a bigger sample of the relevant fossils is at hand, it is perhaps idle to speculate on the possible developmental or functional significance of the unusual pattern.

The relation of the sigmoid sinus to the posterior face of the petrous part of the temporal bone

In *Zinjanthropus*, the groove for the sigmoid sinus lies far forward in the posterior cranial fossa, close to, but not under cover of, the petrous temporal. It lies about one-fifth of the way between the anterior and posterior margins of the cerebellar hemispheres, as seen on the plaster endocast. According to Schepers (1946), in *Australopithecus* the sigmoid sinus descends along the junction of the anterior and middle thirds of the cerebellar hemisphere, whereas, on the endocasts of pongids, the sulcus impression appears about half-way between the anterior and posterior margins of the hemisphere. In a group of pongid crania in which I examined these features, there were some, especially chimpanzees, in which the sigmoid sinus groove was further forward than half-way. In one gorilla cranium it was so close to the posterior face of the petrous bone as to be partly overhung by a spur of bone. In a sample of chimpanzee crania, the groove was more intimately related to the petrous than in gorilla crania. The exact relations are sometimes difficult to determine in pongid crania, because of the gentle slope of the posterior face of the petrous pyramid, in contrast with its high steep slope in the australopithecines and hominines. In modern man, the sigmoid sinus groove is apparently situated about one-quarter to one-fifth of the way from the anterior margin of the posterior cranial fossa. Sometimes, indeed, in modern human crania the groove is so intimately in contact with the posterior face of the petrous pyramid as to be partly embedded in it; thus, in the cranium of a Frenchman in the National Museum of Kenya, the groove is enclosed in bone around three-quarters of its circumference. It would seem that, in the position of the sigmoid sinus, *Zinjanthropus* and *Australopithecus* agree more with the hominine trend than with the pongid trend, but there is a great range of variation in both groups.

CHAPTER VII

THE THICKNESS OF THE CRANIAL BONES

The calvaria of *Zinjanthropus* is a curious blend of extraordinary robusticity in parts and unusual thinness in others. Basically a thin-walled calvaria, the brain-case is modified in parts by (*a*) massive ectocranial superstructures, and (*b*) excessive pneumatisation.

A. Robusticity owing to pneumatisation

One could multiply measurements apace to demonstrate the unusual thickness and robusticity of different parts of the cranium of *Zinjanthropus*. A good area in which to demonstrate this is the region of the parietomastoid and occipitomastoid sutures, since comparative data are available, once again provided mainly by Weidenreich (1943, p. 66).

The first thickness is that of the parietomastoid suture just in front of asterion. In *Zinjanthropus*, this thickness measures 22·2 mm. on the left and 24·1 mm. on the right. For comparison the values in three crania (five sides) of *Homo erectus pekinensis* range from 15 to 18 mm.; and in modern man from 3·5 to 7·0 mm.

The second thickness is that of the occipitomastoid suture 'medial to the mastoid process'. As in anthropoid apes, it is extremely difficult to take this measurement in *Zinjanthropus* because the whole pars mastoidea (not just the mastoid process) is pneumatised. For instance, in *Zinjanthropus*, the parietomastoid suture just in front of asterion is *not* the thickest part (which it is in *H. e. pekinensis*); instead the thickest part is to be found just behind the right mastoid process proper, where readings of 26–29 mm. can be taken. The greatest thickness was measured as nearly as possible at right angles to the general plane of the bone. To obtain readings comparable with those of Weidenreich, the thickness was taken not on the mastoid process itself, as indicated by the position of the mastoid crest, but on the plane surface of the pars mastoidea behind the process. Immediately behind asterion, the thickness of the occipitomastoid suture in *Zinjanthropus* is 21·7 mm. (L) and 19·2 mm. (R); the reading for the corresponding area in *Paranthropus* (Kromdraai type cranium) is ±14·0 mm. and for *Australopithecus* (Sts 19) 14·5 mm. (own measurements). In *Zinjanthropus*, some 15 mm. behind asterion, a greater thickness is attained on the right, the readings being 19·1 and 20·5 mm. The maximum thickness of the occipitomastoid suture in *H. e. pekinensis* (two crania, four sides) ranges from 6·5 to 8·0 mm.; that of modern man from 3 to 6 mm. Thus, the mastoid region in *Zinjanthropus* is appreciably thicker than in *H. e. pekinensis*—even though in the latter it is so thick as to have permitted Weidenreich (1943, p. 66) to state, '...the mastoid portion of *Sinanthropus* is three to four times thicker than that of modern man where the bone is thickest, and more than twice where it is thinnest'.

If, however, one measures the thickness 30–35 mm. behind asterion in *Zinjanthropus*, the readings are only 9·6 (L) and 12·8 (R). Hence the thickening is localised. There is little doubt that the localised, excessive thickness of the mastoid region in *Zinjanthropus* is owing to marked pneumatisation: this is clearly observable through areas of damage in the original specimen. Thickening of this nature aligns the Olduvai cranium with those of pongids, in which—especially in male gorilla crania—the entire mastoid portion may be inflated by a high degree of pneumatisation. Weidenreich draws a definite distinction between such expansion from pneumatisation (as in pongids) and enlargement from bony thickening (as in *H. e. pekinensis*). However, the distinction is not absolute: for in female pongids and in both sexes of the chimpanzee, where pneumatisation is far less pronounced, 'the bone is thick but does not reach such proportions as in *Sinanthropus*' (Weidenreich, 1943, p. 66). It would seem that both factors—pneumatisation and bony thickening—are operating in pongids, australopithecines and early hominines to varying degrees: *H. erectus* stands out from the other two hominoid groups mentioned in the degree to which it has emphasised thickening or massiveness. This early hominine feature, characteristic of the specimens of *H. erectus* from Indonesia, Choukoutien and Olduvai (hominid 9 from high in Bed II, formerly called 'Chellean Man'), is definitely lacking in the australopithecines. The latter group resemble the pongids in combining an essentially thin cranial vault with a degree of pneumatisation varying from moderate (as in *Australopithecus* and chimpanzee) to very marked (as in *Zinjanthropus*, *Paranthropus* and gorilla). The problem of pneumatisation will be discussed again in chapter XI.

B. Robusticity owing to ectocranial superstructures

In several areas, there is a heaping-up of bone on the outer surface of the calvaria. Such excrescences greatly enhance the impression of robusticity in *Zinjanthropus* and create further resemblances to the general pattern of the pongid cranium. Examples are the temporal crests and the product of their coalescence, the sagittal crest; the nuchal crest and its offshoots, the occipitomastoid crests. These have been described in chapter III, C.

C. The thinness of the parietal bones

It has already been pointed out (chapter VI, B, p. 54) that, in the region of the mastoid angle of the parietal, the thickness in *Zinjanthropus* is only 6·5–7·0 mm., in contrast with a maximum of 17·4 mm. in *H. e. pekinensis*.

A greater thickening occurs at the lambdoid angle, where the bone reaches 9·0 mm. in thickness just lateral to the slightly diverging temporal crests. The thickness of the parietal rises from front to back along the sagittal margin, from 5·5 to 10·0 mm. It drops from medial to lateral, being 6·5–7·0 mm. in the centre of the bone (the region of the parietal boss), dropping to 3·0 mm. at the lowest anterior point, that is, the nearest intact point to the sphenoidal angle. As in *H. e. pekinensis*, the thinnest part of the parietal is not the centre of the bone: it is lower down and more anterior. These thickness measurements reveal an extraordinarily thin parietal bone, for so robust and massive a hominoid cranium.

The readings in *Zinjanthropus* fall short of those in *H. erectus*. Thus, *in the mastoid angle region*, where *H. erectus* presents a strong torus angularis, the maximum readings (mm.) are as follows (comparative data from Martin, 1928; Twiesselmann, 1941; Weidenreich, 1943; Vallois, 1958 and Weiner and Campbell, 1964, as well as previously unpublished measurements by myself):

H. e. pekinensis	13·5–17·4	(mean 14·8)
H. e. erectus	14·0	(mean 14·0)
H. e. mauritanicus	12·0	(–)
Neandertal group	4·0–9·0	(mean 7·25)
Modern man	4·5–5·2	(mean 4·85)
S. African australopithecines ($n = 6$)	4·5–12·0[1]	(mean 6·4)
Zinjanthropus	6·5–7·0	(mean 6·75)

Weiner and Campbell (1964) cite thickness measurements 'at asterion': these may not be quite the same as the maximum thickness in the mastoid angle region, as the torus angularis is sometimes best developed a short distance from asterion. Their measurements (mm.) are accordingly included here, although they are listed separately as follows:

Thickness at asterion

Swanscombe (to 1 mm.)	7·0–9·0 (mean 8·0)
Modern Australian ($n = 16$)	Range not cited (mean 7·75)
Modern European ♂ ($n = 200$)	2–7 (mean ?)
Modern European ♂ + ♀ ($n = 400$)	Range not cited (means 4·01 (L), 3·79 (R))

These figures extend the range for modern man beyond the figures cited by Weidenreich (1943),

---

[1] In five australopithecine crania, the thickness ranges from 4·5 to 6·5 mm. (mean 5·6 mm.), but in the Kromdraai cranium the thickness appears to be 12·0 mm.

and give additional emphasis to the smallness of the values in *Zinjanthropus*.

In the region of the parietal boss (tuber parietale), thickness measurements (mm.) are as follows:

H. e. pekinensis	5·0(?)–16·0	(mean 10·8)
H. e. erectus	9·0–12·5	(mean 11·0)
H. e. mauritanicus	7·0	(–)
Neandertal group	6·0–11·0	(mean 9·0)
Modern Australian (*n* = 16)	Range not cited	(mean 9·2)
Modern European ♂ (*n* = 200)	3·5–9·5	(mean ?)
Modern European ♂ + ♀ (*n* = 400)	Range not cited	(means 5·8 (L), 5·6 (R))
Modern man (Weidenreich, 1943)	2·0–5·0	(mean 3·5)
S. African australopithecines (*n* = 6)	4·0–±7·0	(mean 5·8)
Zinjanthropus	6·5–7·0	(mean 6·75)

Near the bregma, readings are:

H. e. pekinensis	7·0–10·0	(mean 8·8)
H. e. erectus	5·5(?)–9·0	(mean 8·4(?))
H. e. mauritanicus	7·0	(–)
Neandertal group	5·0–9·0	(mean 7·7)
Swanscombe (L)	7·0	
Modern man	—	(mean 5·5)
S. African australopithecines (*n* = 7)	3·0–7·7	(mean 5·4)
Zinjanthropus	5·5	(mean 5·5)

In all three areas of the parietal bone, the values for *Zinjanthropus* are far less than those for *H. erectus*, less than those for Neandertal and, generally, intermediate between those for Neandertal and modern man. The thicknesses of the parietal in *Zinjanthropus* are closer to those for modern man (in one area, near bregma, they are identical with the mean) than to those for *H. erectus*. In this respect, the cranial bones of *Zinjanthropus* and other australopithecines resemble those of pongids. For the cranial bones of pongids are generally thin—according to Harris (1926) and Weidenreich (1943, p. 164), thinner even than in modern man—except where they are ballooned by excessive pneumatisation, or adorned by heavy, ectocranial cresting superstructures. As mentioned above, massiveness in *H. erectus*, however, is based on a thickening of the bone substance itself, especially of the outer and inner tables, over and above any effects owing to pneumatisation and superstructures which may concomitantly be present (Weidenreich, 1943, p. 164).

This radical departure of the parietal bones of *Zinjanthropus* from the robust pattern of its other bones led Leakey (1959a, p. 492) to formulate the twelfth of his twenty diagnostic criteria of *Zinjanthropus*—'The relative thinness of the parietals in comparison with the occipitals and the temporals'.

Admittedly, however, the parietals of *Zinjanthropus* lack two of the features which in other bones of the cranium are at least partly responsible for robusticity: there is no indication of pneumatisation which is a thickening factor in its temporal, sphenoid and frontal bones; nor are there any muscular superstructures on the surface of the parietal (except for its sagittal margin) to be compared with the massive and rugged sculpturing on the nuchal surface of its occipital bone.

The thinness of the parietal bones may be considered in relation to two other factors: the age of the individual and the relationship to the temporalis muscle. First, as to age, Getz (1960) has shown that the thickness of the parietal bone tends to increase with age. If this tendency applied to *Zinjanthropus* as well, part at least of his parietal thinness would be a function of his adolescence; with further development and ageing, it might have been expected to become somewhat thicker. There is no reason to expect, however, that further growth would have converted him into a thick-skulled individual. Secondly, the thinness of the *Zinjanthropus* parietals might seem the more surprising because the temporalis muscles have spread right across the surface of the parietals to reach the mid-line and throw up a small crest. During the ontogeny of these crests it has been claimed that 'as the area of attachment of the temporal muscles increases, in keeping with the development of the jaws, the temporal lines move peripherally as the edge of a thin film of new bone' (Zuckerman, 1954, p. 322). Elsewhere in the same study, Zuckerman tells us that this external bony film in a creature like the gorilla may be several millimetres thick (p. 318).

However, according to Scott (1957), the new temporal crest bone forms only at the 'spreading edge' of the muscle, the crest moving before the migrating muscle by a process of resorption and redeposition of bone. In describing the method of formation of sagittal crests, Scott states:

The temporal ridges develop where the fascia covering the temporal muscle is attached to the fibrous outer layer of the periosteum. Along the line of union of the two fibrous layers, the underlying cellular osteogenetic layer of the periosteum forms a bony ridge...As the muscle migrates upward and backward with growth, the bony ridge moves before it by a process of resorption and redeposition of bone until, in the male gorilla or baboon, the two ridges come together along the middle line above the sagittal suture. Here the fasciae covering the temporal muscles of the two sides of the skull meet and run together to the underlying bone, forming a two-layered fibrous septum between the muscles. Ossification extending into this fibrous septum produces a midline sagittal crest (pp. 217–18).

This interpretation varies from that of Zuckerman, inasmuch as it does not recognise a continuous film of ossification spreading across the surface of the bone with the migration of the temporal muscles. We may confidently assume that the mechanism of temporal muscle migration and crest formation in *Zinjanthropus* is similar in histogenetic essentials to that in the gorilla and the baboon. Hence, Scott's view gains support from the very fact that the parietals in *Zinjanthropus* have remained so thin, although in the mid-line, each of the two temporal crests comprising the sagittal crest reaches a thickness of 2–6 mm. Thus, each temporal crest in parts reaches a thickness equal to or greater than the entire thickness of the parietal bone which, in some areas, may be as little as 3–5 mm. Had a crest of such thickness left a comparable film of bone over the surface of the parietal bones, we should have expected the parietals to attain a greater thickness than they have in fact achieved.

Experiments have long been carried out on the relationship between the temporal muscle and the cranial vault, and some of these have a bearing on this subject. During the nineteenth century, experiments involving the surgical removal of one temporal muscle were performed by L. Fick (1857, quoted by Weidenreich, 1941*a*), by P. Lessghaft (notified to the author by Professor D. A. Shdanov of Moscow), and by Gudden (notified by Professor A. Dabelow of Mainz). The observations of R. Anthony (1903), cited by Weidenreich (1941*a*, p. 389), are interesting and perhaps relevant. Anthony removed the temporal muscle of one side from very young dogs, and subsequently compared the altered conditions with those of the normal side. Apart from a deflection of the sagittal crest toward the operated side, the bone on the operated side became *thicker* in the *absence* of the temporalis muscle than on the control side where the muscle was present. Nikitiuk has recently confirmed that the cranium becomes thicker on the operated side in dogs (1965) and has made a similar observation on the rhesus monkey (1964). He has extended the experiment to cats, while Darlington and Lisowski (1965) have recently applied similar experimental procedures to the ferret (*Mustela furo*).

The observation that the operated side becomes thicker offers more support to the idea that the *presence* of the temporal muscle, far from adding an ectocranial film over the surface of the parietal, tends to check the growth of the bone and thus to keep the parietal thin—save for the crest at the 'spreading edge' of the muscle.

Attempts have been made to correlate the thickness of the cranial wall in the course of human evolution with the expansion of the brain-case (Weidenreich, 1941*a*, 1943). Comparing the thickness and cranial capacities of the Indonesian and Chinese crania of *H. erectus* with those of modern human crania, Weidenreich states explicitly, '...there is no doubt that the decrease in the thickness of the wall of the brain case has something to do with the expansion of the cavity' (1941*a*, p. 329). Again, he states later, 'In modern man this region (the parietal tuberosity) is much thinner than in the other groups which, in turn, seems to indicate that the reduction of the cranial wall in the course of human evolution is correlated with the expansion of the braincase' (1943, p. 164).

This view, however, fails to explain that from the australopithecine to the *H. erectus* grades of hominid organisation, both the thickness of the parietals and the volume of the cranial cavity expanded considerably. Such an expansion of the brain—from about 500 c.c. to about 1,000 c.c.—should, on Weidenreich's view, have been accompanied by a marked thinning of the cranium: instead there is an approximate doubling in the thickness of the parietal bone. Clearly some other explanation will have to be sought. For the moment we offer the comment that it is the thickening of the cranial walls in *H. erectus* that demands an explanation, rather than the thinning during post-*erectus* evolution. Both the pongids and the australopithecines are characterised by thin parietal bones; *H. erectus* then seemingly underwent a remarkable specialisation manifested by excessively heavy bones. Weidenreich pointed out that this was by no means confined to the cranial bones: the limb-bones of *H. e. pekinensis* showed marked narrowing of the medullary canal and thickening of the walls in the shaft of the femur (1941 b). He concluded (1943, p. 175) that 'general massiveness is a primitive hominid character which disappears in the course of human evolution'. In the light of our newer knowledge of the australopithecines, even of their most massive member, *Zinjanthropus*, we should be inclined today to say rather that general massiveness is a character of early hominines of the Middle Pleistocene; it is not a feature of the australopithecines nor, as far as we know, of any of the Lower Pleistocene hominids. Somewhere between the Lower and Middle Pleistocene hominids, massive ectocranial superstructures on a thin-walled calvaria gave way to generalised massiveness of the bitabular intrinsic structure of the cranial bones, with superadded ectocranial adornments. It seems that this was part of the general complex of changes which marked at least one transition from the australopithecine to the early hominine level of organisation.

CHAPTER VIII

THE ENDOCRANIAL CAST OF ZINJANTHROPUS

No natural endocast was found with the cranial remains of *Zinjanthropus*. However, so much of the endocranial surface of the calvaria was preserved as to make possible the preparation of a plaster endocast (pl. 28). This was effected by me, with the invaluable expert assistance of Messrs A. R. Hughes and T. W. Kaufman of the staff of the Department of Anatomy, University of the Witwatersrand. Approximately the posterior two-thirds of the brain-case was virtually complete; likewise, the frontal poles and rostral regions. The area between had to be reconstructed. There is little doubt as to the intervening distance, because of a satisfactory approximation between the anterior and posterior calvarial parts. On the basis cranii, the gap includes the anterior part of the middle cranial fossa and most of the anterior cranial fossa; thus, the precise extent of the temporal lobes and the exact position of the temporal poles could not be determined from the surviving cranial bones. A guide to these points is provided by the body of the sphenoid, most of which is present as far forward as the posterior half or more of the hypophyseal fossa on its dorsum, while ventrally a substantial part of the vomer articulates with the sphenoidal rostrum. In higher Primates, the medial aspect of the temporal poles abuts close to the body of the sphenoid. There were available for comparison a large collection of endocranial plaster casts of gorilla, chimpanzee and orang-utan, of cercopithecoids, of *Homo erectus* and other fossil hominines, of modern man, and of australopithecine natural endocasts hitherto discovered, as well as plaster endocasts of those australopithecines where no natural endocast had formed or been found. In the circumstances, it is felt that the plaster endocast of *Zinjanthropus* cannot be far from the mark and cannot deviate volumetrically by more than perhaps 10–20 c.c. in either direction.

A. The cranial capacity of *Zinjanthropus*

The volume of the endocast was determined by displacement of water, after the cast had been varnished (following Welcker). In seven determinations, the displaced water was measured by volume and in three further determinations by weight. The mean of the first seven estimations was 529·7 c.c., the mean of the next three estimations 528·1 c.c.—and the overall mean for ten estimations is 529·2 c.c. For practical purposes, we may call the cranial capacity of *Zinjanthropus* 530 c.c. (Tobias, 1963).

The Olduvai australopithecine falls somewhat short of full adulthood, the third molars being not yet in the fully-erupted position of occlusion. The age of the youth may be about 15 or 16 years. Very little if any further growth of the brain occurs after this age: in man, for instance, brain growth is said to end at about 20 or 21 years of age (Marchand, 1902; Pffister, 1903), although a few workers would extend the period of growth somewhat longer (*see* Zuckerman, 1928). In the chimpanzee, Zuckerman found that 'in the resting stage after the eruption of the first molar', the capacity in males is already 90 per cent of the adult capacity in males and 95 per cent in females (p. 34).

For a comparable dental stage ('milk dentition with the first permanent molars erupting or erupted'), both Schultz (1940) and Ashton and Spence (1958) reported that the endocranial capacity of chimpanzee had reached 94 per cent of adult values; corresponding figures for gorilla are

90 per cent (Ashton and Spence, 1958); orang-utan 92 per cent (Ashton and Spence, 1958) or 91 per cent (Schultz, 1941b); and for modern man 94 per cent (Todd, 1933; Ashton and Spence, 1958). According to Weidenreich (1941a, p. 414), actual cessation of brain growth coincides in man and pongids with the eruption of the third molar. Since *Zinjanthropus* is nearly dentally mature, the value of 530 c.c. may be accepted as the adult cranial capacity.

This is the first of the large-toothed, heavily-muscled forms of australopithecine in which it has been possible to obtain a reliable estimate of the endocranial capacity, none of the *Paranthropus* specimens having proved suitable. It would therefore be useful here to review the evidence bearing on cranial capacity in the australopithecines as a whole.

B. The cranial capacities of the australopithecines
(Fig. 11)

The cranial capacity of the australopithecines has always been a problem of great interest, especially since several workers such as Keith (1948) and Vallois (1954) have recognised a 'Rubicon' of brain size between apehood and manhood. Keith's 'Rubicon' was 750 c.c., while Vallois's was 800 c.c. Other workers, including Straus (1953), Dart (1956) and von Bonin (1963), have emphatically rejected this notion of a cerebral 'Rubicon'. Recent studies (for example, that of Schultz, 1962) have shown that there is no clear dividing line, the largest gorilla cranial capacity (752 c.c.) differing hardly at all from the smallest capacity of an Indonesian *H. erectus* (775 c.c.). Even if we do not recognise a dividing line, it is of interest to see where the cranial capacities of the australopithecines lie in relation to those of other hominoids. It should be stressed, however, that the case for the hominid status of the Australopithecinae does not stand or fall by the brain size, and not too much should be read into the cranial capacities. Here it is relevant to quote the caution sounded by Le Gros Clark (1964):

In some of the discussions resulting from the first discoveries of australopithecine skulls, rather unnecessary emphasis was placed on the exact value of the cranial capacity, apparently on the assumption that upon this character mainly depends the question whether the Australopithecinae are to be regarded as representatives of the Pongidae or whether they are hominids of very primitive type...But (let us once more emphasise) the expansion of the brain to the size characteristic of the *later* Hominidae evidently did not occur until very long after this family had become segregated in the course of evolution from the Pongidae, and a large cranial capacity is thus not a diagnostic feature of the family Hominidae as a whole (pp. 133–4).

A second point should be emphasised at the outset: cranial capacity is not the same as actual brain size, since the cranial cavity accommodates as well as the brain the dura mater and the other coverings of the brain,cerebrospinal fluid, cranial nerves, blood vessels and blood. The relationship between endocranial volume and brain volume has been thoroughly reviewed by Zuckerman (1928): there is no consistent ratio between the two values, and it has long been customary to accept the cranial capacity as an approximation to brain size. For instance, instead of comparing the brain weights of living Primates with their body weights as Connolly (1950) did, Schultz (1941a, 1950a) has expressed cranial capacity in cubic centimetres as a percentage of body weight in grams. In the present study, cranial capacity alone is considered, with no attempt to convert it to brain volume or brain weight.

1. *The cranial capacity of the gracile australopithecines* (Australopithecus)

Six crania of *Australopithecus* are sufficiently well preserved to have made good estimates of the cranial capacity possible (Table 12). Four of these are adult specimens from Sterkfontein and their capacities are 435, 480, 480–520 and 530 c.c. (Broom and Schepers, 1946; Broom and Robinson, 1948; Broom et al. 1950). The fifth is the child skull from Taung which, at about 5 years of age, has a capacity of 500–520 c.c. (Dart, 1926; Broom and Schepers, 1946). The brain capacity which the Taung individual would have attained to, had it lived to adulthood, has been variously estimated at 570, 600, 624 and 625 c.c. (Keith, 1931; Le Gros Clark, 1947, 1955; Dart, 1926, 1956). Newer data provided by Ashton and Spence (1958) have indicated that, by the dental age of the Taung

Table 12. *Cranial capacity of* Zinjanthropus *and other australopithecines* (c.c.)

Zinjanthropus		530	Tobias (1963)
Australopithecus			
(A) Sterkfontein	Sts I	435	Schepers (1946)
	Sts 5	480	Broom *et al.* (1950)
	Sts VII	480–520	Broom *et al.* (1950)
	Sts VIII	530	Broom and Robinson (1948)
(B) Makapansgat	MLD 37/38	480	Dart (1962*b*)
(C) Taung	Taung juv.	500–520	Dart (1926)
			Schepers (1946)
	Taung adult (estimates)	570–600–625	Le Gros Clark (1947)
			Le Gros Clark (1955)
			Dart (1956)
		562	Tobias (1965*e*)
Australopithecus	Adult sample range	435–562	Present study
	Adult mean* ($n = 6$)	498	Present study
Australopithecinae	Adult mean ($n = 7$)	502	Tobias (1965*e*)

* The central value (500 c.c.) has been used for Sts VII and the new estimate (562 c.c.) for the adult capacity of Taung. The resulting means are therefore somewhat lower than those already published (Tobias, 1963), as the latter were based on an estimate of 600 c.c. for Taung.

child, the endocranial capacities of various hominoids have reached 90–94 per cent of the adult mean values. The highest percentages are for man and chimpanzee (94), the lowest for gorilla (90) and orang-utan (91–92). The mean percentage for the four kinds of hominoids is 92·5. On the basis of these figures, I have re-computed the 'adult value' for Taung as 562 c.c. and this value is accepted here. The sixth specimen is the recently discovered tolerably complete cranium (MLD 37/38) from Makapansgat (Dart, 1962*a*). This cranium agrees in its dimensions so closely with the very complete cranium (Sts 5) from Sterkfontein that the same cranial capacity should be attributed to it, viz. 480 c.c. (Dart, 1962*b*, p. 126). Thus, six specimens of *Australopithecus* range in cranial capacity from 435 to 562 c.c., giving an average value of 498 c.c. (Table 12).

2. *The cranial capacity of the robust australopithecines*

Until recently, no single specimen of the large-toothed, heavily-muscled form of australopithecine had been discovered which was sufficiently complete and undistorted to make possible a reliable estimate of the cranial capacity. True, some guesses had been made: the robust dimensions of the teeth, jaws and crania of *Paranthropus* had led to estimates such as 650 c.c. for the Kromdraai specimen (Broom and Schepers, 1946), 700 and 750–800 c.c. for two child crania from Swartkrans and 'probably over 800' and 'probably over 1,000 c.c.' for two Swartkrans adult specimens (Broom and Robinson, 1952). All these estimates agreed in giving the robust ape-man an appreciably bigger brain than the gracile form; in much the same way, the enormous teeth and jaws of *Gigantopithecus* have been held by some to connote a generalised giantism of the whole body of these curious hominoids. Insecurely based as these guesses were, it had virtually become accepted by some workers that the robuster australopithecine was not only bigger-toothed, bigger-jawed and bigger-muscled, but also bigger-brained than the gracile hominid from Taung, Sterkfontein and Makapansgat.

The Leakeys' discovery of *Zinjanthropus* in 1959 provided the first specimen of a robuster australopithecine which is sufficiently complete and well preserved to have enabled an accurate assessment of the cranial capacity to be made. Its capacity of 530 c.c. is the same as that of the largest-brained adult *Australopithecus* from Sterkfontein, but falls short of the value estimated for the adult capacity of the Taung ape-man, namely 562 c.c. Although the Olduvai creature is by far the biggest-toothed and the biggest-muscled of all the robust ape-men thus far discovered, its brain size proves to have been no larger than that of its gracile cousins. While its robusticity is a reflection essentially of its massive dentition and the voluminous jaws and muscles required to accommodate and manipulate these teeth, as far as its cranial capacity goes, its brain could be interchanged with that of the small-toothed, gracile form, *Australopithecus*.

If this is true of the most robust of all the australopithecines, it is probably equally true of the somewhat less robust South African form, *Paranthropus*. In this regard, it is interesting to note that, since the discovery of the small cranial capacity of *Zinjanthropus*, it has become clear that the cheek-teeth (and their supporting paraphernalia of jaws, buttresses and jaw muscles) were disproportionately large in some branches of the Australopithecinae, without any accompanying enlargement of the brain. Robinson has since revised the earlier estimates of the capacity of *Paranthropus* from the high values quoted above to a modest 450–550 c.c., i.e. the same order of size as obtains in the gracile *Australopithecus* (Robinson, 1961, 1962). Thus, in July 1960, Robinson could state of the australopithecines in general, 'The endocranial volume appears to be only about 500 cm.^{3}—I know of no sound evidence at present indicating a brain significantly larger than this. The range is evidently about 450–550 cm.^{3} and therefore well below the pongid maximum of 685 cm.^{3}' (1961, p. 4). In his definition of *Paranthropus* (1962, p. 138), he cited 'an endocranial volume of the order of 450–550 cm.^{3}' as one of its features.

Of course, the single valid estimate of 530 c.c. in one robust australopithecine cannot provide any estimate of the range of capacities. It may yet prove that the range and the mean capacity of the robust forms (*Paranthropus* and *Zinjanthropus*) differ from those of the gracile form (*Australopithecus*), although the ranges may overlap appreciably. Only the discovery of more and better specimens can answer this question. Meantime, there is no reason why at this stage the six capacities of the gracile *Australopithecus* and the single capacity of the robust *Zinjanthropus* should not be combined as a single sample of seven australopithecine capacities. The mean of this combined sample is 502 c.c. (Table 12).

Seen against this small array of data, the Taung child's brain was potentially the biggest of the group. Whichever estimates of its capacity one accepts (500 or 520 c.c.), Ashton and Spence's (1958) figures for growth increment lead one to infer that, as an adult, the Taung australopithecine would have had a cranial capacity beyond any of the reliably determined values so far available.

C. Australopithecine capacities compared with those of other hominoids

In Table 13, data have been assembled for the ranges and means of a number of hominoid groups. The values of the hylobatids and pongids (except the gorillas) have been taken from Vallois's compilation (1954). As in that study, the samples quoted comprise male and female crania combined. In respect of the gorilla values, Vallois cited a sample of 532 gorilla capacities ranging from 340 to 685 c.c. with a mean of 497·8 c.c. Subsequently, Ashton and Spence (1958) published a mean for a further 63 adult male gorilla crania, and Schultz (1962) for a further 58 West African male gorilla crania, the capacities in Schultz's series ranging from 423 to 752 c.c. with a mean of 536·1 c.c. The value of 752 c.c. is exceptional, exceeding the previous maxima of (c.c.): 605 (Gyldenstolpe, 1928), 623 (Weidenreich, 1943), 652 (Harris, 1926), 655 (Bolk, 1925) and 685 (Randall, 1943–4). The value of this 752 c.c. cranium has been added to Vallois's sample of 532, giving a combined ♂+♀ sample of 533, a combined range of 340–752 c.c. and a combined mean of 498·3 c.c. In the second gorilla estimate, the series of 63 crania determined by

ENDOCRANIAL CAST

Table 13. *Ranges and means of cranial capacities of hominoids (c.c.)*

Species	Size of sample	Range	Mean	Reference
Gibbon	86	87– 130	89·3	Vallois (1954)
Siamang	40	100– 152	124·6	Vallois (1954)
Chimpanzee	144	320– 480	393·8	Vallois (1954)
Orang-utan	260	295– 475	411·2	Vallois (1954)
Gorilla ♂+♀	533	340– 752	498·3	Vallois (1954) with Schultz's (1962) cranium of 752 c.c.
Gorilla ♂+♀	653	340– 752	506	Present study, based on Vallois (1954), Ashton and Spence (1958) and Schultz (1962)
Gorilla ♂	400	420– 752	534·8	Schultz (1962)
Australopithecinae	7	435– 562	502·4	Present study
Homo erectus	9	775–1,225	978	Tobias (1963)
Modern man	—	±1,000–2,000	±1,300	Martin (1928), Weidenreich (1943)

Table 14. *Cranial capacity of* Homo erectus *(c.c.)**

Specimen	Capacity	Reference
Homo erectus erectus		
Java I (Trinil)	935 (900)†	Weidenreich (1943)
Java II (Sangiran)	775‡	Weidenreich (1943)
Java III (Sangiran)	c. 880 (900)†	Weidenreich (1943)
Mean (n = 3)	863 (858)†	Tobias (1965a)
Homo erectus pekinensis		
Choukoutien II	1,030	Weidenreich (1943)
Choukoutien III	915	Weidenreich (1943)
Choukoutien X	1,225	Weidenreich (1943)
Choukoutien XI	1,015	Weidenreich (1943)
Choukoutien XII	1,030	Weidenreich (1943)
Mean (n = 5)	1,043	Weidenreich (1943)
Homo erectus of Olduvai		
Olduvai Hom. 9	1,000	Tobias (1965b)
Homo erectus mean (n = 9)	978	Present study

* For later data, see Appendix to this chapter, p. 94.
† It is not clear whether Weidenreich (1943) accepted Weinert's (1928) estimate for *H. e. erectus* I (935 c.c.) which he quotes in the table on p. 108, or his own estimate (900 c.c.) to which he refers in the text on p. 114, stating again on p. 116 that the range for the three crania from Java is 775 to 900 c.c. Again, on p. 116, he quotes for the third *H. e. erectus* cranium both c. 880 and 'close to 900 c.c.'. The mean of 860 c.c. which he quotes on p. 116 could have been derived from *either* 900+775+900 (mean equals 858) *or* from 935+775+880 (mean equals 863).
‡ Given on p. 115 of Weidenreich (1943) as 745.

Ashton and Spence (1958) and Schultz's entire new sample of 58 have been added to Vallois's sample of 532, giving a combined ♂+♀ sample of 653, a combined range of 340–752[1] and an estimated combined mean of 506 c.c. The third set of gorilla values quoted are for 400 ♂ crania, compiled by Schultz (1962) and including both his new series and that of Ashton and Spence (1958): the values for males alone range from 420 to 752 with a mean of 534·8 c.c.

The values for *H. erectus* in Table 13 are based upon three Indonesian specimens, five Choukoutien crania and Olduvai Hom. 9 ('Chellean Man'), details of which are conveyed in Table 14. It will

[1] The range of values was not stated by Ashton and Spence (1958).

be noted that this *H. erectus* series differs from that compiled by Ashton (1950). Unfortunately, Ashton has included the six Ngandong crania under '*Pithecanthropus pekinensis*', of which he states Weidenreich (1943) gave the cranial capacities of eleven specimens. In fact, Weidenreich tabulated the capacities of only five specimens of *P. pekinensis* (although mentioning another two estimates, 850 and 1,300, in the text). When to these five Pekin cranial capacities are added Weidenreich's figures for six Ngandong crania and three Javanese

possible to take the view—with Boule and Vallois (1957)—that 'Ngandong Man has moved sufficiently far away from *Pithecanthropus* to be regarded as a different type' and that he may be placed 'in the great species of *Homo neanderthalensis*, of which he represents simply a special race' (p. 401). Campbell (1962, 1963) places him in *H. sapiens soloensis*. Since there is certainly no agreement with Ashton (1950) that the Ngandong crania belong to *Pithecanthropus pekinensis* (*Homo erectus pekinensis*), I have excluded them from my *H.*

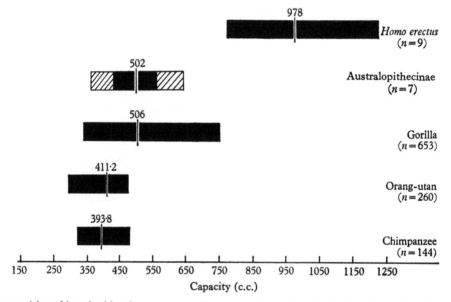

Fig. 11. Cranial capacities of hominoids: the mean, sample range and sample size is given for each of two groups of fossil hominoids and three groups of living hominoids. The hatched area represents the estimated population range for the Australopithecinae (mean ± 3 s.D.'s). (For later data on *H. erectus*, see Appendix on p. 94.)

Pithecanthropus crania, Ashton's total of fourteen crania and his mean of 1,026 c.c. are obtained. Clearly Ashton has included the six Ngandong crania (*op. cit.* p. 715), although Weidenreich (1943, p. 232) indicated that he regarded them as 'intermediate between the *Pithecanthropus* and *Sinanthropus* stage, on the one hand, and Neanderthal types on the other' and, again, spoke of them as representing 'the next evolutionary step in the line leading from *Pithecanthropus* to modern man'. The usual difficulty obtains here that intermediate or transitional forms often cannot be classified under the International Code of Zoological Nomenclature. While Weidenreich's view would be that Ngandong Man ought to be classified as *P. soloensis*, it is

erectus group in Table 14. Likewise, I have excluded the crania of Broken Hill and Hopefield (Saldanha) which Coon (1963, p. 337) would classify as *H. erectus* (*see* p. 94 for later figures).

Most of the data in Table 13 are summarised in Fig. 11.

The mean of seven australopithecine cranial capacities (502·4 c.c.) is virtually the same as the mean for the biggest-brained of the anthropoid apes, the gorilla, which has a mean capacity of 506 c.c. for a large sample comprising crania of both sexes (Table 13). The mean value of the orang (411·2 c.c.) is 90 c.c. smaller than the australopithecine mean, while the mean for chimpanzee is somewhat smaller again (393·8 c.c.).

Nevertheless, the biggest orang and chimpanzee crania (about 480 c.c.) overlap the lower part of the australopithecine range (435+ c.c.).

Ashton's (1950) analysis of the endocranial capacities of the Australopithecinae was based on seven crania. These included the Taung child, unfortunately not adjusted for age, and two crania estimated to have capacities of 650 c.c. which are not accepted in the present study. He compared these values with means for male and female chimpanzee and gorilla, as well as for a combined sample of *H. erectus* and the Ngandong crania. Ashton found that all the australopithecine capacities but one were significantly greater than those of the chimpanzee. None differed significantly from those of male gorilla, while only the two estimated capacities of 650 c.c. were significantly greater than those of female gorilla. *Zinjanthropus* provides one more australopithecine datum well within the gorilla range of cranial capacities.

However man-like (hominine-like) the australopithecines were in other respects, in their cranial capacity and therefore, presumably, their brain-size, they showed no substantial difference from gorillas. Whichever method we use to estimate the range of capacities in the population of australopithecines, the greater part of the estimated range overlaps with the gorilla range, while there is generally substantial overlap with the ranges of the smaller-brained orang and chimpanzee.

This conclusion would not be materially affected by considering male and female crania separately, since we are comparing a combined sample of presumed male and female australopithecines with a combined sample of male and female gorilla. While the means and the standard deviations would be biased by the relative numbers of males and females in each sample, the ranges, with which we are dealing here, are less likely to be so influenced.

On the other hand, the australopithecine mean falls far short of the smallest hominine mean, which, according to my calculations, is 973·7 c.c. for eight crania of Asian *H. erectus* or 978 c.c. for nine crania of *H. erectus* if Olduvai Hom. 9 is included. Ashton's mean for '*Pithecanthropus*' (including six Ngandong crania) was 1,026 c.c. Using the latter figure, he found that all the australopithecine capacities, even including the estimates of 650 c.c., were significantly smaller than those of '*Pithecanthropus*'. There is a gap of 213 c.c. between the largest australopithecine capacity accepted in the present study (562 c.c.) and the smallest known *H. erectus* capacity (775 c.c.). At the same time, the gap between the largest gorilla cranium and the smallest *H. erectus* cranium has been progressively reduced from 152 c.c. (Weidenreich, 1943) to only 90 c.c. (Randall, 1943–4), and now to a mere 23 c.c. (Schultz, 1962). As yet we have no australopithecine brain-case as large as the biggest-brained gorilla cranium. This raises the question of predicting the population range of *Australopithecus* from the sample range, in which 127 c.c. separates the largest and the smallest cranial capacities.

1. *Estimation of the population range in Australopithecinae*

In his definition of the proposed genera, *Paranthropus* and *Australopithecus*, Robinson (1962) ascribes to both an endocranial volume 'of the order of 450–550 cm.^{3}'. This range of only 100 c.c. is clearly too small, as a glance at Table 13 will confirm. Even in the smaller-brained chimpanzee and orang-utan, there are respectively 160 and 180 c.c. between the minima and maxima; while the extent of the range in gorilla—of comparable mean to the australopithecines—is 412 c.c. for both sexes combined or 332 c.c. for male crania alone. Even the small sample of *H. erectus* has a range of 450 c.c. Clearly, the population range in the Australopithecinae must far exceed 450–550 c.c.: even the existing sample range (435–562) is slightly greater. In this section, accepting the adult value of 562 c.c. for Taung, I shall use the sample range 435–562 c.c. as a basis for estimating the probable population range.

If the seven australopithecines whose cranial capacities are recorded in Table 12 are treated as members of a single population, the standard deviation by the conventional formula for S.D.'s of samples is 42·1 c.c. In this event the mean ± 3 S.D. gives a range of 376–629 c.c. However, owing to the smallness of the sample, it is preferable to derive an estimate of the standard deviation for the population, by multiplying the range (127 c.c.) by A_n which, for a sample of $n = 7$, is 0·3698 (Lindley and Miller, 1953, Table 6, p. 7; see also Simpson, Roe and Lewontin, 1960, Table 1A, p. 41): the standard deviation on this basis is somewhat higher, namely, 47·0 c.c. On this estimate, the mean ± 3 S.D. gives the somewhat larger range of 361–643 c.c. The estimated population range of the Australopithecinae above and below the extremes of the sample range is shown as a hatched zone in Fig. 11. On this estimate of the population range, the largest australopithecine value (643 c.c.) is definitely smaller than the second largest gorilla capacity known (685 c.c.), and falls short by 109 c.c. of the largest known gorilla capacity (of a very large sample). The value of 643 c.c. falls short by some 132 c.c. of the smallest known *H. erectus* capacity (Java II of Sangiran, 775 c.c.); and even by 37 c.c. of the central value (680 c.c.) estimated for *H. habilis*, the new early hominine from Bed I, Olduvai Gorge (Tobias, 1964a). The estimated cranial capacity of *H. habilis* is 3·79 S.D.'s larger than the mean australopithecine capacity.

The estimated australopithecine population range of 361–643 is based upon a distribution about the present sample mean of 502 c.c. This would be a valid estimate if the present sample of seven australopithecine capacities were drawn from the middle reaches of the population distribution. Such a value as 650 c.c. which has been estimated for the Kromdraai calvaria (Broom and Schepers, 1946) and for the first Makapansgat calvaria (MLD 1) (Dart, 1948a) would then be improbable. However, through sampling bias, the present australopithecine sample may give a poor indication of the population mean. In the extreme case, if 562 c.c., the highest value in the sample, were also the highest value in the population, the estimated population range of 282 c.c. would yield values of 280–562 c.c.; conversely, if 435 c.c., the lowest value in the sample, coincided with the lowest value in the population, the estimated population range would be 435–717 c.c.

It may tentatively be concluded that, *if* all the australopithecines whose cranial capacities are listed in Table 12 were members of a single population and, *if* the population variability (as expressed by the range) were of the order of six times the estimated standard deviation, then it is highly unlikely that any adult australopithecine cranium still to be discovered would have a capacity smaller than 280 c.c. or greater than 717 c.c. It is of interest to note that the latter value is smaller than the largest gorilla capacity (752 c.c.), the smallest *H. erectus* (775 c.c.), Keith's 'Rubicon' of 750 c.c. and Vallois's of 800 c.c.

We may conclude that the estimates of 1,000 c.c. for the upper limits of the australopithecine cranial capacity (Broom and Robinson, 1952; Dart, 1956) are excessive and highly unlikely.

The 'middle reach' range of 361–643 c.c. (where the extent of the range is 282 c.c.) is the most likely estimate of the upper and lower limits of cranial capacity in the Australopithecinae. Both the conservative and less conservative estimates yield ranges almost entirely within the gorilloid range of capacities. Only if the existing seven australopithecine capacities are a highly biased sample, on the low side of the population mean, is it at all conceivable that the biggest australopithecine brains exceeded (or even reached) the biggest gorilla brains.

2. *The variability of the cranial capacity*

Table 15 shows the ranges between maxima and minima and relates them to the mean for each hominoid group; from this index an estimate of the coefficient of variation has been derived for each good sample or population estimate. No attempt has been made to derive the coefficient for *H. erectus* because of the marked difference in capacity between the available two small sub-samples of *H. e. erectus* and *H. e. pekinensis*.

It is clear from Table 15 that the cranial capacity tends to be more variable, the bigger it is. Thus, a sample of 86 gibbons had a mean capacity of 89·3 c.c. and a range of 43, or 48·1 per cent of the mean. In a sample of 40 siamang, the range of 52 comprised 41·7 per cent of the mean of 124·6 c.c. In the chimpanzee sample ($n = 144$), the range was 160, being 40·6 per cent of the mean of 393·8 c.c.

Table 15. *The variability of hominoid cranial capacities*

Species	Size of sample	Mean capacity (c.c.)	Diff. between minimum and maximum capacities (c.c.)	Range/mean (%)	Estimated coefficient of variation (%)
Gibbon	86	89·3	43	48·1	8·0
Siamang	40	124·6	52	41·7	7·0
Chimpanzee	144	393·8	160	40·6	6·8
Orang-utan	260	411·2	180	43·8	7·3
Gorilla ♂+♀	653	506	412	81·4	13·6
Gorilla ♂	400	534·8	332	62·1	10·4
Australopithecinae	7	502·4	127	25·3	—
Australopithecinae (population estimate)	—	502·4	282	56·1	9·35
Homo erectus	9	978	450	46·2	—
Modern man	—	±1,300	±1,000	±77	12·8

The range in the orang-utan sample (*n* = 260) was 180, being 43·8 per cent of the mean of 411·2 c.c. In these four apes, the extent of the range is just under half of the mean and the estimated coefficient of variation is small (6·8–8·0). When one turns to the gorilla, there is a sudden and dramatic jump to a range/mean index of 81·4 per cent; in other words the range of 412 c.c. is 81·4 per cent of the mean which is 506 c.c. The coefficient of variation for the large gorilla sample is 13·6 per cent.

The wide range and high variability of gorilla cranial capacities may reflect in part a greater degree of sexual dimorphism in gorilla than in other anthropoid apes. This is borne out by the figures quoted by Ashton (1950): his adult chimpanzee females have a mean capacity which is 91·5 per cent of that of the males, whereas the mean capacity of gorilla females is only 85 per cent of that of their males. This difference may in turn simply reflect different degrees of sexual dimorphism in body size. When a sample of male gorilla crania is considered alone, variability is of course less than in the combined sample. The coefficient of variation is 10·4 per cent, as contrasted with the figure of 13·6 per cent in the mixed sample of both sexes. Nevertheless, even in the all-male sample, the variability is appreciably higher than in any of the mixed (♂+♀) samples of smaller-brained apes.

Variability is high, too, in modern man, with an estimated coefficient of variation of 12·8 per cent. One gains the impression that the jump from the moderate-sized hominoid capacities (with means up to about 400 c.c.) to the large-sized hominoid capacities (with means of over 500 c.c.) was accompanied by a large jump in the range of capacities, not only absolutely but relatively. Thus the estimated coefficient of variation jumps from 6·8–8·0 per cent to about 10–13 per cent. The australopithecine sample with its low range/mean index of 25·3 per cent looks clearly out of place in the table; even the population estimate (based on the mean ±3 s.d.) gives an index of only 56·1 and an estimated coefficient of variation of 9·35 per cent. While these values for the australopithecines are larger than those for the smaller-brained pongids, and not very different from those for male gorilla, they are still low as compared with the gorilla sample of both sexes, and the samples of *H. erectus* and *H. sapiens*. Either we are making inferences from a very poor and unrepresentative sample, or the australopithecines are less variable than other large-brained hominoids. The difference in variability is especially striking when we compare Australopithecinae with *Gorilla*; the means are similar, but the variability is not. Some part of the greater variability in *Gorilla* may be owing to a greater degree of sexual dimorphism in *Gorilla* than in the australopithecines.

A further factor possibly making for greater variability in the gorilla series published is that they have included wild-shot specimens and zoo specimens; at least one of the very high gorilla capacities was that of a zoo-reared animal. Aside from these possible sources of variance in *Gorilla*, it is valid to state that (*a*) a real increase in variability with increasing brain size seems to be a feature of the hominoids; and (*b*) we do not yet have enough information to determine whether or not the Australopithecinae were an exception in this respect.

D. Cranial capacity in relation to body size

Schultz (1936, 1941*a*, 1950*a*, 1957) and others have repeatedly stressed that absolute brain size is of less phylogenetic importance than brain size relative to body size: thus the large variability of gorilla cranial capacities may reflect wide differences in the body weight of the gorilla. It is in the interpretation of these cranial capacities that one would have to take relative brain size into consideration. Thus, a 500 c.c. capacity in a gracile or light-weight australopithecine would clearly have a different significance from a 500 c.c. capacity in a heavy-weight gorilla.

A simple way of approaching this problem would be to express the brain weight as a fraction of the body weight, or to divide the body weight by the brain weight (cf. Dubois, 1898; Hrdlička, 1925, etc.). Jerison (1961, 1963) has attempted a more refined analysis. He has demonstrated that, when comparisons are restricted to differences among species and higher taxa, the brain weight or volume provides a reasonable basis for estimating the number of cortical neurones. A further analysis, based upon certain assumptions about the size of the brain, enabled him to resolve brain size into two components, one of which is related to body size and the other of which is interpreted as being associated with improved adaptive capacities. Given certain assumptions, he claims that it is possible to estimate the number of cortical neurones in each of the two components. He has developed a series of equations for the calculation of these neuronal values, given the size of the brain (or an approximation to it, such as cranial capacity) and the size of the body (or an estimate of it). By applying these formulae, he has been able to compute the number of 'extra' neurones that may be interpreted as being 'associated with the evolution and adaptation of brain: behaviour mechanisms in response to the challenge of the environment'. With this second parameter—the number of 'excess neurones'—he has found it possible to differentiate the Primates, including the hominids, on the basis of relative brain development.

Table 16 conveys the results obtained for a series of hominoids. While most of the data in the table are taken from Jerison's (1963) analysis, I have recomputed the values for *Australopithecus africanus* and for *A. (Zinjanthropus) boisei*, on the basis of later data and different estimates of body size. Thus, for *A. africanus*, he based his estimate on a brain size of 500 g. and a body size of 20,000 g. I have recalculated these values for brain sizes of 435, 500 and 560 c.c., being the extremes and a middle value for the sample of *Australopithecus* cranial capacities; the body weight of 20,000 g. seems to me to be somewhat too small an estimate and I have accordingly made the calculations for body weights of 25,000 and 35,000 g., associating the two smaller brain sizes with the smaller body weight and the largest brain size (560 c.c.) with the heavier body weight. The estimates of 'extra neurones' obtained by applying Jerison's formulae to these figures are 3·9, 4·3 and 4·7 thousands of millions respectively; Jerison's figure for *A. africanus* (4·4) fell in the middle of this range of values (Tobias, 1965*a*, *c*).

Again, Jerison used a guess of 600 c.c. for the brain size of *Zinjanthropus*; I have used the value 530 c.c., while adhering to his estimate of body size, namely 50,000 g. This gave a value of 4·2 in place of his estimate of 4·7 thousands of millions.

Despite the altered values, and within the limits of the method and the assumptions on which it is based, my results have corroborated one of Jerison's conclusions, namely, that the australopithecines 'were clearly, if only slightly, in advance of the level of brain evolution achieved by the anthropoid apes of our time' (Jerison, 1963,

ENDOCRANIAL CAST

Table 16. *Estimates of 'extra neurones' in hominoids*

(Modified and added to after Jerison, 1963)

	Brain size (g. or c.c.)	Body size (g. or c.c.)	Estimates of total neurones (in thousands of millions)	Estimates of 'extra neurones' (in thousands of millions)	Reference
Chimpanzee	400	45,000	4·3	3·4	Jerison (1963)
Gorilla A	540	200,000	5·3	3·5	Jerison (1963)
Gorilla B	600	250,000	5·7	3·6	Jerison (1963)
Zinjanthropus	530	50,000	5·2	4·2	Present study
Australopithecus Sts I	435	25,000	4·6	3·9	Present study
Taung (adult)	560	35,000	5·5	4·7	Present study
Homo erectus	900	50,000	7·4	6·4	Jerison (1963)
H. erectus	1,000	50,000	8·0	7·0	Jerison (1963)
H. erectus	775–1,225	50,000	6·8–9·4	5·8–8·4	Present study
H. sapiens	1,300	60,000	9·5	8·5	Jerison (1963)
Varied population means	1,276–1,400	53,000–68,000	9·4–10·0	8·4–8·9	Present study

p. 288). The pongid values range from 3·4 to 3·6 thousand million 'excess neurones'. *H. erectus* could likewise be differentiated from the Australopithecinae on the one hand and from *H. sapiens* on the other: Jerison's figures for *H. erectus* were based on two specimens, one of 900 c.c. and one of 1,000 c.c., which give him values of 6·4 and 7·0. Using the smallest *H. erectus* capacity (775 c.c.) and the largest (1,225 c.c.), I have widened the range to 5·8–8·4 thousand million 'extra neurones'. Similarly, the range for *H. sapiens* is given by Jerison as 8·5; I have calculated outer limits of 8·4–8·9 for population means (Fig. 12).

The results may be summarised as follows:

African great apes	3·4–3·6	
Australopithecines	3·9–4·7	thousand million extra neurones
H. erectus	5·8–8·4	
H. sapiens (various populations)	8·4–8·9	

In passing, the estimated cranial capacity of *H. habilis* gives a 'Jerison value' of 5·3, about midway between the values for the australopithecines and for *H. erectus* (Tobias, 1964b).

These estimates are admittedly based upon many assumptions and, in almost every instance, it has been necessary to make a guess at the body size. However, varying the body size within fairly wide limits does not appreciably alter the results: since it is the smaller size-component of the total brain which is related to body size, there are fairly broad tolerance limits in the method. The underlying neurohistological premises are, of course, basic to any validity the method may possess, and they need to be tested on a greater number of modern specimens, where histological controls can be maintained. In particular, the variable development in different mammals of areas in which large or small neurones predominate needs to be more carefully assessed and perhaps allowed for in the formulae.

Meantime, the first results obtained by this new approach, so elegantly devised by Jerison, are indeed suggestive. For they have confirmed the evidence provided from other sources that *Australopithecus* and *Zinjanthropus* are more advanced hominoids than the apes, but not so advanced as *H. erectus*. Furthermore, if *H. habilis* is included in the series, there appears an interesting stepwise progression in the number of 'excess neurones' with succeeding grades of hominisation.

E. Morphological features of the endocranial cast

1. *Encephaloscopic features*

The impressions of the gyri and sulci are well preserved only in the frontal region. Here, the impressions of the superior, middle and inferior frontal gyri can be identified on both sides,

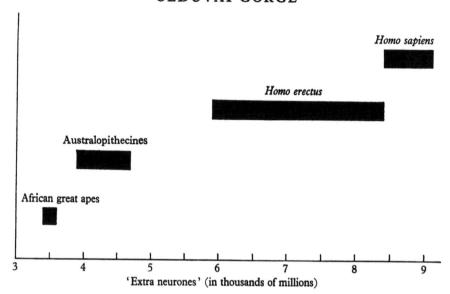

Fig. 12. Range of estimates of 'extra' cortical neurones of hominoids, based upon the formulae of Jerison (1963). 'African great apes' includes chimpanzee and gorilla; 'australopithecines' includes *Australopithecus* and *Zinjanthropus*; *H. sapiens* comprises Neandertal man and modern man.

separated by those of the superior and middle frontal sulci. The impression of the inferior frontal convolution is very prominent, and imparts an angularity to the dorsal contour of the frontal lobe: this feature *Zinjanthropus* shares with Sts II, Sts VII and lower Homininae. Schepers (1950) considered it to be an advanced feature of these brains. Over the parietal, temporal and occipital regions, very little gyral modelling is evident while, over the cerebellum, there is an indication of the horizontal fissure, the deep sulcus separating the superior and inferior surfaces of the cerebellar hemispheres.

As seen in norma ventralis (Fig. 13), the brain-stem is short and squat, turning sharply downwards at the presumed junction of the medulla oblongata and medulla spinalis, well forwards of the hindmost extremity of the brain. Behind the brain-stem can be seen, first, the impression of the marginal and occipital sinuses; secondly, the cerebellar hemispheres and, thirdly, the occipital poles of the cerebral hemispheres. This forward placement of the brain-stem in *Zinjanthropus* agrees well with the position in the endocast of Sterkfontein VIII, though in the endocasts of Sts 5 and VII the brain-stem seems to be placed a little further back (Schepers, 1950). In no single one of a number of adult pongid endocasts does the brain-stem occupy the position it does in *Zinjanthropus*. Figures 13 and 14 demonstrate how much further back, relative to the cerebellar hemispheres and occipital poles, the brain-stem is placed in an adult gorilla, as compared with *Zinjanthropus*. In the adult chimpanzee endocast shown for comparison, the brain-stem is not so far back as in the gorilla, nor yet as far forward as in *Zinjanthropus* and Sts VIII. In the australopithecine endocasts, the brain-stem is almost wholly in front of the posteromedial poles of the cerebellar hemispheres; these hemispheres are large and, behind the brain-stem, extend farther medially than in pongid endocasts. In adult pongids, the brain-stem is not in front of the posteromedial poles of the cerebellar hemispheres, but rather between or medial to them, so that they appear to have been forced apart by the posteriorly-placed brain-stem. Very little of the cerebellum or of the occipital poles of the cerebrum shows *behind* the brain-stem in the pongid endocasts.

Interestingly, this posterior placement is not present in two juvenile gorilla and two juvenile chimpanzee endocasts (Fig. 15). For example, in the endocasts of a juvenile gorilla (in which M^1 is in the process of erupting) and of a juvenile

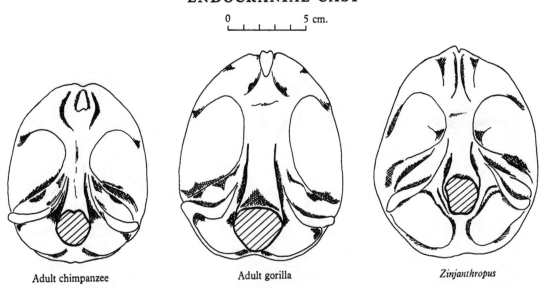

Fig. 13. Dioptographic tracings of norma ventralis (basalis) of endocranial casts of adult chimpanzee, adult gorilla and *Zinjanthropus*. Note the position and arrangement of the brain-stem and cerebellar hemispheres.

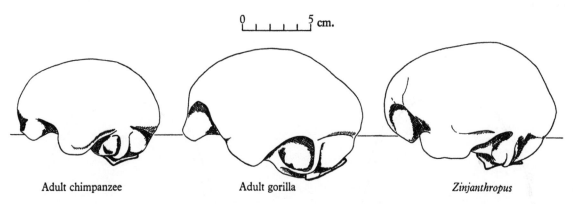

Fig. 14. Dioptographic tracings of norma lateralis of endocranial casts of adult chimpanzee, adult gorilla and *Zinjanthropus*. Note the difference in the rostral, parietal and medullary regions.

chimpanzee (in which M^1 has just erupted), the brain-stem is much further forward, in a position comparable with that in the adult Australopithecinae. This observation, of course, merely confirms the trend discussed by Bolk (1915), for the foramen magnum to be far forwards in juvenile anthropoid apes and progressively to move towards the posterior position of adulthood. Since the brain plays an overwhelmingly important part in the morphogenesis of the calvaria (Wagner, 1935, 1937; Moss, 1954; Mednick and Washburn, 1956; Tobias, 1959c), it seems that we should regard the posterior (or caudal) shift of the brain-stem in apes, not as the consequence of primary osseous developments (affecting the position of the foramen magnum), but as the consequence of morphogenetic changes in the brain. However, we cannot rule out the possibility that postural factors, too, may play a part in moulding the head about a physiological frame of reference, such as the vestibular plane of Delattre (Delattre and Fenart, 1960).

When the endocast of *Zinjanthropus* is compared with those of several *Australopithecus* casts (Sts II, Sts 5 and Taung) (Figs. 16–18), it is seen that the former is both longer and broader than the two specimens from Sterkfontein; on the other hand, it is broader than, though of similar length to, the

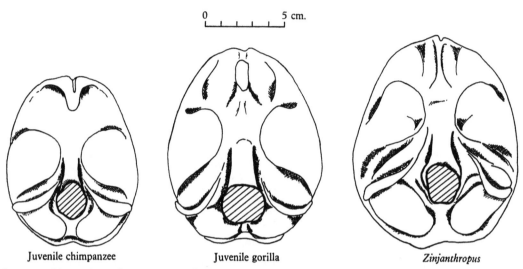

Fig. 15. Dioptographic tracings of norma ventralis (basalis) of endocranial casts of juvenile chimpanzee, juvenile gorilla and *Zinjanthropus*. Note that the resemblances in the cerebello-medullary region are much greater than when the adult pongids are compared.

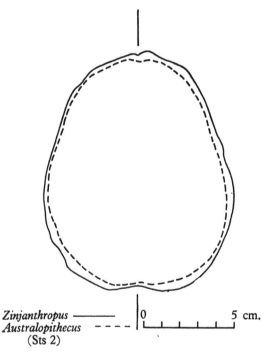

Fig. 16. Outline of norma dorsalis of endocranial cast of *Zinjanthropus* (continuous line) superimposed on that of *Australopithecus*—Sts II (interrupted line). Dioptographic tracing with endocast orientated on the orbito-occipital plane.

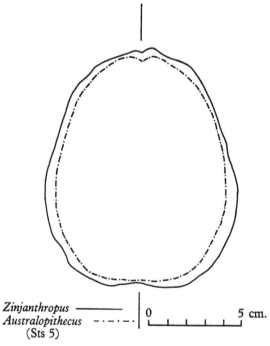

Fig. 17. Outline of norma dorsalis of endocranial cast of *Zinjanthropus* (continuous line) superimposed on that of *Australopithecus*—Sts 5 (dot-and-dash line). Dioptographic tracing with endocast orientated on the orbito-occipital plane.

specimen from Taung. The breadth preponderance is greatest in the parietal lobe area (see especially the comparison with Sts 5, Fig. 17), suggesting that this part of the brain is relatively larger in *Zinjanthropus* than in the other australopithecines.

The tendency towards parietal expansion is manifest also in the side view of *Zinjanthropus* as compared with pongids (Fig. 14) and with the same three endocasts of *Australopithecus* (Fig. 19).

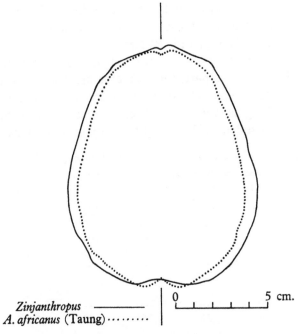

Fig. 18. Outline of norma dorsalis of endocranial cast of *Zinjanthropus* (continuous line) superimposed on that of *Australopithecus* juvenile from Taung (dotted line). Dioptographic tracing with endocast orientated on the orbito-occipital plane.

In each of these figures, the endocast has been drawn in the orbito-occipital plane and the fronto-temporal (Sylvian) notches have been superimposed. The marked vertical expansion of the parietal lobe area of *Zinjanthropus* is clearly apparent in these comparisons.

The degree of advance of the temporal pole in relation to the Sylvian vertical plane cannot be determined, as the part of the middle cranial fossa which lodges the pole is missing. Convolutional and sulcal markings over the parietal and occipital areas of the endocast are hardly represented at all; neither the lunate sulcus nor the parieto-occipital fissure, which Schepers (1946, 1950) found so typically hominine in the other australopithecine endocasts, can be located in the *Zinjanthropus* endocast. Le Gros Clark (1964, p. 135) is emphatic that it is really not possible to identify the lunate sulcus with certainty from the impressions on the casts—which is precisely what Weidenreich (1936a) said about the endocasts of *H. e. pekinensis*, while von Bonin (1963) questions Ariens Kappers's (1929) identification of the lunate sulcus on the endocast of *H. e. erectus* I.

The cerebellum is well underslung, so that it is not to be seen at all when the endocast is viewed in norma dorsalis. In this respect, the endocast of *Zinjanthropus* agrees with those of other Australopithecinae and Homininae and differs from the common arrangement in the Pongidae.

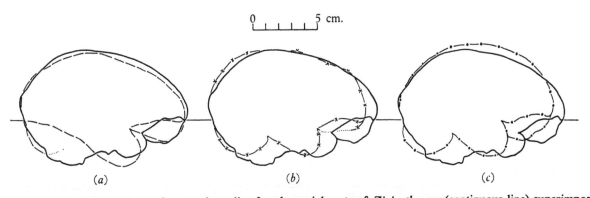

Fig. 19. Dioptographic tracings of norma lateralis of endocranial casts of *Zinjanthropus* (continuous line) superimposed respectively on those of (a) Sts II (interrupted line), (b) the Taung juvenile (×-and-dash line) and (c) Sts 5 (dot-and-dash line).

2. Encephalometric features

Table 17 conveys some of the measurements of the *Zinjanthropus* endocast, in comparison with those of the best-preserved of the other australopithecine endocasts, as measured by Schepers (1946, 1950). These comparisons show that the length of the *Zinjanthropus* endocast (129 mm.) lies within the range for adult *Australopithecus* endocasts (118–132 mm.), whereas the biparietal breadth of 100·3 mm., as we have already indicated, exceeds the australopithecine sample range of 84–98 mm. Biparietal expansion is present in the *Australopithecus* endocasts and Schepers (1950) commented that this expansion represents a highly significant advance. It would seem that this lateral growth tendency has been carried somewhat further in *Zinjanthropus* than in the South African forms, as is indicated both by the absolute biparietal width, and by the biparietal width indices. The large size of the cerebellum of *Zinjanthropus*, to which reference has been made in the account of the posterior cranial fossa, is brought out both by the absolute measurements and by the cerebellar indices. The cerebellum seems to have been better developed in *Zinjanthropus* than in any of the other australopithecine brains.

Von Bonin (1963) has recorded a number of encephalometric lengths and arcs in fossil endocasts, including those of '*Australopithecus*' (presumably Taung) and '*Plesianthropus*' (presumably some or all of the endocasts from Sterkfontein). For the most part, these lengths are based upon von Economo's (1930) encephalometric constants; the lengths and the indices derived from them are intended to throw light mainly on the relative

Table 17. *Encephalic measurements and indices of* Zinjanthropus *and other australopithecines**

Measurements	*Zinjanthropus*	Sts I	Sts II	Sts 5	Sts VII	Sts VIII	Taung
Cerebral length	129·0	118	120†	121	124	132	125
Biparietal width	100·3	88	84	82	90	98	92
Bitemporal width	104·1	100	94	92	100	104	90
Bicerebellar width	91·5	82	—	74	78	75	66
Cerebellum to temporal pole	91·1 (L) 97·5 (R)	80	84	—	—	—	74
Projective height at cerebello-temporal notch	61·5	68	70	—	—	—	70
Cerebellar length	45·0 (L, R)	33	—	35	36	43	38
Indices (%)							
Biparietal width / Cerebral length	77·8	74	70	68	72	74	73
Bitemporal width / Cerebral length	80·7	84	78	76	80	79	72
Biparietal width / Bitemporal width	96·3	88	89	89	90	94	102
Bicerebellar width / Bitemporal width	87·9	82	—	80	78	72	73
Bicerebellar width / Cerebral length	70·9	69	—	61	63	57	53
Cerebellum to temp. pole / Cerebral length	70·6 (L) 75·6 (R)	68	70	—	—	—	59·2
Cerebellar length / Cerebral length	35	28	—	29	29	33	30

* Measurements on South African australopithecine endocasts are taken from Broom and Schepers (1946) and Broom *et al.* (1950), while the indices based on these measurements have for the most part been re-computed.

† Stated as 120 in Broom and Schepers (1946, Table 1) but 130 in Broom *et al.* (1950, Table 2).

development of various parts of the brain. Among the few conclusions drawn are that an increase in the relative height of the cerebrum is 'the most important and the most obvious change that the brain undergoes during these stages of phylogenesis' (i.e. the australopithecine and *H. erectus* stages); that the difference between the dimensions of australopithecine 'brains' and those of *H. erectus* 'brains' is 'not as great as one might have suspected at first'; and that the frontal lobe of the fossil forms 'differs but little in (relative) size from that of modern man' (von Bonin, 1963, pp. 54–5). The relative extent of the parietal and occipital lobes is more difficult to evaluate, since the limiting points are ill defined: thus, while the sagittal extent of the parietal lobe shows no clear-cut trend between australopithecines, *H. erectus*, Neandertalers, Upper Palaeolithic men and modern men, the relative depth of the parietal lobe suggests a trend towards slight increase. However, the landmarks cannot be relied upon and the study was largely inconclusive. Most of these landmarks could not be identified on the plaster endocast of *Zinjanthropus* and I have therefore not attempted this type of analysis here.

F. The pattern of meningeal vascular markings

Figure 20 shows the meningeal vascular markings on the right lateral aspect of the endocranial cast. The basic pattern comprises a short common trunk on the inferior surface of the temporal lobe bifurcating, where the inferior and lateral surfaces join each other, into anterior and posterior trunks. These diverge widely from each other. Thus far, the pattern resembles those described in other australopithecines; differences occur, however, in the further distribution of these trunks. The anterior trunk divides very soon into anterior and posterior branches. The exact distribution of the anterior branch is difficult to determine: it courses over the temporal lobe, heading towards the fronto-orbital region where the interruption occurs between the anterior and posterior parts of the endocast. The anterior branch of the anterior trunk then courses upwards and continues as, or gives off, a large parietal branch which courses

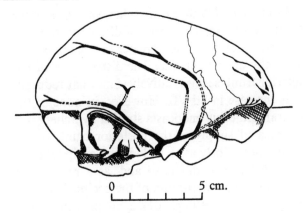

Fig. 20. Dioptographic tracing of norma lateralis of endocranial cast of *Zinjanthropus*, to show the pattern of meningeal vessels on the right side of the brain. The pattern is nearly the same on the left, save that the lower of the two parallel parietal branches comes off the *posterior* trunk of the middle meningeal artery.

diagonally across the inner aspect of the parietal bone, from the vicinity of the pterion in the general direction of the lambda. The posterior branch of the anterior trunk also curves upwards giving rise to a similar diagonal branch parallel to and below the former. These branches of supply of the anterior trunk thus cover an area of the endocast surface which extends probably from the frontal lobe to the part of the brain underlying lambda. The posterior trunk courses backwards close to the junction between the inferior and lateral surfaces of the temporal lobe, supplying fine branches to the temporal lobe. It supplies the lower part of the parietal region by means of a medium-sized branch, then dips over on to the occipital pole of the hemisphere. This right-sided pattern partakes of features of both types I and III of Giuffrida-Ruggeri's (1912) classification.

The pattern on the left is similar, save that the anterior trunk does not seem to have two main branches. Instead the anterior trunk describes a wide arc, convex forwards, to continue as the upper of the two parallel diagonal vessels on the parietal lobe. The lower and bigger of the two diagonal vessels apparently comes off the posterior division on the left, but, as on the right, the two parallel vessels anastomose with each other. The left-sided pattern seems to be closest to Giuffrida-Ruggeri's type IV.

As compared with the vascular pattern in other Australopithecinae, the area of distribution of the posterior ramus is much greater in *Zinjanthropus*, especially on the left side. The distribution of the anterior trunk seems relatively somewhat reduced, as is the case in Sts VII. However, in none of the australopithecine endocasts studied by Schepers is the area of supply of the posterior trunk so extended posteriorly as to include the occipital area. In *Zinjanthropus*, this is so and, as a result, 'the incomplete vascularisation of the occiput', spoken of by Schepers (1950, p. 101), is not present. In this respect, the Olduvai specimen seems to tally with *Paranthropus robustus* of Kromdraai, insofar as the incomplete specimen extends posteriorly: from the vascular markings present at the posterior extremity of the Kromdraai calvarial fragment, it seems that the posterior trunk would have supplied the occipital area, as in *Zinjanthropus*.

When compared with that of the great apes, it is seen that the middle meningeal pattern on the *left* side of *Zinjanthropus* is very similar to that of the gorilla figured by Schepers (1946, fig. 27A, p. 225), while that on the *right* side closely resembles his chimpanzee pattern (fig. 27C of Schepers, p. 225). On both sides, the *Zinjanthropus* pattern differs appreciably from that in Schepers's illustrations of orang (in which the arrangement approaches more closely that in *Australopithecus*). The left pattern in *Zinjanthropus* also resembles that on the parietal endocast of *H. e. pekinensis* (skull 3 from locus H) (Weidenreich, 1936b).

The differences between the two sides in *Zinjanthropus*, as well as the wide diversity among a relatively small number of australopithecine endocasts available for detailed study, show that little store can be set by these vascular patterns as indicators of taxonomic status.

Appendix to Chapter VIII

Since this chapter was written, new data on the cranial capacity of *Homo erectus* have become available. First, with the kind collaboration and help of Dr J. T. Wiebes, Deputy Director of the Rijksmuseum voor Natuurlijke Historie, Leiden, I was able to re-determine the water capacity of the Trinil calvaria (Java I). From a comparable determination on Java II of Sangiran, made with the consent of Professor G. H. R. von Koenigswald and the help of his assistant, Miss H. Cardinaals, I re-computed the total capacity for Java I as 844–54 c.c. or, in round figures, 850 c.c. Dubois's figure of 900 or 935 c.c. was arrived at by multiplying the water capacity of the Trinil calotte by a fraction he derived from a *gibbon* cranium in 1898 (*Proceedings of IVth International Congress of Zoology*, Cambridge, pp. 85–6) and it is this value which has been quoted in the literature ever since!

Secondly, Professor von Koenigswald has recently completed a new reconstruction of the robust cranium of Java IV. Although his results are not yet published, he has generously permitted me to quote the capacity determined on the new reconstruction: it is 750 c.c.

Thirdly, the capacity of the new hominid calvaria from Sangiran (Java V of S. Sartono, *Bulletin of the Geological Survey of Indonesia*, 1964, vol. 1, pp. 2–5, and of T. Jacob, *Anthropologica*, 1964, vol. 6, pp. 97–104) is apparently 975 c.c. (von Koenigswald, 1966, *ex litt.*).

Fourthly, J. K. Woo has estimated the capacity of the new pithecanthropine cranium from Lantian, Shensi Province, as 775–83 c.c., or, in round figures, 780 c.c. (*Scientia Sinica*, 1965, vol. 14, pp. 1032–5).

The data in Table 14, p. 81, should therefore be replaced by the following more up-to-date list:

Homo erectus of Java

	(c.c.)
I (Trinil)	850
II (Sangiran)	775
III (Sangiran)	c. 880 (900)
IV (Sangiran)	750
V (Sangiran)	975
Mean (n = 5)	848

Homo erectus of Choukoutien

	(c.c.)
II	1,030
III	915
X	1,225
XI	1,015
XII	1,030
Mean (n = 5)	1,043

Homo erectus of Lantian

	(c.c.)
I	780

Homo erectus of Olduvai

	(c.c.)
Old. Hom. 9	1,000

Total *H. erectus*

Mean (n = 12)	936

The mean for a dozen estimates or determinations of cranial capacity in *H. erectus* is 936 c.c. and the range is increased to 750–1,225 c.c. The gap between the largest capacity of a gorilla (752) and the smallest capacity of the *H. erectus* sample (750) is thus completely eliminated.

CHAPTER IX

METRICAL CHARACTERS OF THE CALVARIA AS A WHOLE

In Table 18 are recorded a number of measurements of the *Zinjanthropus* cranium, especially metrical characters which have not been dealt with in the preceding sections. Most of these measurements relate to the calvaria alone, but several—such as nasion–basion length and basion–prosthion length—are, strictly speaking, cranial rather than calvarial measurements. To compare with the *Zinjanthropus* data, I have assembled as many measurements on the South African australopithecines as could be culled from Broom *et al.* (1950), Broom and Robinson (1952), Dart (1962*b*) or obtained by personal measurement; as well as some comparable measurements on *Homo erectus*.

Considerable difficulties arise when one attempts to apply measurements devised and defined for the description of modern human crania to ancient hominids, as well as to non-hominid Primates. These problems have been fully discussed by Davidson Black (1931), Washburn (1942) and Weidenreich (1943) among others. Among the most difficult termini of measurements are inion and euryon.

A. Cranial length and the toro-occipital index

In modern hominine crania, the inion and the opisthocranion (furthest occipital point) are different points, the former low and the latter high on the cranium. On the other hand, in some Neandertal crania, in *H. erectus*, in the australopithecines and in many pongid crania, the two points coincide to mark the posterior terminus of the maximum cranial length (glabella–opisthocranion). In modern man, with a small glabellar prominence and an opisthocranion well above the external occipital protuberance, the glabella–opisthocranion length is dependent largely on the length of the brain; in hominoids with a prominent glabella as the median component of the supra-orbital torus, and with an opisthocranion thrown back on to the inionic summit of a large nuchal crest, the glabella–opisthocranion length depends only partly upon the cerebral length, partly on the degree of development of the supra-orbital torus and partly on that of the nuchal crest or occipital torus. Thus what is ostensibly the same measurement—maximum cranial length—reflects totally different biological parameters in modern man and in other extant or extinct hominoids. Even between the two Upper Pleistocene forms of Southern Africa represented by the Broken Hill cranium and that of Boskop, the ratio between maximum external cranial length and maximum internal cranial or cerebral length is grossly different (Tobias, 1960*c*). In the same way, most of the large difference in maximum cranial length (mm.) between *Zinjanthropus* (173) and the reconstructed *Paranthropus* SK 48 (*c*. 170), on the one hand, and *Australopithecus* Sts 5 (146·8) on the other, is owing to the larger size of the supra-orbital torus and nuchal crest in the first two as compared with the last (Table 18).

The endocranial maximum length, as reflected by the maximum cerebral length, is much closer, that of *Zinjanthropus* being 129 mm. and that of Sts 5, 121 mm. This smaller difference in cerebral length obtains, despite the fact that *Zinjanthropus* has a cranial capacity of 530 c.c. and Sts 5 one of 480 c.c. If one compares the maximum cerebral length of *Zinjanthropus* with that of an *Australopithecus* of like capacity, e.g. Sts VIII with a capacity of 530 c.c., the maximum cerebral length differs by only 3 mm., that of Sts VIII being 132 mm. as compared with 129 mm. in *Zinjanthropus*. Unfortunately, none of the *Paranthropus* crania is well enough preserved to give both length measurements.

Table 18. *Measurements of the calvariae of* Zinjanthropus *and other hominids** (*mm.*)

	Zinjanthropus	Paranthropus SK 48	Australopithecus Sts 5	Australopithecus MLD 37/38	H. e. pekinensis	H. e. erectus
Maximum cranial length (L) (glabella–opisthocranion)	173·0	c. 170	146·8	—	188 – 199	176·5– 199
Horizontal projective length (prosthion–inion)	209·0	—	185	—	184 – 197·5	?174 –?182
Minimum frontal breadth (ft–ft) (B^1)	69·4	c. 71	c. 58·0	—	81·5– 91	79 – 85
Maximum breadth on temporal squames (B^t)	116·5	c. 107	99	106	?133 – 139†	?125 – 131†
Maximum biparietal breadth (B)	110·0	—	96·5	100·5	127 – 140	134
Maximum bimastoid breadth (ms–ms)	142·0‡	—	c. 111·5	120‡	?103 –?106	102
Maximum breadth across supra-mastoid crests	139·5‡	—	107	109	?144 –?150	?134 – 140
Biauricular breadth (au–au)	139·0	—	c. 110	108·2	141 – 151	?129 – 156
Biasterionic breadth (ast–ast)	89·2	—	c. 74	79·5	103 – 117	?92 –?120
Interporial breadth	134·4	—	108	107·6	120 –?128	114
Basibregmatic height (H')	c. 98	—	100	95·0	?115	102 –?105
Auricular height (height above F.H. in biauricular plane) (OH)						
Left of sagittal crest	75·0	—	—	—	93·5– 105	89 – 92
Right of sagittal crest	75·0	—	—	—	—	—
Top of sagittal crest	86·0	—	—	—	—	—
Maximum height above basion in coronal plane through basion						
Vault ht. left of sagittal crest	92·0	—	c. 103	c. 94	—	—
Vault ht. right of sagittal crest	90·5	—	—	—	—	—
Top of sagittal crest	103·5	—	—	—	—	—
Maximum height of sagittal crest	13·0	?	—	—	—	—
Height of sagittal crest in standard transverse (auricular) plane	11·0	?	—	—	—	—
Height of sagittal crest in coronal plane through basion	11·5 (L) 13·0 (R)	—	—	—	—	—
Nasion–basion length (LB)	112·5	—	c. 97·5	—	105·5*	108 – 113
Nasion–opisthion length (n–o)	136·5	—	c. 122	—	?144 – 147§	?134 – 144
Height of lambda above F.H.	30·5	—	—	—	—	—
Height of F.H. above inion	8·5	±0*	±0	—	—	—
Height of highest point of nuchal area (lateral to inion) above F.H.	7·1	—	—	—	—	—
Maximum height of temporal squame above F.H.	45·0 (L) 42·0 (R)	—	—	—	29·0– 39·0	—
Foramen magnum length from ectobasion (fml)	26·4	c. 28	c. 31·4	c. 30	—	—
Foramen magnum length from endobasion	25·6	—	c. 31·2	c. 27·2	—	—
Foramen magnum breadth (fmb)	26·1	c. 21·5	c. 24·3	c. 25	—	—
Distance from endobasion to klition	36·5	—	—	—	—	—
Distance from opisthion to klition	56·5	—	—	—	—	—
Width of clivus (at presumed level of occipitosphenoid synchondrosis)	21·0	—	—	—	—	—
Distance between inion and internal occipital protuberance	26·8	—	—	—	27·5– 38·0	15·0– 16·0
Distance between basion and hormion	25	—	31·5	27	—	—

* Values for *H. erectus* are mainly after Weidenreich (1943, 1945). The symbols in brackets are for the most part the notation employed by the Biometric School.

† These values are for the temporoparietal breadth, which is less than the maximum breadth on the temporal squames (Weidenreich, 1943, p. 100).   ‡ Measurements on reconstruction.

§ Given as 144–147 on pp. 106, 119 and 124, and as 144–148 on p. 137 of Weidenreich (1943).

It would seem then that the chief usefulness of the dimension, maximum cranial length, in the australopithecines is to convey metrically the differing degrees of torus and crest formation. If we subtract the maximum cerebral length from the maximum cranial length, we obtain a rough measure of the degree of torus and crest formation. It is proposed to call the difference of the two lengths, the '*toro-cristal length*'. For practical purposes, we may subtract either the maximum cerebral length (as measured on an endocast) or the maximum length of the cranial cavity as measured with one or other of a variety of endocranial calipers, placed to the right and then to the left of the median sagittal plane where the greatest lengths are to be found (Hoadley and Pearson, 1929): the mean of the left and right readings is accepted in the studies cited here. It is assumed that, for our present purposes, there is no material difference between the cerebral and endocranial diameters, although, strictly speaking, such is not the case.

A series of toro-cristal lengths is given for different hominids in Table 19. The only australopithecines which permit both diameters to be assessed are *Zinjanthropus* and *Australopithecus* Sts 5. The difference in absolute size of the toro-cristal length is striking: that of *Zinjanthropus* is 44 mm. This is far and away the greatest value of all the hominids listed in Table 19; only the massive-browed crania of Broken Hill ('Rhodesian Man') and of Hopefield ('Saldanha Man') with toro-cristal lengths of 37 and 35 mm. respectively approach at all near it. The value in *Australopithecus* (25·8 mm.) is less even than those of four *H. erectus* crania, two from Indonesia and two from China, about the same as Choukoutien X and slightly more than that of Choukoutien XI. In other words, the toro-cristal length in *Australopithecus* Sts 5 is within the range for *H. erectus*, whereas that of *Zinjanthropus* is far greater. The toro-cristal lengths of five European and Palestinian Neandertaloid crania range narrowly—from 22 to 25 mm.—which is just below the sample range for six *H. erectus* crania (25–31·5 mm.). On the other hand, the values for the Broken Hill cranium (37 mm.) and the Hopefield calvaria (35 mm.) exceed the sample range for Asian *H. erectus*. The cranial lengths cited in Table 19 for recent human crania are sample means: thus, each toro-cristal length is the difference of the external and internal mean lengths. This figure is, of course, not the same as the mean toro-cristal length, but it provides a close approximation. Several modern human cranial series are available, those cited in the table being from the studies of Hoadley and Pearson (1929) and of Wagner (1935). The toro-cristal lengths for males range from 11·12 mm. in Lapp crania to 19·53 mm. in Australian crania. The corresponding female values are 9·39–17·00 mm. The absolute values of the toro-cristal length in *H. erectus* and the Australopithecinae are thus appreciably greater than those of modern human cranial series, except that, with a mean of 19·53 mm., the Australian male crania may well have overlapped in the upper reaches of their range with the lower part of the sample range of *H. erectus*.

The absolute values of the toro-cristal length reveal little difference between *Australopithecus* Sts 5 and *H. e. pekinensis*; but 25·8 mm. in Sts 5 is the toro-cristal length in a cranium of maximum external length 146·8 mm., whereas 26 mm. in Choukoutien X is in a cranium of length 199 mm. To bring out this distinction, an index has been devised: the *toro-cristal index* is the ratio of the toro-cristal length to the maximum external cranial length expressed as a percentage. In Table 19, the indices for *H. sapiens* have been based upon values derived from sample means. The range of 'mean' toro-cristal indices in *H. sapiens* is from 6·21 per cent in male Lapp crania to 10·65 per cent in male Australian crania with Dynastic Egyptian, Norwegian and Maori male crania all just over 8 per cent. The corresponding figures for female crania are 5·49–9·80 per cent. Thus, in modern man, the supra-orbital torus and nuchal thickness constitute between about 5 and 10 per cent of the total external cranial length.

The index in five Eurasian Neandertaloid crania ranges from 10·58 to 12·95 per cent, with a mean value of 12·07 per cent. These values are intermediate between those of recent man and of the Asian representatives of *H. erectus*. Broken Hill and Hopefield, however, have indices as high as 17·62 and 17·50 per cent respectively, that is,

Table 19. *Toro-cristal length and index in hominids**

Series	n	External cranial length (L) (mm.)	Internal cranial length (L_i) (mm.)	Reference	Toro-cristal length ($L-L_i$) (mm.)	Toro-cristal index $\left(\frac{L-L_i}{L}\%\right)$
Homo sapiens (recent)						
Egyptian ♂ (26th–30th dynasties)	729	185·53	170·44	Hoadley and Pearson (1929)	15·10	8·13
Norwegian ♂	81	187·70	172·28	Wagner (1935)	15·42	8·21
Norwegian ♀	76	179·28	166·09	Wagner (1935)	13·19	7·36
Lapp ♂	63	179·17	168·05	Wagner (1935)	11·12	6·21
Lapp ♀	55	170·96	161·57	Wagner (1935)	9·39	5·49
Maori ♂	16	186·81	171·21	Wagner (1935)	15·60	8·35
Maori ♀	14	180·00	168·04	Wagner (1935)	11·96	6·64
Australian ♂	13	183·30	163·77	Wagner (1935)	19·53	10·65
Australian ♀	10	173·40	156·40	Wagner (1935)	17·00	9·80
Neandertaloids						
Broken Hill	—	210	173	Weidenreich (1943)	37	17·62
Hopefield (Saldanha)	—	200	165	Drennan (1953)	35	17·50
Neandertal	—	199	175	Weidenreich (1943)	24	12·06
La Chapelle	—	208	186	Weidenreich (1943)	22	10·58
Gibraltar	—	193	168	Weidenreich (1943)	25	12·95
Ehringsdorf	—	196	171	Weidenreich (1943)	25	12·75
Tabūn I	—	183	161	Weidenreich (1943)	22	12·02
Mean (n = 5)		195·80	172·20		23·60	12·07
Homo erectus pekinensis						
III		188	156·5	Weidenreich (1943)	31·5	16·76
X		199	173	Weidenreich (1943)	26	13·07
XI		192	167	Weidenreich (1943)	25	13·02
XII		195·5	168	Weidenreich (1943)	27·5	14·07
Mean (n = 4)		193·6	166·1	Weidenreich (1943)	27·5	14·23
Homo erectus erectus						
I		183	153·5	Weinert (1928)	29·5	16·12
II		176·5 ?	148	Weidenreich (1943)	28·5 ?	16·15 ?
Australopithecus (Sts 5)		146·8	121	Broom *et al.* (1950)	25·8	17·57
Zinjanthropus		173·0	129·0	Present study	44	25·43

* The values of external cranial length and internal cranial length for *H. sapiens* are sample means.

higher than any member of the Asian *H. erectus* sample, and about the same as *Australopithecus* Sts 5. No other fossil hominine cranium has such massive toro-cristal appendages.[1] It would be interesting to know if the newly discovered cranium from Petralona in Greece, which in some respects resembles the Broken Hill and Hopefield crania, agrees with them in its toro-cristal length and index as well. The values of the index for four crania of *H. e. pekinensis* range from 13·02 to 16·76 per cent, with a mean of 14·23 per cent; while the two crania of *H. e. erectus* give almost identical values to each other, namely, 16·12 and 16·15 per cent. Finally, the index in *Australopithecus* Sts 5 is 17·57 per cent, just outside the sample range for *H. erectus* of Asia, whilst that of *Zinjanthropus* gives the very high value of 25·43 per cent. The index separates Sts 5 slightly from the *H. erectus* group, but, on a sample comprising one *Australopithecus* specimen, it is impossible to say whether this is significant. On the other hand, the supra-orbital torus and nuchal crest of

[1] Since this was written, I have determined the values for the massive *H. erectus* cranium, Olduvai Hom. 9. With an external cranial length of 206·5 mm. and an internal cranial length of 154·5 mm., it has a toro-cristal length of 52·0 mm., exceeding that of all hominid crania on record. Its index is 25·18 per cent, almost the same as the highest value in Table 19, namely, 25·43 per cent for *Zinjanthropus* (unpublished data).

Zinjanthropus are so massive that, together with the actual thickness of the frontal and occipital bones, they account for over a quarter (25·43 per cent) of the maximum external cranial length!

Variability would undoubtedly be high, since at least two distinct variables contribute to the toro-cristal length. Unfortunately, save for the small fossil samples, no *intraracial* variances (to use the term which Karl Pearson proposed many years ago) of either the toro-cristal length or index are available; only some idea of the *interracial* variability can be obtained from the range of population means of modern human crania cited in Table 19. Nor have I been able to find suitable measurements from which to estimate the toro-cristal lengths and indices of pongids. It is to be expected that sexual dimorphism for toro-cristal length and index would be high both in the gorilla and in the more robust australopithecines like *Zinjanthropus*.

B. The position of euryon and the maximum cranial breadth

Determining the position of euryon as the terminus of the greatest transverse breadth of the cranium occasions no difficulty in recent man. Although varying within wide limits (Huizinga, 1958), the most lateral point on the side-wall of the modern calvaria is most commonly on the parietal bone, or else it is on the temporal squame. A glance at Table 18 reveals that this is true neither of *H. erectus* nor of the Australopithecinae. In *H. erectus*, the greatest breadth occurs in one of two places: either it is the biauricular breadth (as in Choukoutien XII and Java I) or it is Davidson Black's (1931) 'intercrestal breadth' or Weidenreich's (1943) 'maximum intercristal breadth' (as in Choukoutien III, X and XI and Java II and IV). In Table 18, I have called the latter dimension the 'maximum breadth across supramastoid crests' to make its definition unequivocal. In the australopithecines, the bimastoid breadth is the maximum, while the breadth across the supramastoid crests competes with the biauricular breadth for second place. Thus, in *Zinjanthropus*, the bimastoid breadth is 142·0 mm., the breadth across the supramastoid crests 139·5 mm. and the biauricular breadth 139·0 mm. Comparable readings in Sts 5 are *c.* 111·5, 107 and *c.* 110 mm. and in MLD 37/38 they are 120, 109 and 108·2 mm.

The maximum biparietal breadth in *Zinjanthropus* (110) is no less than 32 mm. smaller than the maximum bimastoid breadth, while the corresponding deficit in Sts 5 is 15 mm. and in MLD 37/38 it is 19·5 mm. That is, on each side of the *Zinjanthropus* cranium, the mastoid process protrudes 16 mm. further laterally than the most lateral point on the parietal bone; this gives a rough measure of the additional thickness owing to the excessive pneumatisation of the pars mastoidea. The value for this excess mastoid protrusion in *Australopithecus* is 7·5 mm. on each side in Sts 5, and 9·75 mm. on each side in MLD 37/38. The relationships are exactly opposite in *H. erectus*, where the parietal at its most lateral projection—the lateral salience of the torus angularis—protrudes appreciably further laterally than the pars mastoidea: the excess of maximum parietal width over bimastoid breadth (mm.) is *c.* 25 in Choukoutien III, *c.* 33 in Choukoutien XI and 32 in Java II. Three factors would seem to be responsible for this big difference between the australopithecine and *H. erectus* crania: first, the mastoid processes of *H. erectus* are turned inwards to a strong degree; secondly, the amount of pneumatisation of the pars mastoidea is much less in *H. erectus*; and, thirdly, there is undoubtedly a real expansion of the cranial cavity with an increase especially in the mediolateral dimensions of the parietal bones, more particularly the posterior part of the parietals—so that, of the four margins of the parietal bone, the lambdoid margin shows the greatest differences in chord and arc between the australopithecines and *H. erectus* (*see* Table 2, p. 12). A possible fourth factor is the powerful development of the torus angularis in *H. erectus*, since it is on the most lateral projection of the torus that the terminus of the maximum biparietal diameter is located in these early hominids.

The greater pneumatisation of the australopithecine temporal affects the squamous portion as well as the pars mastoidea. This factor presumably is responsible, in part at least, for the relationship between the maximum breadth on the temporal squames and the maximum biparietal breadth: in all three australopithecine crania, the former exceeds the latter—by 6·5 mm. in *Zinjanthropus*, 5·5 mm. in MLD 37/38 and 2·5 mm. in Sts 5. In *H. erectus*, these measurements cannot be compared in the same way, because unfortunately Weidenreich did not measure the maximum breadth on the temporal squames, as Davidson Black (1931) had recommended; instead he measured a *temporoparietal breadth* between 'the points where the interporial coronal contour meets the squamous suture' (1943, p. 100). This point does not give a breadth coinciding with the maximum breadth on the temporal squames; in fact, Weidenreich admits, 'In primitive hominids such as *Sinanthropus* the greatest breadth still falls below these points, even if the (supramastoid) crest region proper is disregarded'. When the maximum parietal breadth of *H. erectus* is compared with this temporoparietal breadth, the relationship is variable. More frequently the maximum biparietal width just exceeds the temporoparietal breadth—by 1·0 mm. in Choukoutien XI and XII and by 3·0 mm. in Java II; on the other hand, the temporoparietal breadth is in excess by 1·0 mm. in Choukoutien X and by 2·0 mm. in Choukoutien III.

In sum, the australopithecines are characterised by a strong lateral development of the mastoid–supramastoid–auriculare complex; whereas only the more anterior parts of this complex, the supramastoid–auriculare components, display excessive lateral development in *H. erectus*. The difference can be attributed to the very much more marked pneumatisation of the australopithecine temporal, whereas in *H. erectus* the supramastoid crest forms a powerful component of Weidenreich's transverse reinforcing-system of the cranial vault (1943, pp. 159–61).

C. The postero-anterior tapering of the cranial vault

The minimum frontal breadth shows marked restriction: the value in *Zinjanthropus* is 69·4 mm., that in *Paranthropus* (SK 48) *c*.71 and Sts 5 *c*.58·0 mm. These absolute values fall well short of those for the *H. erectus* sample, which range from 79 to 91 mm. When the minimum frontal breadth is expressed as a percentage of the maximum breadth of the cranial vault, an index is obtained which in modern man is known as the *transverse frontoparietal index*. This is an unfortunate name as it is with the maximum cranial breadth that it is clearly intended to compare the minimum frontal breadth and, as we have seen, the maximum cranial breadth may be on the *temporal* squames even in recent man. Hence, the maximum breadth of the vault proper should be used; this excludes such additional projecting superstructures as the supramastoid crest, the root of the zygomatic arch and the entire neighbouring region above the external acoustic meatus (Martin, 1928). The values for the index in the australopithecines are 59·6–*c*.66·4 per cent. The index used by Weidenreich (1943) relates the minimum frontal breadth to the average of four breadths, namely the maximum supramastoid intercristal breadth, the maximum biparietal (torus angularis) breadth, the temporoparietal breadth and the biauricular breadth. The mean of the four breadths is termed the *average 'maximum' breadth*. In Table 20, the values in *H. erectus* for all cranial indices involving the cranial breadth are based upon this average 'maximum' breadth, whereas those cited for the Australopithecinae are based upon the maximum breadth on the temporal squames. The range of transverse 'frontoparietal' index values in *H. erectus* is 58·5–64·1 per cent. If we may accept both the index I have employed and that of Weidenreich as statements of the approximate amount of tapering from the parietosquamosal region of the vault to the postorbital constriction, it seems that the australopithecines and *H. erectus* taper to an equal degree. This ectocranial equivalence finds a parallel in the endocranial proportions, as measured on endocasts by Schepers (1950): the ratio of the 'bi-frontal width' to the 'bi-parietal width' ranges from 80 to 95 per cent in australopithecines, while he quotes values, presumably means, of 91 per cent for *Pithecanthropus* and 93 per cent for *Sinanthropus*. It would seem that between the australopithecines and the pithecanthropines, there is little difference in the degree of tapering of the

METRICAL CHARACTERS OF CALVARIA

Table 20. *Indices of the calvariae of* Zinjanthropus *and other hominids (per cent)**

Index	*Zinjanthropus*	SK 48	Sts 5	MLD 37/38	H. erectus pekinensis	H. erectus erectus
Cranial index ($100L/B^t$)	67·3	62·9	67·5	—	71·4 –72·6†	73·2–79·3†
Altitudinal index ($100H'/L$)	c. 56·6	—	c. 71·5	—	59·6‡	51·2–59·6
Vertical index ($100H'/B^t$)	84·1	—	106·1	89·2	75·6‡†	64·6–78·3†
Auricular ht.–breadth index ($100\ OH/B^t$)	64·4	—	—	—	66·9§–74·0†	57·0–68·5†
Transverse frontoparietal index ($100B'/B^t$)	59·6	c. 66·4	c. 60·1	—	59·7 –64·1†	58·5†
Transverse parieto-occipital index ($100\ ast–ast/B^t$)	76·6	—	—	75·0	78·3 –86·2†	89·0†
Nasion–basion length index ($LB/n–o$)	82·5	—	80	—	72·6‡	76·2–78·6
Occipital length index I	21·8	—	—	—	25·2 –26·1	22·5–26·6

* Values for *H. erectus* mainly after Weidenreich (1943, 1945). The symbols in brackets are for the most part the notation of the Biometric School.

† These indices are based upon Weidenreich's *average 'maximum' breadth*, i.e. the mean of four breadths, the maximum breadth across the supramastoid crests, the maximum biparietal (torus angularis) breadth, the temporoparietal breadth and the biauricular breadth.

‡ Indices based on dimensions of restoration.

§ This value is given by Weidenreich (1943) as 67·2 per cent in his Table XXI, p. 110, but as 62·2 per cent in his Table XXVII, p. 121 and in the text on p. 127. I have re-computed the index from the two basic measurements to give 66·9 per cent (cranium XI).

brain, which, in turn, conditions a similar resemblance in the degree of anterior narrowing of the brain-case. These groups differ but little in these respects from the pongids, both ectocranially and endocranially.

Neandertal and recent man, however, show less tapering: the transverse 'frontoparietal' index ranges in the former from 67·4 to 78·0 per cent (mean 70·9) and in the latter from 65·0 to 76·6 (mean 71·6) (Weidenreich, 1943, p. 122). It seems that the essential jump upwards in the value of the index came at the post-*erectus* stage of hominisation; it clearly reflects a lateral increase in the frontal lobe of the brain, with a consequent filling out of the 'pinched' postorbital constriction.

Another measure of the tapering of the cranium is provided by the transverse 'parieto-occipital' index. The same objection may be taken to the name of this index and, again, the index as computed by Weidenreich (1943, 1945) in *H. erectus* does not have the same denominator as in our estimates for the Australopithecinae. In *Zinjanthropus*, the index has a value of 76·6 per cent and in *Australopithecus* MLD 37/38 a value of 75·0 per cent. That is, the biasterionic breadth is three-quarters of the maximum vault breadth (across the temporal squames). The values (per cent) in the *H. erectus* sample range from 78·3 to 89·0 (mean 83·2); the Neandertalers have a sample range of 72·5–90·6 (mean 81·8) and modern man 76·6–85·6 (mean 82·4). The sample of two australopithecine crania thus lies near the bottom of the range for *Homo*. It is doubtful if this index is of any taxonomic value: Martin (1928) quotes a mean of 83·8 per cent for anthropoid apes. In fact, as Weidenreich (1943, p. 128) has pointed out, 'the index in question is merely an expression of the well-known fact that the oval form of the vault is, in principle, common to anthropoids and to all the types of hominids'.

D. The height of the cranial vault

The bregma point is missing in *Zinjanthropus*, but it is not difficult to estimate its position. The estimate gives a basibregmatic height of c. 98 mm. This lies between the two values for *Australopithecus*, namely, 95 mm. in MLD 37/38 and 100 mm. in Sts 5. This serves to confirm what has earlier been stressed, namely, that the virtual absence of a forehead in *Zinjanthropus* is *not* the consequence of a low-vaulted neurocranium; rather it is owing to the lower hafting of the neurocranium on to the facial skeleton. The values in the

Australopithecinae fall very little short of those in *H. e. erectus*. In Weidenreich's (1945) restoration of *H. e. erectus* IV ('*Pithecanthropus robustus*'), the basibregmatic height was only 102 mm., i.e. 2 mm. more than in the well-preserved Sts 5. In *H. e. erectus* I and II the height was estimated to be 105 mm. The restoration of Choukoutien XI yielded a height of 115 mm. Thus, it seems that the ranges for the Australopithecinae and for *H. e. erectus* must have overlapped. However, an element of uncertainty remains in all of the *H. erectus* crania, since not one of them possesses an intact basion.

The australopithecine sample range of 95–100 mm. overlaps that of the pongids, cited by Weidenreich as 85–100 mm. It falls far short of those of the Neandertalers, namely, 115–131 mm. (mean 125 mm.) and of modern man, 123–141 mm. (mean 134 mm.).

The basibregmatic height is used in the computation of two indices, the *altitudinal* or *cranial height index* ($100H'/L$) and the *vertical index* ($100H'/B^t$). The altitudinal index of *Zinjanthropus* is low ($c.56.6$ per cent—i.e. chamaecranial), as was to be expected from its great cranial length exaggerated by a toro-cristal index of 25 per cent. The value in *Australopithecus* Sts 5 is appreciably higher ($c.71.5$ per cent—i.e. orthocranial), on account of its much smaller cranial length. The values (per cent) in *H. erectus*, too, are low and chamaecranial (51·2–59·6), whereas in the Neandertalers they rise to 60·2–66·8 (mean 63·2) and in modern man to 65·6–77·9 (mean 72·9 according to Weidenreich). In its index of $c.71.5$ per cent, *Australopithecus* Sts 5 is seen to be most akin to modern man. The corresponding endocranial index (cerebral height/cerebral length) is given by Schepers (1950) as 72 per cent, all the other australopithecine endocasts ranging from 61 to 67 per cent. The index is not, of course, exactly the same as the internal height–length index, since the measurements of Schepers were made on endocasts without reference to skeletal planes of orientation. The internal height–length indices as measured on crania have means ranging for a variety of modern human populations from 73·89 to 78·67 for males and from 73·84 to 80·61 per cent for females (Wagner, 1935). Endocranially, as well, it seems that the proportions of *Australopithecus* Sts 5 correspond closely to those of modern man.

Similarly, when the basibregmatic height is compared with the maximum breadth of the cranial vault (*see* 'Vertical index' in Table 20), the value (per cent) for Sts 5 (106·1) is far greater than that of *Zinjanthropus* (84·1), and well within the range for modern man, namely, 86·7–109·2 (mean 100·6). However, in another specimen of *Australopithecus*, MLD 37/38, the index of 89·2 (per cent) is much closer to that of *Zinjanthropus*. Their general resemblance in height–breadth proportions is readily apparent in pl. 9, in which the originals of MLD 37/38 and of *Zinjanthropus* have been photographed side by side. The values in *H. erectus* are lower (64·6–78·3 per cent) since, although the height is somewhat greater than that of australopithecines, the breadth of the vault is appreciably greater. It seems that in the hominising steps from australopithecine to pithecanthropine grades of organisation, the broadening of the calvaria about a broadening brain was a more noteworthy change than the heightening effect. Thus, all early hominids (except Sts 5) fall into the tapeinocranial category ($x - 91.9$): it is only in the further hominising changes from *H. erectus* to *H. sapiens* that the heightening effect became prominent, and metriocranial (92·0–97·9) and acrocranial ($98.0 - x$) calvariae came into being. In this sequence, Sts 5 is exceptional in its acrocranial index of 106·1 per cent; it may well be enquired whether it has not been subject to some distortion in the deposit, as Broom and Robinson surmised for Sterkfontein cranium VII (Sts 71) (1950, p. 25), or whether, in the reconstruction of the cranium—it was exposed in two moieties, the calotte and the rest of the cranium (*see* Broom and Robinson, 1950, pl. 7)—some additional height may not, inadvertently, have been added.

As the basion is missing in all the crania of *H. erectus*, a more accurate gauge of vault height may be furnished by the auricular height. In *Zinjanthropus* this has been measured to the upper surface of the vault *on either side* of the sagittal crest: the reading is 75·0 mm. (Table 18). This measurement in *H. erectus* yields values of 89–

92 mm. in the Indonesian form and 93·5–105 mm. in the Chinese subspecies. When this height measurement is used as the basis for an auricular height–breadth index, the value (per cent) for *Zinjanthropus* (64·4) falls within the sample range for *H. e. erectus* (57·0–68·5), which, in turn, overlaps the sample range for *H. e. pekinensis* (66·9–74·0). Thus, different relationships are found with the two different measures of vault height: when basibregmatic height is employed, the index of *Zinjanthropus* is higher than those of the Asian pithecanthropines; when auricular height is used, the value for *Zinjanthropus* is lower than those of *H. e. pekinensis* and comparable with those of *H. e. erectus*. It might be inferred that the supra-auricular height of the cranial vault in *Zinjanthropus* is not as great as in *H. erectus*, although the depth of the basal portion, from the horizontal plane through auriculare down to the horizontal plane through basion, is appreciably greater in the Olduvai specimen. That is, in *Zinjanthropus*, the plane of the porion is relatively higher up the cranial vault. Thus the height indices provide us with indirect metrical confirmation of what we had already determined by the *porion position–height index* (chapter v, C, pp. 48–9), namely, that the porion in *Zinjanthropus* shows the hominine tendency to rise above the nasion–opisthion baseline, to a greater degree even than in *H. e. pekinensis* and the Neandertalers.

E. Some metrical features of the base of the calvaria

The nasion–basion length is also known as the craniobasal length, or as the basal line of the face (Weidenreich, 1943, p. 99). Its value of 112·5 mm. in *Zinjanthropus* is large compared with *c.* 97·5 mm. in *Australopithecus* Sts 5, and this would be in keeping with the much larger, heavier face of the former (Figs. 1 and 8). The restored *H. erectus* crania yield values of 105·5 mm. in Choukoutien XI, 108 in Java I and 113 in the robust Java IV cranium. The latter value is virtually the same as in *Zinjanthropus*. The Neandertalers range from 98 to 125 mm. (mean 111) and modern man from 90 to 107 mm. (mean 102·7). The mere comparison of absolute values is, however, rather meaningless in view of marked variations in (*a*) the salience of glabella and (*b*) the relation between nasion and glabella, as is mentioned in the discussion on Flower's gnathic index (*see* p. 116). Weidenreich (1943, p. 138) has drawn attention to a shortening of the nasion–basion line in modern man, 'chiefly the result of the deflection of the base'. That is to say, as the basis cranii is deflected upwards toward the interior of the cavity, 'the entire skull rolls up'—thus bringing the nasion and basion nearer to each other. Elsewhere, Weidenreich held that the nasion–basion length depended to a large extent on the size and depth of the face and he preferred to regard the line as the *basal line of the face* (1943, p. 99). If the length of the line is correlated with both the degree of deflection of the cranial base and the massiveness of the face, then *Zinjanthropus* is in an interesting position. In *Zinjanthropus* we have seen that shortening of the cranial base is apparent. As a yardstick of the deflection of the cranial base, porion has risen above the nasion–opisthion baseline even more than in *H. erectus*. From this, one might have predicted a marked degree of shortening of the nasion–basion line, as compared, say, with a pongid standard. However, this degree of shortening is not apparent, presumably because of the massiveness of the zinjanthropine face (*see* chapter x), more especially the glabellar salience and the rostral position of nasion. The relative shortening of the nasion–basion line in *Zinjanthropus* is thus not as marked as one would have expected. Confronted by these seemingly conflicting tendencies, the nasion–basion line of *Zinjanthropus* is 82·5 per cent of the nasion–opisthion length (chapter v, B, p. 46). Sts 5 has a nasion–basion length index of exactly 80 per cent. That is, in both of these australopithecine crania, the nasion–basion/nasion–opisthion ratio lies between the values reported by Weidenreich for hominines (75 per cent or less) and for pongids (85 per cent or more). *Zinjanthropus*, with its massive face, inclines a little more towards the pongids; *Australopithecus*, with a smaller face, has a lower index.

Most of the remaining measurements in Table 18 and indices in Table 20 are dealt with in the appropriate sections elsewhere.

CHAPTER X

THE STRUCTURE OF THE FACE

A. The supra-orbital torus (Pls. 1 and 15)

Dominating the face is a powerful supra-orbital torus which is far more robust than in any of the australopithecines hitherto discovered. To illustrate this point, the vertical thickness of the torus was measured at the highest point on the orbital margin: the value in *Zinjanthropus* (13·6 mm. *left*, 13·7 *right*) is nearly twice as great as in most specimens of *Paranthropus* (7·4 in SK 46; 7·2 in SK 847; 7·8 and 5·6 in SK 48),[1] with the exception of SK 52 which approaches the value in *Zinjanthropus* (12·4). The values in Sts 5 (*Australopithecus*) are 7·8 and 9·6 mm. The values quoted by Weidenreich (1943, p. 30) for the toral thickness in *Homo erectus pekinensis* reveal a range in the middle of the orbital margin of 11·5–17·4 mm. ($n = 5$ crania, 8 sides). If our technique of measuring this vertical thickness of the torus is comparable with his, then three out of five Choukoutien crania have thicker tori even than *Zinjanthropus*. This is in keeping with the overall robusticity of the cranium of *H. erectus*, whereas the australopithecine cranium is an essentially gracile one, though with heavy ectocranial embellishments and tori. Again, if the anteroposterior width of the torus be measured from the mid-point of the anterior surface of the upper margin of the orbit (just lateral to the trochlea) to the temporal crest or line behind, the values in millimetres are as shown in the table in the next column.

Weidenreich's figures for the 'sagittal length' of

	Left	Right
Zinjanthropus	17·8	17·4
Paranthropus SK 48	15·1	14·8
SK 46	14·4	—
SK 847	13·3	—
SK 52	—	14·8
Australopithecus Sts 21	—	13·8[1]
Sts 17	8·8	—
Sts 5	13·3[1]	14·5[1]

the tori in Pekin Man cannot be compared with ours, because he has used a different posterior terminal, namely, 'the point where the frontal tuberosity rises above the supratoral sulcus' (1943, p. 29).

Thus, in both vertical and anteroposterior thickness, the supra-orbital torus of *Zinjanthropus* exceeds those of all known members of *Paranthropus* and *Australopithecus*.

The forehead rises very little above the torus; in fact it would almost be true to say, as Robinson (1961, p. 4) said of *Paranthropus*, there is no trace of a forehead. Instead, immediately above and behind the glabellar region of the torus is a broad triangular hollow, bounded anteriorly by the torus and posterolaterally by the converging temporal crests. This hollow represents the medial part of the *sulcus supratoralis ossis frontalis* of Weidenreich (1943, pp. 29 and 31). In other words, the sulcus has been progressively encroached upon from either side by the advancing temporal crests, which by their union to form a sagittal crest convert the sulcus to a 'sealed off' triangle, to which the name *trigonum frontale* is given. In its supratoral trigonum frontale, *Zinjanthropus* closely resembles two *Paranthropus* specimens, SK 46 and SK 48, though in the Olduvai cranium, the trigone is more clearly defined, because its

---

[1] The unusual, probably juvenile, Swartkrans cranium, SK 54, has virtually no supra-orbital torus at all. This may be owing to its young age—there is no way of determining the precise age, but the vault-bones are very thin, the sutures open and the inferior temporal lines far from the mid-line, their nearest approach to the sagittal suture being 38 mm. (chord distance on anterior part of left parietal). It is not at all certain that this specimen belongs to *Paranthropus*: it may represent a more advanced hominid (? *H. habilis* or *H. erectus*).

[1] At the point of measurement, the temporal line was already diverging markedly posteriorly, thus exaggerating the measurement.

boundaries—the temporal crests and the supraorbital torus—are so much more strongly developed. The frontal trigone in African pongids is much more steeply hollowed, as the torus tends to rise upwards; whereas in the australopithecines, the torus extends forwards but not upwards, the plane of the superior surface of the torus sloping downwards, though not as steeply as the plane of the frontal trigone itself. This is well brought out in a comparison between the median sagittal sections of the *Zinjanthropus* cranium and a female gorilla cranium (Fig. 8, p. 46); for comparison, a median sagittal section of an *Australopithecus* cranium, Sts 5, kindly prepared by Dr J. T. Robinson, is included in the figure. In *Australopithecus*, there is only a slight degree of hollowing of the supratoral area.

The glabellar part of the torus is the most prominent. It forms a broad, rounded, anterior projection extending between one supra-orbital notch and the other. This glabellar portion has two distinct surfaces: an anterior face, rounded from above downwards as well as from side to side, and marked by a number of fairly coarse pits, and a superior face, smooth and relatively unpitted, and slightly concave from side to side. It is this upper surface which delimits the frontal triangle anteriorly. These features are reproduced on a more modest scale in SK 48, but the relevant area is missing from SK 46.

The superciliary part of the torus is of a completely different character from the glabellar moiety in *Zinjanthropus*. This portion of the torus is deeply etched on its posterior surface by the anterior part of the impression for the temporalis muscle, rising to its highest along the temporal crest. The temporal crest thus forms the highest part of the lateral or superciliary portion of the supra-orbital torus on each side. The prominent keeled crests for the anterior parts of the temporalis muscles in the region of the postorbital constriction, exceeding those of even the most muscular *Paranthropus*, had provided Leakey with the thirteenth of his twenty diagnostic criteria of *Zinjanthropus* (1959a, pp. 492–3). Anterior to this part of the crest, there are not two surfaces, a superior and a facial, as in the glabellar region, or as in *Paranthropus*; but only a single anterosuperior surface, sloping steeply downwards to the upper margin of the orbit. The presence of an anterior and a superior surface in the glabellar portion and of only an anterosuperior surface in the superciliary portions gives a curious twisted appearance to the lateral parts of the torus.

At the same time, the highest part of the torus as seen in norma facialis is at the lateral ends of the glabellar portion; from there, the superciliary moiety slopes steeply downwards mediolaterally to the frontozygomatic suture. Thus the lateral ends of the superior orbital margin are far lower than the medial, the difference in height (when the cranium is in the Frankfurt Horizontal) being some 10·5 mm. This downward slope, which contrasts markedly with the position in *Paranthropus* and *Australopithecus*, contributes further to the impression of a twist between the median and lateral components of the crest.

It is presumably this complex of features affecting the superciliary portion of the torus that Leakey was referring to, when he listed as the sixteenth of his twenty diagnostic criteria of *Zinjanthropus*, 'The whole shape and position of the external orbital angle elements of the frontal bone' (1959a, p. 493).

Such an arrangement is not encountered in *Paranthropus*, where even the lateral parts of the torus have upper and anterior surfaces. However, in none of these does the temporal crest appear to have ascended as high up the posterior surface of the orbital process as in *Zinjanthropus*. The steep anteroposterior descent of the lateral part of the torus in the latter seems clearly to be a reflection of the extreme degree of development of the anterior part of the temporalis muscle and its crest. In a series of pongid crania, the lateral part of the torus is seen always to have two surfaces, save in those examples where part of the temporal crest rises to a very superior position: in such crania—and only in the area where the temporal crest attains so high a position—does the torus show a single anterosuperior face. For instance, the cranium of a large male gorilla, recovered by the Witwatersrand University Uganda Gorilla Research Unit from the slopes of the Birunga volcanoes in South West

Uganda (Tobias, 1961), shows an anterior temporal crest which is well developed over the lateral half and poorly developed over the medial half of the superciliary torus. The medial part of the torus has two surfaces, but laterally from the point where the crest rises above the top level of the torus, there is only a single anterosuperior surface. In no pongid cranium examined by me for this feature did the area of torus with only a single anterosuperior face extend further medially than the *lateral half* of the superior orbital margin; generally a smaller area over the superolateral angle of the orbit was the most that could be found with this morphology. Other pongid crania, which lacked any area where the temporal crest rose above the toral level, lacked any part of the torus with a single anterosuperior surface. *Zinjanthropus* is distinctive among all the hominoid crania I have examined in that the area of torus with a single anterosuperior surface extends completely across the top of the orbit to its supero*medial* angle.

This morphological difference between *Zinjanthropus* and the pongids reflects an interesting difference in the degree of development of the anterior part of the temporalis muscle. In *Zinjanthropus*, the temporal crest runs parallel to the orbital margin, almost as far medially as the superomedial angle of the orbit, before deviating posteriorly. The temporalis muscle has utilised a maximum of space anteriorly to gain a purchase. In *Paranthropus* the temporal crest extends about half-way across the top of the orbit from the superolateral angle, before diverging backwards. In most of the pongid crania, however, even in very rugged male gorilla skulls, the temporal crest does not proceed further medially than the lateral one-quarter to one-third of the superior orbital margin before deviating posteriorly. In a large male chimpanzee (Za 94) in the Anatomy Department, the temporal crest does reach the half-way mark across the orbital margin before turning backwards. (As might be expected, the lateral half of the superciliary part of the torus has only a single, anterosuperior face.) In none of the pongid crania examined by me does the anterior temporal crest extend as far medially as in *Zinjanthropus* before turning posteriorly.

This indicates that the anterior part of the temporalis muscle is relatively better developed in *Zinjanthropus* than in the Pongidae, a conclusion for which we find further support in other parts of this study. The peculiar morphology of the *Zinjanthropus* brow ridge, it is suggested, is the result of the unusually powerful development of the anterior fibres of the temporalis muscle and its temporal crest.

B. The orbits and the interorbital area

The orbits as a whole slope downwards laterally. Although overshadowed by a strong torus, they appear to be high, especially on the right, and this impression is confirmed by the measurements (Table 21) and indices (Table 22). When the orbital width is measured from maxillofrontale, orbital indices of 80·6 (L) and 86·0 (R) are obtained. The former value falls into the mesoconch category and the latter into the hypsiconch class. These values compare well with those of a few other australopithecines (86·2 per cent in Sts 5, 84·2–?81·3 per cent in SK 48); they lie close to the mean for a group of *H. e. pekinensis* crania and well within the range of means in modern man (73·9–93·2 per cent—after Martin, 1928). On the other hand, the pongid means are appreciably higher (93·3–113·5 per cent), though lower individual values, ranging down to 76 per cent, are met with in the chimpanzee.

The entire rim of both orbits is intact, save for a short length near the middle of the inferior margin on each side. The blunt upper margin, in keeping with the torus which overtops it, slopes markedly downwards to the frontomaxillary suture which lies in the superolateral angle of the orbit. This upper rim overhangs the orbit, for both the medial and lateral orbital margins are incurved somewhat, below the glabella and the external angular process respectively. Such an incurvation is met with in all of the australopithecines, including the juvenile from Taung; it occurs, too, in the Homininae and Pongidae.

The lateral orbital margin is rounded from within outwards and concave mediolaterally and from above downwards. The frontal process of the

THE FACE

Table 21. *Facial measurements of* Zinjanthropus *and other hominids* (mm.)*†

		Zinjanthropus	Sts 5	SK 48	SK 46	*H. erectus pekinensis* (mean)
Superior facial length (basion–prosthion)	(40)	137	122	—	—	114
Lateral facial length‡ (*fmo*–porion)	(41)	99·8 (L)	c. 78·5	c. 84	—	83
Superior facial breadth (above *fmt*–*fmt*)	(43)	c. 115·4	c. 91·5	99	c. 102	121
Inner biorbital breadth (*fmo*–*fmo*)	(43–1)	99·9	c. 80·6	c. 91·5	c. 88	111
Biorbital breadth (*ek*–*ek*)	(44)	96·2	87	?101	—	111
Nasomalar arc (*fmo*–*fmo*)	(44–1)	105·5	—	—	—	119
Bizygomatic breadth	(45)	c. 168·0	127	146	—	148
Maxillary breadth (*zm*–*zm*)	(46)	c. 121·5	c. 90	c. 95	—	98(?)
Superior facial height (nasion–alveolare)	(48)	111·5	c. 75	c. 80	—	77
Alveolar height	(48–1)					
Nasospinale–alveolare		43·3	—	29·2	29·5	—
Nasospinale–prosthion		42·2	32·3	26·5	26·2	25
Orbito–alveolar height	(48–3)	75·1	c. 52	c. 61·5	—	44
Posterior interorbital breadth (*la*–*la*)	(49)	c. 32·5	—	c. 30	—	30
Anterior interorbital breadth (*mf*–*mf*)	(50)	23·4	16·3	c. 23·5	—	25
Orbital width (*mf*–*ek*)	(51)	40·7 (L) 39·4 (R)	c. 34	c. 38·6 (L) ?40 (R)	—	44
Orbital height	(52)	32·8 (L) 33·9 (R)	c. 29·3	c. 32·5 (L) c. 32·5 (R)	—	36

* Nasal measurements are dealt with in Table 23, maxillo–alveolar and palatal measurements in Tables 26 and 27.
† The symbols and key numbers refer to Martin's (1928) list.
‡ Measured from *frontomalare orbitale* as by Weidenreich (1943) instead of from *ektoconchion* as in Martin's measurement *41*.

Table 22. *Indices of the face of* Zinjanthropus *and other hominoids** (per cent)

		Zinjanthropus	Sts 5	SK 48	*H. erectus pekinensis* (mean)	Modern man†	Great apes†
Superior facial I	(48/45)	c. 66·4	c. 59·1	c. 54·8	52·1	49–57·4	c. 70
Superior facial II	(48/46)	c. 91·8	c. 83·3	c. 84·2	?78·5	—	—
Zygomatico-maxillary	(46/45)	c. 72·3	c. 70·9	c. 65·1	66·2	—	—
Orbital	(52/51)	80·6 (L) 86·0 (R)	c. 86·2	c. 84·2 (L) ?81·3 (R)	81·9	73·9–93·2	93·3–113·5
Interorbital I	(50/43)	20·3	c. 17·8	c. 23·7	22·5	—	—
Interorbital II	(50/44)	24·3	18·7	?23·3	—	18·2–22·2	17·0–25·6
Nasomalar	(44–1/44)	109·7	—	—	107·3	105·9–113·0	—
Zygomatico-suprafacial	(43/45)	c. 68·7	c. 72·0	67·8	81·7	—	—

* The key numbers refer to Martin's (1928) list.
† Means and ranges of means from Martin (1928).

zygomatic bone, which forms this margin, faces largely anteriorly, more so than in other australopithecines, and slopes steeply forwards to meet the body of the zygomatic at the well-rounded inferolateral angle. The width of this process increases from above downwards, as in other australopithecine and hominine crania, but in contrast with most pongid crania examined, in which the greater width is just *above* the zygomaticofrontal suture. Immediately below the suture in *Zinjanthropus* is a marginal process of a form somewhat different from that described in Pekin Man and in some modern human crania (Weidenreich, 1943, p. 87). Instead of being a posterolateral projection of the temporal edge itself, it is a strong double projection immediately behind the edge. The anterior knob of the double projection is very close to the temporal edge; at first glance the edge seemingly runs into it. However, on closer inspection, part of the temporal muscle area is seen to lie just anterior to it. A deep vertical groove separates the anterior from the posterior process: the latter is situated well within the area for the temporalis muscle. Eisler (1912), quoted by Weidenreich, related the *processus marginalis* to the temporalis muscle, as follows: 'The succession of muscle and fascia bundles in the superficial layers of the anterior portion of the temporalis is usually less numerous than in other portions of the muscle, yet there is an additional muscular portion that arose from a strong tract attached to the marginal process of the zygomatic bone and branching out backward and upward.' Although Weidenreich (1943, p. 87) questioned whether the temporalis muscle could actually be responsible for the development of the process, there can be no doubt that the particular form of the process in *Zinjanthropus* is related developmentally to several fascicles of the temporalis muscle. It would seem that, in the strong development of this twin marginal process (or, better perhaps, *postmarginal process*), we have further evidence of the robusticity of the anterior part of the temporalis muscle.

The body of the zygomatic bone slopes so strongly downwards and laterally as to form a distinct shelf at the inferolateral angle of the orbit. The lower orbital rim is thus set on a more anterior plane than the upper, at least as far medially as the zygomaticomaxillary suture. The exact position where the latter suture crosses this margin is missing on both sides, but there must clearly have been an abrupt change of contour at the suture: the more medial part of the inferior margin slopes backwards and then, slightly hollowed, it proceeds upwards towards the blunt-rounded inferomedial angle.

The medial orbital margin is tolerably well preserved. The frontal process of the maxilla has a fairly sharp orbital margin delimiting a large nasolacrimal duct. A small part of the lacrimal bone is present in the right orbit.

A variable part of the walls of the orbit has survived, but this does not include the orbital plate of the ethmoid bone; the posterior portions of each wall are broken some distance in front of the optic foramen.

The interorbital area is very broad and has an inflated appearance. Its great width, together with the unusual shape of the nasal bones, comprised the fifteenth of Leakey's (1959 a) diagnostic criteria for *Zinjanthropus*. In the upper part of the interorbital area, much of the width is composed of the greatly broadened upper ends of the nasal bones, while the frontal processes of the maxilla are narrow; lower down, the nasal bones are attenuated, but the frontal processes much widened. The anterior interorbital breadth (measured between the maxillofrontalia) is virtually identical with that of *Paranthropus* SK 48 of Swartkrans (23·5 mm.), though much greater than in Sterkfontein 5 (16·3 mm.). But it is smaller than the mean for *H. e. pekinensis* (25 mm.). On the other hand, if the width is measured between the lacrimalia (posterior interorbital width—Martin 49), it is about 32·5 mm. in *Zinjanthropus*, and about 30 mm. in SK 48. In other words, as one moves back from the plane of the maxillofrontalia to that of the lacrimalia, the interorbital width in *Zinjanthropus* increases by 9·1 mm. and in *Paranthropus* SK 48 by 6·5 mm. The corresponding mean increase in Pekin Man is 5 mm. (from 25 to 30 mm.). The much greater increase in *Zinjanthropus* is undoubtedly due to the bulging of the air cells in the part of the orbital plate of the

THE FACE

frontal with which the lacrimal articulates. These air cells are clearly seen through cracks and gaps in the orbital wall, and are closely related to the ethmoid air cells behind (where the orbital plate of the ethmoid is missing). This inflation of the bone carries the lacrimal point laterally, thus accounting at once for the much increased *la–la* breadth and for the generally inflated appearance of the region. This is one of many evidences of excessive pneumatisation of the *Zinjanthropus* cranium (see pp. 126–31).

C. The nose

The nasal region is set in a central depressed area of the face, for on all sides the adjacent parts of the face are on a more anterior plane. Above, the glabella; below, the paranasal and subnasal parts of the maxilla; and, on either side, the prominence of the zygomaticomaxillary portions—all leave the piriform aperture and the nasal bones in a hollow recess. In his reconstruction of *Paranthropus*, Robinson (1961) commented that 'the enormously robust cheek bones actually project further forward than does the nose, which is completely flat' (p. 6). In *Australopithecus*, on the other hand, 'the nasal region is slightly raised above the surrounding level of the face' (p. 4). This feature is well brought out in the lateral craniograms (Fig. 1, p. 9).

The only part of the nose of *Zinjanthropus* which is an exception to the foregoing is the nasion itself. It is set in a high and prominent position, very slightly below the most jutting part of the glabellar salience, as can be seen in the median sagittal craniogram (Fig. 8, p. 46). While the lower parts of the nasal bones are well within the central facial hollow, the upper parts are thrown forwards on the slope leading to glabella.

The high anterior position of nasion provided Leakey with the fourteenth of his twenty diagnostic criteria of *Zinjanthropus*, namely, 'the very unusual position of the nasion, which is on the most anterior part of the skull, instead of being behind and below the glabella region' (1959a, p. 493). The uniqueness of this feature was, however, rightly questioned by Robinson (1960, p. 457), who pointed out that the proximity of nasion to glabella was true of *Paranthropus* and at least some crania of *Australopithecus* also. I can corroborate Robinson's statement, especially with regard to *Paranthropus robustus crassidens* (SK 48), in which nasion is unmistakably in an almost identical position to that in *Zinjanthropus*. It must be added, however, that in their original description of SK 48, Broom and Robinson (1952, pp. 10–13) made no mention of this high position of nasion, stating instead, 'The suture between the nasal and the frontal has closed and cannot be traced with certainty, but is probably as we indicate it.' Their accompanying drawing shows a relatively low nasion, well down in the region of *sellion* (Bunak's term for the deepest point in the hollow below glabella). The area is not preserved in SK 46 or the Kromdraai cranium. In Sts 5, the nasion is a little lower, being half-way between sellion and glabella (as depicted in the median sagittal craniogram in Fig. 8, and as shown in the original median sagittal craniogram of Broom and Robinson, 1952, p. 24). The nasals are not preserved or are hidden in Sts I (the type skull from Sterkfontein), nor are they available in Sts VI and VIII. In the incomplete calvariofacial fragment from Makapansgat (MLD 6), as well as in the Taung child, nasion occupies a similar position on the upward slope to glabella.

From these scanty data, we can say that *Zinjanthropus* and *Paranthropus* agree in having nasion almost coincident with glabella, while in *Australopithecus* nasion is a little lower down.

The nasal bones are of unusual form, fanning out in racket-shape to a maximum width of about 13 mm. above, but tapering to a minimum width of under 6 mm. below, a short distance from the inferior extremity of the bones. This curious *superior* expansion of the nasal bones of *Zinjanthropus* is exactly opposite to the general hominoid trend of an *inferior* expansion, as seen in most Pongidae, Homininae and in *Australopithecus*. On the other hand, the condition is duplicated in *Paranthropus* (SK 48) in which, despite a slight degree of frontonasal fusion, the suture between the bones is clearly preserved and outlines a strikingly similar, racket-shaped, superior expansion. This unusual superior expansion and inferior

Table 23. *Nasal measurements of* Zinjanthropus *and other hominids (mm.)**

		Zinjanthropus	SK 48	Sts 5	H. erectus pekinensis (mean)
Nasal width	(54)	31·8	c. 32†	29	30
Nasal height (nasion–nasospinale)	(55)	70·2	c. 55·5	46	52·5
Height of piriform aperture	(55–1)	33·8	22	c. 24	33
Length of nasal bones (chord)	(56)	35·8	?26·8	c. 24·4	20
Length of nasal bones (arc)	(56–1)	37·0	c. 36·0	—	21·5
Length of lateral margin of nasal bone	(56–2)	31·0	c. 30·5	—	22–23‡
Least breadth of nasal bones	(57)	5·8	c. 5·3	—	14–17‡
Greatest breadth of nasal bones	(57–1)	c. 12·7	c. 11·0	—	18
Superior breadth of nasal bones	(57–2)	c. 12·7	c. 7·9	—	17·3–18‡
Inferior breadth of nasal bones	(57–3)	6·7	c. 11·0	c. 17	?15

* The key numbers refer to Martin's (1928) list.

† The greatest width of the nasal opening of SK 48 is described as 22 mm. by Broom and Robinson (1952, p. 12), but it is difficult to see how it could be so small, even allowing for some distortion from crushing. The author's measurement of 32 mm. (with allowance for distortion) is included in the table.

‡ Different mean values are quoted in Tables XVI and XXXII of Weidenreich (1943).

constriction contrasts with the condition illustrated by Martin (1928): both in some modern human and in some pongid crania, there may be a small degree of enlargement above, a constricted area beneath that, and the widest expansion below that along the margin of the piriform aperture. Generally, the inferior breadth is the greatest, whereas in *Zinjanthropus* the inferior breadth (6·7 mm.) is only slightly greater than the least breadth (5·8 mm.), while both are approximately half the greatest width (c. 12·7 mm.), which coincides with the superior width (Table 23).[1] The details in SK 48 differ somewhat from those in *Zinjanthropus*, in that the inferior expansion (c. 11·0 mm.) is far greater than the least breadth (c. 5·3 mm.), exceeds the superior breadth (c. 7·9 mm.), and slightly exceeds the greatest breadth of the upper expansion (10·1 mm.). However, I cannot accept that the inferior breadth of the nasal bones in SK 48 could have been about 19·5 mm. (see Broom and Robinson, 1952, p. 11). The superior expansion of the nasal bones in *Zinjanthropus* may perhaps be seen as part of the general expansion and inflation of the glabellar region. The shape of the nasal bones, in association with the great absolute and relative width of the interorbital area, formed the substance of the fifteenth of Leakey's diagnostic criteria of *Zinjanthropus* (1959a, p. 493).

The two nasal bones are set at a very wide angle to each other: in the expanded part above and in the suprapiriform part below they are angulated to each other at an angle of about 170°, but in the intermediate depressed area the nasal bridge is absolutely flat, the angle between the two nasal bones being 180°. In SK 48 and SK 29, the bridge is seemingly slightly more elevated, the angle in parts reaching about 140–160°. In the Makapansgat specimen (MLD 6) and in Sts 71, the angulation is flat, never becoming less than 160–170°, while it is nearly uniformly 180° in Sts 5.

Another Sterkfontein cranium, Sts 52a, is peculiar in possessing an elevated nasal bridge, in which, in parts, the left and right nasal bones make an angle of less than 90° with each other! This acute angulation affects only the medial half of each nasal bone which is flexed forwards on its lateral half to make a median ridge with the medial half of the opposite nasal bone. The acute angulation is lacking below, moderately developed in the middle third, and prominent and rounded in the upper third near nasion.

---

[1] This contrasts interestingly with *H. e. pekinensis* in which Weidenreich comments, 'One of the most characteristic features of the *Sinanthropus* nasal bone is that there is practically no difference between the upper breadth and the least one which usually is found within the upper moiety of the bone' (1943, p. 73).

Before we leave the nasal bones, another odd feature in *Zinjanthropus* may be mentioned: the lower 4–5 mm. do not follow the general hollow (retroussé) contour of the lower half of the nose, but turn slightly inwards towards their pointed lower end (*rhinion*). Thus, the end of the nasal bones has a slightly beaked appearance, well shown in the median sagittal craniogram (Fig. 8). A similar recurvation seems to have been present in SK 48, and possibly in Sts 5. This feature is said to be usual in modern man (Weidenreich, 1943, p. 74). If we knew more about the relationship between the angles and slopes of the piriform aperture and the conformation of the attached cartilages, it might be possible from this observation to reconstruct fairly faithfully the cartilaginous part of the nose. For the time being, it is deemed worthwhile merely to place the fact on record.

The internasal suture is median in position for the lower 23·5 mm. of the nasal bones; above, the right nasal bone sends out a tenuous process which slightly deflects the suture to the left for a few millimetres. Above that again, the left nasal bone sends out a process which substantially deviates the internasal suture to the right, so that it intersects the frontonasal suture about 2·0 mm. to the right of the median plane (pl. 15). Manouvrier (1893), quoted by Weidenreich, drew attention to the fact that such encroachment of one nasal bone on to the other occurred frequently in modern man. It is encountered, too, in Choukoutien cranium XII, in the Broken Hill cranium, in Krapina C, but not in *Australopithecus* (neither Sts 5, nor MLD 6, nor Taung, as far as can be determined). It cannot be detected in SK 48. The nasofrontal suture ascends towards the mid-line, farther upward than the frontal processes of the maxillae, as occurs in some Neandertal and modern European crania and, especially markedly, in the crania of anthropoid apes.

Part and parcel of the general facial elongation, the nose as a whole and the *apertura piriformis* are markedly elongated. This elongation reflects itself in the low nasal index of 45·3 and the relatively low index of the aperture, 94·1 (Table 24). Other australopithecines are more chamaeprosopic or short-faced: they have correspondingly higher nasal indices (57·7 in SK 48 and 63·0 in Sts 5), and higher indices of the piriform aperture (145·5 in SK 48 and 120·8 in Sts 5). The general nasal elongation of *Zinjanthropus* in comparison with the other two australopithecine forms is clearly apparent in the frontal or facial craniograms (Fig. 21).

The lateral margin of the piriform aperture is sharp above, in the usual hominoid fashion, but becomes thick and rounded below. With the cranium in the F.H., the lateral margin runs slightly forwards to join the floor (or at least the prenasal fossa) well in front of the foot of a perpendicular dropped from rhinion to the nasal floor. Only the roof of the aperture overhangs slightly, because of a modest elevation of the lowest part of the nasal bones from the flat or hollowed surface of the frontal processes of the maxillae. In these respects, *Zinjanthropus* is somewhat intermediate between anthropoid apes and *H. e. pekinensis*.

The floor of the nose is particularly well preserved. There is a distinct *fossa praenasalis* limited in front by a gentle, rounded sill, anterior

Table 24. *Nasal indices of* Zinjanthropus *and other hominoids (per cent)**

	Zinjanthropus	SK 48	Sts 5	H. erectus pekinensis (mean)	Modern man†	Great apes†
Nasal index (54/55)	45·3	c. 57·7	63·0	57·2	38·9– 60·2	36·2–50·1
Index of piriform aperture (54/55–1)	94·1	c. 145·5	c. 120·8	90·8	70 –c. 110	—
B/L index of nasal bones (57–2/56–2)	c. 41·0	c. 25·9	—	75·2	—	—

* The key numbers refer to Martin's (1928) list.
† Ranges of means.

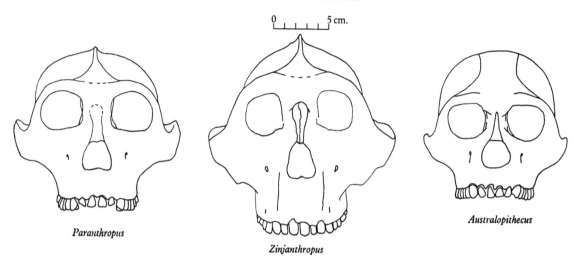

Fig. 21. Craniograms of norma facialis of *Paranthropus*, *Zinjanthropus* and *Australopithecus*. The dioptographic tracing of *Zinjanthropus* was made with the cranium in the F.H.; the missing parts have been restored. The other two craniograms are from those published by Robinson (1961) and are based on the reconstruction and restoration of SK 48 (*Paranthropus*) and on Sts 5 (*Australopithecus*).

to which is the long prognathic slope of the *clivus naso-alveolaris*. The *crista nasalis*, as it descends towards the floor of the nasal cavity, appears to divide into two rounded elevations: the posterior of these swings around to become continuous with the sill; the anterior continues downwards over the lateral part of the subnasal maxilla, in line with the roots of the lateral incisors and, behind them, of the canines.

Posteriorly, the prenasal fossa is limited by another rounded ridge which runs in from the sidewall of the nasal cavity proper. At the point where the left and right *posterior* ridges meet in the midline, there is a strongly developed bifid spinous process, some 2·0–2·5 mm. high. Behind that, the floor falls away into a marked *incisive fossa* in which is the upper opening of the incisive canal, as in the chimpanzee. At the point where the two *anterior* ridges meet in the mid-line is a poorly-developed, bifid spinous process. Between the anterior and the posterior median spinous processes is a linear, bifid *incisor crest*. An interesting feature which distinguishes this cranium from those of most modern men is that it is the posterior end of the incisor crest which is markedly developed as an *anterior nasal spine*, whereas in modern man, the anterior end of this crest forms the anterior nasal spine. The general arrangement of the floor of the nose in *Zinjanthropus* resembles that of the Pongidae, in having a marked incisive fossa behind the prenasal structures, but in none of the pongids is there an anterior nasal spine at all approaching that of the Olduvai creature. Schultz (1958) described a well-marked anterior nasal spine in the cranium of an acrocephalic chimpanzee child. Apart from this abnormal specimen, there is only a faint blunt elevation in many pongid crania, such as those of a chimpanzee and an orang figured by Vogel (1963), but, where recognisable, it lies at the back of the prenasal area, just in front of the incisive fossa, i.e. the corresponding position to that in *Zinjanthropus*. Vogel (1963) has reported the occurrence of small anterior nasal spines as individual variants in *Callicebus* and *Alouatta*, but considers these to be the result of parallel development of no phylogenetic significance.

A similar position to that in *Zinjanthropus* is encountered in *Paranthropus* (SK 12), in which Broom and Robinson (1952, p. 8) reported the presence of 'a rudimentary nasal spine', which is far back in the same plane as the lower part of the nasal bones. If we regard this posteriorly situated spinous process in *Zinjanthropus* and *Paranthropus* as marking the anterior limit of the nasal opening, then we can say, with Broom and Robinson, that in these forms, the lower border of the nasal opening is far behind the plane of the sides of the nasal opening; whereas in anthropoid apes the lower border is almost in the same plane as the sides. The blunt elevation which, in some pongid crania, passes as the homologue of the anterior nasal spine, is thrown far forward in the apes, especially the chimpanzee and the orang-utan, so that it tends to form a single median elevation at the medial ends of both the anterior and the posterior limiting ridges of the fossa (or sulcus) praenasalis. Hence, despite some features which the floor of the zinjanthropine nose shares with that of the chimpanzee, there are several points of clear distinction. The high anterior nasal spine is distinctly hominine, while the posterior placement of that spine indicates (*a*) an orientation of the nasal opening proper which is closer to that of man; and (*b*) a tendency for the prenasal fossa to 'drop over' from the floor of the nose to become part of the naso-alveolar clivus as in modern man. In modern Bushman crania, for instance, although a fossa or sulcus is often present, it is so commonly a definite part of the clivus rather than of the nasal floor that it has for decades been known to workers in the field of Southern African craniology as the *subnasal fossa* (or *sulcus*), rather than the prenasal fossa (or sulcus)!

In sum, the shape of the nasal margin and floor of *Zinjanthropus* is in general reminiscent of that of the chimpanzee, while at the same time showing a number of departures which are clearly hominine in character.

D. The maxillary and zygomatic bones

The greater part of the bodies of the maxillae of *Zinjanthropus* are satisfactorily preserved, as are the alveolar processes, with the exception of the posterior extremities, the maxillary tuberosities. The palatine processes are perfectly preserved, save their most posterior parts just in front of the palatomaxillary suture. Fairly extensive reconstruction has been necessary in the upper part of the maxillary body, the frontal process and the zygomatic process. Even after reconstruction several gaps remain, from which fragments of bone are missing; however, these gaps in no way affect the interpretation of facial anatomy, since the adjacent parts articulate perfectly.

1. *The infra-orbital foramen*

The very large and ramifying maxillary antrum which pervades the body of the bone will be described in the section on Pneumatisation (chapter XI). The lower opening of the infra-orbital canal is preserved on both sides: it is a single foramen, as in all other australopithecines in which it is preserved, with the exception of Sts 52a in which three foramina occur on each side. It lies well below the orbital margin in *Zinjanthropus*, the shortest distance being 30·0 mm. on the left, and 28·9 mm. on the right (where the lower orbital margin descends to a somewhat lower level). Corresponding distances in *Paranthropus* are 18·0 mm. (SK 12), 23·0 mm. (SK 48), 22·0 mm. (SK 46) and 20·0 mm. (Kromdraai), whilst, in *Australopithecus*, the distance is shorter, being 17·0 mm. in Sts I, 12·3 and 13·8 mm. on the left and right sides of Sts 5, and 12·2 mm. in MLD 6. The appreciably greater distance in *Zinjanthropus* is yet another reflection of the facial elongation of the Tanzanian australopithecine as compared with the South African forms.

2. *The size of the maxilla*

Some of the measurements and indices of the maxilla are given in Tables 21 and 22. All betray a great facial enlargement in *Zinjanthropus* as compared with other australopithecines. Thus, the maxillary breadth of 121·5 mm. (measured between left and right zygomaxillaria) is about one-third as much again as the breadths in Sts 5 and SK 48 (90 and 95 mm. respectively). The orbito-alveolar height (measured as the projection height from the inferior orbital margin to the alveolar margin between M^1 and M^2) is 20 per cent as much again as in *Paranthropus* and 50 per cent or more as much again as in *Australopithecus* from Sterkfontein, the values being 75·1 mm. in *Zinjanthropus*, 61·5 in *Paranthropus* (SK 48), 52 and 44 in *Australopithecus* (Sts 5 and MLD 6, the last mentioned not being quoted in Table 21). The alveolar height shows the same enlargement: the upper terminal for this measurement is *nasospinale*, the identification of which occasioned some difficulty. In the preceding section of this chapter, it was mentioned that a small nasal spine occurs where the two anterior ridges from the crista nasalis come together, and a large nasal spine (regarded as the definitive anterior nasal spine) where the two posterior ridges come together. For purposes of measurement, the smaller anterior spine was accepted as nasospinale, because it seems to mark the junction of the floor of the nose and the naso-alveolar clivus more clearly than does the larger spine further back. (Only Sts 52a from Sterkfontein has a clear delineation between nasal floor and naso-alveolar clivus; in this respect it resembles the Swartkrans maxilla which Robinson (1953b) has attributed to Telanthropus or *H. erectus*.) The lower terminal of the alveolar height is variously given as *prosthion* or *alveolare*: following the teaching of the Biometric School, I do not consider that these terms apply to the same point. Prosthion is the *most anteriorly-projecting* point on the process of bone between the maxillary central incisors; alveolare is the *lowest* point on the same bony process. There may be several millimetres between them. Hence, in Table 21, measurements to both lower terminals are cited. Again, the values for *Zinjanthropus* are 33½–50 per cent greater than those for the South African australopithecines. The very prognathous Sts 5 has a greater alveolar height (32·3 mm.) than do two crania of *Paranthropus* (26·2, 26·5 mm.), the three falling far short of the 42·2 mm. of *Zinjanthropus*.

Although these metrical comparisons show that *Zinjanthropus* has a face which is very large in all directions, the major emphasis is clearly on heightening or elongation. This is revealed by the superior facial indices, in which the nasion–alveolare height is compared either with the bizygomatic breadth or with the maxillary breadth. For both indices, *Zinjanthropus* yields values well in excess of other australopithecines which, in turn, exceed the means for Pekin Man (Table 22).

The maxillary bone itself is responsible for much of the facial enlargement in *Zinjanthropus*. We have already seen the preponderance of the orbito-alveolar height and the alveolar height, both of which dimensions are based on the maxilla. The great width of the maxilla is revealed, not only by the absolute value of the maxillary breadth, but by the zygomaticomaxillary index. This shows that the maxillary breadth of *Zinjanthropus* is 72·3 per cent of its bizygomatic breadth, which index is greater than in any other hominoid cited in Table 22 (70·9 in Sts 5, 65·1 in SK 48 and a mean of 66·2 per cent in *H. e. pekinensis*).

The contribution of the maxillae to the piriform aperture and to the floor of the nose, as well as the nature of the transition from nasal floor to naso-alveolar clivus, are discussed in the preceding section on the Nose. Here it may be noted that, if the nasospinale were regarded as lying between the two *posterior* nasal spines separating the prenasal fossa from the fossa for the incisive canal, then the distance from nasospinale to prosthion in *Zinjanthropus* would be 50·0 mm. instead of 42·2 mm.! For comparison, it is of interest to recall that the distance between the akanthion (not the nasospinale) and 'the inferior alveolar border at the prosthion' in the massive-jawed cranium of Broken Hill man is 38 mm. (Pycraft, 1928).

3. *Prognathism*

The subnasal maxilla in *Zinjanthropus* slopes markedly forwards, immediately in front of the prenasal fossa, for at least two-thirds of the height of the naso-alveolar clivus; then, at the level of the anterior tooth-roots, it curves over practically into a vertical plane; the teeth emerge in the same vertical plane, there being no trace of prodontism or dental prognathism. This arrangement, clearly seen in the median sagittal craniogram (Fig. 8, p. 46) and in the lateral craniogram (Fig. 1, p. 9), contrasts rather sharply with that of other australopithecines. In SK 46, although there is a noticeable change in the surface relief of the naso-alveolar clivus at the level where the anterior tooth-roots are reached from above, there is hardly any change in the contour, and the prognathic slope continues to the alveolar margin. There is even less change evident in SK 48 and in the Kromdraai fragment. In *Australopithecus* (Sts 5), although a slight change of contour is apparent, the prognathic slope is so marked that the alveolar process juts forward right down to its margin. These differences are clearly seen in the lateral craniograms of the three forms of australopithecine (Fig. 1, p. 9). Even more striking is the view from above (Fig. 2, p. 10): when all three types of australopithecine cranium are aligned on the F.H. it is seen that the maxilla juts forward from the facial plane to the least extent in *Zinjanthropus*, to the greatest extent in *Australopithecus*, and to an intermediate degree in *Paranthropus*. So vertical is the alveolar process of the Olduvai maxilla that, in norma verticalis, no teeth can be seen in the dioptographic tracing, whereas they are visible in both of the other forms (Fig. 2).

To give metrical expression to the degree of prognathism, several different indices and angles are commonly recorded (Laffont and Aaron, 1961). The facial angle (i.e. the angle between nasion–prosthion and the F.H.) is 78° in *Zinjanthropus*. According to Robinson (1956, p. 14), this angle measures 75° in *Paranthropus* (SK 46) and only 50° in *Australopithecus* (Sts 5). This confirms the impression conveyed by our craniograms that *Zinjanthropus* is less prognathous than either of the other two; however, as Robinson points out, Sts 5 should not be taken as reflecting the average conditions in *Australopithecus*, for it is clearly the most prognathous specimen of *Australopithecus* at present known. Unfortunately, in no other adult specimen of this form is it possible to obtain an accurate estimate of the degree of prognathism.

The *Zinjanthropus* value of 78° and the *Paranthropus* angle of 75° fall well outside the values for anthropoid apes (41–56°, range of means; 30–68°, range of individual measurements). The australopithecine readings are a little low compared with the mean for *H. e. pekinensis* (84·5°), and for a group of Neandertalers (86·8°), but can be accommodated within the range of means for modern man (76·8–89·2°) and well within the range of individual values (70·0–99·0°). The reading of 50° in Sts 5 is within the pongid range and outside the hominine range. In the classical system of categories, Sts 5 would be classified as hyperprognathous (x–69·9°), but *Zinjanthropus* and *Paranthropus* as prognathous (70–79·9°).

A second angular measurement is the angle of alveolar prognathism or alveolar profile angle, that is, the angle which the nasospinale–prosthion line makes with the F.H. This value in *Zinjanthropus* is $62\frac{1}{2}°$; the mean in *H. e. pekinensis* is 72°; the value in Broken Hill man is 85°; while the range of means in modern man is 62·8–86°, according to Martin (1928). In series of anthropoid apes, the means range from 31 to 47° and individual values from 10 to 55°. Thus the value in *Zinjanthropus* is well outside the pongid range, both of means and of individual values; on the other hand, it lies close to the lowest modern human mean (62·8° in a group of 38 north-east African Negroes), and well within the range of human absolute values (49–100°). No other australopithecine values for this angle have been placed on record, but a measurement made by me on the median sagittal craniogram of the highly prognathous Sts 5 gives an angle of 37°, well within the pongid range and well outside the hominine range. Hence, by this index, Sts 5 would be classified as ultra-prognathous (x–59·9°), and *Zinjanthropus* as hyperprognathous (60·0–69·9°).

A third angular measurement of prognathism is the nasal profile angle, i.e. the angle between the nasion–nasospinale line and the F.H. Using the forwardly-placed nasospinale as defined previously, I obtained a reading in *Zinjanthropus* of 93·5°! The angle being greater than a right angle indicates that nasospinale is actually *behind* the foot of a perpendicular from nasion to the nasal floor. This value places *Zinjanthropus* in the *hyperorthognathous* category. The reasons are not far to seek: first, nasion lies in an unusually anterior position near the most salient point on the glabellar eminence, and, secondly, the nasal region below nasion is markedly flattened behind the facial plane. These readings confirm Robinson's statement that 'Owing to the flatness of the nasal region in the pre-hominids, most of this prognathism is maxillary prognathism' (1956, p. 14). The value of 93·5° for the nasal profile contrasts with 74° on Sts 5 (as measured on the median sagittal craniogram), with a mean of 89° in *H. e. pekinensis*, 85° in Broken Hill man, and 80·5–90·3° as the range of means in modern human populations, or 73–100°

as the range of individual values in modern man. On the other hand, the range of pongid means is 51–69° or, of individual values, 43–76°. Thus, the value of 74° in *Australopithecus* (as represented by Sts 5) lies within the extremes of the pongid and hominine ranges, where these two ranges overlap marginally; while the value of 93·5° in *Zinjanthropus* is well outside the pongid range and well within the hominine range, in fact on the opposite ('ultra-human') side of the hominine range from that which approaches the Pongidae.

To summarise, on facial profile angles, *Zinjanthropus* is hyperorthognathous in its nasal profile, but hyperprognathous in its alveolar profile, giving an overall prognathous classification. Sts 5 is prognathous in nasal profile, ultraprognathous in alveolar profile, with an overall effect which is hyperprognathous.

One further index of prognathism is Flower's gnathic index, the ratio of basion–prosthion to basion–nasion. The value in *Zinjanthropus* is 121·3 and in Sts 5, based on measurements on a good cast, 125·1. In other words, this index shows up only slightly the different degrees of alveolar prognathism in the two crania. The reasons are clear: basion–nasion in *Zinjanthropus* includes the great glabellar bulge, near the front of which nasion is situated, but this feature is not a part of the basion–nasion measurement in *Australopithecus*, in which (a) there is a relatively weak glabellar protrusion, and (b) nasion is well below and, therefore, well behind, the glabellar bulge and, hence, nearer the basion. Again, in *Zinjanthropus*, basion–prosthion includes a dental arcade and palate of near-*Gigantopithecus* dimensions which are thrown downwards, further from basion, by the vertical facial lengthening—neither of which features Sts 5 enjoys. It is not surprising that, when two such biological non-comparables are matched mathematically, the result fails to do justice to the differences clearly revealed by the profile angles of the face. This is a good example of the common biometrical fallacy of making comparisons between the biologically non-comparable, against which warning voices have often been sounded (e.g. Le Gros Clark, 1955; Mednick and Washburn, 1956; Tobias, 1960c).

4. *The architecture of the infra-orbital area*

Leakey (1959a, eighteenth criterion) drew attention to certain peculiarities of the malar-maxillary area of *Zinjanthropus*, while Robinson (1960) claimed that some of these features occurred in one or two *Paranthropus* crania. We shall therefore take a closer look at the anatomy of this area.

Two powerful ridges characterise the lateral parts of the body of the maxilla in most pongid crania. The more medial of these is a strong rounded ridge which borders the piriform aperture for most of its height, giving way above to a distinctly flattened or hollowed anterior face of the frontal process of the maxilla. Below, this ridge gives off a medially-directed offshoot to limit the fossa or sulcus praenasalis anteriorly; but the main ridge continues somewhat obliquely over the large root of the canine. Weidenreich (1943, p. 77) regarded this ridge as nothing more than the *jugum alveolare* of the canine. Lateral to it—and therefore on the lateral face of the body of the maxilla—is a deep depression extending from the line of the infra-orbital foramen or foramina in front, as far as the root of the zygomatic process of the maxilla behind. A circumscribed area in this depression constitutes the so-called *canine fossa*. Immediately behind the depression and fossa, the zygomatic process attaches to the body of the maxilla; from the lower border of the attachment in most pongid crania, a second strong ridge extends downwards over the roots of M^1 or between the roots of M^1 and M^2. This posterior ridge, likened by Robinson (1960) to a 'flying buttress', limits the depression and canine fossa posteriorly.

A similar distinct anteroposterior sequence of structures is clearly seen in most South African australopithecine crania, except that the posterior buttress, leading from the root of the zygomatic process, is on a much more anterior plane. Instead of passing over the roots of M^1 or between M^1 and M^2, it clearly leads down to P^4. As a result, the depression between the anterior and posterior ridges is reduced from a broad canine fossa to a relatively narrow vertical sulcus situated directly above P^3. In the Kromdraai face, for instance, the anterior ridge (canine jugum) is full and rounded medially, but ends laterally as a prominent blunt keel, in a vertical line with the medial margin of the infra-orbital foramen. Then follows a narrow sulcus, twisted on itself, so that the upper part in line with the foramen faces anteriorly, while the lower part in line with P³ faces laterally. Behind that again is the posterior ridge which flattens out above P⁴. SK 46 tallies in almost every detail, including the keeled lateral edge of the anterior ridge. SK 48 is similar, though the keeling is absent. Very similar features are seen in the Sterkfontein fossils, TM 1512 and Sts 5, and in the Makapansgat fragment, MLD 6 (Dart, 1949*a*).

Zinjanthropus differs markedly. Instead of two ridges with a sulcus between them, there is only a single, broad, rounded ridge, the anteromedial part of which leads down to both the canine and P³. The root of the zygomatic process lacks an independent buttress or posterior ridge, but flows smoothly into the posterolateral part of the single ridge, above P³/P⁴. The intermediate area boasts no hollow or sulcus, but is inflated like the adjacent parts of the maxilla. In other words, it seems that both anterior and posterior ridges have coalesced into a highly distinctive, broad, rounded, compound buttress, overlying the roots of canine and third premolar. This curious departure seems to be the result of two dental features of *Zinjanthropus*: first, the buccolingual enlargement of the P³ and concomitant expansion of its roots, and secondly, the tendency towards lateral or buccal displacement of P³ (*see* section on the Shape of the Dental Arcade, pp. 132–7). These features seem to have led to so great an enlargement of the related jugum alveolare of the premolar, as to have occasioned its fusing with the jugum of the canine. Into the fused juga flows the unflanged, unkeeled root of the zygomatic process.

A further factor emerges when the Australopithecinae are compared with the Pongidae. It seems that the malars are hafted on to the face far more posteriorly in the apes (in the area of M¹ or M¹/M²), whereas in the australopithecines, the area of hafting is situate more anteriorly (in the area of P⁴ or even P³/P⁴); or conversely, the face may be regarded as being tucked in under the hafting to a greater extent in the australopithecines. This process of tucking in of the face has been carried much further in *Paranthropus* than in *Australopithecus* and has reached its furthest development in *Zinjanthropus*, as is evidenced by the lesser degree of prognathism, the flatness of the face and, possibly, the obliteration of the fossa or sulcus between the anterior and posterior ridges. Thus, we may detect in the unusual malar-maxillary anatomy of *Zinjanthropus* an adjustment between two seemingly opposite tendencies: first, the tendency towards dental enlargement with its implied 'unfolding' of the lower face, and secondly, the opposite process of tucking in of the face with progressive reduction of prognathism. *Zinjanthropus* illustrates a crucial stage in this odontofacial adjustment.

Although a well-developed canine fossa or sulcus is lacking in *Zinjanthropus*, a small depression or sulcus leads downwards for a short distance from the infra-orbital foramen. However, this *maxillary sulcus* extends no further than about a centimetre below the foramen and does not even reach the level of the *incisura malaris* (Weidenreich's name for the notched lower border of the medial attachment of the zygomatic process). The sulcus in *Zinjanthropus* is comparable with the upper part of the canine sulcus or fossa in various crania of *Paranthropus*, but it does not twist below on to the lateral surface of the jugum. It is not even as distinct a hollow as occurs in the Broken Hill cranium, in which there is but a slight depression below the infra-orbital foramen.

This maxillary sulcus, well developed in the South African australopithecines and poorly represented in *Zinjanthropus*, corresponds closely to the condition of maxillae III and V of *H. e. pekinensis* (Weidenreich, 1943, p. 78; figs. 143, 153) in which 'a "*sulcus maxillaris*" replaces the "*fossa canina*" of modern man'. Weidenreich describes this feature in Pekin Man as a 'novelty', although he stresses that in the chimpanzee the form of the fossa may closely resemble this sulcus. He adds: 'It may be superfluous to stress the fact that the sulcus is certainly not caused by the nerves and vessels which the foramen infra-orbitale admits to the malar region though the sulcus takes its origin there, but must be considered a kind of fold between the two prominences, the canine jugum, on the one side, and the zygomatic process, on the other.' Although this is precisely the explanation arrived at here from our study of the australopithecines, it must be admitted that the very narrow form of sulcus immediately below the foramen looks remarkably like a vascular groove in its uppermost part (the only part which exists in *Zinjanthropus*). In several chimpanzee crania, there are undoubted grooves flowing out of the foramen, traversing the upper sulcus-like depression, and then entering the lower part of the depression which looks like a canine fossa *sensu stricto*, lateral to the canine jugum. In one chimpanzee cranium (Za 1071), some of the vessels (and possibly nerves) appear to have re-entered the maxillary bone close to the apex of the premolar roots. An orang cranium (Za 93) shows clearly-defined vascular grooves running down the sulcus maxillaris. It must be concluded that two factors are operative in the moulding of this part of the maxilla: the upper or infra-orbital part of the maxillary sulcus is related to infra-orbital nerves and vessels, even when the anterior buttress (or canine jugum) is modestly developed, or slight and keeled as in the Kromdraai *Paranthropus* and SK 48; the lower part or canine fossa depends on the relative development of the anterior and posterior buttresses (i.e. the canine jugum and the root of the zygomatic process). In *Zinjanthropus*, by the forward movement of the zygomatic process relative to the naso-alveolar part of the face or, alternatively, the tucking in of the lower face beneath the zygomatic yoke, as well as by the great size and lateral placement of the premolars and their roots, the lower part or canine fossa is completely obliterated.

5. *Inframalar features*

The lower border of the zygomatic process of *Zinjanthropus* is entirely different from those of Kromdraai, SK 48, SK 46 and Sts 5. In all of these it runs downwards at a sharp angle, as a straight or slightly curved edge, into the alveolar process; whereas in *Zinjanthropus* the malar pillar is well-rounded, forming a distinct '*incisura malaris*' (Weidenreich) or '*incurvatio inframalaris frontalis*' (S. Sergi), almost exactly comparable with those described and figured in maxillae III and V of *H. e. pekinensis*. The uncurved oblique ascent of the zygomatic process in the other australopithecines resembles in its slope the arrangement in the crania of La Chapelle-aux-Saints, Monte Circeo, Saccopastore II, and other Neandertal crania (Sergi, 1947, 1960 a, b). In most of the latter group of crania, with a relatively shallow alveolar ridge, the sloping edge of the zygomatic process descends to the alveolar margin; whereas, in the australopithecines, with their deep alveolar ridges, it lodges on the maxilla some distance above the alveolar margin. In *Zinjanthropus*, as in *H. e. pekinensis*, the inframalar notch seemingly cuts into the lower border of the process, which attaches to the body of the maxilla high above the alveolar margin. Sergi has given much attention to the conformation of the maxilla in this area: by comparing the well-rounded, inframalar incurvation of *H. e. pekinensis* with that of modern man, he showed that this character of modern man (the *inflexion type* of maxilla) was present, too, in 'the most ancient type of palaeanthropi'. The present study permits one to extend this distribution back to the Australopithecinae because of its presence in *Zinjanthropus*. It may be noted, too, that a similar inframalar notch characterises orang-utan crania. On the other hand, Sergi recognised the *extension type* of maxilla with little or no canine fossa, nor inframalar incurvation, as characteristic of the 'typical Würmian neanderthalians', 'a different type coming between ancient and modern man'. However, the present study, by drawing attention to the *extension type* of maxilla in SK 48, Kromdraai and other Australopithecinae, shows that this maxillary tradition was present long before the time of the Würmian Neandertalers. The straight pillar-like lower edge to the zygomatic process occurs as well in gorilla and chimpanzee crania. This study shows, also, that a canine fossa and an inframalar incurvation are not necessarily correlated with each other: *Zinjanthropus* has a perfect example of an inframalar notch but no canine fossa; SK 48 has a modest canine fossa but no incurvation. Strictly speaking, these two maxillae could be accommodated neither in the extension type nor in the inflexion type of maxilla of Sergi. Different factors must govern the two morphological features. We have already discussed the factors governing the absence of a canine fossa in *Zinjanthropus*. Some light may perhaps be shed on the relief of the inframalar notch by a consideration of the masseteric impression just lateral to it.

6. *The attachments of the masseter muscle*

Zinjanthropus is characterised by a very rugose, strongly developed and partly excavated impression for the masseter muscle, a feature which provided Leakey (1959a, p. 493) with the nineteenth of his twenty diagnostic criteria of the Olduvai hominid. The lower border of the zygomatic bone is preserved in its entirety on the right; on the left it is entirely missing, from the sutural surface of the zygomatic process of the maxilla to the sutural surface of the zygomatic process of the temporal. The anterior part of the masseteric impression on the right encroaches beyond the zygomaticomaxillary suture on to the zygomatic process of the maxilla; the encroachment extends on to the maxilla for 10 mm. at the anterior margin of the impression, 12·5 mm. in the centre and as much as 17·7 mm. at the posterior margin of the impression. The anteromedial extremity of the masseteric impression projects downwards as a strong hook-like process. Not only does this process mark the lateral limit of the smooth inframalar notch, but it helps to form the notch. This process leads laterally to an oval expanded area, some 19 mm. long and 11 mm. wide (across the width of the lower border of the zygomatic bone). The oval area, which is for the most anterior part of the superficial laminae of masseter, is hollowed, surrounded by a low ridge, and traversed from before backwards by the zygomaticomaxillary suture; it faces lateralwards, slightly downwards and slightly forwards, so a small part of it is visible from the front (pl. 2). There follows an intermediate constricted area, overhung by a strong lateral lip. This lip marks the most prominent point on the body of the zygomatic bone, since, from it, the surface of the bone falls steeply away antero-inferiorly to the zygomaticomaxillary suture, and retreats posterosuperiorly towards the temporal process.

Posterior to the intermediate constriction is an expanded oval area, about 17 mm. by 6·5 mm., facing mainly laterally. This posterior oval is lined by a low ridge which, on the medial aspect of the oval, makes a strong lip directed downwards and backwards, while laterally, there is a moderate lip, stopping short at a higher level than the medial lip. These two lips fuse posteriorly, closing off the posterior oval impression some 16·5 mm. from the zygomaticotemporal suture. The two oval impressions and the intermediate constriction bespeak powerfully developed superficial laminae of the masseter, as in the gorilla (Raven, 1950). Such robusticity of the masseter is to be seen as part of a generalised hypertrophy of all the muscles of mastication, rather than a specific masseteric hypertrophy, such as occurs not infrequently in modern man (cf. Maxwell and Waggoner, 1951; Guggenheim and Cohen, 1959, 1961).

The deep lamina of the muscle, which in man and the apes attaches to the deep surface of the zygomatic arch (Sonntag, 1924), has produced a deeply-excavated hollow impression just medial to the intermediate constriction. A powerful bundle of the more anterior fibres of the deep lamina of masseter must have attached there. Some deep fibres manifestly attached still further forward, to the posterior surface of the zygomatic process of the maxilla. The more posterior fibres of the deep lamina have left only a weak impression, which turns around the inferior border of the arch, just behind the posterior oval impression for the superficial laminae of masseter, and expands there into a fairly smooth impression on the inferolateral aspect of the temporal process. This smooth expansion for the most posterior of the deep fibres ends just in front of the zygomaticotemporal suture. From the intact left zygoma, there is no evidence that the most posterior fibres of masseter extended across the suture on to the temporal bone. The deep lamina was clearly not as powerfully developed as the more superficial part; in particular, the posterior fleshy attachment of the deep lamina—which is on the smooth inferolateral impression—is small and weak in the adolescent *Zinjanthropus* in comparison with its large size in gorilla.

The masseteric impressions are not nearly as well developed in other australopithecines. In SK 52 and in SK 847, there is a slight downturning of the medial end of the zygomatic bone, close to the suture, but, whereas in the former the impression for masseter is confined to the zygomatic bone, in the latter it encroaches for as much as 8·5 mm. on to the maxilla. In SK 48, there is neither downturning nor encroachment. In SK 46, there is a different arrangement: a sharp flange occurs along the lower border of the root of the zygomatic process extending on to the zygomatic bone. The flange is undercut on its medial surface by the masseteric fossa, as far medialwards as the zygomaticomaxillary suture, but apparently not further medially. Only the most medial 7·5–8·0 mm. of the border is rounded and not flanged, at the anchorage of the process just above P⁴. The masseteric impression is slight in *Australopithecus*: in Sts 5, for instance, it is less marked than in Broken Hill man or even the Springbok Flats (Tuinplaas) skull. Sts 21 has very little more impression, while only in Sts 52a does the impression encroach slightly anterior to the zygomaticomaxillary suture.

The general pattern of the attachment in *Zinjanthropus*—two oval areas connected by a constricted isthmus, and an inferolateral posterior impression for the fleshy origin of the deep posterior fibres—duplicates fairly faithfully the conditions of the masseteric impressions in the gorilla, of which a large number of crania have been examined by me in the Powell-Cotton Collection, as well as in the British Museum (Natural History), the National Museum of Kenya, and the Anatomy Department of the University of the Witwatersrand. But there is one important difference: in the gorilla, the muscle spreads widely along the zygomatic arch, the posterior smooth impression being situate on the zygoma of the temporal bone, immediately in front of the zygomatic tubercle and even encroaching on to the anterior surface of the tubercle. In *Zinjanthropus*, the whole muscle is crowded forwards on to the zygomatic and maxillary bones, the posterior smooth impression lying just in front of the zygomaticotemporal suture. There is an anterior trend in the attachment and, furthermore, the anterior part of the muscle is most strongly developed. In the great majority of gorilla crania examined, the masseteric impression stops just behind the zygomaticomaxillary suture; in only two out of twenty male gorilla crania (and in no female gorilla cranium) does the masseteric area encroach as much as 1–2 mm. beyond the suture. Encroachment is commoner in the chimpanzee crania examined, the masseteric impression commonly extending 3–5 mm. beyond the suture, and in one example attaining a distance of 8 mm. beyond the suture. In no ape cranium does the muscular impression extend as far forwards as in *Zinjanthropus*—10·0–17·7 mm. beyond the suture.

7. *The masseteric fibres in relation to the mandibular ramus and the zygomatic arch*

The form and bulk of the masseter muscle are closely related to the extent of its area of attachment, on the one hand to the zygomatic arch and, on the other, to the mandibular ramus (Scott, 1954, 1957). Scott has demonstrated that the breadth of the mandibular ramus is related to the degree of development of the muscles of mastication. The latter factor, in turn, is related to the size of the molars. He has therefore devised an index which compares the combined mesiodistal length of the lower permanent molars (M) with the breadth of the mandibular ramus (R). According to Scott (1957), the mean M/R index in hominines ranges from 70 to 107 per cent, that of chimpanzee and gorilla from 61 to 62 per cent. He cites the value in one of the Sterkfontein mandibles ('Plesianthropus') as 80 per cent, while that of *H. erectus* (Telanthropus) from Swartkrans is 88 per cent. For contrast, the Heidelberg mandible, with its considerable breadth of ramus, has a value of 70 per cent: it combines modestly sized teeth with an extremely broad mandibular ramus, which it is reasonable to infer is related to a strongly developed masseter. In *Zinjanthropus*, we have a creature with very large maxillary molars; its mandibular molars would, if anything, have been somewhat bigger, to judge by Robinson's figures for other australopithecines (1956). At the same time, its masseter muscle was manifestly very well developed, from which we may infer that the ramus would have been very broad.

If the value of the M/R index in *Zinjanthropus* had been the same as that of *Australopithecus* of Sterkfontein (80 per cent) and the combined mesiodistal length of its lower molars had exceeded that of its upper molars in the same proportion as did the mean lengths of lower and upper molars of *Paranthropus*, we should expect the lower molar length of *Zinjanthropus* to have been 53·2 mm. and the mandibular ramus to have been 66·5 mm. broad. If we had used a lower index, nearer to that of the pongids, say 70 per cent, as in the Mauer mandible, the rameal breadth would have been 76·0 mm. (*see* pl. 42). The mean breadth in the gorilla is 85·0 mm. The rameal breadth in 'Plesianthropus' (quoted by Scott, 1957) is 56·0 mm.; while the minimum rameal widths in various mandibles of *Paranthropus* are 57·0 mm., 57·5 mm. and about 69 mm. So the value of about 76 mm. in *Zinjanthropus* would not be unreasonable.

The foregoing analysis leads to two general conclusions: the mandibular ramus of *Zinjanthropus* must have been very broad, while the masseter was clearly placed well forward at its zygomatic origin. This might suggest an especially oblique direction of the outer fibres of masseter in *Zinjanthropus*. Since the inner fibres usually run vertically downwards or even anteriorly downwards, it is likely that, with the marked anterior placement of their origin, the inner fibres could not have formed a strong sheet if they followed the customary direction. Only the more posterior of the inner fibres would have had a sufficiently posterior purchase on the zygomatic arch, to have run downwards and forwards and still to have implanted upon the mandibular ramus. If, however, with the coming into play of the M^3's, the attachment of masseter had extended posteriorly on to the zygomatic process of the temporal bone, the posterior deep part of the muscle (which constitutes the hindmost fibres of the masseter) would have been able to develop further. The possibility of a change in direction of fibres cannot, however, be overlooked.

That there is a relationship between the extent of origin of the masseter and the number of molar teeth upon which it is operating is suggested by the following observations. In a lateral craniogram of *Zinjanthropus*, if two perpendiculars be erected to the line of the occlusal plane, one through the distal margin of M^2 and the other through the mesial margin of M^1, the two perpendiculars between them cut off a length of the zygomatic arch (as well as zygomatic body and zygomatic process of maxilla) which exactly defines the anterior and posterior limits of the masseteric impression. That is, the origin of the masseter is co-extensive with the two molar teeth which were in occlusion in *Zinjanthropus*: for the youth died before his erupting M^3's were in occlusion. If the same exercise be carried out on lateral craniograms of adult *Paranthropus* and *Australopithecus*, the perpendiculars to the occlusal plane cut off a zygomatic area which extends well back on to the zygoma of the temporal bone. This intersected area tallies with the extent of the masseteric origin in these two australopithecines.

It should be noted, in relation to the above exercise, that the occlusal plane of *Zinjanthropus* differs markedly from those of *Paranthropus* and *Australopithecus* (*see* Fig. 1, p. 9). In *Zinjanthropus*, the occlusal plane slopes markedly upwards and backwards, at an angle of $11\frac{1}{2}°$ to the F.H. In the other two australopithecine forms, the plane of occlusion is parallel to the F.H. over the molar part of the arcade, but from P^4 to the front teeth turns upwards, at an angle of 7·5° in *Australopithecus* and 17° in *Paranthropus*. In the latter two forms, the perpendiculars should be erected on the posterior, horizontal portion of the occlusal plane.

It is likely, therefore, that when the M^3's of *Zinjanthropus* had come into occlusion, there would have been a posterior extension of the origin of the masseter. Such a 'migration' or expansion of the masseteric origin is fully in keeping with the conclusions of Symons (1954), who showed that continual adjustments must be made to the attachments of the masticatory muscles during skeletal growth. He pointed out, too, that different mechanisms for such shifts obtain in the case of fleshy muscle attachments and tendinous attachments. It is easier to envisage the mechanism in the first group, to which belong the most posterior fibres of the deep masseter: 'By an interstitial growth of the periosteum, different rates of lengthening at different regions allow the periosteum to shift or "slip" relative to the bone carrying the muscle attachments with it, so maintaining their constant spatial relationships' (p. 76). Several studies of such shifts (e.g. that of Baume and Becks, 1953, on the medial pterygoid and masseter of the rhesus monkey) have related to shifts in the *insertion* of the muscles; similar shifts, or rather expansions, seem likewise to affect the *origin* of the masseter, just as they do the origin of the temporalis muscle (Washburn, 1947; Zuckerman, 1954).

Zinjanthropus must be regarded as still in the phase of skeletal growth; his M^3's were not yet in occlusion and the full development of his masticatory muscles could not have been completed. When the M^3 came into occlusion in this type of creature, it is likely that the posterior part of the muscle extended back over the zygomaticotemporal suture for some distance along the zygoma. It seems unlikely that it would have attached as far back as in modern man, because of the apparent relationship between the occlusal plane and the projection of the masseteric origin and the molars on to that plane. If the occlusal plane itself altered with the coming into occlusion of M^3, then further adjustments of the masseteric origin might have occurred.

8. *The palate*

The form and size of the dental arcade and the palatal, maxillo-alveolar and dental arcadal indices are discussed in the section on the Dentition. A few other features of the bony palate remain to be discussed here. One of its most remarkable features in *Zinjanthropus* is its great depth, as measured with the palatometer devised by van Schaik (1958). The depth opposite the mesial part of M^2 is 21·5 mm. Comparable figures in other australopithecines are 15·5 mm. in SK 48, 14·0 mm. in SK 46 and *c.* 9·0 mm. in the Kromdraai type cranium. From M^2 to the incisive foramen, the palate shows very little tendency to decrease in depth, the readings being 20·0 mm. opposite M^1 and 16·5 mm. at the level of P^4. From the incisive foramen forwards, the palate shelves steeply downwards to the alveolar margin. This late anterior shelving occurs too in several of the Sterkfontein *Australopithecus* skulls, especially Sts 5 which is the most prognathous, and it is encountered in many prognathous recent hominine crania. On the other hand, in *Paranthropus*, the palate shelves sharply downwards from the molar teeth forwards, so that, by the level of the incisive foramen, it is relatively shallow: there remains a short gentle slope to the alveolar margin. In this palatal feature, *Zinjanthropus* is closer to *Australopithecus* than to *Paranthropus*.

The fossa for the incisive canal is distinct and large (pl. 29). From its anterior bevelled margin, a deep, median, intermaxillary sulcus extends to within a short distance of the alveoli of the central incisors. It is flanked by two faint, slender premaxillary–maxillary sutures passing forwards and laterally from the incisive foramen and then disappearing some distance before the alveoli are reached.

The *greater palatine foramen* is clearly visible opposite the alveolus of the last molar, in the suture between maxillary and palatine bones. From it, deeply incised grooves for the greater palatine nerves and vessels run forward over the inner aspect of the alveolar ridge, towards the incisive canal (pl. 20). On each side, the main groove is double, the deeper (usually arterial) groove running nearer the alveoli. On the right this deeper groove is bridged by bone for a short distance, between the levels of M^2 and M^1. There is no trace of a torus maxillaris, either median or lateral (cf. Campbell, 1925).

The junction between the palatine processes of the maxillae and the horizontal laminae of the palatine bones is represented only in part; several areas of bone in this vicinity are missing. The area on either side of the median palatine suture is elevated into a small but distinct *torus palatinus medianus* (Campbell, 1925), extending from the posterior nasal spine as far forwards as the palatine bones are preserved (14·0 mm.), but not continuing forwards as a *torus maxillaris medianus*. The greatest width of the torus is 5·6 mm. A similar torus is present in most specimens of *Paranthropus*, but is missing from the *Australopithecus* palates, including Sts 5 and MLD 37/38. In the Kromdraai cranial fragment, the torus seems to have overstepped the palatomaxillary suture on to the palatine process of the maxilla as a true torus maxillaris medianus, at least for a short distance.

THE FACE

The maximum width of the torus in the Kromdraai cranium is about 9·0 mm. It is unnecessary here to review the large literature and varied views on this feature, which is relatively common in modern man and rare in primates other than man. Suffice it to re-state Scott's view, reconciling several earlier theories: 'It would appear that torus palatinus like torus mandibularis... and parietal torus is a morphological feature inherent in human heredity but which requires for its development an environment in which there is a heavy use of the masticatory apparatus' (1957, p. 212). As such, palatine torus, Scott suggests, would be only one of a number of morphological traits which may exist in a species in latent form, and which develop only with the acquisition of specific functional behaviour. Such latent variability would appreciably extend the adaptive potentialities of a given species. It is of interest to speculate whether the skeletal stigmata of muscular development (e.g. crests) in the more robust australopithecines fall into this category and are based upon a genetic potential with powerful functional evocation.

9. *The zygomatic bone*

The zygomatic bone is large, in keeping with the other components of the face. From the infraorbital to the masseteric margins, the minimum distance is 39·5 mm., the maximum 44·0 mm. The distance in a single zygomatic of *H. e. pekinensis* is 31·5 mm. and in the Broken Hill cranium 27·5 mm. (Weidenreich, 1943, p. 83). In SK 48, the corresponding minimum and maximum values are 30·5 and 36·8 mm. The greatest distance (height) between the lowest point of the body of the zygomatic and the highest point of the frontal process (on the right) is 71·3 mm. in *Zinjanthropus* as compared with 65 mm. in *H. e. pekinensis* II and only 54 mm. in the Broken Hill cranium. The range of means for some modern races is 44·5–46·5 mm., the largest individual value amounting to 59 mm. Thus the single Pekin specimen is well outside the modern human range and this is even truer of the long-faced *Zinjanthropus*. However, values for other australopithecines are closer to those of Pekin and modern man: for instance, in SK 48 it is 59·5 mm., while in Sts 5 it is about 43 mm.

(measured on a cast). The zygomatic preponderance of *Zinjanthropus* over other hominoids is clearly not as overwhelming as is the maxillary.

As in Pekin Man, in *Zinjanthropus* the anteroinferior part of the malar surface shows a prominence or malar tuber, close to the zygomaticomaxillary suture. It forms the most anterior and the lowest part of the bone, participating in the formation of the anterior oval for the superficial masseter. The infra-orbital margin is rounded and leads to a slight, rounded hollow in the inferolateral angle of the orbit. The malar face looks anteriorly for a short distance between the zygomaticomaxillary suture and the malar tuber; then, at the tuber, it swings sharply backwards to face mainly laterally and somewhat anterosuperiorly. The tuber is in the line of change of direction of the *facies malaris*. A similar flexion of the facies is present in other australopithecine crania (e.g. SK 48, Sts 5).

The frontal process faces mainly anteriorly, so there is a twist between it and the more laterally directed body of the zygomatic. The process is moderately concave from above downwards, but swells out to meet the frontal bone at the frontozygomatic suture. The special form of marginal (or better, postmarginal) tubercle on the temporal surface of the frontal process is discussed in the section on the Orbit (chapter x, B, p. 108). The temporal process is short and relatively slender, bearing on its inferolateral aspect the smooth facet for the posterior fleshy fibres of the deep masseter. The sutural surface is intact on the right (pl. 5).

E. Facial measurements and indices and calvariofacial indices

Facial measurements and indices of *Zinjanthropus* in comparison with those of other hominoids have been given in Tables 21 and 22. Most of these metrical features are discussed in the sections dealing with the Orbits and Interorbital Area, the Maxilla and the Zygomatic, the Palate and the Dental Arcade. Nasal measurements and indices are dealt with separately in Tables 23 and 24. There remain to be discussed the following metrical features.

1. *Size of the face*

The superior facial height measured from nasion to alveolare (the *lowest*, not the most anterior, point on the spur of bone between the upper two central incisor sockets) is enormous in *Zinjanthropus*. The value of 111·5 mm. is more than 30 mm. greater than the estimated heights in SK 48 (*c.* 80 mm.) and in Sts 5 (*c.* 75 mm.), and the mean height in *H. e. pekinensis* (77 mm.). Several Neandertaloid crania have greater heights than the latter specimens, e.g. La Chapelle (86 mm.) and Broken Hill (95·2 mm.). While the entire face of *Zinjanthropus* is built on a large scale, two factors undoubtedly contribute a large part of the facial lengthening: first, the marked 'verticalising' of the face with the tucking-in of the lower part of the upper face; and secondly, the great depth of the subnasal maxilla in association with very long dental roots. This marked depth of the subnasal maxilla is conveyed either by the nasospinale–alveolare or the nasospinale–prosthion dimensions, as discussed on pp. 113–14.

While the subnasal maxillary features contribute substantially to the facial lengthening, the orbito-alveolar height of 75·1 mm. shows that little or no extra height is gained from the region of the maxillary antrum in *Zinjanthropus* as compared with SK 48. Thus, the difference between the alveolar height and the orbito-alveolar height in both crania is 30–35 mm.

The breadth of the face is large, though not as strikingly so as is the height. The maxillary breadth of 121·5 mm. in *Zinjanthropus* exceeds that of SK 48 (95 mm.) by 26·5 mm., while the bizygomatic breadth of 168 mm. in *Zinjanthropus* exceeds that of SK 48 (146 mm.) by 22 mm. The value in SK 48 is exceeded by the *mean* value in *H. e. pekinensis* (148 mm.), and by the values in La Chapelle (153 mm.) and Broken Hill (147 mm.). Little additional breadth is added to the face of *Zinjanthropus* by the zygomatic arches, the difference residing essentially in the maxillary bone itself. This reflects itself in the higher value of the zygomaticomaxillary index (per cent) in *Zinjanthropus* (72·3) than in SK 48 (65·1), while a very modest flaring of the zygomatic arches in Sts 5 results in an index of 70·9, almost the same as in *Zinjanthropus*.

The greater predominance of facial height as compared with facial breadth in *Zinjanthropus* is reflected in the superior facial indices. In the first of these the superior facial height is related to the bizygomatic breadth. The value of 66·4 per cent in *Zinjanthropus* exceeds that of other australopithecines and hominines, though Broken Hill (64·6) and Tabūn I (60·8) are not far off. Weidenreich (1943, Table XXXIII) quotes a figure of *c.* 70 per cent as the mean for this index in pongid crania. In this respect, therefore, *Zinjanthropus* lies in an intermediate position between the hominids and pongids, though other australopithecines lie well within the hominine range.

The second of the superior facial indices relates the superior facial height to the maxillary breadth. Here, too, *Zinjanthropus* is pre-eminent with an index of 91·8 per cent, the nearest contendant being Broken Hill with an index of 86·3 per cent. Values for pongids are not available to me.

The face tapers strongly upwards towards the hafting with the narrow frontal part of the braincase. This is revealed by the zygomatico–suprafacial index: in this ratio, the numerator is the superior facial breadth, measured just above the frontomalare temporale, while the denominator is the bizygomatic breadth. In *Zinjanthropus* and *Paranthropus*, this value (per cent) is low (68·7, 67·8); it is somewhat higher in Sts 5 (72·0), while the mean in *H. e. pekinensis* is 81·7 and the values in a number of fossil hominine crania range from 82·8 to 95·4. Values for anthropoid apes and modern man are not available to me. A further indication of the superior facial tapering in *Zinjanthropus* is provided by a comparison between the superior facial breadth and the bimaxillary breadth: only in *Zinjanthropus* is the former smaller than the latter (115·4, 121·5). In SK 48 the values (per cent) are 99 and 95 respectively, in Sts 5 91·5 and 90, in *H. e. pekinensis* 121 and 98 (means), in Broken Hill 140 and 110, La Chapelle 128 and 110, Gibraltar 118 and 103, Skhūl V 120 and 110. Two factors may contribute to this unique relationship in *Zinjanthropus*: the strongly down-curved lateral part of the supra-orbital torus would tend to diminish

THE FACE

Table 25. *Calvariofacial indices of* Zinjanthropus *and other hominoids (per cent)*

	Zinjanthropus	Sts 5	SK 48	H. erectus pekinensis (mean)	Modern man*	Great apes*
Longitudinal craniofacial index (sup. facial l./max. cr. l.)	79·2	83·1	—	58·8	45–60	over 100†
Transverse craniofacial index (bizyg. br./max. cr. br.)	c. 144·2	128·3	c. 136·4	105·7	87–102·2	115·7–154·4
Zygomaticofrontal index (min. front. br./bizyg. br.)	c. 41·3	c. 45·7	c. 48·6	64·7	66·2–91·1	—
Frontobiorbital index (min. front. br./sup. facial br.)	60·1	c. 63·4	c. 71·7	—	—	—

* Ranges of means from Martin (1928).
† Weidenreich (1943).

the superior facial breadth, whilst the marked bilateral expansion of the maxillae would further contribute towards the upper facial tapering.

2. *Calvariofacial indices*

In Table 25, a few calvariofacial indices are compared. When facial elongation is measured against the total cranial length, it is seen that *Zinjanthropus* and *Australopithecus* (Sts 5) have indices (79·2, 83·1 per cent) appreciably greater than in the Homininae, but falling well short of the values in the Pongidae. The marked differences in the degree of longitudinal expansion of the cranial vault are probably responsible for these differences. On the other hand, the transverse craniofacial indices of the australopithecines are far higher than those of the Homininae and fall well within the pongid range. The key variable here would seem to be the degree of lateral flaring of the zygomatic arches which is related to the development of the temporalis muscles.

The zygomaticofrontal indices of the australopithecines are far lower than in the hominines, on account of both the relatively unexpanded anterior part of the calvaria and the flaring of the zygomatic arches.

CHAPTER XI

THE PNEUMATISATION OF THE ZINJANTHROPUS CRANIUM

The cranium of *Zinjanthropus* is heavily pneumatised. All the paranasal sinuses represented show excessive development and spread into adjacent parts not regularly pneumatised in hominids. In the ninth of his twenty diagnostic criteria of *Zinjanthropus*, Leakey (1959a, p. 492) had commented on 'the very great pneumatosis of the whole of the mastoid region of the temporal bones, which even invades the squamosal elements'. Each part of the cranium will be dealt with here in turn.

A. The maxillary sinus

Through damage to the posterior aspect of the maxillae, almost the entire volume of both maxillary sinuses is exposed from behind (pl. 18B). This excellent exposure provides an unusually good view of the antrum on each side.

The maxillary sinus is vast and partly bilocular. There are a partial septum on the right and a more complete transverse septum on the left, approximately at the level of the lower edge of the root of the zygomatic process. The partial septum cuts off a lower compartment which, inferiorly, and especially anteriorly, is largely multilocular. This lower compartment corresponds to the *recessus alveolaris* of the maxillary antrum in gorilla and orang (Wegner, 1956). In *Zinjanthropus*, the roots of M^2 have broken partly through into the alveolar recess. The roots of M^3 are set in the more solid bone of the maxillary tuberosity and only the mesial edge of the mesial root of M^3 is set in the area thus far pneumatised. To a slight extent, the alveolar recess has invaded the hard palate behind the incisive canal (*recessus palatinus*), thus tending to intervene between the nasal floor and the roof of the mouth. Anterior to the incisive canal, pneumatisation has gone much further, the alveolar recess extending right around, under the naso-alveolar junctional area (the top of the naso-alveolar clivus), almost to the mid-line in front. These conditions obtain as well in gorilla, chimpanzee and orang crania, though in adults the invasion of the palate is carried much further (Cave and Haines, 1940; Wegner, 1956; Cave, 1961).

The superior compartment of the maxillary sinus is not loculated and extends laterally into the zygomatic bone (*recessus zygomaticus*). It is traversed by a distinct bony tube, for which the name *tuba infra-orbitalis* may be suggested, to convey the infra-orbital nerve and vessels from the infra-orbital canal to the infra-orbital foramen (pl. 19). The infra-orbital foramen is situate 29–30 mm. below the inferior orbital margin, so the nerve and vessels must have pursued an oblique, nearly vertical, course to that position, ensheathed in a delicate sleeve of bone.

The upper part of a similar bony canal for the infra-orbital nerve and vessels is widely exposed in the right maxillary sinus of one of the *Paranthropus* crania from Swartkrans (SK 847). It occurs, too, in the crania of anthropoid apes, in which it may comprise a double tube, leading to double infra-orbital foramina (cf. Wegner, 1956). Sometimes, as in chimpanzee I. 44 (♀) in the Powell-Cotton collection at Birchington, the bony canal is not closed right around to form a tube, but remains partly open as a gutter (*canalis infra-orbitalis*), the lips of which are continued by intra-antral sheets of bone to the nearest wall of the sinus, either superiorly (i.e. continuous with the infra-orbital canal as generally understood) or anteriorly. Canals are sometimes seen in other positions within the maxillary sinuses of pongids, such as a canal for the descending palatine artery and anterior palatine nerve, in those crania in which the maxillary sinus has spread to embrace the pterygopalatine canal. Canals for the superior alveolar nerves are common in hominoids, including modern man, although they vary in the degree to which they are attached to the adjacent walls of the sinus. The infra-orbital foramen in modern man is generally so high on the face that the infra-orbital canal pursues an almost horizontal course just below (and sometimes completely open into) the floor of the orbit.

Above and medially, the maxillary sinus extends up the frontal process of the maxilla (*recessus anterior*), even passing medially around the nasolacrimal duct (*recessus lacrimalis*), to come into close contiguity with part of the frontal sinus system. Continuity between the two systems does not appear to occur directly, since I could not probe from one into the other. These features duplicate those of gorilla crania.

Little is known of the ramifications of the maxillary sinus in *Homo erectus pekinensis*, save that it has a broad, lateral expansion into the zygomatic bone, but does not penetrate the palate nor engulf the nasolacrimal canal (Weidenreich, 1943, p. 167). In *H. e. erectus* IV, Weidenreich gained the impression that the maxillary sinus extended further medially towards the palate than in *H. e. pekinensis* and modern man.

In the adolescent *Zinjanthropus*, the sinus does not extend back into the maxillary tuberosity, as I found it doing in a number of pongid crania; in this respect, *Zinjanthropus* resembles *H. e. erectus* IV (Weidenreich, 1945, p. 29) and modern man, in whom the maxillary tuberosity is a thick and massive bone wall. In many pongid crania studied in London and at Birchington, the maxillary tuberosity was found to be well 'inflated', the pneumatic recess extending back towards but not into the pterygoid formation. In all pongid crania where section or damage permitted the precise conditions to be determined, the extension of the sphenoidal sinus into the root of the pterygoid formation (*vide infra*) is distinct from the extension of the maxillary sinus into the maxillary tuberosity. They form parts of totally different sinus systems (cf. the gorilla— Cave, 1961). In *Zinjanthropus*, only the more medial part of the pterygomaxillary contact area is present, where the pterygoid process abuts against the maxillary tuberosity. More laterally, the lower part of the maxillary tuberosity only is preserved. There is no sign that the pterygoid recess of the sphenoidal sinus extended forwards into the maxillary tuberosity.

In sum, the maxillary sinus of *Zinjanthropus* is voluminous and ramifying, extending in the form of recesses into the pars alveolaris (*recessus alveolaris*), the frontal process (*recessus anterior*), the palatine process (*recessus palatinus*), the lacrimal bone and nasolacrimal duct (*recessus lacrimalis*), and the zygomatic bone (*recessus zygomaticus*). It was not possible to determine whether, in addition, it pneumatises the palatine bones, the maxilloturbinals and the ethmoidal bones, but it does *not* extend into the pterygoid processes, nor is there any other evidence of an intrasphenoidal encroachment by the maxillary sinus, such as not uncommonly occurs in gorilla crania (Cave, 1961). With the possible exception of the latter extensions, the maxillary sinus of *Zinjanthropus* tallies fairly closely with that of gorilla, of which Cave (1961) could say, 'The maxillary sinus, ontogenetically the first and topographically the most extensive, of the series, is unmistakably the most "dominant" of the gorilla sinuses' (p. 369). There is little doubt that the maxillary sinuses of *Zinjanthropus*, like its frontal sinuses, would with age have extended further—for instance, in the palatine processes and pars alveolaris, and into the maxillary tuberosities—as has been recorded for man (Jovanovic, 1958) and pongids (Cave, 1961).

B. The frontal sinus

The frontal sinus system, too, is extensive in *Zinjanthropus* and is honeycombed with septa and locules. The highest (or most posterior) extent of the frontal sinus is apparent on the left through a defect in the outer table, a short distance from the back of the preserved part of the frontal squame (pl. 15B). This furthest point of the sinus is 54 mm. (chord distance) from nasion, and would perhaps correspond with the *superior* (or *dorsal*) *diverticulum* of the gorilla frontal sinus, described by Cave (1961) as 'burrowing paramedially upwards and backwards into the floor, and towards the apex, of the trigonum frontale' (p. 363).

Some idea of the lateral extent of the frontal sinus is gleaned from an exposure at the outer angle of the frontal endocranial wall, where the frontal squame begins to turn backwards. This small exposure of sinus is interposed between the outer and inner tables, at the postorbital constriction. It corresponds to the *posterolateral diverticulum* of the gorilla frontal sinus (Cave, 1961). This posterolateral diverticulum seems to occupy the triangular gap between three surfaces—the medially-situated anterior cranial fossa, the inferiorly-situated orbit and the laterally-situated temporal fossa opposite the postorbital constriction.[1] The continuity of this recess with the frontal air-sinus is suggested on the right, but is not apparent on the left. Anterior to the postorbital constriction, the frontal sinus seems to extend even further laterally, into the zygomatic process, almost to the frontozygomatic suture (cf. *lateral diverticulum* of the frontal sinus in gorilla—Cave, 1961): it does not appear to transcend that suture.

The large size of the frontal sinus compares well with conditions in *H. erectus* of Indonesia, and in gorilla and chimpanzee; but contrasts with the small frontal sinuses of *H. e. pekinensis* and of the orang-utan. Little is known of the frontal sinuses in other australopithecines. Those of Sts 5 are illustrated and referred to in Broom *et al.* (1950, pp. 23–4 and fig. 5, p. 22), but no details are given. Cranium VII from Sterkfontein is reported as having 'a very large median frontal sinus' (p. 26), which is illustrated in Broom's fig. 8 (p. 26). The frontal sinuses of the type skull from Sterkfontein are likewise referred to without any details by Broom and Schepers (1946, p. 52). I have not been able to locate any published reference to the frontal sinuses of *Paranthropus*.

[1] The presence of pneumatisation in this position would seem to support Weidenreich's view that the sinuses represent 'void' chambers, 'not essential for the transmission of stress and strain in the architectonic structure of the skull' and that, 'consequently, they develop where a large incongruity exists between the frameworks of two adjacent systems of organs' (Weidenreich, 1943, p. 165 and 1924, 1941*a*).

C. Pneumatisation of the naso-orbital region

The precise relationship of the frontal sinus to the ethmoidal bone and sinus system cannot be determined in *Zinjanthropus*, as the ethmoid is missing. This is a pity as the characteristic arrangement in *Gorilla* is, according to Cave (1961), the direct contrary of conditions in *Homo*. In *Gorilla*, there is an *inferior* (or *anterior*) *diverticulum* of the frontal sinus, which extends into the lacrimo-ethmoidal region, absorbs the anterior ethmoidal air cell, and pneumatises the lacrimal bone and a variable extent of the ethmoid. After a primary pneumatisation of the *initially narrow* naso-orbital region of the gorilla cranium by the ethmoidal sinuses, the region is secondarily pneumatised by the frontal and maxillary sinuses acting 'in mutual antagonism' to each other. This secondary pneumatisation of the naso-orbital area leads to the reduction or even total obliteration of the initial ethmoidal air cells. In *Homo*, on the other hand, an *initially wide* naso-orbital region is pneumatised by an abundance of ethmoidal cells, which 'continue to proliferate and to extend topographically throughout life' (Cave, 1961, p. 370). One might say that the pneumatising potential of the ethmoidal air cells is far greater in man than in pongids. They may even invade the middle nasal concha (turbinal) creating therein what were earlier called 'bone-cysts', but what we should today regard as pneumatic spaces or air cells (Radoievitch and Jovanovic, 1959). This 'dominance' of the ethmoidal air cells was stressed earlier by Weidenreich (1943), who said, 'Independent cellulae ethmoidales in larger numbers have developed only in man. In this respect he differs not only from the anthropoids but also from all the catarrhine and platyrrhine apes in which these cells are either completely lacking or restricted to one or two cells, the spaces they occupied being taken over by the frontal or maxillary sinus' (p. 166). Jovanovic (1958) has claimed that in modern man the frontal sinus may pneumatise the *crista galli* (8 per cent of crania) and the perpendicular plate of the ethmoidal bone (3 per cent), and even the nasal bone and the frontal process of the maxilla (3 per cent). Nevertheless, it seems

PNEUMATISATION

clear that the frontal sinus of *Homo* is not nearly as 'dominant' a sinus, to use Cave's terminology, as it is in *Gorilla*, while the converse is true of the ethmoidal sinus system of *Homo*.

Although the ethmoid bone is missing in *Zinjanthropus*, there is no doubt of the extensive pneumatisation of the naso-orbital region. The inflated appearance of the anterior part of the medial orbital wall and the great posterior interorbital width measured between the lacrimalia (32·5 mm.) have already been stressed in chapter x, B (p. 108). This degree of inflation had been achieved by the age of adolescence—for the M^3's of *Zinjanthropus* were in process of erupting. I am unable to reach a definite conclusion about whether the naso-orbital region of the *Zinjanthropus* cranium was 'initially narrow' as in *Gorilla* or 'initially wide' as in *Homo*; the fact that, by an age of, say, 15–17 years, the region had become as wide as it is in the fossil cranium would tend to speak in favour of an initially wide naso-orbital region. However, the general extent of the entire sinus system of *Zinjanthropus* suggests a rapidity of pneumatisation which would weaken this line of reasoning. Secondly, without breaking or sectioning the specimen (and even then, much of the relevant area is missing), it seems impossible to say whether the naso-orbital inflation of *Zinjanthropus* was owing to pneumatisation by an abundance of proliferating ethmoidal air cells, or to secondary pneumatisation by either the frontal or maxillary sinus, or to both. It is therefore impossible to conclude whether, in its pattern of growth and extension, the pneumatisation of the naso-orbital region resembles the pattern in man or that in gorilla.

D. The sphenoidal sinus

The whole floor of the middle cranial fossa is underlaid by an air-cushion of spaces which extend laterally from the sphenoidal air-sinus. The body of the sphenoid is broken across in the midst of the sella turcica, revealing a profusion of air locules. A vertical septum in a sagittal plane lies to the right of the mid-line and separates the left from the right sphenoidal sinuses. The left sphenoidal sinus is very large and is partly divided into two by a transverse septum. The right sinus is itself spacious and is also divided into upper and lower compartments by transverse horizontal septa. These compartments have several other subdivisions; one large recess passes to the right, even further laterally than the foramen ovale. The sinuses invade the root of the pterygoid process massively, extending downwards for a short distance, though not invading more than the upper part of the laminae themselves. The air cells extend inside the greater wing of the sphenoid as far as the extreme upper limit of the preserved part of the alisphenoid on the right. Basally, the sphenoidal sinus extends into the sphenoidal rostrum, the lateral walls of which are widely separated: so that the medial parts of the two alae of the vomer inserted on the rostrum are unusually far apart. The total width at the attached base of the vomer is 5·4 mm., of which 3·6 mm. is the width occupied by the sphenoidal rostrum with its contained air cell.

This spread of the sphenoidal air-sinus well beyond the limits of the body of the sphenoid is paralleled by conditions in Sts I (the type skull from Sterkfontein—Broom and Schepers, 1946, p. 52), Sts VIII (Broom, Robinson and Schepers, 1950, p. 33), *H. e. erectus* IV, the Ehringsdorf and Galilee crania, but *not* those of *H. e. pekinensis*; it is characteristic also of many pongid crania.

Posteriorly, the sphenoidal air-sinus in *Zinjanthropus* extends just beyond the presumed position of the spheno-occipital synchondrosis. This is true as well of Sts 19 (cranium VIII of Sterkfontein), in which the posterior extremity of the sinus reaches to within 18·5 mm. of basion. Neither in this cranium, nor in any of a number of pongid crania examined for this feature, does the sphenoidal sinus extend far beyond the synchondrosis: instead, the central part of the basioccipital, between the outer and inner tables, presents a finely spongy texture in sectioned pongid crania. Speaking of the sphenoidal sinus in *Gorilla*, Cave (1961, p. 369) states: 'The sphenoidal sinus, although a major member of the series, is significantly less "dominant" in the pneumatisation of the skull than either the frontal or the maxillary sinus. Its pneumatising activity is confined to the basis cranii (i.e. to the body, great wings and pterygoid processes of the sphenoid) outside which limit it never obtrudes.' Strictly speaking, this passage from Cave does not allow of the spread of the sphenoidal sinus posteriorly into the basi-occipital; yet such extension does occur, not only in *Zinjanthropus* and *Australopithecus*, but in other hominid and pongid crania (*see* chapter VI, pp. 59–60). The posterior extension of the sphenoidal sinus in *Gorilla* is well illustrated in fig. 3 of Hofer (1960). Furthermore, there is no trace in *Zinjanthropus* of the expansion of the maxillary sinus posteriorly into the body of the sphenoid, a feature which Cave (1961) described as not uncommon in gorilla crania. The overall impression is that the sphenoidal sinus of *Zinjanthropus* is slightly more 'dominant' than it is in *Gorilla*.

E. Pneumatisation of the temporal bone

Pneumatisation of the *pars mastoidea* is marked and extends well beyond the limits of the mastoid process itself. The coarse air cells of the pars mastoidea are well seen in *Zinjanthropus* through damage, especially to the left temporal bone (pls. 24 and 25). They cause the whole pars mastoidea to be puffed out, so that this area projects most laterally on the skull (fig. 5). The pneumatisation extends forward into the root of the zygomatic process and upward for some distance within the squama. The entire height of the squama is, in fact, pneumatised as far as the line of junction between the endocranial surface and the bevelled sutural surface for the squamosal suture. The degree of pneumatisation of the squama is much less marked in *Zinjanthropus* than in *Gorilla*, in which it causes a characteristic change of external contour of the vault, very close to the line of the squamosal suture (Hofman, 1926–7). On the other hand, in some crania of *H. e. pekinensis*, the squama is pneumatised as far upwards as 'the lower margin' of the squamosal suture (Weidenreich, 1943, p. 168); this degree of pneumatisation is apparently not encountered in modern man.

Extensive pneumatisation of the pars mastoidea occurs in all the australopithecines, being especially well marked in *Paranthropus* (such as Kromdraai, SK 48 and SK 46), but it is also a feature of *Australopithecus* (such as Sts 5, Sts VIII and MLD 37/38).

In *H. erectus*, although Davidson Black (1931) spoke of the 'peculiarly restricted character of the temporal pneumatisation' in Choukoutien III, Weidenreich (1943) showed that in at least some crania, and especially Choukoutien V, there is a fairly wide extension of air cells into the pars mastoidea behind the mastoid process, into the temporal squama and into the petrous pyramid. Some crania, especially skull III, display relatively poor pneumatisation.

In modern man, Calogero (1959) has shown that the degree of development of mastoid pneumatisation and that of the paranasal sinuses do not necessarily parallel each other, as though different factors controlled the formation of the two sets of cavities. However, subjects having strong temporal pneumatisation likewise have well-developed sinuses. Calogero suggests a possible neuro-endocrine explanation in the latter instance, involving pituitary hyperfunction or diencephalic disease.

F. Developmental aspects of pneumatisation

In *Zinjanthropus*, the degree of pneumatisation of the maxilla, frontal, sphenoid and temporal bones is greater than in any previously described australopithecine or hominine. It is tempting to associate this factor with the other respects in which *Zinjanthropus* is extreme, namely, the size of the teeth and jaws, and the development of his masticatory and nuchal muscles. Cave (1961) has shown that accelerations of sinus growth in *Gorilla* can be correlated with growth spurts in the development of the facial skeleton occasioned principally by the successive eruption of the deciduous and permanent dentitions. He speaks of 'the over-riding physiological necessity of obviating undue cranial-mass weight by a pneumatic lightening of cranial components'. If this correlation holds good for *Zinjanthropus*, it remains true that, despite a voluminous literature on the pneumatisation of the cranium, we still do not know enough of the factors influencing the development of air spaces in the skull to determine whether epigenetic events could account for the postulated association between heavy teeth and jaws and extreme pneumatisation, or whether it is necessary to postulate an association between separate and distinct genetic effects. Moreover, the frontal sinus (and the maxillary) display 'a propensity for steadily persistent expansion even after cranial growth is complete and the full permanent dentition has been acquired (i.e. throughout adult life into old age)' (Cave, 1961). As mentioned above, Weidenreich, too, has stressed the additional or non-lightening function of the sinuses, namely, as 'void rooms' not essential for the transmission of stress and strain in the architectonic structure of the skull. Negus (1965), also, has spoken of 'an unwanted space remaining laterally in the maxilla'—and added, 'this empty space is known as the maxillary sinus' (p. 145). We might then recognise a *functional* and a *spatial* aspect to pneumatisation, just as we may have to acknowledge both *genetic* and *epigenetic* or functional factors in its ontogeny.

CHAPTER XII

THE DENTAL ARCADE AND THE PALATE

Although over fifteen partial or complete australopithecine maxillae have thus far been described from Africa, *Zinjanthropus* is only the second australopithecine specimen to have presented the world with a complete adult maxillary dentition. The first was a fine *Australopithecus* maxilla from Sterkfontein, Sts 52a. The presence of all sixteen teeth in the alveolar process of the Olduvai specimen, as well as of almost the entire palate, has made possible a very detailed study of the teeth and supporting structures. In the present chapter, the following aspects will be treated: the shape of the dental arcade and related features of the palate; the arrangement of the teeth in the arcade.

A. The shape of the dental arcade, alveolar process and palate

The maxillary dental arcade of *Zinjanthropus* presents an evenly arched, parabolic curve (pls. 20 and 29). The incisors and medially-placed canines are arranged in a low arch, extending between the right and left anterior premolars. From immediately behind the canines, the premolar–molar series continues with no trace of a diastema. The premolar-molar teeth lie in two slightly diverging rows, which are nearly straight save that the distal ends of the series are somewhat inturned. In other words, the arcade is very similar in shape to those of other australopithecines and of the Homininae (Le Gros Clark, 1955; Robinson, 1956). It differs markedly from the typical pongid pattern, in which the canine and post-canine teeth form approximately straight rows, parallel or even slightly divergent anteriorly, and separated from the anterior teeth by a distinct diastemic interval in the overwhelming majority of instances (Le Gros Clark, 1952b).

Many attempts have been made to give metrical expression to the shape of the dental arcade and its related alveolar arch and bony palate. Broadly speaking the diverse methods, definitions and nomenclature can be classified under three headings:

(a) *measurements on the bone*—taken *outside* the alveolar arch, generally called *maxillo-alveolar* dimensions;

(b) *measurements on the bone*—taken *inside* the alveolar arch: these are correctly called *palatal* dimensions;

(c) *measurements on the dental arch*, designed to give metrical expression to the shape of the dental arcade, irrespective of the shape or size of the alveolar arch and the palate. We shall briefly consider each of these approaches in turn.

1. *Maxillo-alveolar dimensions and index*

Of the many early studies on palatal and maxillary size and shape (*see* review in Campbell, 1925), Flower's (1881, 1885) proposals have received the most support over a lengthy period. He favoured including the alveolar arch as part of the bony palate. Thus he used external or maximum measurements to derive what the International Monaco Congress (1906) called the 'maxillo-alveolar index'. The termini of the measurements proposed by Flower (1881) are as follows:

Maxillo-alveolar length—'from the alveolar point in front' (interpreted here as the prosthion, not the alveolare) 'to the middle of a line drawn across the hinder border of the maxillary bones (the maxillary "tuberosities")'.

Maxillo-alveolar breadth—'between the outer borders of the alveolar arch immediately above the middle of the second molar tooth'.

Although a variety of names has been applied to these measurements,[1] the definitions of Flower

---

[1] The term 'maxillo-alveolar length' was adopted officially by the International Congress of Prehistoric Anthropology and Archaeology at Monaco in 1906; it is synonymous with the

were adopted *inter alia* by Campbell (1925) in his work on Australians, by Martin (1928), by Shaw (1931) in his contribution on *The Teeth, the Bony Palate and the Mandible in Bantu Races of South Africa*, by Jacobson (1966) in a more recent study on larger samples of the same population, and by Weidenreich (1943). These definitions are followed here and the term 'maxillo-alveolar' is used to designate them.

The maxillo-alveolar length of *Zinjanthropus* is 86·1 mm., the maxillo-alveolar breadth 81·8 mm., and the resulting index is 95·0 per cent. This places *Zinjanthropus* in the dolichuranic category.

2. Palatal dimensions and indices

The alveolar ridge itself varies considerably in thickness, quite apart from variations in the hard palate as such. Hence, there is an obvious place for a set of inner or palatal dimensions, as well as the outer or maxillo-alveolar dimensions. Many workers therefore use the palatal length and breadth measured *within the alveolar ridge* to derive a *palatal index*. It is clearly these inner dimensions which should be called 'palatal'.

The *palatal length* is defined as the distance from *orale* to *staphylion*.

Two different methods of measuring the *palatal breadth* have been proposed. Olivier (1960) measures the breadth between the lingual surfaces of the second molars, while Montagu (1960) uses the inner aspects of the alveolar margins opposite the mesiodistal mid-point of the second molars. Since there is no general agreement on which of the two techniques to employ for palatal breadth, both are recorded for *Zinjanthropus*. Fortunately, the teeth, palate and alveolar ridge of *Zinjanthropus* are so well preserved as to permit the measurement of a variety of internal and external dimensions, and the computation of several different indices. The palatal measurements of *Zinjanthropus* are recorded in Table 26 and the resulting indicial

'maxillary length' of Flower (1881) and the 'palato-maxillary length' or 'palato-alveolar length' of Turner (1884). The use of 'palatal length' for this dimension, as advocated by Campbell (1925) and followed by Shaw (1931), is inadvisable, as the term 'palatal' is generally applied to an *internal* palatal dimension (see Table 26). In the same way, 'maxillo-alveolar breadth' and 'maxillo-alveolar index' are to be preferred to any of the corresponding synonyms.

values are 48·3 per cent (breadth from alveolar margins) and 50·2 per cent (breadth from M^2). These values place the palate in the leptostaphyline category.

3. Dental arch dimensions and indices

Several different sets of measurements have been used in measuring dental arches. Moorrees (1957) speaks of an 'odontometric method' and an 'anthropometric method'.

(a) Odontometric method

Following a simplified version of the method proposed by Korkhaus (1939), Moorrees (1957, pp. 123–4) defines his 'odontometric' measurements as follows:

Dental arch length (odontometric)—'the distance between a tangent to the labial surfaces of the median incisors and a plane through the arch breadth line (M1–M1) perpendicular to the occlusal surface'.

Dental arch breadth (odontometric)—'the distance between the intersection of the main occlusal grooves of the first molars'.

(b) Anthropometric method

According to Moorrees (1957, p. 124), the anthropometric measurements are defined as follows:

Dental arch length (anthropometric)—'the distance between the tangent to the labial surfaces of the central incisors and a plane tangent to the distal surfaces of the second molars, perpendicular to the occlusal plane'.

Dental arch breadth (anthropometric)—'the greatest distance between the buccal surfaces of the second molars'.

(c) Arcadal index of Laing

Laing (1955) has devised a *length–breadth index of the dental arcade* or, more succinctly, an *arcadal index*, which is slightly different from the anthropometric method as set forth by Moorrees (1957). She measures the *maximum arcadal length* from 'the mid-point of a line joining the most anterior points on the tips of the incisor teeth' to 'the midpoint of a line joining the most posterior points of the third molar teeth'. The termini of her

Table 26. *Palatal and arcadal dimensions and indices of* Zinjanthropus

Maxillo-alveolar length (Flower) (From prosthion to the point where a tangent to the posterior borders of the maxillary tuberosities crosses the median plane)		86·1 mm.
Maxillo-alveolar breadth (Flower) (The distance between the lower borders of the outer surface of the alveolar arch, immediately above the middle of M^2)		81·8 mm.
Maxillo-alveolar index (Flower)	(dolichuranic)	95·0 %
Palatal length (orale–staphylion)		79·1 mm.
Palatal breadth (a) (Endomolare–endomolare: i.e. the distance between the inner aspects of the alveolar margin, opposite the middle of M^2)		38·2 mm.
Palatal breadth (b) (The shortest distance between the lingual faces of the M^2's themselves)		39·7 mm.
Palatal index (a) (Using breadth between alveolar margins; cf. Montagu)	(leptostaphyline)	48·3 %
Palatal index (b) (Using breadth between M^2's; cf. Olivier)		50·2 %
Arcadal length (Laing) (From the mid-point of a line joining the most anterior points on the tips of the incisor teeth to the mid-point of a line joining the most posterior points of the third molars)		82·4 mm.
Arcadal breadth (Laing) (The distance between the buccal surfaces of the second molars)		80·4 mm.
Arcadal index (Laing)		97·6 %
Distance between buccal faces of P^3's		63·0 mm.
Distance between buccal faces of M^3's		83·2 mm.
Distance between lingual faces of P^3's		30·3 mm.
Distance between lingual faces of M^3's		41·8 mm.

maximum breadth are on the buccal surfaces of the second molar teeth.

Thus, of the three techniques for determining the dimensions of the dental arch, the odontometric method uses M^1, the anthropometric method M^2 and Laing's arcadal index M^3, for determining the posterior terminus of the dental arch length. This is not the place to attempt to evaluate the relative merits of the three techniques. In *Zinjanthropus*, I have determined Laing's arcadal dimensions and index (Table 26): they are 82·4 mm. (arcadal length), 80·4 mm. (arcadal breadth) and 97·6 per cent (arcadal index).

(d) *Index of tapering* (*Robinson*)

In order to obtain an indication of the degree of tapering or otherwise, between the anterior and posterior parts of the dental arcade, Robinson (1956) measured the distance between the buccal faces of the right and left P^3, as well as that between the right and left M^3. Similarly, he measured the distance between the lingual faces of the right and left P^3, and of the right and left M^3. By comparing these several measurements, he obtained a metrical statement of the degree of divergence between the margins of the dental arcade. Corresponding measurements for *Zinjanthropus* are recorded in Table 26.

4. *Comparisons between* Zinjanthropus *and other hominoids*

The only other data available for the Australopithecinae are those of Robinson (1956, pp. 12–14). Unfortunately, he has apparently confused the indices based on internal and external measurements of the palate and alveolar ridge. To compare with

Table 27. *Maxillo-alveolar index of* Zinjanthropus *and other hominoids*

Zinjanthropus	95·0	Present study
Paranthropus (SK 46)	96·0	Robinson (1956)
Paranthropus (SK 48)	c. 94·0	Robinson (1956)
Paranthropus (Kromdraai adult)	c. 92·0	Robinson (1956)
Australopithecus (Sts 5)	90·0	Robinson (1956)
Australopithecus (Sts 53)	c. 96·0	Robinson (1956)
Homo erectus erectus IV (unrestored)	104·0	Weidenreich (1945)
H. e. erectus IV (restored)	116·0	Weidenreich (1945)
H. e. pekinensis (reconstructed)	107·6	Weidenreich (1943)
La Chapelle	101·2	From Weidenreich (1943)
Skhūl V	106·0	From Weidenreich (1943)
Broken Hill	116·2	From Weidenreich (1943)
Australian ♂ + ♀ (n = 134)	95·2–124·5* 108·9 (mean)	Campbell (1925)
Bantu ♂ + ♀ (n = 91)	101·5–120·5* 110·5 (mean)	Shaw (1931)
Bantu ♂ (n = 334)	87·5–142·5* 112·2 (mean)	Jacobson (1966)
Bantu ♀ (n = 95)	92·5–128·0* 110·3 (mean)	Jacobson (1966)
Modern races	108·2–126·0† 94·0–154·0*	Martin (1928)
Anthropoid apes	52·1–106·0* 76·1 (mean)	Weidenreich (1943)
Gorilla ♂	52·1– 73·7* 64·7 (mean)	Martin (1928)
Orang-utan ♂	66·3– 85·7* 74·8 (mean)	Martin (1928)
Chimpanzee ♂	70·8– 90·4* 80·8 (mean)	Martin (1928)

* Range of individual values.
† Range of population means.

his own data, he cites Weidenreich's (1943) results on comparative series for the *palatal index*. Weidenreich's figures clearly refer to the true or internal palatal index, based on orale–staphylion and endomolare–endomolare, as is evident both from his figures (1943, p. 149) and from the fact that he quotes the maxillo-alveolar index separately. Robinson compares these internal palatal indices of Weidenreich with his own 'palatal index' determinations on *Australopithecus* and *Paranthropus*: but it seems clear both from his results and from the fact that he bases his index on 'maximum dimensions' (p. 13), that what Robinson has, in fact, determined is the external or maxillo-alveolar index. On the basis of comparing these two non-comparables, he is even led to feel that Weidenreich's figures for *Homo erectus pekinensis* are 'insecurely based' (p. 13)! The confusion has apparently arisen because of Campbell's (1925) attribution of the term 'palatal' to the external or maxillo-alveolar dimensions, an attribution which was followed by Shaw (1931) and evidently Robinson (1956). If any further attempts are made to standardise anthropological terms internationally, a start could well be made with the term 'palatal': it should be urged that this term be confined to the anatomical entity, the palate, which lies wholly *within* the alveolar arch. Another term—maxillo-alveolar—is needed for measurements *outside* the alveolar arch.

The detailed comparative data available for the maxillo-alveolar index appear in Table 27; while Table 28 summarises these data. On the basis of these figures, the Australopithecinae, including *Zinjanthropus*, have a relatively narrower and longer arcade than the Homininae, only a few hominine

Table 28. *Ranges of maxillo-alveolar indices in hominoid groups (per cent)*

Gorilla	52·1–73·7
Orang-utan	66·3–85·7
Chimpanzee	70·8–90·4
Australopithecinae	90·0–96·0
Fossil *Homo*	101·2–116·2
Modern *Homo*	87·5–154·0

individuals at the lower extreme of their ranges having maxillo-alveolar indices within the australopithecine sample range. On the other hand, the australopithecine palates and alveolar ridges are not as long and narrow as in the pongids, only the upper extreme of the chimpanzee range falling just within the bottom of the australopithecine sample range. On this index, the Australopithecinae occupy a clearly intermediate position between the Pongidae and Homininae. Furthermore, the very few determinations of this index do not show any distinction between *Paranthropus* (92–96 per cent) and *Australopithecus* (90–96 per cent), the two values for the latter sample being the greatest and smallest of the australopithecine sample range.

The palatal index of *Zinjanthropus* (48·3 per cent) is much lower and more ape-like than in any hominine group. Thus, the mean (per cent) for *H. e. pekinensis* is 75·1, the value for Skhūl V is 75·0, La Chapelle 80·6, Broken Hill 84·6, and the indices for modern human races 63·6–94·6. On the other hand, according to Weidenreich (1943), the pongids have indices ranging from 34·5 to 62·5 per cent (the latter extreme being the value for a single male chimpanzee). This extraordinarily low palatal index in *Zinjanthropus* is owing not so much to the rather small endomolar breadth of 38·2 mm. (though the mean in *H. e. pekinensis* is 39·0 mm.), as to the excessively long orale–staphylion length of 79·1 mm. Among fossil hominines, the lengths are 52 mm. (mean) in *H. e. pekinensis*, 57·7 in Broken Hill, 60 in Skhūl V and 62 in La Chapelle. Among living hominines, values for palatal length range from 37·0 to 63·0 mm. in various samples of male Bantu-speaking negroids (Galloway, 1941; De Villiers, 1963; Jacobson, 1966), with sample means of 47·0, 48·6 and 49·0, while figures for females range from 39·5 to 57·0, with means of 45·3 and 47·0. The mean for adult Australians is 51·5 mm., ranging up to an individual maximum of 59·5 mm. The palatal narrowing and lengthening of *Zinjanthropus* are related to the broadening and lengthening of the alveolar ridge with its contained immense teeth; thus the external (maxillo-alveolar) and internal (palatal) indices of *Zinjanthropus* convey two very different pictures. In breadth, the ectomolar dimension is of course greater than the endomolar breadth by *twice* the thickness of the buccolingually expanded alveolar ridge; in length, the prosthion–postmaxillary dimension is bigger than orale–staphylion by only *one* thickness of alveolar process, and this is at the thinnest (anterior) part of the process. Hence, the internal (palatal) index is very low (within the pongid range), while the external (maxillo-alveolar) index is much higher (lying between pongid and hominine values).

Laing's Arcadal Index for *Zinjanthropus* (97·6 per cent), when compared with the few available data in the literature, is lower than the range for a group of 24 Bushmen, 100–136 per cent (Laing, 1955). The value for a single chimpanzee is given by Laing as 80 per cent, those for the Talgai cranium 95, the Grimaldi youth 100, a Tasmanian 106, an Australian aboriginal 110, a Předmostí male 110, La Chapelle 116, Broken Hill and Předmostí female 120, and the mean for modern English crania (120). These sparse results with the Arcadal Index are expectedly very close to those with the Maxillo-alveolar Index: they show that *Zinjanthropus* falls at or close to the lower limit of the hominine range.

All three indices point to *Zinjanthropus* having a longer and narrower palate than do the Homininae. The external proportions (which include the alveolar ridge) are intermediate between those of Pongidae and Homininae; the internal proportions, which are available for only a single, very unusual australopithecine, namely *Zinjanthropus*, lie well within the pongid range.

The last measurements in Table 26 show that the outer (buccal) width of the arcade increases from P^3 to M^3 by divergence of the tooth rows from 63·0 to 83·2 mm., an increase of 20·2 mm. At the same time, despite the increasing breadth of the posterior teeth, the inner (lingual) width of the arcade increases from 30·3 to 41·8 mm., an increase of

DENTAL ARCADE AND PALATE

11·5 mm. The corresponding P³–M³ increases for a *Paranthropus* specimen, SK 46, are 19·2 mm. on the buccal aspect and 11·4 mm. on the lingual surface. The cheek tooth-rows thus diverge to an almost identical degree in *Zinjanthropus* and SK 46. The tendency of the M³'s to turn inwards somewhat is brought out by comparing the bidental distances between the buccal faces of each pair of teeth from P³ to M³. In *Zinjanthropus*, these figures are (mm.):

P³–P³ 63·0
P⁴–P⁴ 68·1—5·1 ⎫
M¹–M¹ 74·8—6·7 ⎬ greater than the preceding
M²–M² 80·4—5·6 ⎪
M³–M³ 83·2—2·8 ⎭

These figures demonstrate that, although the diverging tendency continues as far as the third molars, the degree of divergence decreases somewhat from M¹ to M² and most markedly from M² to M³, so that the increase in buccal breadth from M² to M³ is much less than between any other two successive teeth in the premolar–molar series. The M²'s and especially the M³'s are beginning to reverse the curve of divergence, thus converting the arcade from a hyperbola into a parabola.

B. The front of the alveolar process

Robinson (1956, p. 14) pointed out that the shape of the dental arcade differs very little between *Australopithecus* and *Paranthropus*. On the other hand, the shape of the alveolar margin in the canine–incisor region 'differs appreciably in the two forms'. According to Robinson, in *Australopithecus*, the alveolar margin in this region is curved forwards, as is the dental arcade, whereas in *Paranthropus*, although the dental arcade is curved, the alveolar margin runs more or less straight across from one canine alveolus to the other. Figure 22 is based upon fig. 1(a) and (b) of Robinson (1956) and shows (i) the straight alignment of anterior tooth sockets in *Paranthropus*, in contrast with the curved arrangement of the crowns of the front teeth, and (ii) the curved alignment of both sockets and crowns in *Australopithecus*. In comparison with these two forms, *Zinjanthropus* occupies an intermediate position. The posterior mar-

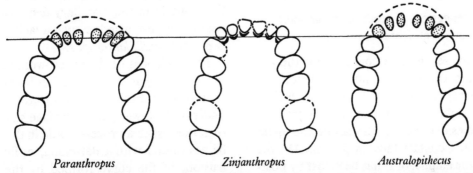

Fig. 22. The shape of the dental arcade in *Paranthropus*, *Zinjanthropus* and *Australopithecus*. The figures for *Paranthropus* and *Australopithecus* have been adapted from those of Robinson (1956), to approximately the same size as the tracing of *Zinjanthropus*: the diagrams are thus for comparison of shape alone, not of relative size. The three arcades are orientated on the prelacteon line. In *Paranthropus* and *Australopithecus*, the dotted areas represent the sockets, the interrupted line the labial face of the tooth crowns. In *Zinjanthropus*, the outlines of the anterior tooth crowns are shown, as well as the lingual parts of the sockets shown as dark crescents lingual to the tooth crowns. Note the variable relationship of the anterior tooth-crowns and sockets to the prelacteon line.

gins of the alveoli, which are shown in Fig. 22, are arranged in a moderate curve, not as straight as in the *Paranthropus* illustrated by Robinson, nor as convex as in the depicted *Australopithecus*. However, the shapes drawn by Robinson represent extremes: when I re-examined the original specimens at the Transvaal Museum, Pretoria, I noted that of four maxillae of *Australopithecus*, three had the highly convex arch depicted (Sts 5, Sts 17 and Sts 52a), while one, T.M. 1511, had a low curve, very similar to that of *Zinjanthropus*. In addition, the second adult palate of *Australopithecus* from Makapansgat (MLD 9) shows a high convexity

(Dart, 1949b). Out of seven maxillae of *Paranthropus*, five have the straight contour depicted (SK 11, SK 12, SK 13, SK 46 and SK 48), while two (SK 52 and SK 55) have a degree of curvature fully comparable with that of *Zinjanthropus*. While agreeing with Robinson that the *modal* patterns of maxilla in these small samples of the two South African australopithecine forms differ, it is necessary for us to draw attention to the variability of both groups. Thus, in a small proportion of members of both groups (one out of five specimens of *Australopithecus* and two out of seven specimens of *Paranthropus*), the more gently curved intermediate form of alveolar process occurs. In this respect, then, the dental arcade and alveolar margin of *Zinjanthropus* fall within the range of variation of both *Australopithecus* and *Paranthropus*.

C. The arrangement of teeth in the arcade: evidence of dental crowding

The canines are set in close contact with the anterior premolars, there being a small interproximal attrition facet between the two teeth on each side. Likewise, the right lateral incisor is in direct contact with the mesial surface of the canine and again an interproximal contact facet is present. On the left, much of the enamel of I^2 has been lost by postmortem damage, including the relevant area on the distal face; nevertheless, a facet on the mesial face of the left canine testifies that the two teeth had been in the same intimate contact on the left as on the right. Thus, there is not the slightest trace of a pre- or postcanine diastema. Even the minute gap which Robinson (1956, pp. 14–15) occasionally detected between the lateral incisor and canine of some australopithecines is not in evidence in *Zinjanthropus*.

Both the left and right canines and the left and right third premolars show slight labial displacement. This can be seen clearly on both the photograph (pl. 20) and the dioptographic tracing (Fig. 22), but the minute degree of displacement is insufficient to mar the smooth parabolic curvature of the dental arcade. At the same time, the canines show slight rotation, especially on the left, so that the lingual surface faces distolingually, i.e. towards the third premolar (pl. 29). Both the displacement and the rotation suggest a degree of crowding and a shortage of space. Robinson (1956, pp. 16–17) has drawn attention to several instances of dental crowding in the australopithecines, as well as other features suggesting a shortage of space, mainly in the big-toothed form, *Paranthropus*. Oppenheimer (1964) has recently suggested that 'the only plausible explanation for the crowded dentition of the *Australopithecines* is the substitution to a significant degree of tools and weapons for teeth and jaws'.

In a biometrical study of crowding and spacing in the *mandibular* teeth of modern man, Moorrees and Reed (1954) have shown that 'a lack of association between tooth size and arch size leads to crowding and spacing' (p. 87). If similar conditions obtain in the maxillary dentition, it is pertinent to enquire whether such a lack of association is present in *Zinjanthropus*.

The evidences of dental crowding are localised in *Zinjanthropus* to the junction between the front tooth-row and the cheek tooth-row. The tucking-under of the lower part of the face has already been remarked: such a process would in the first instance be associated with a flattening, or failure of protrusion, of the curve formed by the front tooth sockets, in contrast with the highly-arched curve present in most of the more prognathous (and prodontic) australopithecines. The flattening of the anterior dental curve in *Zinjanthropus* is clearly associated with the relatively orthognathous face, as Robinson has already indicated for *Paranthropus*. Such a morphogenetic process of facial flattening and tucking-under, affecting the anterior teeth, might well have come into conflict with a separate process of dental enlargement affecting the cheek-teeth. The opposing morphogenetic tendencies might be expected to set up competition for the available space in the area in which the two growth potentials overlap, i.e. the canine–premolar zone. This might lead to a lack of association between tooth size and arch size *in this area*, with consequent localised manifestations of dental crowding.

CHAPTER XIII

THE PATTERN OF DENTAL ATTRITION AND OCCLUSION, WITH COMMENTS ON ENAMEL HYPOPLASIA

A. Attrition of individual teeth

Although the M^3's of *Zinjanthropus* are not fully erupted, the other teeth are already markedly worn with one or more areas of exposed dentine on every one. Such early wear is characteristic of the Australopithecinae. The degree of wear may be compared with that in South African australopithecine maxillae, of which two specimens of *Australopithecus* (Sts 52a and Sts 37) and four specimens of *Paranthropus* (the Kromdraai specimen, SK 13, SK 52 and SK 49) have the M^3's not fully erupted, some being a little more advanced and some at a slightly earlier stage of eruption than *Zinjanthropus*: in all these maxillae, the amount of attrition is far less than in *Zinjanthropus*. The degree of attrition in the latter has already levelled the cusps and left numerous areas of dentine exposure (pl. 29). The exposed dentine is black in colour, in contrast with the yellowish-grey enamel. Areas of exposure are thus particularly clearly delimited.

The incisors have a substantial strip of exposed dentine, extending for the full mesiodistal diameter of the teeth, save for the mesial and distal enamel walls (pl. 32). Wear is slightly greater on the right I^1 than on the left, the incisal surface being hollowed on the former, but straight to convex on the latter. The canines show the typically hominid form of wear from the tip (cf. Le Gros Clark, 1950a), which has resulted in a nearly flat occlusal surface with an irregularly biconvex area of dentine exposure, somewhat larger on the right canine than on the left.

The buccal cusp of P^3 has a biconvex island of exposure, very similar in shape to that on the canine (pls. 33 and 34). On the right P^3, there is a second pinpoint islet of exposure distad of the main exposure; on the left, attrition on the buccal cusp is slightly more advanced, so that the distal islet has become confluent with the main buccal cusp exposure. On the lingual cusp the opposite condition obtains, namely, that the right P^3 has a large island of exposure and the left P^3 only a pinpoint exposure. The right P^3 has lost all the enamel off its lingual face by post-mortem damage, and it is at first difficult to determine whether some of the enamel on the occlusal surface of the lingual cusp has not come away with the enamel lost from the lingual surface. From a careful study under magnification of the edges of enamel mesiad and distad of the defect, it seems most unlikely that the large island of dentine exposure on the lingual cusp was the result of post-mortem damage. It seems to be a typical exposure following attrition in life. It follows that the right P^3 is more worn lingually than buccally, whereas the left P^3 is more worn buccally than lingually. In general, the right P^3 agrees with the right anterior teeth in being more worn as a whole than the left P^3; this is confirmed by the fact that the left P^3 clearly shows the central primary fissure as well as traces of both buccal fissures, whereas on the right P^3, only the faintest vestiges of the central fissure and of the distobuccal fissure can with difficulty be detected.

The left P^4 shows a pattern directly opposite to that of the left P^3 (pl. 33): it is more worn on the lingual cusp with a resulting large island of dentine exposure, and much less worn on the buccal cusp with a pinpoint islet of exposure. On the other hand, the right P^4 as a whole is much less worn than the left, retains more of its fissure pattern and has two subequal islets of exposure on the buccal and lingual cusps. The heavily worn lingual cusp of the left P^4 is markedly hollowed, so that there is a decided step from the high plateau of the buccal cusp to the deep valley of the lingual. Such a concavity is not present on the right.

The left M^1 has a similar marked hollowing, with a large exposure of dentine, over the distolingual cusp (the hypocone); around the distobuccal margin of the island, the enamel appears to have become chipped and broken away, probably in life (pls. 33 and 37). The corresponding cusp of the right M^1 has only a pinpoint islet of exposure. On the other hand, over the mesial half of the occlusal surface, the left M^1 is less worn: there is a moderate island of exposure on the mesiobuccal cusp (the paracone), a small islet on the mesiolingual cusp (the protocone), and a third tiny islet between these two. On the right M^1, attrition on the mesial half of the occlusal surface is more advanced and the islands of exposure have become confluent to produce an irregular, dumb-bell-shaped area, extending from protocone to paracone. On both sides, the distobuccal cusp (the metacone) has a pinpoint of dentine exposure. To sum up on the M^1: the right M^1 is as a whole more worn than the left; on the right M^1, wear is maximal on the mesial half, and especially mesiobuccally; on the left M^1, wear is maximal distolingually.

Both M^2's are worn and fractured, enamel having been lost from the mesiobuccal margin, part of the occlusal surface adjacent and a small part of the mesio-occlusal margin, in the left M^2 (pl. 37). The right M^2 lacks a large part of the mesial half of the crown, both enamel and dentine having fractured away, probably by post-mortem damage. In addition, a large area on the mesial half of the buccal face and a small area on the lingual half of the mesial face lack enamel alone. The distal half of the occlusal surface of the right M^2 shows signs of fracturing. It is very likely that at least these fractures occurred in life, since the fractured enamel is in position, with some suggestion of post-fracture healing reaction. Aside from these indications of ante- and post-mortem damage, a large island of dentine is exposed on the protocone, bigger on the right than on the left M^2. There is a pinpoint of exposed dentine on the left paracone; through loss of tissue the position on the right paracone remains indeterminate. The distal cusps have no dentine exposure, despite the signs of fractured enamel.

B. The pattern of attrition and occlusion

The highly irregular pattern of attrition is striking. In advanced wear, Robinson showed that the maxillary teeth of *Australopithecus* wear down most strongly lingually, those of *Paranthropus* most strongly buccally, a difference which he attributed to the lower dental arcade being respectively a little narrower and a little wider than the upper arcade in the two forms. The wear on *Zinjanthropus* teeth is perhaps not sufficiently advanced to permit a ready classification into either of Robinson's categories. But, even as far as it has gone, it suggests either an uneven anatomical relationship between the left and right sides of the upper and lower postcanine tooth rows, or irregularity of function, or both. Thus, there is a general tendency towards greater wear on the right front and cheek teeth than on the left; exceptions to this generalisation are provided by the buccal cusp of P^3, the lingual cusp of P^4 and the hypocone of M^1, on all of which the reverse obtains. Furthermore, from P^3 to M^2 on each side, there is a change in the relative emphasis of attrition on buccal and lingual aspects. On the left, P^3 is more worn buccally, but P^4, M^1 and M^2 are more worn lingually; conversely, on the right, P^3 is more worn lingually, P^4 is equally worn lingually and buccally; M^1 and M^2 are rather more worn buccally. This change of emphasis represents a gradient of maximum wear from left to right passing distad along the premolar–molar tooth-rows. From this it might be inferred that the mandibular tooth-rows were not well aligned against the maxillary, either structurally or functionally. The absence of centric occlusion in normal mastication would seem to follow from this inference, but the clear evidence for mal-alignment applies only to the premolar-molar series. The lack of symmetry between the patterns of wear on the dental arcades of the two sides and the apparent absence of centric occlusion find a parallel in the asymmetry of the temporalis muscles, the evidence of the temporal and sagittal crests pointing to a preponderance of the right muscle over the left (chapter III, C, p. 19).

It may be concluded that the diet of *Zinjanthropus* was of an extremely coarse or gritty nature, for his teeth to be showing such advanced attrition even before the third molars had erupted fully.

Although the M³'s were not yet in position, it is patent that the occlusal plane is not flat, but is curved downwards, the lowest region of the convexity being at about the opposed surfaces of M¹ and M². This curve of Spee is characteristic of the Australopithecinae and the Homininae (Robinson, 1956, p. 18).

C. The state of the enamel

Well-marked areas of hypoplastic enamel have been identified on most of the teeth of *Zinjanthropus*. On the medial incisors, these take the form of a horizontal strip devoid of some of the surface enamel, on the labial surfaces (pl. 32). The strip is 2·2 mm. in cervico-occlusal extent; its lower margin 0·8 mm. from the incisal edge; its upper margin is 6·0 mm. from the summit of the curved cervical line. A convenient vertical fracture, with loss of nearly half the enamel on the labial face of the right I¹, makes it possible to measure the loss of enamel in the area of hypoplasia: above and below the strip the enamel is 1·0 mm. thick; in the strip, it is 0·9 mm. thick. Yet, this minute loss of 0·1 mm. enamel can be both visually and palpably appreciated. The strip is somewhat irregular in outline, and several tiny projections of surface enamel protrude downwards from the upper (cervical) edge of the strip. Within the strip, a series of wave-like imbrications (*perikymata*) are clearly discernible, more obviously so than on the adjacent intact enamel surface. The adjacent border of intact enamel is slightly puffed out, especially along the cervical border of the strip. On the lingual face, a few pits, or small, transverse, linear areas occur at about the same level as the strip on the labial face (pl. 32). These appear to form an incomplete line of hypoplastic enamel, suggesting that some factor interfered with the calcification of the crown on both labial and lingual surfaces during the early developmental period.

The labial face of I² has a crescentic area of hypoplastic enamel, tapering mesially and distally, and having the greater cervico-occlusal extent (2·4 mm.) near the mesial surface. Some 1·3 mm. of apparently normal enamel intervenes between the hypoplastic area and the incisal edge and some 3·6 mm. between it and the cervical line. Perikymata are again more readily detectable in the hypoplastic area, though clearly apparent over the whole labial face. As with the medial incisors, the lateral incisors also show transverse linear areas or pits on the lingual face.

The labial face of the canines is again marked by manifestations of hypoplastic enamel (pls. 32 and 35). A particularly well-impressed feature occurs on the left canine: this takes the form of a deep transverse groove, concave incisally, on the mesial part of the labial face. The groove is limited cervically by a marked swelling which overhangs it; on either side of the groove, this swelling thins out. The incisal verge of the groove is bevelled and, in turn, the bevel is flanked incisally by a smaller swelling. The two swellings become continuous with each other around the distal limit of the groove; mesially, however, the two swellings are separated by a narrow, cleft-like mesial continuation of the groove. The mesial part of the cervical swelling is interrupted further incisally by a small area of chipped enamel, showing perikymata, and comparable with the areas on the labial faces of the incisors. Similar areas or pits are seen, too, on the more distal part of the labial face of the left canine. A somewhat similar deep hypoplastic groove is figured by Colyer (1936, fig. 794, p. 584) on the right I¹ of a captive baboon.

The right canine shows no similar groove, but there is a comparable swelling, giving way incisally to a hollowed area of hypoplastic enamel (pl. 35). In addition to these features near the incisal edge, the labial face of both canines bears two additional rows of pits or hypoplastic areas, one about 2·25 mm. and the other 3·0 mm. from the cervical line. Thus, the canines testify to at least three phases during which the development of these teeth was impaired.

The third premolars (pl. 35) have a distinct strip of hypoplastic enamel about 2·0 mm. in cervico-occlusal extent and extending right around from the buccal face to the distal face and, as discontinuous areas or pits, on to the lingual face. A further series of hypoplastic areas lies mesiobuccally at a lower level (4·2 mm. from the cervical line).

The fourth premolars show only a few traces of hypoplasia: a few pits on the buccal half of the mesial face and on the lingual face.

The first molar shows no signs of hypoplastic enamel at all: as it is generally the first permanent tooth to come into occlusion, it is possible that any hypoplasia which affected it has already been removed by attrition. On the other hand, as it is also the first permanent tooth to begin calcifying and to complete enamel formation, it is possible that these processes might have been largely accomplished by the time of onset of the disturbances responsible for the hypoplasia.

The second molar has curiously irregular and pitted enamel over the buccal and distal faces, contrasting markedly with the smooth enamel surface on the first molar (pl. 39). Similar pitted enamel occurs on the lingual face close to the occlusal margin. On the lingual face of the protocone is an ill-defined roughened area suggesting hypoplasia.

The enamel on the buccal face of M^3 is rugose and irregular, sometimes faintly pitted, with markedly wavy perikymata (pl. 39). The lingual and occlusal faces are highly crenulated: this may or may not be related to the hypoplastic process.

The total picture is one of generalised hypoplasia which affected *Zinjanthropus* intermittently during his developmental period. Such generalised effects have commonly been attributed in the past to temporary impairments of nutrition, such as one might ascribe to the exanthematous fevers. However, the manifestly lengthy duration of these episodes—such that two and more millimetres of enamel were laid down in that time—rather speaks against these fevers with their specific and circumscribed duration. We must think rather of a longer-lasting impairment such as gastro-enteritis, which Mr G. G. Baikie of the Witwatersrand University Dental School has suggested to me.

It is of interest to record that comparable lines or strips of hypoplastic enamel on the labial faces of incisors are common on Robinson's Sterkfontein sample (*Australopithecus*), but not in that from Swartkrans (*Paranthropus*) (1956, p. 28). On the other hand, hypoplastic pits occur on several molars of *Paranthropus*, on one of which there are present, in addition, carious pits (Broom and Robinson, 1952, p. 78). Colyer (1936) has reported similar hypoplastic strips or grooves in *Gorilla, Pan, Hylobates, Papio, Cercocebus* and *Cebus*, but he makes the point that such manifestations in animals from the wild state are relatively rare. The more severe grades of hypoplasia occur in captive animals and in man.

At the time when Colyer published his important work (1936), hypoplastic conditions were usually attributed to malnutrition. Colyer did not accept this view, saying: '...it is my experience, and I think must be a common experience of dental practitioners, that many individuals who have been exposed to conditions which must inevitably cause malnutrition show an unexpected freedom from hypoplasia of the enamel' (p. 596).

Instead, he invoked an explanation based on heredity, suggesting that 'where in certain individuals there is a latent hereditary tendency to vary in the direction of defective enamel formation, this tendency may be quickened into activity by some departure from normal environment such as a change producing malnutrition, and hypoplasia may be due to a large extent, and possibly in the majority of cases, to the interaction of endogenous and exogenous causes'.

While the hereditary factor is now well confirmed, at least in some cases of hypoplasia (e.g. Shear, 1954), the environmental precipitants which have been inculpated have altered from time to time. Following the doubts about malnutrition as a causal agent, the exanthematous fevers were incriminated; but the disparity between the short, well-defined duration of these conditions and the obvious length of time during which, in many instances, enamel formation has been impaired, has led this explanation to fall into desuetude. Longer-acting impairments of vital activities, such as might be occasioned by gastro-intestinal infections, may be considered more likely causes of hypoplasias, especially if such conditions occur at times when the body is advancing from one developmental plateau to another, with consequent metabolic readjustments. Thus, it may be suggested that the genotype, the developmental stage of the organism and adverse environmental circumstances are jointly responsible for such a pattern of hypoplasia as *Zinjanthropus* shows.

Whatever the cause, the position of the areas of hypoplasia may enable one to determine the approximate ages at which the disturbances occurred. Using the standards for modern man worked out by Logan and Kronfeld (1933), Massler and Schour (1946) and Fanning (1961), and estimating from Robinson's data on other australopithecines the approximate total (unworn) height of the *Zinjanthropus* teeth, we are able to say that the belt of hypoplasia nearest the occlusal surface corresponds with what, in a modern human child, would be the 'early childhood ring' ($\pm 2\frac{1}{2}$ years); the two other strips of hypoplasia tally with the ring of enamel formed during the latter part of the period of early childhood (± 4 years), and with the 'later childhood ring' ($\pm 4\frac{1}{2}$–5 years). If similar developmental phases applied to *Zinjanthropus*, it is tempting to suggest that, at three such periods, *Zinjanthropus* may have been subject to systemic upsets, which had the effect of impairing enamel formation for a time. The actual ages were probably a little younger in the australopithecine than those obtaining for modern hominine children.

CHAPTER XIV

THE SIZE OF INDIVIDUAL TEETH, ABSOLUTE AND RELATIVE

The most striking feature of the dentition of *Zinjanthropus* is the enormous size of the premolars and molars. All of the cheek-teeth are significantly greater even than those of the crassident South African australopithecine, *Paranthropus*, and, in some dimensions, a few of the teeth approach and even surpass some of the *Gigantopithecus* teeth of China. In Table 29, the dimensions of each of the sixteen teeth of *Zinjanthropus* are given.

A. Notes on methodology and terminology

In this account, we have used the term *mesiodistal crown diameter* (or simply M.D.) in preference to the term 'length', for some workers (e.g. Olivier, 1960) have designated this dimension 'maximum breadth'. In fact, as Moorrees (1957, pp. 79–80) has pointed out, almost as many workers have called this measure crown breadth as have called it crown length!

Similarly, we have used the term *labiolingual crown diameter* (L.L.) for anterior teeth and *buccolingual crown diameter* (B.L.) for cheek-teeth, instead of 'breadth' or 'maximum thickness'. Where both anterior and cheek-teeth are reflected in a single table (e.g. Table 29), the term buccolingual is preferred.

To measure these two dimensions, a Helios

Table 29. *Crown dimensions (in mm.) and indices of individual teeth of* Zinjanthropus

	Crown diameters		Module	Crown area	Shape index	Shape index
	Mesiodistal	Buccolingual	(M.D.+B.L.)/2	M.D.×B.L.	(B.L./M.D.)%	(M.D./B.L.)%
I¹ (L)	10·0+	8·0	9·0+	80·0+	80·0−	125·0
I¹ (R)	?	7·7	?	?	?	?
I² (L)	?	?	?	?	?	?
I² (R)	6·9	7·5	7·2	51·7	108·7	92·0
C (L)	8·7	9·7	9·2	84·4	111·5	89·7
C (R)	8·8	9·9	9·35	87·1	112·5	88·9
P³ (L)	10·9	17·0	14·0	185·3	156·0	64·1
P³ (R)	10·9	17·0*	14·0†	185·3†	156·0†	64·1†
P⁴ (L)	11·8	18·0	14·9	212·4	152·5	65·6
P⁴ (R)	12·0	17·6	14·8	211·2	146·7	68·2
M¹ (L)	15·2	17·7	16·45	269·0	116·4	85·9
M¹ (R)	15·2	17·7	16·45	269·0	116·4	85·9
M² (L)	17·2	21·0*	19·1†	361·2†	122·1†	81·9†
M² (R)	17·2*	21·0*	19·1†	361·2†	122·1†	81·9†
M³ (L)	15·7	21·4	18·55	336·0	136·3	73·4
M³ (R)	16·3	20·5	18·4	334·1	125·8	79·5

* Through enamel loss, these dimensions could not be precisely determined; they have been estimated.
† These indices have been computed from the *estimated* value of the relevant tooth dimension(s).

Vernier sliding caliper was used, with points especially sharpened to a fine taper, in order to enable them to be inserted between teeth in a jaw. The instrument was calibrated in millimetres and tenths of millimetres, as well as tenths and hundredths of an inch.

The mesiodistal crown diameter measured here was accepted as the maximum mesiodistal dimension with the two points of the instrument making contact with the tooth in a 'horizontal' plane, that is in a plane parallel to the occlusal plane. Moreover, the tips of the two points of the instrument lay in a plane as nearly as possible parallel to the mesiodistal axial plane of the tooth; this approximates to the position prescribed by Moorrees (1957, p. 78), namely, 'parallel to the occlusal and labial surfaces'. Since the labial (or buccal) surface is often highly curved, it is perhaps more precise to define the mesiodistal diameter with respect to the mesiodistal axial plane of the tooth. As pointed out by Moorrees, such a definition does not depend upon the position of the teeth in the dental arch; it does not follow Remane's (1930) 'mean length', which utilised the points of contact with neighbouring teeth as termini for this dimension. A glance at pl. 20 shows that the maximum mesiodistal diameter as defined here would differ appreciably from Remane's 'mean length' in P^4, M^1 and M^2, and probably several other teeth. The method used here is identical with that described and lucidly illustrated by Korenhof (1960, pp. 24–6).

The labiolingual or buccolingual crown diameter used here was the greatest distance between the labial (or buccal) and lingual surfaces of the tooth crown, with the arms of the instrument tangent to the respective surfaces and parallel to the mesiodistal axial plane of the tooth. As Moorrees puts it (1957, p. 80), this is the greatest distance between the two surfaces 'in a plane perpendicular to that in which the mesiodistal diameter was measured'. This definition of Moorrees does not, however, make clear whether the contact points of the two arms of the instrument must lie in the same buccolingual plane; in my method they do not necessarily do so. By placing each arm tangentially along the respective surface, I have measured the perpendicular distance between two planes parallel to that in which M.D. was measured, but the actual points of contact of the left and right arms with the tooth surfaces are not necessarily directly opposite each other, i.e. in the same buccolingual plane perpendicular to the M.D. plane. In fact, only in a markedly skewed tooth would there be any noticeable difference in the L.L. or B.L., as measured by the 'tangent method' and as measured with the contact points in the same plane: in that event, the 'tangent method' would yield a slightly higher reading than the other method. The 'tangent method' employed here is identical with that described and figured by Korenhof (1960, pp. 25–6).

Crown height has been measured only on M^3, the other teeth showing too much occlusal attrition to permit this measurement to be taken. The definition used was that of Moorrees (1957, p. 80), namely, 'the distance between the tip of the mesiobuccal cusp and the deepest point of the cemento-enamel junction on the vestibular side measured along a line parallel to the long axis of the tooth'.

In comparing the sizes of different teeth in the same jaw, as well as in making comparisons between different forms, e.g. *Paranthropus* and *Australopithecus*, Shaw (1931) and Robinson (1956) have relied upon the module

$$\tfrac{1}{2}(\text{'Length'} + \text{'Breadth'}).$$

The value of the module has been doubted by Pedersen (1949), Moorrees (1957) and Leakey (1960b). Different geometric shapes with the same module may, nevertheless, be of widely differing surface area. However, since Robinson has provided modules of every australopithecine tooth measured by him, modules have been computed for *Zinjanthropus* as well, and are included in Table 29. In addition, 'crown area' has been determined, as the product of the mesiodistal and labiolingual (or buccolingual) crown diameters, as used, for instance, by Weidenreich (1937) and Moorrees (1957): although somewhat crude and inaccurate, crown area determination thus provides a rough indication *for comparative purposes only* of the space occupied by the tooth in the dental arch.

Finally, as a rough measure of crown form, the *crown shape index* has been determined as follows:

$$\frac{\text{buccolingual crown diameter}}{\text{mesiodistal crown diameter}} \times 100.$$

Some workers have expressed the shape index as 'length'/'breadth': hence, to facilitate comparison, I have computed a second crown shape index as follows:

$$\frac{\text{mesiodistal crown diameter}}{\text{buccolingual crown diameter}} \times 100.$$

For ease of reference, the two shape indices are referred to in text and tables as B.L./M.D. and M.D./B.L. respectively.

Each tooth was measured on at least two separate occasions by the same observer and with the same instrument. Furthermore, to achieve some standardisation and basis for comparison, a number of original australopithecine teeth measured by Robinson (1956) were re-measured by me and a high level of agreement was recorded. Generally, my measurements were within 0·1–0·3 mm. of Robinson's, showing that our techniques must have been essentially similar.

The most important comparisons were with the hundreds of australopithecine tooth measurements recorded by Robinson (1956), as well as those recorded by Dart (1962c). Since Robinson completed his superb, compendious study of the australopithecine dentition, further teeth have been recovered from the breccia, so the available samples have become still bigger. With the kind permission of the Director of the Transvaal Museum (Dr V. FitzSimons), I have measured all previously unpublished and apparently unmeasured teeth of *Australopithecus* and *Paranthropus*; to these have been added further australopithecine teeth from Makapansgat. The total number of teeth reasonably considered to be australopithecine, from five sites in the Republic of South Africa and three sites in the Republic of Tanzania, are shown in the next column. The figure for Olduvai includes only the sixteen maxillary teeth of the type specimen of *A. (Zinjanthropus) boisei*, although other isolated teeth may need to be added to this tally on further study.

Taung	24
Sterkfontein	162
Kromdraai	39
Swartkrans	311
Makapansgat	55
Garusi	2
Peninj (Lake Natron)	16
Olduvai (*Zinjanthropus*)	16
	625

Not all of these 625 australopithecine teeth are so complete as to permit accurate measurement. Furthermore, some are the rootless crowns of immature teeth, the dimensions of which may not validly be compared with those of fully-formed and erupted teeth. Again, a proportion consists of deciduous teeth. The number of australopithecine mature, permanent teeth, of which metrical characters are now available, is as follows:

I^1	8	I_1	10
I^2	12	I_2	8
$\underline{C}$	19	$\underline{C}$	20
$\overline{P^3}$	34	$\overline{P_3}$	23
P^4	31	P_4	22
M^1	30	M_1	33
M^2	31	M_2	26
M^3	27	M_3	28

Maxillary teeth 192   Mandibular teeth 170

Total australopithecine permanent teeth, 362.*

* The measurements of several additional teeth given by Robinson (1956) have been excluded here on grounds of their incompleteness or immaturity.

By any standards, this is an excellent sample, even though it includes teeth of more than one taxon. Thus, the number of teeth available for either of the South African australopithecine taxa apparently exceeds the number belonging to any dryopithecine taxon (Simons and Pilbeam, 1965).

To compare the measurements of the *Zinjanthropus* teeth with those of the existing taxa of australopithecines, I have divided all the known specimens between the two taxa, *Australopithecus* (= *A. africanus*) and *Paranthropus* (= *A. robustus*). The taxon *A. africanus* includes the specimens from Taung, Sterkfontein, Makapansgat and Garusi; while *Paranthropus* (*A. robustus*) includes specimens from Kromdraai and Swartkrans. For each taxon, sample means ($\bar{x}$), standard deviations (s), standard errors and coefficients of variation (v) have been computed for each dimension and each index of each tooth. Table 30 gives the results of

SIZE OF INDIVIDUAL TEETH

Table 30. *Metrical characters of maxillary permanent teeth of* Australopithecus (A. africanus) *and* Paranthropus (A. robustus)*

Tooth	Metrical character	Australopithecus (A. africanus)					Paranthropus (A. robustus)				
		n	Range	$\bar{x}$	s	V	n	Range	$\bar{x}$	s	V
I^1	Mesiodistal diameter	3	9·3–10·0	9·6±0·24	0·41	4·31	8	8·3–10·8	9·5±0·31	0·88	9·29
	Labiolingual diameter	2	8·2–8·3	8·3±0·06	0·09	1·07	6	7·3–7·8	7·6±0·08	0·20	2·60
	Module	2	8·8–8·85	8·8±0·03	0·04	0·05	6	8·1–9·2	8·5±0·19	0·45	5·35
	Shape index (M.D./L.L.)	2	112·1–115·9	113·9±2·38	3·37	2·95	6	106·4–142·1	124·1±5·75	14·08	11·35
I^2	Mesiodistal diameter	6	5·8–7·3	6·7±0·24	0·59	8·81	8	5·9–9·0	7·2±0·38	1·09	15·20
	Labiolingual diameter	5	5·6–7·0	6·5±0·27	0·61	9·48	7	6·3–7·6	6·9±0·18	0·48	6·98
	Module	5	5·7–7·1	6·6±0·28	0·62	9·35	7	6·2–8·1	7·1±0·21	0·55	7·86
	Shape index (M.D./L.L.)	5	100·0–111·5	104·8±2·21	4·93	4·71	7	90·8–123·3	104·9±4·55	12·03	11·47
C	Mesiodistal diameter	7	8·8–9·9	9·5±0·14	0·37	3·87	13	8·1–9·3	8·5±0·10	0·36	4·21
	Labiolingual diameter	6	8·7–9·9	9·5±0·19	0·47	4·97	13	8·4–11·1	9·5±0·22	0·81	8·50
	Module	6	9·1–9·9	9·5±0·15	0·37	3·90	13	8·3–10·2	9·0±0·16	0·58	6·47
	Shape index (M.D./L.L.)	6	93·6–108·0	99·9±2·32	5·68	5·80	13	83·0–97·7	89·9±0·85	3·06	3·40
P^3	Mesiodistal diameter	17	8·5–9·4	8·9±0·07	0·29	3·31	18	9·0–10·8	9·9±0·12	0·50	5·06
	Buccolingual diameter	17	10·7–13·9	12·4±0·15	0·80	6·42	17	13·1–15·3	14·1±0·19	0·78	5·50
	Module	17	9·7–11·5	10·6±0·11	0·47	4·45	17	11·3–12·9	12·0±0·14	0·58	4·82
	Shape index (M.D./B.L.)	17	65·4–81·3	71·8±1·06	4·39	6·11	17	64·3–74·5	70·2±0·80	3·31	4·71
P^4	Mesiodistal diameter	11	7·8–10·5	9·1±0·26	0·85	9·33	20	9·2–11·8	10·6±0·16	0·71	6·71
	Buccolingual diameter	11	12·0–13·9	12·9±0·18	0·58	4·54	20	13·7–16·5	15·4±0·17	0·75	4·89
	Module	10	10·3–12·2	11·0±0·16	0·51	4·64	20	11·5–13·9	13·0±0·15	0·68	5·21
	Shape index (M.D./B.L.)	10	56·5–79·2	71·1±2·13	6·74	9·48	20	63·0–75·2	68·9±0·74	3·31	4·80
M^1	Mesiodistal diameter	15	11·9–13·2	12·6±0·10	0·38	3·04	16	13·1–14·6	13·8±0·12	0·49	3·57
	Buccolingual diameter	14	12·8–15·1	13·8±0·17	0·63	4·53	16	12·7–16·6	14·5±0·21	0·86	5·93
	Module	14	12·7–14·1	13·2±0·11	0·43	3·25	16	12·9–15·5	14·1±0·15	0·60	4·22
	Shape index (M.D./B.L.)	14	86·5–97·7	90·9±1·06	3·96	4·35	16	86·7–106·1	95·1±1·23	4·93	5·19
M^2	Mesiodistal diameter	16	12·7–15·1	13·8±0·20	0·79	5·76	15	13·6–15·7	14·5±0·19	0·72	4·95
	Buccolingual diameter	18	13·8–17·1	15·4±0·24	1·00	6·48	15	14·2–16·9	15·9±0·22	0·86	5·44
	Module	16	13·5–16·1	14·6±0·20	0·80	5·47	15	13·9–16·3	15·2±0·18	0·69	4·55
	Shape index (M.D./B.L.)	16	83·1–107·1	89·7±1·54	6·14	6·85	15	81·4–98·0	91·7±1·21	4·68	5·11
M^3	Mesiodistal diameter	12	11·6–15·2	13·2±0·32	1·11	8·39	16	12·8–17·0	15·0±0·28	1·12	7·49
	Buccolingual diameter	11	14·6–17·9	15·5±0·31	1·04	6·72	16	15·9–18·2	16·8±0·19	0·77	4·57
	Module	11	13·2–16·5	14·3±0·32	1·06	7·36	16	15·0–17·4	15·9±0·19	0·76	4·81
	Shape index (M.D./B.L.)	11	77·7–93·7	85·3±1·52	5·04	5·91	16	70·3–95·5	89·0±1·62	6·47	7·27

* Where samples are 14 or over, the standard deviations have been computed by the method of deviations; where the samples are below 14 in number, the standard deviations have been estimated from the sample range.

this analysis for maxillary teeth. The analysis of the mandibular teeth is not presented here, since we have no mandibular teeth of the type specimen of *Zinjanthropus*.

Similarly, I have compiled comparative tables of absolute measurements and of indices for some 66 maxillary and 136 mandibular teeth of *Homo erectus* (including *Pithecanthropus, Sinanthropus, Atlanthropus, Telanthropus*, Mauer and Montmaurin). Table 31 gives the results of this analysis for maxillary teeth of *H. erectus*.

In computing the standard deviations, I have followed this procedure: where the sample for any tooth numbers fourteen or over, I have employed the conventional formula based upon deviations from the mean. For smaller samples, however, I have obtained an estimate of the standard deviation from the range: the range is multiplied by a_n, a factor which varies with the sample size. The standard table of a_n values given by Lindley and Miller (1953, p. 7, Table 6) was employed (*see also* Simpson *et al*. 1960, p. 41, Table 1A). The standard deviation estimated from the range is generally, though not always, higher than the S.D. obtained by the method of deviations. Table 32 compares the S.D.'s obtained by the two methods for all metrical characters for which the sample size was thirteen or fewer. Irrespective of which was the larger, however, for all samples of thirteen or fewer, only the S.D. estimated from the range was employed.

B. Dimensions of the incisors

(Tables 29, 33 and 34)

In absolute M.D. crown diameter, the I^1 of *Zinjanthropus* falls near the top of the range for eight I^1's of *Paranthropus* (8·3–10·8 mm.), being near or possibly equal in M.D. to the Swartkrans

Table 31. *Metrical characters of maxillary permanent teeth of* Homo erectus*

Tooth	Metrical character	n	Range	$\bar{x}$	s†	V
I¹	Mesiodistal diameter	4	9·8–10·8	10·3 ± 0·24	0·49	4·72
	Labiolingual diameter	5	7·5–8·1	7·7 ± 0·11	0·26	3·34
	Module	4	8·7–9·4	9·0 ± 0·16	0·32	3·49
	Shape index (M.D./L.L.)	4	124·1–144·0	132·6 ± 4·85	9·69	7·31
I²	Mesiodistal diameter	4	8·0–10·0	8·6 ± 0·49	0·97	11·26
	Labiolingual diameter	5	8·0–10·4	8·6 ± 0·46	1·03	11·94
	Module	4	8·1–10·2	8·7 ± 0·50	1·00	11·43
	Shape index (M.D./L.L.)	4	94·1–101·2	98·2 ± 1·73	3·45	3·52
C	Mesiodistal diameter	10	8·5–10·5	9·38 ± 0·21	0·65	6·93
	Labiolingual diameter	10	9·8–11·9	10·5 ± 0·22	0·68	6·53
	Module	10	9·1–10·7	9·9 ± 0·16	0·50	5·08
	Shape index (M.D./L.L.)	10	79·8–101·1	90·0 ± 2·17	6·87	7·62
P³	Mesiodistal diameter	7	7·4–9·2	8·4 ± 0·25	0·67	7·96
	Buccolingual diameter	7	10·5–12·8	12·0 ± 0·32	0·85	7·06
	Module	7	8·9–11·0	10·2 ± 0·29	0·76	7·43
	Shape index (M.D./B.L.)	7	66·1–71·9	69·4 ± 0·80	2·12	3·06
P⁴	Mesiodistal diameter	12	7·2–8·9	8·0 ± 0·15	0·52	6·51
	Buccolingual diameter	12	10·3–12·5	11·5 ± 0·19	0·67	5·83
	Module	12	8·7–10·7	9·7 ± 0·17	0·60	6·14
	Shape index (M.D./B.L.)	12	65·8–75·2	69·8 ± 0·84	2·90	4·15
M¹	Mesiodistal diameter	11	10·0–13·1	11·6 ± 0·29	0·98	8·42
	Buccolingual diameter	10	11·7–14·4	13·1 ± 0·28	0·88	6·72
	Module	10	10·9–13·3	12·3 ± 0·25	0·78	6·37
	Shape index (M.D./B.L.)	10	81·0–96·6	87·7 ± 1·60	5·05	5·76
M²	Mesiodistal diameter	9	10·2–13·6	11·3 ± 0·38	1·15	10·16
	Buccolingual diameter	9	12·2–15·2	13·0 ± 0·34	1·01	7·75
	Module	9	11·4–14·4	12·1 ± 0·34	1·01	8·31
	Shape index (M.D./B.L.)	9	79·1–100·0	86·5 ± 2·35	7·04	8·13
M³	Mesiodistal diameter	10	8·7–10·8	9·7 ± 0·22	0·68	7·00
	Buccolingual diameter	10	10·4–14·0	11·8 ± 0·37	1·17	9·90
	Module	10	9·5–12·4	10·8 ± 0·29	0·93	8·59
	Shape index (M.D./B.L.)	10	77·1–89·5	82·6 ± 1·28	4·03	4·88

* The pooled sample includes Asian fossils (*Pithecanthropus* and *Sinanthropus*) and African fossils (*Atlanthropus*, including Ternifine and Rabat).

† As the sample in every instance numbers below 14, all the standard deviations in this table have been estimated from the sample ranges.

(*Paranthropus*) tooth with the greatest M.D. diameter (SK 68, 10·8 mm.). It exceeds the *Paranthropus* mean (9·45 mm.) by 0·63 S.D.'s (Table 33), and hence is not significantly greater in M.D. diameter. Likewise, the M.D. of *Zinjanthropus* I¹ is slightly greater than that of three maxillary incisors from Sterkfontein (*Australopithecus*) with M.D. 9·3, 9·5 and 10·0 mm. (Fig. 23; Table 30).

The lateral incisor is small in M.D. diameter, as may be determined from the intact right I². Its M.D. of 6·9 mm. falls within the range for *Paranthropus* (5·9–9·0 mm.), only 0·24 S.D.'s below the mean for eight specimens, namely 7·2 mm. (Table 33); its value also falls within the range for six specimens of I² belonging to *Australopithecus* (5·8–7·3 mm.), a little above the mean value of 6·7 mm.[1] (Table 30).

On the other hand, the L.L. diameters of the

[1] The mean value of 6·6 mm. given by Robinson (1956) in the table on his p. 29 is based on three specimens of I² from Sterkfontein; my mean figure of 6·7 is based on a sample comprising these three teeth, together with an additional I² from Sterkfontein and two I²'s from Makapansgat (*Australopithecus*).

SIZE OF INDIVIDUAL TEETH

Table 32. *Standard deviations of hominid teeth as computed by two different methods* (S.D. 1—*estimate by the method of deviations*; S.D. 2—*estimate from the sample range*)*

Tooth	Metrical character	*Australopithecus* (= *A. africanus*)		*Paranthropus* (= *A. robustus*)		*Homo erectus*	
		S.D. 1	S.D. 2	S.D. 1	S.D. 2	S.D. 1	S.D. 2
I^1	Mesiodistal diameter	0·36	0·41	0·70	0·88	0·52	0·49
	Labiolingual diameter	0	0·09	0·17	0·20	0·24	0·26
	Module	0	0·04	0·36	0·45	0·29	0·32
	Shape index (M.D./L.L.)	2·69	3·37	11·70	14·08	8·34	9·69
I^2	Mesiodistal diameter	0·54	0·59	1·03	1·09	0·93	0·97
	Labiolingual diameter	0·61	0·60	0·45	0·48	1·00	1·03
	Module	0·58	0·62	0·73	0·55	0·86	1·00
	Shape index (M.D./L.L.)	4·17	4·93	11·50	12·03	3·61	3·45
C	Mesiodistal diameter	0·41	0·37	0·37	0·36	0·55	0·65
	Labiolingual diameter	0·45	0·47	0·81	0·81	0·76	0·68
	Module	0·39	0·37	0·57	0·58	0·54	0·50
	Shape index (M.D./L.L.)	4·91	5·68	4·92	3·06	6·45	6·87
P^3	Mesiodistal diameter	0·29	—	0·50	—	0·57	0·67
	Buccolingual diameter	0·80	—	0·78	—	0·79	0·85
	Module	0·47	—	0·58	—	0·66	0·76
	Shape index (M.D./B.L.)	4·39	—	3·31	—	1·87	2·12
P^4	Mesiodistal diameter	0·75	0·85	0·71	—	0·60	0·52
	Buccolingual diameter	0·58	0·60	0·75	—	0·67	0·67
	Module	0·51	0·60	0·68	—	0·60	0·60
	Shape index (M.D./B.L.)	6·74	7·37	3·31	—	3·01	2·90
M^1	Mesiodistal diameter	0·38	—	0·49	—	1·03	0·98
	Buccolingual diameter	0·63	—	0·86	—	0·95	0·88
	Module	0·43	—	0·60	—	0·89	0·78
	Shape index (M.D./B.L.)	3·96	—	4·93	—	4·98	5·05
M^2	Mesiodistal diameter	0·79	—	0·72	—	1·09	1·15
	Buccolingual diameter	1·00	—	0·86	—	0·91	1·01
	Module	0·80	—	0·69	—	0·90	1·01
	Shape index (M.D./B.L.)	6·14	—	4·68	—	6·79	7·04
M^3	Mesiodistal diameter	1·16	1·11	1·12	—	0·62	0·68
	Buccolingual diameter	1·03	1·04	0·77	—	1·03	1·17
	Module	1·06	1·06	0·76	—	0·84	0·93
	Shape index (M.D./B.L.)	4·80	5·04	6·47	—	3·55	4·03

* S.D. 2 is quoted only in respect of samples of fewer than fourteen teeth.

Zinjanthropus incisors are nearly equal to or even bigger than the greatest values yet published for australopithecine maxillary incisors (Fig. 24). The values (mm.) for I^1 (8·0, 7·7) lie at the top of and beyond the range for six examples of *Paranthropus* (7·3–7·8) and exceed the mean of 7·6 by 2·10 S.D.'s (left) and 0·60 S.D.'s (right).

The *Zinjanthropus* values fall slightly short of those of two I^1's of *Australopithecus* (8·2, 8·3 mm.). The L.L. value for the undamaged right I^2 of *Zinjanthropus* (7·5 mm.) is again close to the top of the range for seven *Paranthropus* teeth (6·3–7·6 mm.) and 1·27 S.D.'s above the mean of 6·9 mm. (cited by Robinson, 1956, p. 29, as 6·8 mm.); while it exceeds the top of the range (5·6–7·0 mm.) and the mean (6·5 mm.) for five I^2's of *Australopithecus* from Sterkfontein and Makapansgat (I^2 of *Australopithecus* has a *smaller* mean L.L. than that of *Paranthropus*, though I^1 of *Australopithecus* has a *greater* mean L.L. than that of *Paranthropus*; Table 30).

Table 33. *Metrical characters of* Zinjanthropus *teeth as compared with those of* Paranthropus (=A. robustus)

Tooth	Metrical character	Value in Zinjanthropus (A)	Paranthropus mean (B)	Deviation of Zinjanthropus value from Paranthropus mean (A–B)	Paranthropus s* (C)	Standardised deviation of Zinjanthropus value $\left(\frac{A-B}{C}\right)$
I¹ (L)	Mesiodistal diameter	10.0+	9.45	+0.55	0.88	+0.63
	Labiolingual diameter	8.0	7.58	+0.42	0.20	+2.10
	Module	9.0+	8.49	+0.51	0.45	+1.13
	Shape index (M.D./L.L.)	125.0	124.06	+0.94	14.08	+0.07
I¹ (R)	Labiolingual diameter	7.7	7.58	+0.12	0.20	+0.60
I² (R)	Mesiodistal diameter	6.9	7.16	−0.26	1.09	−0.24
	Labiolingual diameter	7.5	6.89	+0.61	0.48	+1.27
	Module	7.2	7.06	+0.14	0.55	+0.25
	Shape index (M.D./L.L.)	92.0	104.92	−12.92	12.03	−1.07
C (L)	Mesiodistal diameter	8.7	8.55	+0.15	0.36	+0.42
	Labiolingual diameter	9.7	9.54	+0.16	0.81	+0.20
	Module	9.2	9.04	+0.16	0.58	+0.28
	Shape index (M.D./L.L.)	89.7	89.93	−0.23	3.06	−0.07
C (R)	Mesiodistal diameter	8.8	8.55	+0.25	0.36	+0.69
	Labiolingual diameter	9.9	9.54	+0.36	0.81	+0.44
	Module	9.35	9.04	+0.31	0.58	+0.53
	Shape index (M.D./L.L.)	88.9	89.93	−1.03	3.06	−0.34
P³ (L and R)	Mesiodistal diameter	10.9	9.94	+0.96	0.50	+1.92
	Buccolingual diameter	17.0	14.15	+2.85	0.78	+3.65
	Module	14.0	12.04	+1.96	0.58	+3.38
	Shape index (M.D./B.L.)	64.1	70.16	−6.06	3.31	−1.83
P⁴ (L)	Mesiodistal diameter	11.8	10.58	+1.22	0.71	+1.72
	Buccolingual diameter	18.0	15.37	+2.63	0.75	+3.51
	Module	14.9	12.97	+1.93	0.68	+2.84
	Shape index (M.D./B.L.)	65.6	68.85	−3.25	3.31	−0.98
P⁴ (R)	Mesiodistal diameter	12.0	10.58	+1.42	0.71	+2.00
	Buccolingual diameter	17.6	15.37	+2.23	0.75	+2.97
	Module	14.8	12.97	+1.83	0.68	+2.69
	Shape index (M.D./B.L.)	68.2	68.85	−0.65	3.31	−0.20
M¹ (L and R)	Mesiodistal diameter	15.2	13.77	+1.43	0.49	+2.92
	Buccolingual diameter	17.7	14.51	+3.19	0.86	+3.71
	Module	16.5	14.14	+2.36	0.60	+3.93
	Shape index (M.D./B.L.)	85.9	95.13	−9.23	4.93	−1.87
M² (L and R)	Mesiodistal diameter	17.2	14.53	+2.67	0.72	+3.71
	Buccolingual diameter	21.0	15.87	+5.13	0.86	+5.97
	Module	19.1	15.20	+3.90	0.69	+5.65
	Shape index (M.D./B.L.)	81.9	91.65	−9.75	4.68	−2.08
M³ (L)	Mesiodistal diameter	15.7	14.96	+0.74	1.12	+0.66
	Buccolingual diameter	21.4	16.83	+4.57	0.77	+5.93
	Module	18.5	15.89	+2.61	0.76	+3.43
	Shape index (M.D./B.L.)	73.4	89.02	−15.62	6.47	−2.41
M³ (R)	Mesiodistal diameter	16.3	14.96	+1.34	1.12	+1.20
	Buccolingual diameter	20.5	16.83	+3.67	0.77	+4.77
	Module	18.4	15.89	+2.51	0.76	+3.30
	Shape index (M.D./B.L.)	79.5	89.02	−9.52	6.47	−1.47

* For the incisors and canines, where the samples number 13 or fewer, the S.D. has been estimated from the sample range in each instance; for the premolars and molars, where the samples are 14 or over, the S.D. has been computed by the method of deviations.

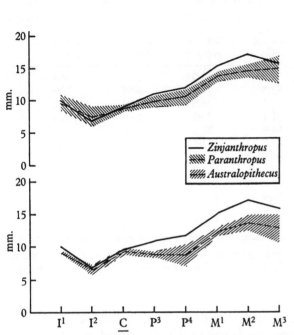

Fig. 23. Mesiodistal diameters of maxillary teeth of *Zinjanthropus* (solid line); *Paranthropus* (upper shaded area, showing mean and sample range); and *Australopithecus* (lower shaded area, showing mean and sample range). The figures for South African australopithecines are based on Robinson (1956).

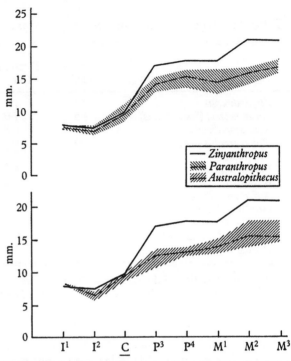

Fig. 24. Buccolingual diameters of maxillary teeth of *Zinjanthropus* (solid line); *Paranthropus* (upper shaded area, showing mean and sample range); and *Australopithecus* (lower shaded area, showing mean and sample range). The values for South African australopithecines are based on Robinson (1956).

If we follow Robinson (1956, pp. 29–30) in comparing the modules of I^1 and I^2, we find that in *Zinjanthropus* the module of I^2 is 80 per cent of the module of I^1 (this value would be somewhat lower if we had the true M.D. of either of the first incisors). According to Robinson, the ratio of the 'modules' of the means for *Paranthropus* is 82 per cent, for *Australopithecus* 72 per cent,[1] for *H. e. pekinensis* 90 per cent, and for Shaw's (1931) Bantu series, Drennan's (1929) Bushmen and Campbell's (1925) aborigines, approximately 85 per cent in each instance (Table 34). From these figures, it looks, first, as though australopithecine values are in general a little lower than those of hominines; secondly, that in the degree of reduction of I^2 in comparison with I^1, *Zinjanthropus* lies between *Paranthropus* and *Australopithecus*, though perhaps slightly nearer to *Paranthropus*. However,

[1] Two additional I^2's from Makapansgat have raised the *Australopithecus* sample from three to five I^2's, and the sample mean module from 6·4 to 6·6. In consequence, the I^2/I^1 ratio for *A. africanus* is raised from 72 to 75·1 per cent (Table 34).

the samples are still very small, and caution should be exercised both in recognising trends and in drawing conclusions from them.

For comparison, estimates of the I^2/I^1 ratios of gorillas have been derived from the valuable data published by Ashton and Zuckerman (1950), by computing the module of the means from the mean dimensions published by them, and by comparing the modules thus obtained. (Their data for chimpanzee and orang-utan cannot be used for this purpose, because, while M.D. dimensions are given for young and old adults separately, the L.L. diameter is given only for a combined sample of young and old adults.) In their method of measuring, Ashton and Zuckerman give two M.D. diameters, 'the maximum transverse dimension of each face' (labial and lingual). Since neither of these M.D. diameters corresponds precisely with the generally-accepted M.D. diameter, such as that employed in the present study, I could obtain only an approximation by using whichever of their M.D.

Table 34. *Modules and crown areas of incisors and I^2/I^1 ratios in* Zinjanthropus *and other hominoids**

	I^1 module (mm.)	I^2 module (mm.)	I^2/I^1 module ratio	I^1 crown area (mm.^{2})	I^2 crown area (mm.^{2})	I^2/I^1 crown area ratio
Zinjanthropus	9·0+	7·2	80·0−	80·0+	51·7	64·6−
Paranthropus (Swartkrans)	8·5	7·0	82·3	71·4	49·0	68·6
Australopithecus	8·8	6·6	75·1	78·0	44·2	56·7
Homo erectus pekinensis	9·0	8·2	91·1	79·3	66·4	83·7
Australian	8·6	7·3	84·9	74·3	53·1	71·5
Bantu Negroid	8·0	6·8	85·0	63·2	46·1	72·9
Bushman	7·4	6·4	86·5	53·9	40·2	74·6
American White	8·0	6·2	77·5	63·0	38·4	61·0
Gorilla ♂ (young adult)	12·8	10·7	83·6	161·3	113·8	70·6
Gorilla ♂ (old adult)	11·5	9·8	84·9	131·5	95·3	72·4
Gorilla ♀ (young adult)	11·6	9·0	76·9	134·1	80·3	59·9
Gorilla ♀ (old adult)	10·1	8·3	81·7	101·2	68·2	67·4

* With the exception of the values for *Zinjanthropus*, all of the values in this table are sample means or ratios of means. For the derivation of the gorilla values from the data of Ashton and Zuckerman (1950), see the discussion in the accompanying text.

diameters was the greater in computing the modules. The resulting I^2/I^1 index is 83·6 per cent for young adult male gorillas ($n=11$), 84·9 per cent for old adult males ($n = 10-14$), 76·9 per cent for young adult females ($n = 10-13$) and 81·7 per cent for old adult females ($n = 7-11$) (Table 34).

It would seem that the ratio of I^2/I^1 modules provides little information of taxonomic value in the Hominoidea in which it has thus far been computed. Thus the range of *means* for Australopithecinae is 75–82 per cent, for Homininae 78–91 and for Pongidae 77–85. When crown area is used instead of the module (Table 34), the sorting of hominoids is no better: the range of means of the I^2/I^1 crown area ratio for Australopithecinae is 57–69, for Homininae 61–84 and for Pongidae 60–72. Thus, in the relative sizes of the front teeth, there is considerable overlap among different groups of hominoids. It is interesting to note that *H. e. pekinensis* has relatively larger lateral incisors than any other hominoid group: this provides extreme values of both mean Module Ratio (91·1 per cent) and mean Crown Area Ratio (83·7 per cent). However, the sample comprises only two lateral incisors and four medial incisors. The trend towards enlargement of the lateral incisors is apparent as well in two other specimens assigned to *H. erectus*: the Rabat I^2 has a module of 8·25 mm. and the Sangiran IV I^2 a module of no less than 10·2 mm. Unfortunately, in neither specimen is the I^1 available, so it is not possible to compute the I^2/I^1 ratios. However, if the modules of these two I^2's are added to those of the two Choukoutien I^2's, the combined *H. erectus* sample of four lateral incisors has a mean module of 8·7 mm. Unless Sangiran IV has an enormously expanded I^1, this would suggest that the overall mean I^2/I^1 ratio for *H. erectus* may prove to be even higher than 91·1 per cent for the mean Module Ratio. On the other hand, the *Australopithecus* sample ($n = 5$ I^2's and 2 I^1's) comprises some of the smallest lateral incisors, with a mean module nearly as small as that of Bushmen (6·4 mm.) and that of American whites (6·2 mm.). *Australopithecus* provides the lowest values in the Hominoidea for both mean Module Ratio (75·1 per cent) and mean Crown Area Ratio (56·7 per cent).

C. Dimensions of the canines
(Tables 29, 33, 35–7)

The size of the canines is of interest, both absolutely and in relation to the size of the more distal teeth. As in other australopithecines and in hominines, the canine of *Zinjanthropus* is a small tooth,

SIZE OF INDIVIDUAL TEETH

contrasting sharply in its dimensions with those of pongids. In Table 35, its mesiodistal and labiolingual dimensions, module and crown area, are compared with those of other hominoids.

The M.D. diameters of the *Zinjanthropus* canines (8·7, 8·8 mm.) fall well within the *Paranthropus* (Swartkrans) range of 8·1–9·3 mm. (excluding SK 92 and the aberrant SK 27 with a M.D. diameter of 10·6 mm.) and are close to the mean (8·5 mm.) of thirteen Swartkrans canines. They are small in comparison with the M.D. diameter of *Australopithecus* canines, the range for seven specimens of which is 8·8—9·9 mm. and the mean 9·5 mm. (this sample includes six canines from Sterkfontein and one from Makapansgat) (Table 30). Thus, the M.D. diameters of the *Zinjanthropus* canines fall at the bottom of the range for *Australopithecus*, but at the mid-value of the range for *Paranthropus* (Fig. 23).

On the other hand, the L.L. diameters of the Olduvai specimens (9·7, 9·9 mm.) fall at the *top* of the range (8·7–9·9 mm.) for six canines of *Australopithecus* and above the mean (9·5 mm.) for this Sterkfontein–Makapansgat group. Falling at the bottom of the M.D. range of *Australopithecus*, it is clear that the *Zinjanthropus* canine has a different shape from that of *Australopithecus* (see chapter XV). The L.L. diameters of the *Zinjanthropus* canines fall well within the range for *Paranthropus* (8·4–11·1 mm. for thirteen examples excluding the aberrant and unerupted SK 27 and the unerupted SK 92) and just above the mean of 9·5 mm. (Fig. 24).

The modules of the *Zinjanthropus* canines (9·2, 9·35 mm.) fall within the range for both (Fig. 25), are a little bigger than the *Paranthropus* mean (9·0 mm.) and a little smaller than the *Australopithecus* mean (9·5 mm.). The same is true for the crown area (Fig. 26).

In contrast, *H. erectus* has an extraordinarily large canine, its mean module and mean crown area exceeding the means of all other hominid populations in Table 35. Indeed, its mean M.D. diameter (9·4 mm.) surpasses the M.D. diameters of all thirteen Swartkrans maxillary canines but is exceeded by those of four *Australopithecus* upper canines out of seven. Again, the mean L.L. diameter of the *H. erectus* canines (10·5 mm.) is surpassed by only one out of thirteen *Paranthropus* teeth and not by any *Australopithecus* teeth. As a result, the mean module (9·9 mm.) of *H. erectus* canines is exceeded by the module of only one out of thirteen *Paranthropus* canines and none out of six *Australopithecus* teeth, while its mean crown area of 98·2 mm.^{2} is exceeded by that of one out of thirteen *Paranthropus* canines and none out of six *Australopithecus* teeth.

The entire range of means for a selection of modern hominine populations (Table 35) is appreciably lower than the canine dimensions of the Australopithecinae. Thus, the mean M.D. diameters (mm.) range from 7·5 to 8·4 (from Bushmen to Australian aboriginals) in comparison with an australopithecine range of means of 8·5–9·5; and the mean L.L. diameters from 7·8 to 9·0, while the australopithecine means are 9·5. Likewise, the mean modules of modern hominine upper canines range from 7·7 to 8·7 mm., those of australopithecine series from 9·0 to 9·5, whilst the mean crown areas of modern hominines range from 58·5 to 75·6 mm.^{2}, those of australopithecine series from 81·7 to 90·5. If, however, one considers *individual* australopithecine teeth, there are a few at the lower extreme of the range of variation which fall among the modern hominine mean values. Of thirteen *Paranthropus* maxillary canines, one isolated tooth (SK 95) has a crown area of 68·0 mm.^{2}; and three others fall in the lower seventies,[1] i.e. they are smaller than the *mean* crown area among Australians; however, none of the seven *Australopithecus* maxillary canines is smaller than the Australian mean.

The mean dimensions in the pongids exceed all others cited in Table 35, with the exception that a sample of sixteen female chimpanzees yielded a mean L.L. diameter of 9·0 mm., smaller than the australopithecine mean of 9·5 mm.

Zinjanthropus thus resembles *Paranthropus* and *Australopithecus* in having relatively small canines.

---

[1] For present purposes, we have accepted Robinson's (1956) attribution of these and other isolated Swartkrans teeth to *Paranthropus*; it should, however, be borne in mind that a second hominid (*Telanthropus*, subsequently regarded as *Homo erectus*) is present in the same deposit. The identification of some of these isolated teeth may therefore have to be re-considered and it may even prove impossible to assign every tooth with confidence to one or other taxon.

Table 35. *Crown dimensions, modules and crown areas of canines of* Zinjanthropus *and other hominoids**

	Crown diameters		Module (mm.)	Crown area (mm.²)
	Mesiodistal (mm.)	Labiolingual (mm.)		
Zinjanthropus (L)	8·7	9·7	9·2	84·4
(R)	8·8	9·9	9·35	87·1
Paranthropus (Swartkrans) (n=13)	8·5	9·5	9·0	81·7
Australopithecus (Sterkfontein and Makapansgat) (n=6)	9·5	9·5	9·5	90·5
Homo erectus				
Sangiran IV (n = 2)	9·5	11·8	10·7	112·1
Choukoutien (n = 6)	9·5	10·2	9·9	96·9
Rabat (n = 1)	9·5	10·0	9·75	95·0
Ternifine (n = 1)	8·7	10·0	9·35	87·0
Total H. erectus (n = 10)	9·4	10·5	9·9	98·2
Australian	8·4	9·0	8·7	75·6
East Greenland Eskimo	7·8	8·4	8·1	65·5
Aleut	7·9	8·3	8·1	65·6
Japanese	7·7	8·2	7·9	63·1
Bantu Negroid	7·6	8·2	7·9	62·3
Bushman	7·5	7·8	7·7	58·5
American White	7·6	8·0	7·8	60·8
Gorilla ♂	20·7	16·4	18·5	339·5
Gorilla ♀	14·5	11·3	12·9	163·9
Chimpanzee ♂	13·6	10·9	12·3	148·2
Chimpanzee ♀	11·3	9·0	10·1	101·7
Orang-utan ♂	16·0	13·8	14·9	220·8
Orang-utan ♀	12·5	9·5	11·0	118·7

* All figures in the table, save those for *Zinjanthropus*, Rabat, Ternifine and *H. e. erectus* IV, are sample means or ratios of means; all populations cited (except the pongids) are represented by combined male and female samples. The figures for *H. e. erectus* IV are means of the left and right teeth (after Weidenreich, 1945).

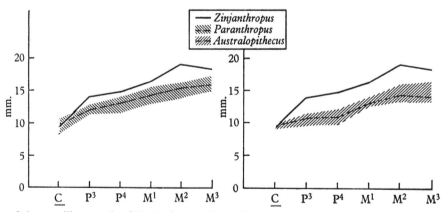

Fig. 25. Modules of the maxillary teeth of *Zinjanthropus*, *Paranthropus* (left) and *Australopithecus* (right). The thick solid line represents the values in *Zinjanthropus*. For each of the other two australopithecines, the interrupted line represents the sample mean and the shaded area the sample range from absolute minimum to absolute maximum values, based on the measurements recorded by Robinson (1956).

SIZE OF INDIVIDUAL TEETH

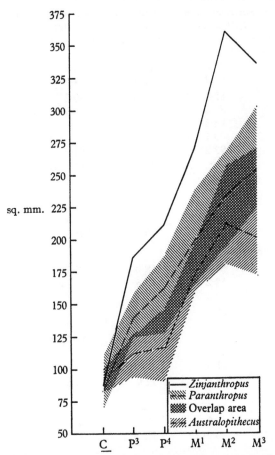

Fig. 26. Crown areas of maxillary teeth of *Zinjanthropus* compared with the ranges in *Paranthropus* and *Australopithecus*. The thick solid line represents the values in *Zinjanthropus*. For each of the other two forms, the interrupted line represents the mean and the shaded area the sample range from absolute minimum to absolute maximum values. The crown areas have been computed as the product of mesiodistal and labiolingual diameters recorded by Robinson (1956).

D. Canine–premolar ratios (Tables 36 and 37)

The small size of the zinjanthropine canine is most strikingly borne out by comparing the size of the canine with that of the teeth distal to it. This point is of taxonomic importance and, among his original list of 20 points diagnostic of the proposed new genus, Leakey (1959a) included 'the relatively greater reduction of the canines in comparison with the molar–premolar series than is seen even in *Paranthropus*, where it is a marked character'. Robinson (1960) contested this claim and adduced as evidence the ratios of the canine and P^3 modules, 'as the marked change of proportion between anterior and cheek teeth occurs in *Paranthropus* between canine and P^3'. The ratio between the modules (per cent) in *Zinjanthropus* was stated by Robinson to be 64·9, while that for three specimens of *Paranthropus* was said to range from 61·8 to 78·8. In his rebuttal, aside from discounting the use of the module, Leakey pointed out that 'the ratio between the canine and the premolar alone cannot have any bearing upon the relation of the canine size to the total molar–premolar series, unless the premolar bears a constant relation to the total postcanine series'.

In the ensuing discussion, the canine–premolar relationship will be considered, as well as the perfectly valid suggestion that a premolar may occupy varying proportions of the cheek tooth row. To Leakey's points, one might add that a comparison of the *modules* alone of the two teeth, $\underline{C}$ and P^3, may mask a trend for the postcanine teeth to be enlarged much more in one dimension, e.g. buccolingual diameter, than in the other. We shall therefore compare $\underline{C}$ and P^3 and also $\underline{C}$ and P^4 in M.D. and B.L. crown diameters, as well as in module and crown area. In Table 36, these values for *Zinjanthropus* are compared with mean and individual values for *Paranthropus* and *Australopithecus* samples. It should be emphasised that the 'mean' values quoted for *Australopithecus* and *Paranthropus* in the table are the ratios of the means for $\underline{C}$ and P^3 (or P^4) rather than the mean of individual ratios, since many teeth in each sample are isolated finds. However, there are five *Paranthropus* maxillae and four *Australopithecus* maxillae, in each of which both $\underline{C}$ and P^3 are present, as well as, in a few instances, P^4: the dental measurements on such specimens have been listed separately in Table 36, so that the actual $\underline{C}/P^3$ and $\underline{C}/P^4$ ratios in several individuals are available. Only a few measurements have been cited from the maxilla SK 83, since both $\underline{C}$ and P^3 are so worn that the exposed area of dentine covers the entire occlusal surface, with apparent pulp exposures on each.

Both the $\underline{C}/P^3$ and $\underline{C}/P^4$ ratios of the *Zinjanthropus* maxilla are lower than in any other australopithecine yet described. This is true whether we compare the canine with each premolar by the

Table 36. *Comparison of maxillary canine and premolar dimensions in Australopithecinae**

Specimen	Canine	P³	C̲/P³ × 100 %	P⁴	C̲/P⁴ × 100 %
A. *Mesiodistal crown diameter (mm.)*					
Zinjanthropus (L)	8·7	10·9	79·8	11·8	73·7
(R)	8·8	10·9	80·7	12·0	73·3
Paranthropus (mean)	8·5	9·9	85·9	10·6	80·2
SK 55 (L)	8·3	9·5	87·4	—	—
SK 55 (R)	8·2	9·5	86·3	—	—
SK 48	8·2	9·2	89·1	—	—
SK 65 (L)	8·3	9·0	92·2	10·0	83·0
SK 83	8·9	—	—	11·6(?)	76·7(?)
Australopithecus (mean)	9·5	8·9	106·7	9·1	104·4
T.M. 1512	8·8	8·9	98·9	—	—
Sts 52a (L)	9·8	8·7	112·6	9·1	107·7
Sts 52a (R)	9·9	8·6	115·1	9·3	106·5
MLD 11/23	9·9	9·1	108·8	9·4	105·3
B. *Buccolingual crown diameter (mm.)*					
Zinjanthropus (L)	9·7	17·0	57·1	18·0	53·9
(R)	9·9	17·0	58·2	17·6	56·25
Paranthropus (mean)	9·5	14·1	67·4	15·4	61·7
SK 55 (L)	8·6	13·1	65·7	—	—
SK 55 (R)	8·9	13·5	65·9	—	—
SK 83	10·3	13·9	74·1	16·7(?)	61·7(?)
SK 48	8·8	13·7	64·2	—	—
SK 65 (L)	10·0	14·0	71·4	15·4	64·9
Australopithecus (mean)	9·5	12·4	76·6	12·9	73·6
T.M. 1512	9·4	11·9	79·0	—	—
Sts 52a (L)	9·9	12·8	77·3	—	—
Sts 52a (R)	9·7	12·8	75·8	13·3	72·9
MLD 11/23	9·9	12·3(?)	80·5(?)	12·6(?)	78·6(?)
C. *Module (mm.)*					
Zinjanthropus (L)	9·2	14·0	65·7	14·9	61·7
(R)	9·35	14·0	66·8	14·8	63·2
Paranthropus (mean)	9·0	12·0	75·0	13·0	69·2
SK 55 (L)	8·45	11·3	74·8	—	—
SK 55 (R)	8·55	11·5	74·3	—	—
SK 48	8·5	11·45	74·2	—	—
SK 65 (L)	9·15	11·5	79·6	12·7	72·0
Australopithecus (mean)	9·5	10·6	89·6	11·0	86·4
T.M. 1512	9·1	10·4	87·5	—	—
Sts 52a (L)	9·85	10·75	91·6	—	—
Sts 52a (R)	9·8	10·7	91·6	11·3	86·7
MLD 11/23	9·9	10·7(?)	92·5(?)	11·0(?)	90·0(?)
D. *Crown area (mm.²)*					
Zinjanthropus (L)	84·4	185·3	45·5	212·4	39·7
(R)	87·1	185·3	47·0	211·2	41·2
Paranthropus (mean)	81·7	139·7	58·5	163·0	50·1
SK 55 (L)	71·4	124·5	57·3	—	—
SK 55 (R)	73·0	128·3	56·9	—	—
SK 48	72·2	126·0	57·3	—	—
SK 65 (L)	83·0	126·0	65·9	154·0	53·9
Australopithecus (mean)	90·5	110·7	81·7	117·2	77·2
T.M. 1512	82·7	105·9	78·1	—	—
Sts 52a (L)	97·0	111·4	87·1	—	—
Sts 52a (R)	96·0	110·1	87·2	123·7	77·6
MLD 11/23	98·0	111·9(?)	87·6(?)	118·4(?)	82·8(?)

* The 'means' cited for *Paranthropus* and *Australopithecus* are the ratios of the means for C̲ and P³ (or P⁴), *not* the means of individual ratios.

M.D. crown diameter, the B.L. crown diameter, the module or the crown area. That is to say, the canine dimensions form a smaller percentage of the premolar dimensions in *Zinjanthropus* than in any *Paranthropus* or *Australopithecus* specimens yet recorded. This study therefore supports Leakey's claim that, in the Olduvai fossil, the canines are more reduced relative to the postcanine teeth than they are in *Paranthropus*.

We have illustrated this point by comparing two dimensions and two indices of the canine with those of both P³ and P⁴, much as Robinson has done with the module alone (1956, p. 148) to show a distinction between *Paranthropus* and *Australopithecus*. The distinction between the latter two groups is clearly demonstrated in our Table 36, in respect not only of the module, but also of each crown diameter and of the crown area. Thus, in round figures, we may compare the Olduvai fossil (Z) with the means for the other two groups (designated by their initial letters, P and A) as in Table 37.

Table 37. *Maxillary canine–premolar percentage ratios in* Zinjanthropus *(Z),* Paranthropus *(P) and* Australopithecus *(A)**

		C/P³	C/P⁴
M.D. crown diameter	Z	80	73·5
	P	86 (86–92)	80 (77–83)
	A	107 (99–115)	104 (105–108)
B.L. crown diameter	Z	57·5	55
	P	67 (64–74)	62 (62–65)
	A	77 (76–81)	74 (73–79)
Module	Z	66·25	62·5
	P	75 (74–80)	69 (—)
	A	90 (87–93)	86 (87–90)
Crown area	Z	46·25	40·5
	P	59 (57–66)	50 (—)
	A	82 (78–88)	77 (77–83)

* Figures quoted for *Zinjanthropus* are the means of left and right tooth ratios. For *Paranthropus* and *Australopithecus*, there are quoted the *ratio of the means* for all specimens of C and P³ (or P⁴), including isolated teeth; and, in parentheses, the range of ratios for individual maxillae in each of which C and P³ (or P⁴) are present. All figures for *Australopithecus* and *Paranthropus* are rounded off to the nearest whole number.

In all the comparisons in Table 37, *Paranthropus* lies intermediately between *Zinjanthropus* (with the lowest values) and *Australopithecus* (with the highest values). In three out of four metrical characters, *Paranthropus* is definitely nearer in C/P³ ratio to *Zinjanthropus* than to *Australopithecus*; in the comparison based on B.L. crown diameter, however, the distance (9·5 per cent) between *Zinjanthropus* and the *Paranthropus* 'mean' is virtually the same as that between the *Paranthropus* and *Australopithecus* 'means' (10·0 per cent). It is in the contrast between the buccolingual dimensions of C and P³ that *Zinjanthropus* is most sharply distinguished from the South African australopithecines. When comparisons are made of the C/P⁴ values, in all four instances *Zinjanthropus* is closer to the *Paranthropus* mean than the latter is to the *Australopithecus* mean.

E. Ratios of premolar dimensions to premolar–molar chords (Tables 38 and 39)

It was mentioned above that Leakey had made the pertinent point that the use of the C/P³ ratio implied a constant size relationship between P³ and the rest of the tooth row. This has been put to the test in *Zinjanthropus* and in as many other australopithecine maxillae as have the premolar–molar series entire. Some specimens have only P⁴–M³ intact and separate ratios have therefore been calculated, relating the dimensions of the premolars to the P³–M³ chord and to the P⁴–M³ chord (Tables 38 and 39).

Table 38. *Percentage ratios of premolar dimensions to P³–M³ chord*

	P³		P⁴	
	M.D.	B.L.	M.D.	B.L.
Zinjanthropus (mean)	15·19	23·69	16·58	24·80
SK 46	14·82	23·57	16·52	26·07
SK 13	15·40	20·77	16·59	23·85
SK 83	14·60	23·88	19·93	?28·69
Sts 17	15·55	23·33	16·64	23·87

Within each vertical column of Tables 38 and 39, the values are remarkably constant, the fluctuations being generally not more than 1–3 per cent of the chord. From this we may infer that the premolar dimensions of this series of australopithecines bear a size relationship to the cheek-tooth

Table 39. *Percentage ratios of premolar dimensions to P^4–M^3 chord*

	P^3		P^4	
	M.D.	B.L.	M.D.	B.L.
Zinjanthropus (mean)	18·05	28·14	19·71	29·48
SK 46	17·40	27·67	19·39	30·61
SK 13	18·12	24·44	19·52	28·07
SK 83	17·10	27·97	23·34	?33·60
SK 11	—	—	20·96	29·94
SK 49	—	—	21·70	31·95
Sts 17	18·42	27·62	19·70	28·27
Sts 53	—	—	?18·14	?

row which is only slightly variable. Robinson's procedure of relating canine dimensions to P^3 and P^4 dimensions may thereby be considered validated.

F. Dimensions of the premolars
(Tables 29, 33 and 40)

In Table 40, the mesiodistal and buccolingual crown diameters, module and crown area of the *Zinjanthropus* premolars are compared with those of other hominoids.

In mesiodistal diameter, both P^3 and P^4 of *Zinjanthropus* fall at the top of the range for

Table 40. *Crown dimensions, modules and crown areas of premolars of* Zinjanthropus *and other hominoids**

	P^3				P^4			
	Crown diameters				Crown diameters			
	Mesiodistal (mm.)	Buccolingual (mm.)	Module (mm.)	Crown area (mm.²)	Mesiodistal (mm.)	Buccolingual (mm.)	Module (mm.)	Crown area (mm.²)
Zinjanthropus (L)	10·9	17·0	14·0	185·3	11·8	18·0	14·9	212·4
(R)	10·9	17·0†	14·0†	185·3†	12·0	17·6	14·8	211·2
Paranthropus (Swartkrans and Kromdraai)	9·9	14·1	12·0	139·7	10·6	15·4	13·0	163·0
Australopithecus (Sterkfontein, Makapansgat and Garusi)	8·9	12·4	10·6	110·7	9·1	12·9	11·0	117·2
Homo erectus								
Sangiran IV ($n = 2$)	8·3	12·4	10·4	103·5	8·35	12·3	10·3	101·5
Choukoutien ($n = 4, 9$)	8·3	11·9	10·1	98·8	7·9	11·4	9·7	90·1
Rabat ($n = 1$)	8·5	?12·0	?10·25	?102·0	8·0	11·0	9·5	88·0
Total *H. erectus* ($n = 7, 12$)	8·4	12·0	10·2	101·0	8·0	11·5	9·7	95·3
Australian	7·8	10·3	9·1	80·3	7·2	10·1	8·7	72·8
E. Greenland Eskimo	7·5	9·2	8·4	69·0	6·8	9·2	8·0	62·6
Aleut	7·1	9·3	8·2	66·1	6·6	9·1	7·9	60·7
Japanese	7·3	9·5	8·4	69·3	6·8	9·3	8·1	63·2
Bantu Negroid	7·2	9·0	8·1	64·8	7·0	9·1	8·1	63·7
Bushman	6·8	8·6	7·7	58·5	6·5	8·5	7·5	55·3
American White	7·2	9·1	8·2	65·5	6·8	8·8	7·8	59·8
Gorilla ♂	11·4	15·8	13·6	180·1	11·2	15·5	13·3	173·6
Gorilla ♀	10·7	14·3	12·5	153·0	10·3	14·1	12·2	145·2
Chimpanzee ♂	7·9	10·1	9·0	79·8	7·5	10·1	8·8	75·7
Chimpanzee ♀	7·5	9·5	8·5	71·3	7·1	9·5	8·3	67·5
Orang-utan ♂	10·0	13·0	11·5	130·0	9·7	13·1	11·4	127·1
Orang-utan ♀	9·0	11·7	10·3	105·3	8·8	12·0	10·4	105·6

* All figures in the table, save those for *Zinjanthropus*, Rabat and Sangiran IV, are sample means or ratios of means; all samples (except the pongids) are combined male and female samples. The figures for Sangiran IV are the averages of left and right teeth (after Weidenreich, 1945).

† Estimated dimension, or index based upon an estimated dimension.

SIZE OF INDIVIDUAL TEETH

Paranthropus (Table 30), whereas in buccolingual diameter, both teeth far exceed the range for *Paranthropus* (Figs. 23 and 24). Thus, the M.D. diameter of the *Zinjanthropus* P³ (10·9 mm.) falls just outside the range for eighteen *Paranthropus* P³'s (9·0–10·8 mm.) and exceeds the *Paranthropus* mean of 9·9 mm. by 1·92 S.D.'s; it is well in excess of the *Australopithecus* mean of 8·9 mm. ($n = 17$) and range of 8·5–9·4 mm. The B.L. diameter of the *Zinjanthropus* P³ (17·0 mm.) is far greater than the top of the *Paranthropus* range (13·1–15·3 mm.) and exceeds the *Paranthropus* mean of 14·1 mm. by 3·65 S.D.'s—a significant difference; it greatly exceeds the *Australopithecus* mean (12·4 mm.) and range (10·7–13·9 mm.). It is clear that, in comparison with *Paranthropus*, the *Zinjanthropus* third premolars are greatly expanded in a buccolingual direction (though not so markedly in the mesiodistal dimension), and therein lies the main basis for the much lower C/P³ indices discussed in the previous section.

The M.D. diameters of the *Zinjanthropus* P⁴ (11·8 and 12·0 mm.) lie at or close to the top of the *Paranthropus* range (9·2–11·8 mm.) and, of course, well above the mean (10·6 mm.), which they exceed by 1·72 S.D.'s (left) and 2·00 S.D.'s (right). *A fortiori*, the M.D. of the Olduvai P⁴ exceeds the mean (9·1 mm.) and range (7·8–10·5 mm.) of *Australopithecus*. As with P³, the main distinction of the *Zinjanthropus* P⁴ resides in its tremendous buccolingual expansion: its B.L. diameters of 18·0 and 17·6 mm. lie well outside the range (13·7–16·5 mm.) of twenty *Paranthropus* P⁴'s and exceed the mean (15·4 mm.) by 3·51 S.D.'s (left) and 2·97 S.D.'s (right), a significant difference. Likewise, they exceed the mean (12·9 mm.) and range (12·0–13·9 mm.) of eleven *Australopithecus* P⁴'s (Figs. 27 and 28).

The module of the *Zinjanthropus* P³ (14·0 mm.) is appreciably greater than those of seventeen *Paranthropus* P³'s (mean 12·0, range 11·3–12·85 mm.) and of seventeen *Australopithecus* P³'s (mean 10·6, range 9·7–11·5 mm.). Likewise, the crown area of the *Zinjanthropus* P³ (185·3 mm.²) is well in excess of those of *Paranthropus* (mean 139·7, range 124·5–159·6 mm.²) and of *Australopithecus* (mean 110·7, range 93·1–126·9 mm.²) (Figs. 25 and 26).

A comparable or even greater excess is evident when the module and crown area of the *Zinjanthropus* P⁴ are compared with those of the South African forms. The module (14·9, 14·8 mm.) exceeds those of *Paranthropus* (mean 13·0, range 11·5–13·9 mm.), the excess over the mean being 2·84 S.D.'s (left) and 2·69 S.D.'s (right); and it is far greater than those of *Australopithecus* (mean 11·0, range 10·35–12·2 mm.). The crown areas of the *Zinjanthropus* P⁴'s (212·4, 211·2 mm.²) appreciably surpass the crown areas of twenty *Paranthropus* P⁴'s (mean 163·0, range 127·4–187·5 mm.²) and of eleven *Australopithecus* P⁴'s (mean 117·2, range 102·5–145·9 mm.²).

All the hominine teeth recorded in Table 40, including those of *H. e. pekinensis*, fall below the australopithecine values in dimensions and indices: although the mean values for a small group of premolars of Pekin Man overlap the *lowest* values for *Australopithecus* premolars in all respects, except the B.L. diameter and the module of P⁴. It seems that buccolingual expansion of the premolars is a distinctive feature of all australopithecines, even the relatively medium-toothed group (*Australopithecus*). This expansion is greater in P⁴ than in P³, and is carried to its most marked degree in *Zinjanthropus*.

A similar tendency towards buccolingual broadening of the third and fourth premolars is apparent in gorilla and orang-utan, though not in chimpanzee (Figs. 29 and 30).

Robinson (1956, pp. 65–6) has stressed the comparative sizes of P³ and P⁴ as a taxonomically important feature. It is clear from our Table 40, that P⁴ in *Zinjanthropus* is bigger in every respect (M.D., B.L., module and crown area) than P³. In this regard, it is comparable with the mean size relations of P⁴ to P³ in the *Paranthropus* group. In *Australopithecus*, however, the mean M.D. and the mean B.L. of P⁴ are only 0·2 and 0·5 mm. greater respectively than those of P³, with the result that the mean module and mean crown area of P⁴ are only slightly greater than those of P³. In *H. e. pekinensis*, *H. e. erectus* and *Meganthropus* the tendency is reversed (von Koenigswald, 1955, 1958), as it is in all the extant hominine samples recorded. It is interesting to note in passing that, to

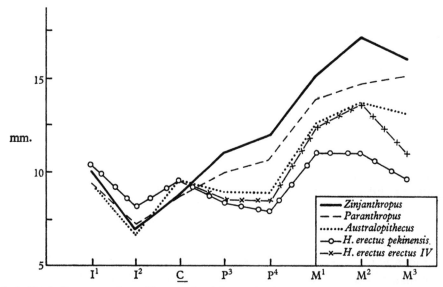

Fig. 27. Mesiodistal diameters of maxillary teeth of *Zinjanthropus*, compared with means for *Paranthropus*, *Australopithecus*, *H. e. pekinensis*, and the absolute values for *H. e. erectus* IV.

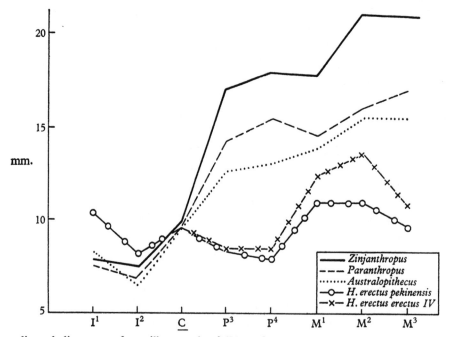

Fig. 28. Buccolingual diameters of maxillary teeth of *Zinjanthropus*, compared with means for *Paranthropus*, *Australopithecus*, *H. e. pekinensis*, and the absolute values for *H. e. erectus* IV.

SIZE OF INDIVIDUAL TEETH

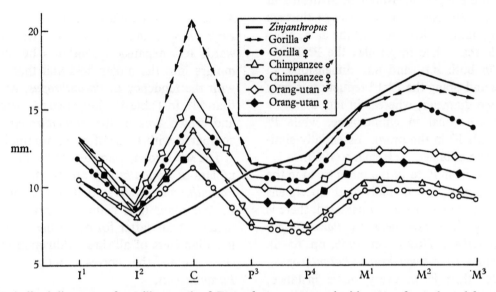

Fig. 29. Mesiodistal diameters of maxillary teeth of *Zinjanthropus* compared with means for male and female Pongidae. The pongid measurements are based on those recorded by Ashton and Zuckerman (1950).

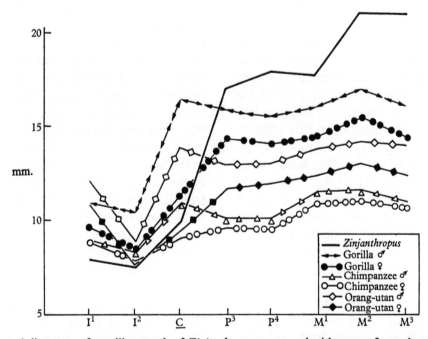

Fig. 30. Buccolingual diameters of maxillary teeth of *Zinjanthropus* compared with means for male and female Pongidae. The pongid measurements are based on those recorded by Ashton and Zuckerman (1950).

judge by module and crown area, P^4 is smaller than P^3 in all of the pongids measured by Ashton and Zuckerman (1950), except in the series of thirteen female orang-utan. A closer look at the data, however, reveals that while in gorillas the $P^3:P^4$ reduction is in both M.D. and B.L. dimensions, in chimpanzee and orang the $P^3:P^4$ reduction is only in M.D. crown diameter, whereas in B.L. diameter P^3 and P^4 are equal in chimpanzee, while P^4 slightly exceeds P^3 in the orang. Essentially similar results were obtained by Remane (1960).

The increase from P^3 to P^4 is a *mean* trend in the Australopithecinae; when individual specimens containing both premolars are considered, there is an interesting difference between *Paranthropus* and *Australopithecus* (Robinson, 1956, pp. 65–6). In five maxillae from Swartkrans (*Paranthropus*), P^4 is greater than P^3 in every single instance; Robinson has plotted the 'length' (M.D.) and 'breadth' (B.L.) of the P^3 and P^4, joining up the points for the two premolars in each maxilla (p. 65, fig. 19). All the lines thus drawn for five Swartkrans (*Paranthropus*) maxillae slope uphill from P^3 to P^4; that is, in every single instance, $P^4 > P^3$ though the inclination (i.e. the size disparity) varies slightly from one jaw to another. When *Zinjanthropus* is plotted on the same co-ordinates, the P^3–P^4 slope is equal to the steepest of the *Paranthropus* curves: this means that in *Zinjanthropus* the degree of increase from P^3 to P^4 is approximately equal to the top of the sample range of increase for *Paranthropus*. On the other hand, out of five Sterkfontein jaws, in two P^4 is greater than P^3 (with a very steep curve of increase!), whereas in the other three, P^4 is markedly smaller than the corresponding P^3. In the small hominine sample measured by Robinson (seven Zulu maxillae), $P^3 > P^4$ in every instance. We may thus agree with him that the *Australopithecus* group with its variable behaviour is intermediate between the *Paranthropus* sample and the hominine sample: in this regard, *Zinjanthropus* groups itself with *Paranthropus* and those specimens of *Australopithecus* in which $P^4 > P^3$. The australopithecine tendency for a further increase, especially in B.L. crown diameter, from P^3 to P^4 characterises *Zinjanthropus* to a marked degree.

G. Dimensions of the molars

At least two significant features of the size of *Zinjanthropus* molars are that the tendency towards B.L. expansion continues beyond the premolars into the molar field and that, in contrast with the tendency in *Paranthropus*, M^3 is smaller than M^2. In Table 41, the crown diameters, module and crown area of the *Zinjanthropus* molars are compared with those of other hominoids.

The M.D. crown diameters of the first and second molars of *Zinjanthropus* are greater than those of *Paranthropus*, whereas that of the third molar of *Zinjanthropus* falls within the range of M.D. diameters of the latter form; on the other hand, the B.L. diameters of all three Olduvai molars on each side exceed the corresponding dimensions in *Paranthropus*.

More particularly, the M.D. diameter of M^1 (15·2 mm.) exceeds those of sixteen *Paranthropus* M^1's (mean 13·8, range 13·1–14·6 mm.) and of fifteen *Australopithecus* M^1's (mean 12·6, range 11·9–13·2 mm.). Again, the B.L. diameter of the *Zinjanthropus* first molar (17·7 mm.) is appreciably greater than those of *Paranthropus* (mean 14·5 mm. which it exceeds significantly by 3·71 S.D.'s; range 12·7–16·6 mm.) and of *Australopithecus* (mean 13·8, range 12·8–15·1 mm.). The disparity is somewhat greater in respect of the B.L. than of the M.D. diameter. Robinson (1956, p. 151) has provided standard deviations for the molar dimensions of the South African australopithecines, but I have employed here the S.D.'s which I have computed from larger samples and by the pooling of data for teeth from different australopithecine sites. The S.D.'s used here are in general slightly smaller than Robinson's, which were based on somewhat smaller samples of teeth from only one site for each taxon (Sterkfontein for *Australopithecus* and Swartkrans for *Paranthropus*). On the newly estimated S.D.'s, the dimensions of the *Zinjanthropus* M^1 exceed the *Paranthropus* means by 2·92 S.D.'s (for M.D. diameter) and 3·71 S.D.'s (for B.L. diameter) (Table 33). These figures tend to confirm that the B.L. excess is greater than the M.D. excess. The M^1 dimensions of *Zinjanthropus* exceed the *Australopithecus* means by 6·84 S.D.'s for mesio-

SIZE OF INDIVIDUAL TEETH

Table 41. *Crown dimensions (mm.), modules (mm.) and crown areas (mm.^{2}) of molars of* Zinjanthropus *and other hominoids**

	M^1				M^2				M^3			
	Crown diam.			Crown	Crown diam.			Crown	Crown diam.			Crown
	M.D.	B.L.	Module	area	M.D.	B.L.	Module	area	M.D.	B.L.	Module	area
Zinjanthropus (L)	15·2	17·7	16·45	269·0	17·2	21·0†	19·1†	361·2†	15·7	21·4	18·55	336·0
(R)	15·2	17·7	16·45	269·0	17·2†	21·0†	19·1†	361·2†	16·3	20·5	18·4	334·1
Paranthropus												
(Swartkrans and Kromdraai)	13·8	14·5	14·1	200·1	14·5	15·9	15·2	230·9	15·0	16·8	15·9	252·0
Australopithecus												
(Sterkfontein, Makapansgat and Taung)	12·6	13·8	13·2	173·4	13·8	15·4	14·6	212·6	13·2	15·5	14·35	205·8
Homo erectus												
Sangiran IV (n = 1–2)	12·2	13·65	12·9	166·5	13·6	15·2	14·4	206·7	10·8	14·0	12·4	151·2
Choukoutien (n = 6–9)	10·9	12·5	11·7	136·3	10·9	12·7	11·8	138·4	9·6	11·5	10·6	110·4
'Sinanthropus officinalis' (n = 1)	12·8	13·7	13·25	175·4	—	—	—	—	—	—	—	—
Rabat	—	—	—	—	11·5	13·0	12·25	149·5	—	—	—	—
Ternifine	12·0	14·4	13·2	172·8	—	—	—	—	—	—	—	—
Total H. erectus (n = 9–11)	11·6	13·1	12·25	150·2	11·3	13·0	12·15	147·4	9·7	11·8	10·8	115·7
Australian	11·4	12·8	12·1	145·9	10·9	13·1	12·0	142·8	10·0	12·3	11·2	123·0
E. Greenland Eskimo	10·7	11·6	11·2	124·1	10·2	11·5	10·9	117·3	9·6	11·1	10·3	106·6
Aleut	10·2	11·3	10·7	116·1	9·9	11·3	10·6	112·0	9·1	10·7	9·9	98·6
Japanese	10·2	11·3	10·7	115·3	9·7	11·4	10·5	110·6	8·7	10·6	9·7	92·2
Bantu Negroid	10·3	11·0	10·7	113·3	10·0	11·5	10·8	115·0	9·5	11·0	10·3	104·5
Bushman	9·9	10·6	10·3	104·9	9·7	10·6	10·2	102·8	8·2	10·3	9·3	84·5
American White	10·7	11·8	11·3	126·3	9·2	11·5	10·4	105·8	8·6	10·6	9·6	91·2
Gorilla ♂	15·1	15·8	15·5	238·6	16·3	16·8	16·6	273·8	15·5	16·1	15·8	249·5
Gorilla ♀	14·1	14·4	14·2	203·0	15·0	15·4	15·2	231·0	13·6	14·4	14·0	195·8
Chimpanzee ♂	10·3	11·4	10·8	117·4	10·2	11·6	10·9	118·3	9·3	10·9	10·1	101·4
Chimpanzee ♀	9·7	10·8	10·3	104·8	9·7	11·0	10·4	106·7	9·2	10·7	9·9	98·4
Orang-utan ♂	12·2	13·7	12·9	167·1	12·2	14·2	13·2	173·2	11·6	13·9	12·8	161·2
Orang-utan ♀	11·5	12·4	12·0	143·4	11·4	13·0	12·2	148·1	10·5	12·4	11·4	129·4

* All figures in the table, save those for *Zinjanthropus*, Sangiran IV, '*Sinanthropus officinalis*', Rabat and Ternifine, are sample means or ratios of means; all samples (except the pongids) are combined male and female samples. The figures for Sangiran IV are for the right teeth (M^2 and M^3) and the average of right and left teeth (M^1) (after Weidenreich, 1945).

† Estimated dimension, or index based upon estimated dimensions. Enamel loss from these teeth prevented precise measurement of the indicated dimensions; the diameters have been estimated.

distal and 6·19 S.D.'s for buccolingual crown diameters (Figs. 27 and 28).

Similarly, the mesiodistal crown diameter of the second upper molar of *Zinjanthropus* (17·2 mm.) far exceeds those of fifteen *Paranthropus* maxillary molars from Swartkrans and Kromdraai (mean 14·5, range 13·6–15·7 mm.) and of sixteen *Australopithecus* teeth from Sterkfontein and Makapansgat (mean 13·8, range 12·7–15·1 mm.). The M.D. diameter of the *Zinjanthropus* M^2 exceeds the mean for *Paranthropus* by 3·71 S.D.'s and for *Australopithecus* by 4·30 S.D.'s (Fig. 27).

The buccolingual diameter of the left M^2 measures 20·9 mm. as the tooth is at present. As a small amount of enamel is missing from the most protuberant part of the buccal face, the diameter has been estimated as 21·0 mm. It may, however, have been a point or two greater. This is mentioned to show that only a minute fraction of the B.L. diameter has had to be estimated. The right

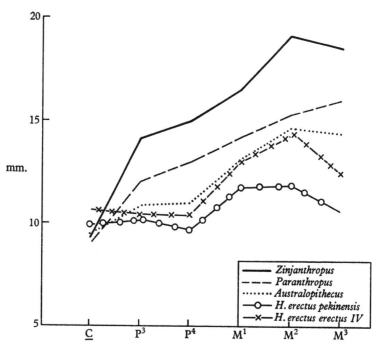

Fig. 31. Modules of maxillary teeth of *Zinjanthropus*, compared with means for *Paranthropus*, *Australopithecus*, *H. e. pekinensis*, and with absolute values for *H. e. erectus* IV.

M^2 in its present damaged state measures as much as 20·3 mm. in B.L. diameter, although the whole of the bucally-protruding paracone lacks enamel from its buccal face. It has therefore been deemed reasonable to estimate the original B.L. diameter of this tooth at the same value as on the left—i.e. 21·0 mm. Neither tooth could have been less than this value, which is close to the greatest dimension of any *Zinjanthropus* tooth (only the B.L. diameter of the left third molar exceeding this value and reaching 21·4 mm.). The value of 21·0 mm. is greatly in excess of the corresponding dimension in the *Paranthropus* series of fifteen M^2's (mean 15·9, range 14·2–16·9 mm.) and in the *Australopithecus* series of eighteen M^2's (mean 15·4, range 13·8–17·1 mm.) (Figs. 24 and 28). The B.L. diameter of the *Zinjanthropus* M^2 exceeds the mean for *Paranthropus* by 5·97 S.D.'s and for *Australopithecus* by 5·60 S.D.'s (Fig. 27).

Finally, the M.D. crown diameters of the *Zinjanthropus* M^3's (15·7, 16·3 mm.) fall within the range for sixteen third maxillary molars of *Paranthropus* (mean 15·0, range 12·8–17·0 mm., S.D. 1·121), the excess of the *Zinjanthropus* values over the *Paranthropus* mean being only 0·66 S.D.'s (left)

and 1·20 S.D.'s (right). However, they exceed the M.D. diameter for twelve *Australopithecus* third upper molars (mean 13·2, range 11·6–15·2 mm., S.D. 1·164) by 2·25 S.D.'s (left) and 2·79 S.D.'s (right) (Figs. 23 and 27).

In B.L. diameter, however, the *Zinjanthropus* M^3's (21·4, 20·5 mm.) are greatly in excess of the *Paranthropus* sample of sixteen M^3's (mean 16·8, range 15·9–18·2 mm., S.D. 0·77), the *Zinjanthropus* values being 5·93 S.D.'s (left) and 4·77 S.D.'s (right) bigger than the *Paranthropus* mean. Needless to say, the Olduvai fossil surpasses by far the B.L diameter of the *Australopithecus* sample of eleven M^3's (mean 15·5, range 14·6–17·9, S.D. 1·04), the disparity being 5·67 S.D.'s (left) and 4·81 S.D.'s (right), despite the large variability of the sample (Figs. 24 and 28).

In consequence of these excessive dimensions, the modules and crown areas of the *Zinjanthropus* molars are far greater than those of other australopithecines.

In particular, the module for M^1 (16·45 mm.) exceeds those for *Paranthropus* (mean 14·1, range 12·9–15·5 mm.) and for *Australopithecus* (mean 13·2, range 12·65–14·15 mm.), the excess over the means being 3·85 S.D.'s and 7·67 S.D.'s respectively. The module for M^2 (19·1 mm.) surpasses the values for *Paranthropus* (mean 15·2, range 13·95–16·3 mm.) and for *Australopithecus* (mean 14·6, range 13·45–16·1 mm.), the excess over the means being 5·65 S.D.'s and 5·63 S.D.'s respectively. Thirdly, the modules of the *Zinjanthropus* M^3's (18·55, 18·4 mm.), although smaller than the module for M^2 in the same fossil, are none the less greater than those for M^3 in *Paranthropus* (mean 15·9, range 15·0–17·4 mm.) and in *Australopithecus* (mean 14·35, range 13·2–16·55 mm.) (Figs. 25 and 31). The differences from the *Paranthropus* mean are 3·50 S.D.'s (left) and 3·30 S.D.'s (right), and from the *Australopithecus* mean 3·96 S.D.'s (left) and 3·87 S.D.'s (right).

In like fashion, the crown area of the first maxillary molar of *Zinjanthropus* (269·0 mm.²) exceeds those of *Paranthropus* (mean 200·1, range 166·4–239·0 mm.²) and of *Australopithecus* (mean 173·4, range 159·5–199·3 mm.²). The crown area of the *Zinjanthropus* M^2's (361·2 mm.²) substantially exceeds those of *Paranthropus* (mean 230·9, range 194·5–265·3 mm.²) and of *Australopithecus* (mean 212·6, range 180·8–258·2 mm.²). Finally, the crown areas of the *Zinjanthropus* M^3's (336·0, 334·1 mm.²), despite the relatively small M.D. crown diameter of this tooth, surpass those of *Paranthropus* (mean 252·0, range 224·2–302·6 mm.²) and of *Australopithecus* (mean 205·8, range 171·7–272·1 mm.²) (Figs. 26 and 32).

All the *Zinjanthropus* molars, therefore, are larger than their counterparts in the two known taxa of southern African australopithecines.

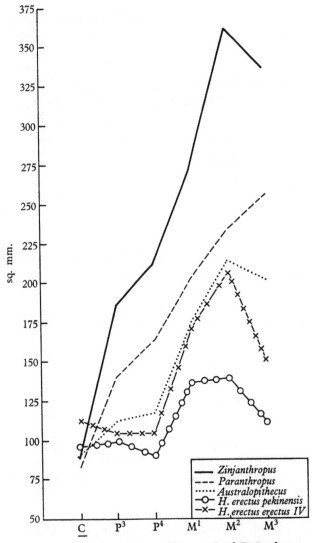

Fig. 32. Crown areas of maxillary teeth of *Zinjanthropus*, compared with means for *Paranthropus*, *Australopithecus* and *H. e. pekinensis*, and with absolute values for *H. e. erectus* IV.

H. Ratios of molar dimensions to premolar–molar chords (Tables 42 and 43)

It may be of interest to compare the proportions of the tooth-row occupied by each of the *Zinjanthropus* molars with those of other australopithecines, as was done for the premolars. The M.D. and B.L. diameters of the molars are compared in some instances with the P^3–M^3 chord and in all examples with the P^4–M^3 chord, because some specimens lack the P^3.

Within each vertical column, there is a remarkable constancy in the percentage ratios, whether the dimensions are related to the full chord (P^3–M^3) or to the partial chord (P^4–M^3) and whether the specimens belong to *Zinjanthropus*, *Paranthropus* (SK) or *Australopithecus* (Sts). Thus, in the sample of Australopithecinae represented in Tables 42 and 43, the M.D. crown diameter of M^1 is 21·11–22·68 per cent of the full chord or 24·91–27·22 per cent of the partial chord; that of M^2 is 23·15–26·43 per cent of the P^3–M^3 chord or 27·41–31·03 per cent of the P^4–M^3 chord; while the M.D. diameter of M^3 is 21·96–27·86 per cent of the full

Table 42. *Percentage ratios of molar dimensions to P³–M³ chord*

	M¹		M²		M³	
	M.D.	B.L.	M.D.	B.L.	M.D.	B.L.
Zinjanthropus* (L)	21.26	24.76	24.06	29.37	21.96	29.93
(R)	21.11	24.58	23.89	29.17	22.64	28.47
SK 46	?	?27.14	26.43	28.93	27.86	29.29
SK 13*	21.17	23.54	24.49	25.83	25.12	27.57
SK 11	?	26.68	24.50	27.68	25.34	27.01
SK 83	?22.68	27.32	?25.09	28.87	26.80	28.35
Sts 17	21.70	24.05	23.15	27.49	—	—

* M³ not fully in occlusion.

Table 43. *Percentage ratios of molar dimensions to P⁴–M³ chord*

	M¹		M²		M³	
	M.D.	B.L.	M.D.	B.L.	M.D.	B.L.
Zinjanthropus* (L)	25.33	29.50	28.67	35.00†	26.17	35.67
(R)	25.00	29.11	28.29	34.54†	26.81	33.72
SK 46	?	?31.87	31.03	33.96	32.70	34.38
SK 13*	24.91	27.70	28.81	30.39	29.55	32.43
SK 83	?26.56	31.99	?29.38	33.80	31.39	33.20
SK 11	?	31.74	29.14	32.93	30.14	32.14
SK 49	27.22	29.59	28.80	32.35	26.33	34.52
Sts 17	25.70	28.48	27.41	32.55	—	—
Sts 53	26.76	28.57	29.14	32.77	27.78	33.33

* M³ not fully in occlusion.
† Based upon estimated diameter.

chord and 26·17–32·70 per cent of the partial chord. The greatest range—which may be taken as a rough measure of variability—is in the mesiodistal diameter of the third molars: the reduced M³'s of *Zinjanthropus* are at the lower end of the range, only one out of five *Paranthropus* maxillae having as shortened an M³ (SK 49).

The B.L. crown diameter of M¹ is 23·54–27·32 per cent of the P³–M³ chord or 27·70–31·99 per cent of the P⁴–M³ chord; that of M² is 25·83–29·37 per cent of the full chord or 30·39–35·00 per cent of the partial chord; and finally, the B.L. diameter of M³ is 27·01–29·93 per cent of the longer chord and 32·14–35·67 per cent of the shorter chord.

The narrow limits within which these percentages vary show that, as far as available australopithecine specimens go, each cheek-tooth from P³ to M³ occupies a relatively constant proportion of the postcanine dental row. It is therefore a valid procedure in this population to judge of the size of one tooth, e.g. a front tooth, by comparing its dimensions with those of any postcanine tooth or combination of teeth.

I. Relative molar size

The most striking feature of the relative molar size in *Zinjanthropus* is the reduction of M³ as compared with M². The reduction affects the M.D. diameter, the trend from M¹ to M³ being 15·2 → 17·2 → 16·0 (average of left and right molars). This is precisely the trend shown by the mean M.D. diameters of *Australopithecus*, in which the relevant mean values from M¹ to M³ are 12·6 → 13·8 → 13·2. In each instance, the reduction from M^2 to M^3 is about half the increment from M^1 to M^2. In *Paranthropus*, however, the mean trend is towards a further increase in M.D. diameter between M^2 and M^3, the relevant mean values being $13 \cdot 8 \rightarrow 14 \cdot 5 \rightarrow 15 \cdot 0$.

In buccolingual diameter, *Zinjanthropus* again agrees with *Australopithecus* in showing an increase from M^1 to M^2, while M^3 remains about equal to M^2. The exact values for *Zinjanthropus* are difficult to determine because of damage to both M^2's: but the trend is approximately:

$$17 \cdot 7 \rightarrow 21 \cdot 0 \rightarrow 20 \cdot 95.$$

The comparable mean values in *Australopithecus* are:
$$13 \cdot 8 \rightarrow 15 \cdot 4 \rightarrow 15 \cdot 5.$$

In contrast again, *Paranthropus* shows a definite progressive increase in mean B.L. crown diameter:

$$14 \cdot 5 \rightarrow 15 \cdot 9 \rightarrow 16 \cdot 8.$$

Zinjanthropus thus differs from *Paranthropus* and agrees with *Australopithecus* in two respects: a mesiodistal diminution and a buccolingual plateau from M^2 to M^3.

In fact, the degree of mesiodistal reduction in *Zinjanthropus* is even greater than the mean reduction in *Australopithecus* and, on the left side, even approaches the mean M^2–M^3 reduction in *H. e. pekinensis*. Frisch (1965, p. 78) has recently pointed out that '*Zinjanthropus*, with his extraordinarily large molars, shows more reduction, both morphological and metrical, in M^3 than either previously known australopithecine'. If the change in mean M.D. diameter from M^2 to M^3 is expressed as a percentage of the mean M.D. diameter of M^2, the following figures result:

Paranthropus (mean)	$+ 3 \cdot 45 \%$
Australopithecus (mean)	$- 4 \cdot 35 \%$
Zinjanthropus (right)	$- 5 \cdot 23 \%$
Zinjanthropus (left)	$- 8 \cdot 72 \%$
H. e. pekinensis (mean)	$-11 \cdot 93 \%$
H. e. erectus IV	$-20 \cdot 59 \%$

In his study of the reduction of the third molar, Frisch (1965) compared the M.D. length of M^3 with that of M^1, devising thus another index of reduction of M^3. He cited figures for *Paranthropus* and *Australopithecus* based on the measurements of Robinson (1956) and for *Zinjanthropus* based on the original set of dental measurements published by Leakey (1959a). I have re-computed the indices, using the larger australopithecine samples now available and the newer measurements of the *Zinjanthropus* teeth published herein. The other results cited below are taken from Table XXIII of Frisch (*op. cit.* p. 73).

	M.D. of M^1	$\dfrac{\text{M.D. of } M^3 \times 100}{\text{M.D. of } M^1}$
Paranthropus	13·8	108·6
Zinjanthropus	15·2	105·3
Australopithecus	12·6	104·8
Sangiran IV	12·3	87·8
Choukoutien	10·9	88·1
East Greenland Eskimo	10·7	89·7
Japanese	10·2	85·3
American White	10·3	86·4

These indices confirm that the M^3 of *Paranthropus* is larger in M.D. diameter relative to M^1 than in *Zinjanthropus* and *Australopithecus*.

Both the M.D. and B.L. trends just mentioned result in a smaller M^3 module and crown area in both *Zinjanthropus* and *Australopithecus*, as compared with those of M^2. For *Zinjanthropus*, the modules from M^1 to M^3 are:

$$16 \cdot 45 \rightarrow 19 \cdot 1 \rightarrow 18 \cdot 475.$$

In *Australopithecus*, the comparable mean values are:
$$13 \cdot 2 \rightarrow 14 \cdot 6 \rightarrow 14 \cdot 35,$$

while *Paranthropus* shows a rising mean trend:

$$14 \cdot 1 \rightarrow 15 \cdot 2 \rightarrow 15 \cdot 9.$$

If we express the change in mean module from M^2 to M^3 as a percentage of the mean module of M^2, we see that the degree of reduction of the *Zinjanthropus* M^3's is somewhat greater than the mean reduction of the *Australopithecus* sample:

Paranthropus (mean)	$+4 \cdot 61 \%$
Australopithecus (mean)	$-1 \cdot 71 \%$
Zinjanthropus (left)	$-2 \cdot 88 \%$
Zinjanthropus (right)	$-3 \cdot 67 \%$
H. e. pekinensis (mean)	$-10 \cdot 17 \%$
H. e. erectus IV	$-13 \cdot 89 \%$

The crown area shows a similar close resemblance between molar size relations in *Zinjanthropus* and *Australopithecus*, while *Paranthropus* stands out from the others.

Thus far, we have compared the trend in *Zinjanthropus* with the *mean* trend in the other two taxa. However, as Robinson has pointed out (1956, pp. 95–6), the facts are seen more clearly if comparisons are made between the sizes of M^2 and M^3 on the same side in the same individual. Of six Swartkrans (*Paranthropus*) specimens cited by him, to which we may add the original Kromdraai specimen, the module of M^3 exceeds that of M^2 in six

(1950) and by Remane (1960) show a similar relationship among the molars only in male gorillas ($M^2 > M^3 > M^1$); in all the other pongid samples (female gorilla, male and female chimpanzee, male and female orang-utan), M^3 is appreciably reduced so that the sequence $M^2 > M^1 > M^3$ results (Figs. 33 and 34). In these pongids, as in the Australopithecinae, the major reduction is in the mesiodistal diameter. No pongid group shows the

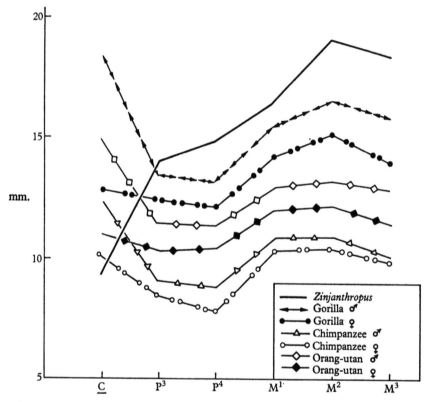

Fig. 33. Modules of maxillary teeth of *Zinjanthropus* compared with mean modules in male and female Pongidae. The pongid measurements are based on those recorded by Ashton and Zuckerman (1950).

cases, while the two are of equal size in the seventh specimen. Not a single *Paranthropus* specimen thus far described shows the M^2–M^3 relationship which obtains in *Zinjanthropus*. Again, of six Sterkfontein (*Australopithecus*) specimens in which M^2 and M^3 are present on the same side, the module of M^3 is smaller than that of M^2 in five instances, and slightly larger in the sixth specimen.

The degree of reduction of M^3 in *Zinjanthropus* and *Australopithecus* is slight, so that M^3 remains the second biggest tooth. The mean values for Pongidae measured by Ashton and Zuckerman

striking mean sequence $M^3 > M^2 > M^1$ found in the *Paranthropus* group. In the very large sample of teeth attributed to *Gigantopithecus*, again the M^3's are smaller in mean M.D. and mean B.L. diameters than the M^2's: this is true for both the 'large type' (?male) and the 'small type' (?female) of these teeth (Woo, 1962). An increase in size of the third maxillary molar as compared with that of the second should perhaps be regarded not as a primitive hominoid feature, but as a specialisation undergone by one branch of the Australopithecinae. Robinson has very tentatively hinted

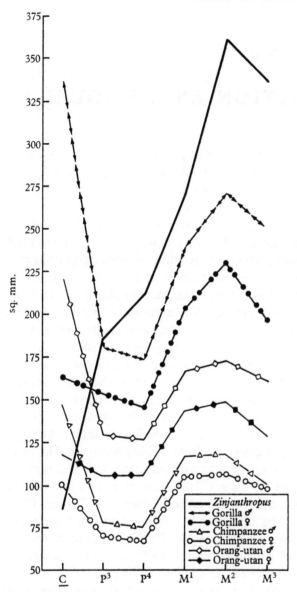

Fig. 34. Crown areas of maxillary teeth of *Zinjanthropus*, compared with means for male and female Pongidae. The pongid values have been computed from those recorded by Ashton and Zuckerman (1956).

would be crucial in resolving this problem. In some of the Miocene hominoids of East Africa (*Proconsul*, *Limnopithecus*), the maxillary third molars are shorter in mesiodistal crown diameter than are the corresponding second upper molars. Thus, in four specimens of *Limnopithecus macinnesi*, in which all three upper molars are present, $M^3 < M^2$ in both M.D. and B.L. diameters (Clark and Leakey, 1951). Again, in three specimens of *Proconsul nyanzae* (which Simons and Pilbeam, 1965, have recently proposed to re-classify as *Dryopithecus* (*Proconsul*) *nyanzae*), the mesiodistal diameter of M^3 is in each instance shorter than that of M^2, although the B.L. diameter of M^3 exceeds that of M^2 in all three maxillae (Clark and Leakey, 1951). Among the Miocene and Pliocene hominoid remains, maxillary molars are relatively scarce compared with the abundance of mandibular teeth: thus, Simons and Pilbeam (1965, p. 132) have pooled a sample of teeth from several taxa which, on their proposed new classification, all belong to *Dryopithecus*. The combined sample includes teeth of *D. fontani*, *D. (Proconsul) nyanzae* and *D. (Sivapithecus) sivalensis*: the mean values for ten M^1, thirteen M^2 and ten M^3 clearly indicate the mean size sequence $M^2 > M^3 > M^1$. From his re-allocation of specimens to *Ramapithecus punjabicus* (formerly designated *R. brevirostris*), Simons (1964a) has illustrated parts of the palate and upper dental arcade of this Mio-Pliocene hominoid which he believes is an early member of the Hominidae (1961, 1963, 1964b): his illustration indicates an M^3 which is appreciably smaller than the adjacent M^2. When all the evidence of these Mio-Pliocene hominoids, *Gigantopithecus*, the Australopithecinae, fossil and recent Homininae, and living pongids, is taken into consideration, it certainly seems that *Paranthropus* is exceptional in showing the sequence $M^3 > M^2 > M^1$ in the majority of specimens. It is suggested that this pattern is a specialised departure, rather than a primitive feature, and has thus far been shown to characterise only *Paranthropus*, but not *Australopithecus* or *Zinjanthropus*.

at this possibility (1956, pp. 148–9), only to dismiss it as unlikely 'in view of the fact that all the available evidence points to a reduction of the face in this form as well as in early hominids in general'.

The evidence of pre-Pleistocene Anthropoidea

CHAPTER XV

THE SIZE OF THE DENTITION AS A WHOLE

A. Flower's Dental Index

This index, devised by Sir William Flower (1879, 1885), relates the size of the teeth, as expressed by the premolar–molar chord, to the size of the cranium, as represented by the basion–nasion length. The premolar–molar chord is the distance from the most mesial point on the maxillary P^3 to the most distal point of the maxillary M^3, all five cheek-teeth being *in situ*. Flower originally suggested the following classification based on the values of this index:

42	microdont
43	mesodont
44 and upwards,	megadont.

However, in later usage, the following categories came to be adopted:

< 42·0	microdont
42·0–43·9	mesodont
44·0–45·9	megadont
> 46·0	hypermegadont.

As Pedersen (1949) and Moorrees (1957) have stressed, the value of this index is small, while Van Reenen (1961) has pointed out that the index has definite limitations: it cannot be used until the third molars have erupted; it can be applied only on crania; it is not a true measurement of tooth size, but a ratio between *two* variables; and, finally, a number of variables other than tooth size can affect the premolar–molar chord measurement (e.g. the curve of Spee, irregularities in the form of the arch and interstitial attrition with age).

Nevertheless, since *Zinjanthropus* is the first australopithecine in which the Dental Index can be determined, it was deemed of interest to place the results on record. Although the M^3's are not yet in the position of wear, they are sufficiently erupted to permit a reasonable determination of the P^3–M^3 chord to be made. The chord is 71·5 mm. on the left (M^3 almost in position) and 72·0 mm. on the right (M^3 a little higher than on the left). The basion–nasion length as determined on our reconstruction is 108·7 mm.

This gives *Zinjanthropus* a Dental Index of 65·8 per cent on the left and 66·2 per cent on the right, and places it in an extremely hypermegadont category. No comparable indices are available for other members of the Australopithecinae. In *Homo erectus erectus* IV, the chord is 55·5 mm. and the basion–nasion line, measured on the reconstructed calvaria, is 113·0 mm., giving an index of 49·1 per cent. In the reconstructed *H. e. pekinensis* cranium, the chord is 46·5 mm., while the mean basion-nasion length is 105·5 mm., giving an index of 44·1 per cent (measurements from Weidenreich, 1945).

Values for small samples of anthropoid apes come closer to those of *Zinjanthropus*. The mean for six orang-utan of both sexes was 55·2 per cent, that for six gorilla of both sexes 54·1 and that for six chimpanzee of both sexes 47·9 (Flower, 1885).

The highest mean values (per cent) for recent races of man are 48·1 for 13 Tasmanians, 46·1 for 100 Australian crania, 45·5 for 17 Andamanese and 44·2 for 21 Melanesians (Flower, 1885; Campbell, 1925). The range of individual values (per cent) in Campbell's 100 Australian crania was 41·3–50·0, while that for Shaw's (1931) 100 crania of South African Bantu-speaking negroids was 37·0–46·0. A wider range was obtained by Jacobson (1966) on 238 male Bantu crania, namely 34·5–50·0, the mean being 42·9; while his range for 67 female Bantu crania was 38·0–49·5 (mean 43·86).

B. 'Tooth material'

Howes (1954) used the sum of the M.D. diameters of maxillary and mandibular incisors, canines, premolars and first molars, as an expression of tooth material. Van Reenen (1961) has adapted this approach to interracial comparisons by accepting the minimum and maximum values of American Whites (163·5–187·0 mm.) as the arbitrarily-defined limits of a microdont category. The mesodont category has a range of the same extent, i.e. 187·1–210·6 mm., and the megadont category 210·7–234·2 mm. Presumably values above 234·2 mm. would be hypermegadont.

Unfortunately, Van Reenen's categories are based on maxillary and mandibular dentition, and the latter is not available in *Zinjanthropus*. However, Van Reenen does quote values for maxillary tooth material alone: in the following table (Table 44), the value for *Zinjanthropus* is compared with *mean* values taken from Van Reenen's Table II (1961, p. 349).

Zinjanthropus has the highest value of those cited in Table 44, though a rough estimate of the values for a gorilla sample, based on the means recorded by Ashton and Zuckerman (1950), gave 162·0 for males and 140·2 mm. for females (Van Reenen's figure of 60·7 mm. for gorillas of unstated sex is clearly wrong).

It should be pointed out, however, that this 'tooth material' estimate may give misleading results. First, it omits the second molar. In modern man, where most commonly $M^2 < M^1$, this may be no serious problem, but in fossil hominids the biggest tooth is often the second molar and relative dental size may be misjudged if this tooth is omitted. Secondly, 'tooth material' is based only on

Table 44. *Maxillary 'tooth material' (sum of mesiodistal crown diameters of left and right I^1–M^1) of* Zinjanthropus *and other hominids (mm.) (partly after Van Reenen)*

	Maxillary 'tooth material'
Zinjanthropus	127·7
Paranthropus	119·2
Australopithecus	109·4
Homo erectus pekinensis	110·2
Australian	103·8
East Greenland Eskimo	96·4
Bantu	96·4
Bushman	91·4
American White	95·4

M.D. diameters; it will therefore not reflect the true position where there is a strong tendency to enlargement of the B.L. diameters. The sum of crown areas might give a truer picture of the amount of tooth material. Thirdly, as mentioned above, it summates maxillary and mandibular M.D. dimensions and is therefore inapplicable to finds of isolated fossil jaws. A modification of the index would be desirable, in a direction of greater flexibility, so that, according to circumstances, upper or lower or both dentitions could be used, and M^2 and even M^3 could be included if present. This would require different sets of standard categories of micro-, meso- and megadontism to be worked out for each set of circumstances.

CHAPTER XVI

THE CROWN SHAPE INDEX OF THE TEETH

The postcanine teeth tend to be enlarged buccolingually far more than mesiodistally. Metrical expression can be given to this tendency by the calculation of the crown shape index, in which the M.D. crown diameter is expressed as a percentage of the B.L. diameter. In Table 45, the M.D./B.L. shape indices of *Zinjanthropus* are compared with those of a number of other hominoids. Two distinct trends seem to characterise the hominoids listed: one is a tendency towards relative buccolingual expansion from $\underline{C}$ to the premolars, resulting in most groups in progressively lower mean indices from before backwards; the other is a similar gradient starting at M^1 and decreasing through to M^3.

The canine–premolar trend shows the most marked drop from $\underline{C}$ to P^3; this is true in Australopithecinae (including *Zinjanthropus*), Homininae and Pongidae. The decrement in shape index from

Table 45. *Shape indices* (M.D./B.L.) *of maxillary permanent teeth of* Zinjanthropus *and other hominoids (per cent)**

	$\underline{C}$	P^3	P^4	M^1	M^2	M^3
Zinjanthropus (L)	89·7	64·1	65·6	85·9	81·9†	73·4
(R)	88·9	64·1†	68·2	85·9	81·9†	79·5
Paranthropus	89·9	70·2	68·9	95·1	91·7	89·0
Australopithecus	99·9	71·8	71·1	90·9	89·7	85·3
Homo erectus						
Sangiran IV ($n = 1$–2)	80·5	67·3	68·4	89·4	89·5	77·1
Choukoutien ($n = 4$–9)	92·9	70·1	69·8	87·0	85·8	83·2
'*Sinanthropus officinalis*' ($n = 1$)	—	—	—	93·4	—	—
Ternifine ($n = 1$)	87·0	—	—	83·3	—	—
Rabat ($n = 1$)	95·0	70·8	72·7	—	88·5	—
Total *H. erectus* ($n = 7$–12)	90·0	69·4	69·8	87·7	86·5	82·6
Homo sapiens						
Australian	93·4	75·7	71·3	89·0	83·2	81·3
East Greenland Eskimo	92·9	81·5	73·9	92·3	88·7	86·5
Aleut	94·6	76·2	73·1	91·2	87·2	85·6
Bantu Negroid	95·7	76·3	73·4	93·2	90·0	84·5
Bushman	96·2	79·1	76·5	93·4	91·5	79·6
American White	95·0	79·1	77·3	90·7	80·0	81·1
Pongidae						
Gorilla ♂	126·1	71·8	72·5	95·6	97·0	96·2
Gorilla ♀	128·9	74·3	72·6	97·9	97·4	94·4
Chimpanzee ♂	123·3	77·8	74·0	90·3	88·0	85·3
Chimpanzee ♀	124·5	78·3	74·6	89·8	88·2	86·0
Orang-utan ♂	115·6	76·4	73·4	89·0	85·9	83·5
Orang-utan ♀	131·1	76·9	73·3	92·8	87·7	84·7

* All figures in the table, save those for *Zinjanthropus*, Sangiran IV, '*Sinanthropus officinalis*', Ternifine and Rabat, are sample means or ratios of means; all samples (except the pongids) are combined male and female samples. The figures for Sangiran IV are the averages of left and right teeth, save the figures for M^2 and M^3, which are for right teeth alone.

† Index based on estimated dimension.

P^3 to P^4 is slight throughout these hominoids, but there are several exceptions in which P^4 has a slightly higher index than P^3. *Zinjanthropus* is one of them. Sangiran IV and Rabat are further examples, so that the overall picture in *Homo erectus* shows a mean shape index of P^4 which is a shade higher than that of P^3. The mean values for the male gorilla sample of Ashton and Zuckerman (1950) provide another instance. However, when one considers single specimens of *Paranthropus* and *Australopithecus*, in which both maxillary premolars are present and measurable on at least one side, the *Zinjanthropus* pattern finds a parallel in several examples of both forms. Of seven *Paranthropus* maxillae, the shape index of P^4 is higher than that of P^3 in three cases (SK 52 left and right, SK 65 and SK 845), the indices are equal in one (SK 57) and they conform to the general hominoid trend of a decrease in shape index from P^3 to P^4 in two specimens (SK 13 left and right, Kromdraai). Of five maxillae of *Australopithecus*, P^4 has a higher shape index than P^3 in two specimens (Sts 52a left and right, Sts 54) and a lower index in three (T.M. 1511, Sts 17, Sts 42). Thus a decrease from P^3 to P^4 in the broadening tendency relative to length characterises not only *Zinjanthropus*, but also a good proportion (five out of twelve) of available australopithecine specimens from South Africa.

The absolute values of the indices for P^3 and P^4 of *Zinjanthropus* are extremely low. For P^3, they are lower than any mean value in the P^3 column of Table 45, while for P^4, the left tooth of *Zinjanthropus* has the lowest value in the P^4 column, while the right P^4 of *Zinjanthropus* is the second lowest and is virtually equal to the mean of the left and right P^4's of Sangiran IV. The P^3 index of 64·1 per cent is just lower than the lowest values for *Paranthropus* (range 64·3–74·5 per cent) and for *Australopithecus* (range 65·4–81·3 per cent): that is to say, the relative broadening tendency reaches its most marked development in the *Zinjanthropus* P^3. The P^4 indices (per cent) of 65·6 and 68·2 fall within the range for both *Paranthropus* (63·0–75·2) and *Australopithecus* (56·5–79·2), the relative broadening tendency reaching its acme in a remarkable tooth from Sterkfontein, Sts 42, with a shape index of 56·5 per cent (Figs. 35 and 36).

The M^1–M^3 trend is clearly manifested in all the groups tabulated, with the exception of two. The gradient of buccolingual broadening starts out from a squarish M^1 with a mean shape index of just under 100 per cent; shows a modest drop to M^2 and a further drop to M^3. This trend is clearly shown by all the hominids in the table, save for Sangiran IV

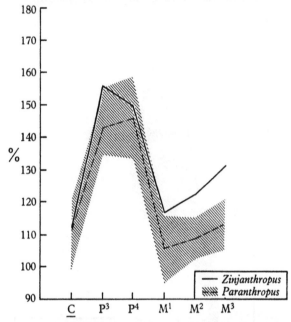

Fig. 35. Crown shape indices of maxillary teeth of *Zinjanthropus* and *Paranthropus*. The thick solid line represents the values in *Zinjanthropus*; the interrupted line the mean for *Paranthropus* and the shaded area the sample range. The indices for *Paranthropus* have been computed from the measurements recorded by Robinson (1956), by expressing the buccolingual diameter of each tooth as a percentage of the mesiodistal diameter.

in which the shape indices of M^1 and M^2 are practically the same, and the American Whites, whose index rises from 80·0 in M^2 to 81·1 per cent in M^3.[1]

Among the pongids, the trend is shown equally clearly by chimpanzee and orang-utan, the values closely resembling those for hominids. But gorilla departs from the trend, especially in the males: in the latter group the mean M^1 index is the smallest,

[1] It should be stressed, however, that the latter figures are not the means of the indices, but the indices of the means, for shape indices were not calculated in the original study of American Whites (Black, 1902); one was therefore obliged to compare the mean buccolingual diameter with the mean mesiodistal diameter, in order to arrive at an approximation to the mean shape index. Both the values for M^2 and M^3 may therefore be 1–2 per cent out in either direction.

the M² the greatest and the M³ intermediate, all three values being very similar and virtually the lowest in the table. That is to say, all three upper molars in male gorillas have a squarish form. In the female gorilla series, M¹ and M² are practically the same in mean shape index, approximating even more closely to a square, while M³ shows a slightly increased degree of buccolingual expansion.

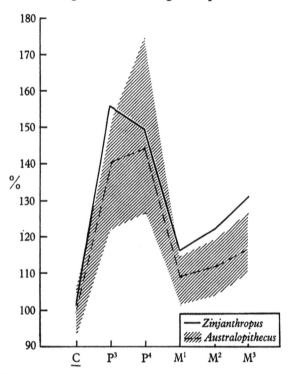

Fig. 36. Crown shape indices of maxillary teeth of *Zinjanthropus* and *Australopithecus*. The thick solid line represents the values in *Zinjanthropus*; the interrupted line the mean and the shaded area the sample range for *Australopithecus*. The indices are based on measurements recorded by Robinson (1956).

Zinjanthropus shows the M¹–M³ trend clearly; but, in comparison with the mean values for the South African australopithecines, the M¹ starting point is a buccolingually broader tooth. The gradient thus reaches much lower values in M³ of *Zinjanthropus* than of *Paranthropus* and *Australopithecus*.

The shape index of the *Zinjanthropus* M¹ (85·9 per cent) is almost the lowest of the M¹ indices recorded in Table 45, only Ternifine M¹ having a smaller index (83·3 per cent). It is just smaller than the lowest values (per cent) in sixteen *Paranthropus* M¹'s (range 86·7–106·1) and in fourteen *Australopithecus* M¹'s (range 86·5–97·7). The index of the *Zinjanthropus* M² (81·9 per cent) likewise falls close to the smallest value in fifteen *Paranthropus* M²'s (range 81·4–98·0) and below the smallest index in sixteen *Australopithecus* M²'s (range 83·1–107·1). The position is somewhat different in respect of M³: first, the shape indices of the left and right teeth in *Zinjanthropus* differ from each other by 6·1 per cent; secondly, the range in sixteen *Paranthropus* M³'s is extended by one tooth (SK 49 right) from a lower limit of 82·7 to one of 70·3 per cent! Excluding this somewhat atypical M³, the range for fifteen *Paranthropus* M³'s is 82·7–95·5 per cent, all of which values exceed those of both M³'s of *Zinjanthropus*. The range for eleven *Australopithecus* M³'s is 77·7–93·7 per cent: this range accommodates the right M³ of *Zinjanthropus* (79·5) but not the left (73·4). It may be concluded that the relative broadening tendency in the third maxillary molar of *Zinjanthropus* is about equal to or just greater than the maximum relative buccolingual broadening in the M³'s of *Australopithecus* and *Paranthropus*. The M³ of Sangiran IV has a comparable shape index (77·1 per cent) while, of living hominines, the Bushmen have the lowest value (79·6 per cent).

We have shown previously that the postcanine teeth of *Zinjanthropus* show remarkably high absolute buccolingual measurements. It may now be concluded that these are not simply part of a general enlargement of the cheek-teeth in all directions, but that the buccolingual expansion has been disproportionately great, as compared with the enlargement in a mesiodistal direction. The main growth peculiarity of the greatly expanded cheek-teeth of *Zinjanthropus* has been a buccolingual exaggeration of the crowns.

CHAPTER XVII

THE MORPHOLOGY OF THE TEETH

A. The maxillary incisors

1. *The central incisors*

Both I¹'s are present and damaged, most of the enamel being lost from the mesial and distal faces, as well as from the distal part of the labial face of the right I¹. The teeth are worn, the right more than the left, so that the height from the cervical line to the incisal edge is 8·1 mm. on the left and 7·4 mm. on the right. Thus the morphological features on only about the cervical two-thirds of the crown remain for examination.

The lingual surface is slightly shovel-shaped; the bases of the faintly raised, mesial and distal marginal ridges flank the lingual fossa. A small pit lies near the base of each marginal ridge, close to the mesial and the distal margins of the tooth respectively. Between the two marginal ridges are the cervical parts of two lingual depressions ('mesiale und distale Vertikalgrube' of Remane, 1960), separated in the midst of the lingual fossa by a slight cervico-incisal elevation ('mittleren Hauptleiste').

At the base of this cervico-incisal elevation, that is, on the upper part of the lingual face, is a rounded protuberance: we shall follow Robinson (1956, p. 23) in calling this the *gingival eminence*, and it corresponds to the 'basale Verdickung der mittleren Hauptleiste' which Remane (1960, p. 782) synonymises with *tuberculum linguale* or *tuberculum dentale*. The prominence is named 'dental tubercle' by Sicher (1949, p. 210). The marginal ridges converge as they pass upwards towards the gingival eminence, running into either side of the eminence: no ridge continues over the eminence (pl. 32).

The labial face is marked by the distinct hypoplastic strip discussed earlier (chapter XIII, C). The smoothness of the transverse curve of the labial face is interrupted distad by a faint vertical ridge, flanked by two shallow, vertical, labial grooves. The groove along the distal aspect of the ridge is somewhat deeper and more marked than that on the mesial aspect. As far as one can judge from the rest of the preserved part of the labial face, it is relatively flat. Both in this respect, and in the presence of labial grooves, the *Zinjanthropus* central incisor is more like those of *Australopithecus* than of *Paranthropus*: the latter form has a smoothly curved labial face and no grooves.

2. *The lateral incisors*

The right I² is completely intact, though worn, but the left I² has lost all enamel on the labial face and most on the mesial and distal faces, only the lingual surface having intact enamel.

The lingual surface seems to have been slightly shovel-shaped. The marginal ridges are apparent, and are clearly seen 'in section' on the incisal surface, where it is apparent that both enamel and dentine entered into the formation of the mesial and distal marginal ridges, especially the mesial one. At the base of the distal marginal ridge is a distinct pit, appreciably larger than the corresponding pit on I¹: the mesial, distal and cervical margins of the pit are slightly undercut, giving it strong relief on these aspects. A similar though smaller pit lies at the base of the mesial marginal ridge.

The gingival eminence in I² is more fully rounded and prominent than in I¹ and, on the right, it almost justifies the name lingual tubercle. The marginal ridges converge upwards on to its sides. Faint finger-like processes of enamel pass downward from the eminence into the midst of the lingual fossa, as in a few of the Swartkrans teeth and in *Homo erectus pekinensis*; they form a slight vertical elevation, as on I¹. The elevation is flanked by two vertical lingual depressions, which in turn are limited by the marginal ridges. The marginal ridges appear to meet each other over the gingival eminence, as in *Paranthropus* and *Australopithecus*, but the ridge is not well defined in the area of meeting. Nevertheless, the impression is conveyed that the lingual fossa is circumscribed by a continuous rampart (pl. 32).

The distal part of the incisal surface of the *Zinjanthropus* I^2 curves smoothly into the distal face on each side, whereas the mesial part of the incisal surface meets the mesial face at an acute angle. This effect is heightened by the incisal part of the mesial face being somewhat flattened for interproximal contact with I^1. The flattened area is close to the incisal surface and marks the level of maximum mesiodistal diameter; from this level of the tooth, the mesial and distal margins taper to the cervical line. The distal face is slightly vertically curved. These relations align this tooth with Robinson's type B lateral incisor, which is the commoner type in his Swartkrans (*Paranthropus*) sample and apparently the only type in the Sterkfontein (*Australopithecus*) sample. (Type A is represented by two teeth from Swartkrans which are almost identical with upper central incisors but which, on circumstantial grounds, Robinson has classified as lateral incisors, 1956, p. 26.)

As in the type B lateral incisor, the incisal surface of the *Zinjanthropus* I^2 slopes upwards distad, to turn smoothly over into the distal face. Since this slope of the incisal surface, even in a somewhat worn tooth, does not parallel the slope of the cervical enamel line on the labial face—in fact, the cervical line tends to slope upwards *mesiad*—it follows that the greatest height of the crown is at the mesial end of the incisal edge. Thus, on the right, the height is 5·9 mm. at the mesial end of the labial face and 5·2 mm. at the distal; on the left, despite the missing enamel, the outline of the cervical line is clear—the height is 6·3 mm. at the mesial end and 5·3 mm. at the distal.

The labial face has no grooves, resembling in this respect *Australopithecus*, whereas *Paranthropus* I^2's tend to develop shallow labial grooves. The surface is more or less evenly curved transversely, but the summit of the curve is somewhat nearer the mesial surface than the distal. Hence, as in *Paranthropus*, the greatest labiolingual diameter is nearer the mesial face. The vertical curve of the labial face is well rounded and even, save for the interruption produced by a hypoplastic area.

3. *The maxillary incisors: general*

Unfortunately, the author did not have the opportunity of studying the roots of the teeth before they were re-inserted in their alveoli. However, from root exposures here and there—and especially in the maxillary antra—as well as from skiagrams, it has been possible to learn much about root-form. As in the Swartkrans *Paranthropus*, the incisor roots of *Zinjanthropus* increase in diameter for some distance above the cervical enamel line, giving a maximum mesiodistal diameter at least 5–8 mm. above the cervical line. Then, the root tapers rather abruptly to its termination, curving slightly distally, so that a small angle appears between crown and root, as seen in labiolingual view. The abrupt taper gives way to a more cylindrical apical third, which ends with a terminal slight taper. Measurements on skiagrams are notoriously inaccurate through distortion: different views of the root of the central incisor gave length measurements of 21·0 and 23·4 mm. for the left I^1 and 21·0 and 23·8 mm. for the right I^1. The root seems clearly to have been of the order of 20 mm. or a little more, as in the Swartkrans *Paranthropus*.

In general morphology, the *Zinjanthropus* incisor crowns resemble those of *Australopithecus* and *Paranthropus*: there is a slight to moderate degree of shovel-shape; the lingual swelling corresponds to Robinson's simple, uncomplicated, gingival eminence rather than to a fully-developed lingual tubercle, although it is better developed on the lateral incisors than on the centrals. The features on the labial surface of both central and lateral incisors are more reminiscent of *Australopithecus* than of *Paranthropus*, but the comparative samples are small and little significance can be read into this observation on its own.

Metrically, the central incisors have a mesiodistal crown diameter which is somewhat larger than the mean for either of the two South African forms, and a labiolingual diameter which is large in comparison with that of *Paranthropus* but not as great as in two specimens of *Australopithecus*. The lateral incisors of *Zinjanthropus* are small in M.D. diameter, falling well within the ranges for both South African taxa, but they show a labiolingual enlargement greater than in any specimens of *Australopithecus* and about equal to the *Paranthropus* tooth which is greatest in L.L. diameter. *Zinjanthropus* is a little closer to *Paranthropus* than to *Australopithecus*, in the degree of reduction of I² in comparison with I¹: again one must stress the exiguous samples of front teeth available for comparison.

B. The maxillary canines

Both are intact and worn from the tip, with islands of dentine exposure. The most striking feature of the canines is their marked asymmetry. This serves to distinguish them from most of the maxillary canines of *Paranthropus* and *Australopithecus* to which Robinson (1956, p. 48) ascribes a 'more or less symmetrical appearance', and to relate them rather to the more markedly asymmetrical mandibular canines. Maxillary canine symmetry and mandibular canine asymmetry are deemed by Robinson (1956) to be such a clear-cut distinction as to have been one of the main factors which led him to re-classify as mandibular the very large, isolated, Sterkfontein canine, Sts 3, earlier regarded as maxillary (Broom *et al.* 1950), despite the fact that its dimensions pointed rather to its being a maxillary canine. Nevertheless, the *Zinjanthropus* canines which are *in situ* demonstrate that marked asymmetry may occur in an australopithecine upper canine, as does the unworn maxillary canine from Makapansgat which shows similar asymmetry.

The occlusal surface of the *Zinjanthropus* C̲ comprises two distinct planes as seen in labial or lingual view (pl. 35). The mesio-occlusal edge of the canine lies in contact with the distal face of I², a short distance above the occlusal surface of I². From that edge, the occlusal surface of the canine slopes steeply *downwards* distad for some 1·5 mm. Then there is an abrupt angular transition and the entire occlusal surface from that line distally slopes gently upwards, so that the distal end of the occlusal surface makes contact with the P³ some 0·9 mm. above the occlusal surface of the premolar. Although the slope is not as marked, it is nevertheless reminiscent of the sloping occlusal surface on the somewhat enigmatic Sterkfontein canine, Sts 3.

Further evidence of asymmetry is seen in the nature of the 'section' presented by the worn occlusal surface: the surface tapers both mesiad and distad, but the degree of tapering is much more marked distally. In fact, the distal part of the tooth is drawn out, as though into an accessory distal cusplet. This is well seen in the photographs and drawings of the occlusal surface of the tooth (pls. 29, 32, 33 and 35).

Asymmetry marks the features remaining on the lingual face of both canines. The gingival eminence (lingual or dental tubercle) is strongly developed and has a full and swollen appearance. Mesially, there is a remnant of a moderately developed mesial marginal ridge. Immediately behind this mesial ridge is a pit, which is better preserved on the less worn left canine than on the more worn right. It seems reasonable to assume that this pit represents the upper or cervical extremity of a mesial lingual groove, such as occurs in the maxillary canines of *Paranthropus* and *Australopithecus*. The pattern is more reminiscent of that in *Australopithecus*, however, for the mesial groove remnant in *Zinjanthropus* at its most basal extremity remains far forward or mesial in position, showing no tendency to pass distad or converge towards the distal lingual groove. The upper ends of the grooves remain widely separated, by 5·6 mm. on the left and 4·9 mm. on the right. This arrangement, whereby the two lingual grooves remain more or less parallel to each other and widely separated, is typical of *Australopithecus* from Sterkfontein and Makapansgat; it contrasts with the arrangement in *Paranthropus* from Swartkrans, in which the lingual grooves are more deeply incised and converge sharply on to the gingival eminence.[1] If the *Paranthropus* arrangement had obtained in the *Zinjanthropus* canines, at

---

[1] Two of the fifteen maxillary canines from Swartkrans (SK 4 and SK 27) do not conform to the typical *Paranthropus* pattern; in the former, there is no clearly defined distal lingual groove, 'nor are the ends of the grooves nearest the cervical line as close together as is usual' (Robinson, 1956, p. 43). In SK 27, apart from being metrically atypical, the canine has an atypical lingual face: 'Neither of the lingual grooves is clearly defined but are merely small depressions half-way down the crown and are widely separated (7 mm.).' Although atypical, these two *Paranthropus* specimens tend to approach the *Australopithecus* pattern.

their present level of wear the upper extremities of the mesial and distal lingual grooves would have been close together near the centre of the gingival eminence. The features mentioned on the *lingual* face of the *Zinjanthropus* canines resemble those of *Australopithecus*, as did those on the *labial* face of the incisors.

Thus far, we have spoken of the mesial marginal ridge and the mesial lingual groove of the *Zinjanthropus* C̱. Distally, there is a powerful distal marginal ridge, the occlusal 'section' of which forms the distal prolongation already mentioned in the account of the occlusal surface (pl. 32). This distal ridge is cut off mesially by a deeply-incised remnant of the distal lingual groove, which remains as a short length of actual groove, not a pit, despite the fact that the distal part of the tooth is more worn than the mesial. The distal groove not only is more deeply incised, but passes further cervically than does the mesial groove. Thus, the distal groove reaches within 3·3 mm. of the cervical line on the left and 3·8 mm. on the right (comparable measurements on TM. 1512 and TM. 1527 from Sterkfontein are 4·0 and 2·0 mm.), whereas the mesial grooves stop short 4·5 and 5·0 mm. from the cervical line. There are no prominent swellings between the bases of the lingual grooves and the cervical line; this resembles the position in *Australopithecus*, whereas such swellings are present in *Paranthropus*. Again, there is no forward or mesial curving of the distal groove and thus no convergence towards the mesial groove. So deep and distinct is this groove and so prolonged the distal portion of the tooth behind it, that what we have called the distal marginal ridge was clearly a distal cusplet, similar to those found moderately developed on the distal face of mandibular canines of *Paranthropus* and well developed on the mandibular canines of *Australopithecus*. In this regard there is a close resemblance between the canines of *Zinjanthropus* and the worn isolated canine from Sterkfontein, Sts 3. Since *Zinjanthropus* demonstrates that such marked asymmetry and distal cusplet formation can occur on *maxillary* canines of an australopithecine, the position of Sts 3 should once more be reconsidered.[1]

The labial face of the canine is disposed in two planes, a smaller portion facing anteriorly and a larger part facing laterally with a well-rounded curvature uniting the two parts (pl. 35). On the distal, laterally facing part of the labial surface are several faint vertical grooves, separated by tenuous linear elevations. The cervico-incisal curvature is well marked on the mesial half of the labial face, though interrupted by the strong hypoplastic features which were described in detail earlier. In the distal half of the labial face, the cervico-incisal curvature is poorly developed (pl. 32); there is, however, a transverse eminence parallel to and immediately below the cervical enamel line, reminiscent of Weidenreich's 'cingular band' on the upper canines of *H. e. pekinensis* (1937). Further incisally on this part of the labial surface, there is a distinct enamel line continuous with the hypoplastic features already mentioned: at this line, the contour of the surface turns fairly sharply lingualwards. On the most distal part of the labial face, there is a strong swelling of white enamel, which has no parallel on the most mesial part of the tooth. This swelling is on the labial aspect of the distal prolongation of the tooth; it is, in fact, the labial surface of the distal cusplet.

The cervical line is distinct and the enamel bulges out sharply from the root, as in Swartkrans canines. The line rises mesiodistally on the labial face, reaching the summit of its curve at about the junction between the anteriorly and laterally

---

[1] The large isolated canine, Sts 3, was originally described as a maxillary canine (Broom *et al.* 1950), largely, according to Robinson (1956, p. 48), on account of its size. Later, Robinson (1956) re-classified the isolated Sts 3 canine as a mandibular canine on the basis of its mesiodistal asymmetry, the presence of a distinct distal cusplet, and one or two other features which he ascribed exclusively to mandibular canines, as well as on an indirect argument based on the relative sizes of upper and lower canines. However, the maxillary canines of *Zinjanthropus* show both asymmetry and a distal cusplet, demonstrating that in at least this australopithecine specimen, these features are not the exclusive preserve of mandibular canines. Furthermore, the maxillary canine from Makapansgat (*see* Robinson, 1956, p. 45, fig. 12C) likewise shares both features with Sts 3 and *Zinjanthropus*, while two of the maxillary canines of *H. e. pekinensis* figured by Weidenreich (1937, Plate 4, figs. 38 and 39) show a semblance of these features, which was pointed out, at least in respect of the tooth in Weidenreich's fig. 38, by Broom *et al.* (1950, p. 41). Hence, the morphology of Sts 3 is compatible with its being an upper canine: its measurements strongly support that view. It is suggested that Robinson's re-classification of Sts 3 as a mandibular canine be once more reconsidered (Tobias, 1965*d*).

facing parts of the labial surface. Mesially and distally from this apogee, the cervical line descends markedly. On the mesial face, the buccal and lingual cervical lines meet at a sharp angle, furthest from the apogee; on the distal face, they meet in a rounded or blunted curve, the cervical line not descending as far on the distal face as on the mesial. This descent of the cervical line reduces the mesial and distal surfaces of the crown to two relatively small, rounded, triangular areas, of which the distal protrudes more strongly.

The root expands for a short distance above the neck, then tapers very gently to a blunt tip. There is a slight distal deflection, which becomes most apparent about half-way up the root where a gentle change in direction occurs.

In sum, the canines of *Zinjanthropus* have a lingual face very like that of *Australopithecus*. They are not at all like the highly distinctive maxillary canines of *Paranthropus*. On the labial face, the sudden bulge of the enamel just below the cervical line is a reminder of *Paranthropus*, and not of *Australopithecus* nor of the Homininae in general. The overall impression is of a canine marked by asymmetry and by a distal cusplet, features which have hitherto been ascribed exclusively to the mandibular canines of australopithecines, and especially to those of *Australopithecus*. In mesiodistal diameter, it falls within the range for the Swartkrans sample, but is small in comparison with the maxillary canines of Sterkfontein; on the other hand, in labiolingual dimension, it falls at the top of the *Australopithecus* range but well within the *Paranthropus* range. Thus, while its crown shape index is almost identical with the mean shape index of *Paranthropus*, it is appreciably lower (i.e. the tooth is broader in a labiolingual direction) than the mean for *Australopithecus* (Table 45). The module and crown area of the *Zinjanthropus* canines are comfortably accommodated within the ranges for the other two forms. *Zinjanthropus* thus agrees with *Paranthropus* in showing no relative labiolingual reduction of the canine, a trend towards which is apparent in the canines of *Australopithecus* (Robinson, 1956, pp. 52–3).

The smallness of the *Zinjanthropus* canine, in comparison with the robust size of the postcanine teeth, is well brought out by comparisons of $\underline{C}$ with P^3 and P^4, since the latter teeth bear a fairly constant size relationship to the premolar–molar tooth row in the australopithecines. From such comparisons, the $\underline{C}/P^3$ and $\underline{C}/P^4$ ratios of the *Zinjanthropus* teeth are lower than those of any australopithecine dentition yet described: this is true whether we compare the canine with each premolar by the M.D. crown diameter, the B.L. crown diameter, the module or the crown area. Thus, in the Olduvai fossil, the canines are more reduced relative to the postcanine teeth than they are even in *Paranthropus*. In most comparisons, *Zinjanthropus* lies closer to *Paranthropus* than the latter does to *Australopithecus*; only in the comparison between the B.L. diameters of $\underline{C}$ and P^3 does *Paranthropus* stand midway between *Zinjanthropus* and *Australopithecus*. This follows from the inordinate degree of buccolingual expansion which the *Zinjanthropus* P^3 has undergone.

C. The maxillary premolars

1. *The anterior premolars* (P^3)

Both are present and moderately worn, with exposure of dentine on both cusps. The left P^3 lacks a small area of enamel close to the cervical line on the buccal face; the defect does not extend up to the occlusal face. The limit of the cervical line is, however, still clearly marked on the dentine surface (pl. 35). The right P^3 has suffered post-mortem loss of most of the enamel on the lingual face, from occlusal surface to cervical line; again, the limit of the cervical line is still clearly marked on the dentine (pl. 36).

On the occlusal surface, the base of the central primary fissure is still clearly apparent on the left, but only a faint trace is present on the right. This fissure demarcates the buccal from the lingual cusp. Nothing can be said of the relative development of the two cusps in height, but in B.L. diameter, the buccal cusp is more extensive. It measures 8·6 mm. from the buccal edge of the occlusal surface to the central fissure on the left and 8·25 mm. on the right, whereas the lingual cusp measures 5·4 mm. from the lingual edge to the fissure on the left and an estimated 5·7 mm. on the right. The lingual cusp seems to have been slightly mesiad of the buccolingual centre line of the tooth, as in both *Paranthropus* and *Australopithecus*, whereas the buccal cusp seems to have been on this centre line, as is common in the Sterkfontein premolars. Apart from the central fissures, traces of both the mesiobuccal and distobuccal grooves are apparent on the left and of the distobuccal alone on the right; no trace of the lingual grooves remains on the occlusal surface.

On the buccal face, aside from the succession of hypoplastic lines, faint vertical depressions represent mesial and distal buccal grooves, the mesial being more distinct on the left, the distal on the right. The extraordinarily well-developed buccal grooves and marginal elevation about the buccal face of P^3 in the Sterkfontein premolars are not present in *Zinjanthropus*. However, as in the Sterkfontein specimens, the buccal face is narrower near the cervical line than at the occlusal plane, and P^3 is buccolingually broader near the cervical line with the beginnings of a marked taper (truncated at the occlusal surface). The enamel on this face forms an upwards directed peak between the two buccal roots (pl. 35).

The buccal half of the mesial face projects slightly more mesiad than the lingual half: this mesially projecting portion makes the main contact with the canine 0·9 mm. above the occlusal surface of the premolar. The cervical line on the mesial face forms a rounded peak.

In contrast with the mesial face, the *lingual* half of the distal surface projects more distad and makes contact with P^4 (pl. 20). The cervical line on this face is bluntly rounded with a downwards convexity, not peaked as on the mesial face.

On the lingual face, the cervical line is smooth and horizontal. The enamel bulges markedly in a lingual direction from the cervical line, then flattens, sloping buccally towards the occlusal face. The bulge is purely enamel, as may be seen on the right P^3: the dentine exposed on the lingual face slopes slightly *buccally* from the cervical line and does not bulge lingually. The broken edges of enamel mesial and distal to the area of loss confirm that the bulge is due to enamel alone: from a very thin layer of enamel near the cervical line, it thickens abruptly to reach a maximum thickness of 1·6 mm. distal to the deficiency and 1·5 mm. mesial to it, at a level approximately 3·0 mm. below the cervical line (pl. 36).

Both teeth have two buccal roots and a single lingual one, as in fifteen out of twenty Swartkrans P^3's but in only one out of thirteen[1] Sterkfontein P^3's (Robinson, 1956, p. 59). Robinson recognised two types of triple-root system in his sample of maxillary anterior premolars from Swartkrans: in one, the two buccal roots are fused for about half of their length, while the mesiobuccal root is connected to the lingual root by a thin plate for most of their length. In the other type the roots are free, although the root trunk or common stem extends for one-third to one-half of the total root length from the cervical line. It is not possible to detect any connecting plate on the skiagrams, but the two buccal roots are clearly free, suggesting that the root pattern in *Zinjanthropus* conforms to Robinson's second type of three-rooted premolar.

Although the differences between the morphological features of *Australopithecus* and of *Paranthropus* are not marked, in most of the foregoing descriptive details the *Zinjanthropus* P^3 accords more closely with that of *Paranthropus*, although in a few respects it is reminiscent of the premolars of *Australopithecus*.

Metrically, the Olduvai premolars are far greater than those of *Australopithecus*, in both M.D. and B.L. crown diameters. The M.D. diameter is just greater than the greatest M.D. diameter of a *Paranthropus* tooth; but the B.L. diameter of the anterior premolars from Olduvai far exceeds that of the largest australopithecine maxillary premolar. It is nearly 2·0 mm. bigger than the top of the range for *Paranthropus* and just over 3·0 mm. bigger than the biggest value for *Australopithecus*. Most of this buccolingual expansion seems to have resulted from excessive development of the buccal cusp, which throws this part of the premolar out of alignment with the adjacent parts of the dental arcade (Fig. 22 and pl. 20). This is well brought out by the crown shape index (63·1 per cent), which is lower than any mean value in the hominoids and falls short of the individual minimal values for both South African australopithecine taxa. The module

[1] Given as one out of fifteen P^3's by Robinson (1954e, p. 271).

and the crown area of the *Zinjanthropus* P³ exceed the highest values for both *Paranthropus* and *Australopithecus* by appreciable margins. Clearly, P³ of *Zinjanthropus* is an excessively large tooth.

2. *The posterior premolars* (P⁴)

Both are in place, intact and moderately worn, though less so than P³. More of the fissure pattern and even the posterior fovea are preserved.

On the occlusal face, the central fissure clearly separates the two cusps. The buccal cusp is more or less on the buccolingual centre line of the tooth, but the lingual cusp is slightly mesiad of this line. Again, the relative heights of the two cusps cannot be determined, but the surface extent of each may be gauged, as before, by measuring the distance on the occlusal face from the buccal and lingual edges to the central fissure. In marked contrast with P³, the buccal and lingual cusps have the same dimension—8·1 mm. on the left P⁴ and 7·5 mm. on the right P⁴. That is to say, the diameter of the buccal cusp of P⁴ (8·1, 7·5) is slightly less than that of the buccal cusp of P³ (8·6, 8·25), whereas the diameter of the lingual cusp of P⁴ (8·1, 7·5) is appreciably greater than that of the lingual cusp of P³ (5·4, 5·7), so that in P⁴ the two cusps are of equal diameter. Thus, P³ and P⁴ are heteromorphic in manifesting directly opposite trends: in the former the buccal cusp is greatly expanded in a B.L. direction, but not the lingual cusp: in the latter, the buccal cusp is expanded slightly less in B.L. diameter, but the lingual cusp is greatly expanded in this direction. The different relative development of the buccal and lingual halves of the two premolars is clearly apparent when the teeth are inspected in the dental arcade, for these opposite trends result in a relative displacement of the premolars from the smooth contour of the arcade—P³ being displaced relatively buccally and P⁴ relatively lingually (pl. 29).

The fissure pattern is roughly H-shaped, the distobuccal and distolingual arms being strongly developed (pl. 33). The mesiobuccal limb is present, though less clearly marked; on the right, it is seemingly represented by two parallel fissure remnants. There is no trace of a mesiolingual limb. One or two faint suggestions of secondary grooves are detectable, but at this level of attrition it is not possible to discern any clear-cut talon or cuspule, both of which features characterise the P⁴ of *Paranthropus*. Nevertheless, the occlusal surface of P⁴ is somewhat more squarish and molariform than that of P³, largely because of the distobuccal talon-like protrusion of the P⁴, as well as the greater length of P⁴ as compared with P³ (pl. 33). These features of the *Zinjanthropus* posterior premolar draw it somewhat closer to that of *Paranthropus* than to that of *Australopithecus*, in which the occlusal surface of P⁴ closely resembles that of P³ (Robinson, 1956, p. 63).

On the buccal face, the enamel bulges immediately below the cervical line, but not as markedly as in P³; in P⁴ the bulge is more evenly rounded cervico-occlusally, the taper towards the occlusal surface being less abrupt. The base or upper extremity of the distal buccal groove is well defined; the part of the buccal face distal to it is large and protuberant distobuccally, as in most Swartkrans specimens. There is only the faintest indication of the base of the mesial buccal groove, as in Swartkrans premolars. As in the latter type and, in contrast with the premolars of *Australopithecus*, the buccolingual diameter is appreciably greater near the enamel line (18·0, 17·6 mm.) than at the occlusal surface (16·0, 14·8 mm.).

There is a slight mesial projection on the buccal half of the mesial surface. The mesial face is excavated slightly on the left, but not on the right. There is a round or oval area of interproximal contact with P³ on the mesial surface about midway between the buccal and lingual faces. The cervical line descends in a broad convexity on this face.

The buccal part of the distal surface protrudes more than the lingual part and makes contact with the paracone of M¹. On this face, the cervical line descends slightly to a broad-angled peak over the lingual root; it is not as low as on the mesial face.

The cervical line on the lingual face is horizontal (pl. 36). The lingual face is smoothly rounded and featureless, there being not even a trace of a distal lingual groove. This is the regular arrangement of the lingual face in *Australopithecus*, but occurs, too, in a minority of *Paranthropus* P⁴'s.

Three roots are present on the *Zinjanthropus*

P^4's and they are clearly separated, as in eight out of ten Swartkrans P^4's. In contrast, there is a single lingual root on all ten Sterkfontein teeth. It is not possible to detect on the skiagrams whether the mesial buccal root in *Zinjanthropus* is connected to the lingual root, as in one type of three-rooted posterior premolar of *Paranthropus*.

The bulk of its morphological features align the P^4 of *Zinjanthropus* more closely with that of *Paranthropus*, although in a few respects, such as the lingual face, there is once more a suggestion of an affinity with *Australopithecus*.

Metrically, the main feature distinguishing the *Zinjanthropus* P^4 is its tremendous buccolingual expansion, its B.L. diameters of 18·0 and 17·6 mm. exceeding the top of the *Paranthropus* range by 1·1–1·5 mm. and the top of the *Australopithecus* range by 3·7–4·1 mm. The crown shape index, however, shows that the buccolingual broadening tendency manifested by P^4 of *Zinjanthropus* is not the most extreme for this tooth, being exceeded by that of several teeth of both South African forms. Although not so exaggerated, the M.D. diameters, too, are large, lying at and just beyond the maximum for *Paranthropus*. Hence, the module and crown area of the Olduvai P^4's far exceed those of the South African australopithecines.

When the ratio of P^3/P^4 is considered, *Zinjanthropus* approximates to *Paranthropus*, whether the comparison is made between M.D. or B.L. diameters, modules or crown areas. In fact, the increase in size from P^3 to P^4 in *Zinjanthropus* is about equal to the greatest degree of increase in any of the South African specimens where both premolars are present in the same maxilla. This steepness of the P^3–P^4 curve in the East African form might be interpreted in at least two different ways. Robinson's view, repeatedly expressed in his publications over the last dozen years (e.g. 1952*b*, 1954*d*, 1956, etc.), is that the australopithecines were undergoing dental reduction, but that this process was retarded in the *Paranthropus* cheek-teeth (e.g. 1954*d*, p. 328). On this basis, the relative sizes of P^3 and P^4 in *Zinjanthropus* might suggest that in this form P^4 had not yet begun to reduce to the smaller size characteristic of the Homininae, whereas P^3 had reduced its size somewhat, especially in respect of the lesser development of the lingual cusp and of the distal, talon-like part of the tooth. On the other hand, the same facts could point to a reverse trend, namely, a specialising expansion of the cheek-teeth. We have earlier found reason to regard at least one other feature of the *Paranthropus* dentition as a specialisation from the common hominoid pattern, namely, the greater size of M^3 than M^2. It seems very probable that the general enlargement of the cheek-teeth is a specialised 'field' phenomenon, which has not only spread throughout the molar field, but has affected the premolar field. Following Butler's (1939) 'field concept', as interpreted and elaborated for hominids by Dahlberg (1949) and Moorrees (1957), we may say that the molar tendency to expansion has affected most the more variable of the two elements in the premolar field, namely P^4, and only to a lesser degree the more stable element, P^3. It is suggested here that the greater size of P^4 in *Zinjanthropus* may thus be seen as a manifestation of a greater degree of molarisation of the P^4. It may be mentioned, in passing, that the *lower* posterior premolar of the robust mandible from Peninj, near Lake Natron, about 80 kilometres north-east of Olduvai Gorge, is highly molarised; in its worn state it strongly resembles a molar tooth (*see* illustrations in Leakey and Leakey, 1964; Tobias, 1965*c*). The view supported here is that the enlarged cheek-teeth of *Zinjanthropus* and *Paranthropus* do not represent the retention of a primitive feature, but constitute a specialisation. This matter will be reverted to in the general discussion.

D. The maxillary molars

1. *The first molars*

Both first molars are present, intact and fairly markedly worn, with striking hollow-wear on the hypocone of the left M^1.

The shape of the occlusal surface is that of a nearly equal-sided parallelogram, with the long axis running from mesiobuccal to distolingual (pl. 37). There are four cusps, the structure of which is still fairly clear. The mesiolingual cusp (the protocone) is partly limited buccally by the remnant of a fissure passing mesiad from the central fovea. The less worn left M^1 shows this well; there is only the faintest trace of it on the right. Distally, the protocone is partly limited by a transverse remnant of the oblique fissure, which demarcates it from the hypocone: this transverse remnant of the oblique fissure runs across the lingual margin of the occlusal face and is isolated by a short distance from the rest of the oblique fissure, which runs to the posterior fovea. The protocone of the left M^1 has a remnant of a deep Carabelli groove on the occlusal face, close to the mesiolingual corner of the tooth. A fainter though still definite groove is found in the same position on the more heavily worn right M^1. Such a Carabelli feature is widespread in australopithecine maxillary molars (Robinson, 1956).

In the *Zinjanthropus* M^1, the oblique crest (trigon crest) connects the distobuccal corner of the protocone to the metacone, the flanking fissures being better preserved on the left than on the right. At this level of wear, the oblique crest is continuous. There is no suggestion of any interruption by a fissure crossing the crest, as is commonly the case in *Paranthropus*, still less of the formation of two small cuspules replacing the crest, a formation dubbed by von Koenigswald 'double metaconulus' and said to characterise the type specimen of his *Hemanthropus peii* and *Paranthropus* from Swartkrans (von Koenigswald, 1957). Although it cannot be avowed that, in the unworn state of this tooth, the oblique crest was not divided, in its preserved simple formation, it strongly resembles the morphology of M^1 in *Australopithecus*.

The metacone is well defined. On its lingual border is the deep posterior fovea with the oblique fissure leading mesiad from it. On its mesial border is a buccal groove, extending from the central fovea and, after a short interruption, incising the buccal edge of the occlusal face. At the distobuccal angle, a slight projection extends distad from the metacone (pl. 37).

The paracone is partly limited by the central fovea at its distolingual angle and by two fissures extending buccally and mesiad from the fovea. At this level of wear, an anterior trigon crest connects the protocone and paracone, but the area mesial to the crest, which might be expected to feature an anterior fovea cut off from the central fovea, is worn to a marked degree with islands of dentine exposure which, on the right, are confluent. The buccal limb of the central fovea is deep and distinct.

The hypocone is somewhat distinct from the trigon on the right tooth; that on the left is markedly hollowed with broken and damaged enamel, resulting in a wide exposure of dentine in a 'stepped down' area. The hypocone is delimited by the very deep posterior fovea and the oblique fissure leading from it, across to the lingual edge. A small interruption in the oblique fissure permits the enamel of the hypocone to become continuous briefly with that of the protocone.

From the central fovea, a triradiate, or perhaps even quadriradiate fissure sends arms buccally, mesially, lingually and distally; while from the posterior fovea, a transverse fissure and an oblique fissure radiate. The oblique fissure circumscribes the hypocone and passes on to the lingual face.

The occlusal edge of the buccal face features a V-shaped nick or groove (pl. 39), which represents the continuation of the groove separating the paracone and metacone on the occlusal surface. The nick is about midway along the mesiodistal extent. The enamel above the nick rises to a peak between the two buccal roots: the peak is a little sharper on the right than on the left. Immediately below the cervical line, the enamel flares buccally, but not as much as in the premolars; then it tapers slightly towards the occlusal surface. Thus, the tooth has a greater buccolingual diameter near the cervical line than at the occlusal surface, the respective B.L. dimensions being 17·7 and 16·3 mm. on the left and 17·7 and 16·2 mm. on the right. The taper is therefore gentler than in the premolars, in keeping with the more modest enamel bulge near the cervical line in M^1 (pl. 39). Above the nick, the buccal groove continues upwards on the buccal surface, tending to fade out near the cervical line without a basal pit, as in *Australopithecus* but in contrast with *Paranthropus*. Immediately mesial to the nick is a small swelling, delimited mesially by another buccal groove on the buccal face of the paracone: this little swelling is so clearly demarcated as possibly to have constituted a small *paramolar cusp* or *buccostyle* in the unworn tooth (pls. 37 and 39). A similar feature occurs, too, in the first maxillary molars from Sterkfontein, Taung and Makapansgat (i.e. *Australopithecus*) but not in those from Swartkrans (i.e. *Paranthropus*). It is to be distinguished from the *molar tubercle of Zuckerkandl* which (at least in deciduous upper first molars) sometimes appears as a hemispheric accentuation of the cervical ridge or buccal cingulum in the mesial half of the buccal surface (Sicher, 1949, p. 234). Korenhof (1960) has discussed in detail the diversity of names and descriptions applied to the cusps, tubercles, 'buccostyles' and other features which he regards as remnants of the buccal cingulum. Reviewing the data on the South African australopithecines, as described and illustrated by Robinson (1956), he concludes 'that buccal remains of the cingulum are not a rare occurrence in the Australopithecinae, although they are present in a weakly developed aspect only' (1960, p. 324). The buccostyle or paramolar tubercle on the *Zinjanthropus* M^1 may perhaps be viewed as a faint vestige of the group of buccal cingular remnants. As Frisch (1965) has stressed, there is a trend toward complete reduction of the cingulum in all hominoid evolutionary lines. The trend has gone furthest in the Hominidae; the buccal cingulum is lost first, both in upper and lower molars, and from M^1 sooner than from the other molars. In the light of these generalisations, the very slightness of the buccostyle in the *Zinjanthropus* M^1 is not altogether surprising.

Near the cervical line on the mesial surface is a depression, so the occlusal part of the mesial face protrudes more mesiad than does the cervical part (pl. 35). These features occur in both South African forms of australopithecine.

The distal face is convex and has no depression near the cervical line. The crown is shorter mesiodistally near the cervical line than at the occlusal surface.

On the lingual face (pl. 36), the base of the lingual groove can be seen, ending in a small pit or depression 3·4 mm. on the left and 4·5 mm. on the right below the cervical line; the pit lies just mesial of a peak in the cervical line. This peak is shallow and points to a groove on the lingual face of the lingual root. Immediately below the peak but just above the lingual groove and pit is a small marked swelling of enamel, causing a local protrusion in a lingual direction: it would appear to be a local development from the cingulum (Frisch, 1965), such as Robinson has reported on the lingual face of the Sterkfontein M^1's and especially the M^2's. It is, however, more localised in *Zinjanthropus* than in *Australopithecus*. No comparable feature is reported in *Paranthropus*.

The skiagrams reveal two smaller buccal roots and a stout lingual root with bifid tip and slender, twin, root canals. The lingual root has a moderately deep groove down the lingual face, as in *Australopithecus* from Makapansgat.

From the foregoing description, the *Zinjanthropus* M^1 shows a number of morphological resemblances to *Australopithecus* rather than to *Paranthropus*.

It is clear from the dimensions of M^1 that the tendency towards relative buccolingual enlargement continues distad of the premolars, for, while the first molar of *Zinjanthropus* is an overall large tooth exceeding the biggest specimen of *Paranthropus* in both crown diameters, the buccolingual excess over the *Paranthropus* mean is 3·71 and the mesiodistal 2·92 S.D.'s. This is brought out as well by the crown shape index, the value of which (85·9 per cent) is lower than any of the means for hominoids and the lowest individual values for *Paranthropus* (86·7 per cent) and for *Australopithecus* (86·5 per cent). Both the module and the crown area of the *Zinjanthropus* M^1 far exceed the largest values for *Paranthropus* and *Australopithecus*: thus, the module of 16·45 mm. exceeds that of the largest estimated *Paranthropus* M^1 by 0·95 mm. or 6·1 per cent, whilst the crown area of 269 mm.^{2} exceeds the highest estimated *Paranthropus* value by 30 mm.^{2} or 12·6 per cent.[1]

---

[1] It is interesting to note from this example that the use of the module alone tends to minimise the size disproportion; this is important because Robinson (1960) has claimed that the teeth of *Zinjanthropus* are only 8·4 per cent bigger than 'the only male specimen (of *Paranthropus*) with good teeth which has P^3–M^1 preserved in sequence and is of the same dental age as the Olduvai specimen'. It is not at all clear how Robinson arrived at this figure of 8·4 per cent. He does not state whether it was obtained by comparing the P^3–M^1 chords of the two specimens, or, as Robinson usually employs the module to compare sizes, whether he compared the sums of the modules of the two premolars and the first molar, or whether 8·4 per cent is the mean of the differences between the modules of the different teeth. In any event, it is not clear why the comparison should be limited to

2. The second molars

Both are in position in the *Zinjanthropus* maxilla, both are somewhat worn and damaged, especially in the mesial half. Fortunately, however, the roots are exposed in the maxillary sinuses.

The occlusal face is worn and much damaged, making the identification of cusps difficult (pl. 37). The tooth as seen from this surface is somewhat skewed, though not as markedly as M^1, in contrast with *Paranthropus* in which M^2 is more skewed than M^1. There is a slight degree of reduction of the distobuccal angle, but it is definitely not as marked as in *Paranthropus*. The four basic cusps are seemingly present and the posterior fovea is clear, with traces of the oblique fissure circumscribing the hypocone. Transverse limbs and a distal limb also radiate from the posterior fovea. On the right, the hypocone is still a low, rounded tubercle, but it is more worn on the left.

The metacone is a distinct, low, rounded tubercle on the left, with a clear oval wear facet on its summit, but, as yet, no exposed dentine (pl. 37). It is encroached upon by the buccal transverse limb emanating from the posterior fovea, and also by a secondary fissure running obliquely distobuccally from the main buccal groove, which limits the metacone mesially. Another few traces of secondary grooves are apparent on the metacone, towards the posterior fovea. There is a tendency to the formation of cuspules ('distoconuli' of Remane, 1960) along the posterior margin and one fairly small extra cusp can be made out distal to the posterior fovea though, through attrition, its limiting fissures are only slightly apparent. Such an extra cusp is present, too, in *Australopithecus*. Again, as in M^1, a slight projection extends distad from the distobuccal angle of the metacone.

The more mesial structures—the central fovea, the oblique crest, Robinson's U-shaped fissure, including the paramesial groove—cannot be identified through a combination of damage and fracture lines, wear and dentine exposures.

On the buccal face of the *Zinjanthropus* M^2, enamel is missing from the mesial one-third on the left and the mesial half on the right (pl. 39). In addition, on the right M^2, the enamel has been lost (post mortem) along an irregular strip near and parallel to the cervical line. The pitting of the surviving buccal enamel has already been referred to, but at least one of the pits is deep and is apparently a morphological entity (pl. 39 L). This pit lies about half-way along the mesiodistal extent, 4·5 mm. from the cervical line, and close to a deep nick on the bucco-occlusal edge. The nick is an exaggeration of the nick in a corresponding position on the first molar. The margin of the pit has a clear thickened edge. These features are lacking on *Paranthropus* M^2's. The enamel bulges a short distance below the cervical line, as in *Australopithecus* and in contrast with *Paranthropus*, then it tapers towards the occlusal surface. Thus, the maximum B.L. diameter of the occlusal surface is some 3–4 mm. less than the B.L. diameter near the cervical line; through enamel loss and damage, the exact diameters cannot be ascertained.

A pair of vertical buccal grooves flanks a slight paramolar cusp-like feature, which would be regarded as a remnant of the buccal cingulum on the criteria of Korenhof (1960) and Frisch (1965). The more mesial buccal groove is narrower and deeper and has the aforementioned pit in it; it is in line with the cervical peak of enamel. The more distal buccal groove is broader, shallower and hence less defined. It is separated from the mesial groove by a blunt, vertical, faint ridge, which is not as prominent as the comparable feature on M^1. This ridge passes upwards towards the cervical line, but when about 2·0 mm. from the line swings mesiad to enclose a tiny pinpoint pit 1½ mm. from the cervical line (pl. 39 L). Mesially, the faint ridge becomes continuous with another faint vertical elevation, which forms the anterior lip of the mesial groove. The distal faint ridge is reminiscent of the poorly developed buccostyle or paramolar cusp on M^1. The whole complex suggests the morphology of the *Australopithecus* second molars from Sterkfontein and that in the calvariofacial fragment from Makapansgat; nothing similar is found on the *Paranthropus* M^2's.

P^3–M^1, since the most dramatic size disparity between *Zinjanthropus* and the biggest corresponding tooth of *Paranthropus* is in M^2, the crown area of which in *Zinjanthropus* exceeds that of the largest tooth of *Paranthropus* by 95·9 mm.² or 36·1 per cent!

The mesial and distal halves of the cervical line on the buccal face form two somewhat uneven curves, gently concave upwards, meeting in a slight though definite peak. The peak does not lie in the customary position between the two buccal roots, but mesiad of the separation, against the distal face of the mesial buccal root (pl. 39).

The mesial face slopes markedly mesiad from cervical line to occlusal edge, in direct contrast with the distal face. The area of most mesial projection is at the occlusal edge, over the lingual half of the paracone, where contact with M^1 is established. Some interproximal attrition is apparent at this contact, the convex distal face of M^1 fitting into a slight concavity on the mesial face of M^2.

On the distal face of the *Zinjanthropus* M^2, there is a depression or faint groove just below and parallel to the cervical line; then the enamel bulges rather abruptly distad for a few millimetres. Below that, there is an abrupt change of contour, the distal surface tapering mesiad as far as the occlusal edge. Thus the most distal point on the tooth is at the summit of the bulge, along the line marking the change of contour. The contact facet with the incompletely erupted M^3 is at this level (pl. 39).

The lingual face is almost intact on both sides (pl. 38). The enamel bulges gently from the cervical line downwards, to reach a maximum bulge nearly 4·0 mm. below the cervical line. The upper half of this 4·0 mm. is somewhat depressed as a faint groove paralleling the cervical line. Below the swelling, this face tapers buccally to a fairly marked degree, as far as the occlusal surface, thus contributing to the reduction in B.L. diameter of the occlusal face mentioned above. The upper, lingually tapering part of the surface meets the lower, buccally tapering part at a fairly sharp angle, so that there is an abrupt change of contour.

Close to the occlusal margin on this face, the enamel is somewhat irregular and pitted, with several vertical grooves extending from the occlusal face. The most central groove is the continuation of the oblique fissure and there is a deep pit just as it turns over on to the lingual face (pl. 38). Mesial to this groove is another pitted groove on the lingual face of the protocone: it is the remnant of a Carabelli groove. Distal to the oblique fissure on the less worn left M^2 is a series of irregular vertical grooves separating several enamel tubercles, similar to the Carabelli structures on the lingual face of the protocone of the *Zinjanthropus* M^3 (see below). On M^2, these structures are well marked over the lingual face of the mesial half of the *hypocone*; but attrition over the protocone is so much more pronounced than over the hypocone (pl. 38) that it is no longer possible to detect whether the Carabelli structures present on the protocone were well developed too. Normally, the much-discussed Carabelli structures adorn the protocone (Korenhof, 1960; Frisch, 1965); but on the M^2 and M^3 of *Zinjanthropus* (see below), they seemingly 'overflow' on to the hypocone to a certain extent. It is likely, then, that M^2 possessed well-developed remnants of the protoconal cingulum, which extended distally on to the hypocone. The cingular remnants on the lingual face are reminiscent of those in *Australopithecus*, and, for the state of wear, are more definite than in *Paranthropus*. The cervical line, as on the buccal face, forms two low concavities meeting at a peak over the mesial face of the distal root.

Through the maxillary sinus, one sees a broad, robust lingual root and two well-separated buccal roots.

Once more, we see that in a number of its morphological features, the M^2 of *Zinjanthropus* agrees closely with those of *Australopithecus* rather than with those of *Paranthropus*.

In absolute size, M^2 is greatly in excess of the largest teeth of *Paranthropus* and *Australopithecus*. Its M.D. diameter of 17·2 mm. exceeds the greatest value for *Paranthropus* by 1·5 mm. and for *Australopithecus* by 2·1 mm. It exceeds the respective means by 3·71 and 4·30 S.D.'s. Its metrical predominance, however, is most developed in its B.L. diameter, which exceeds the greatest value for *Paranthropus* by 4·1 mm. and for *Australopithecus* by 3·9 mm. It surpasses the respective means by 5·97 and 5·60 S.D.'s. The buccolingual preponderance is brought out strikingly by the crown shape index (81·9 per cent), which falls short of all the hominoid means except that for American Whites (80·0 per cent, Black, 1902); it is close to or less than the lowest values for

Paranthropus (81·4 per cent) and *Australopithecus* (83·1 per cent). Both the module and the crown area are far in excess of the maxima for the two South African forms. Its module of 19·1 mm. is 2·8 mm. or 17·1 per cent greater than that of the largest *Paranthropus* homologue, and 3·0 mm. or 18·6 per cent greater than that of the largest *Australopithecus* M^2. The crown area of 361·2 mm.^{2} is almost as great as the maximum rectangle or crown area recorded by Weidenreich for a *Gigantopithecus* upper molar (specimen 3), namely 378 mm.^{2} (wrongly given as 308 mm.^{2} in Table 10 of Weidenreich, 1945). It exceeds the maximum *Paranthropus* value by 95·9 mm.^{2} or 36·1 per cent and the greatest *Australopithecus* value by 103·0 mm.^{2} or 39·9 per cent.

3. *The third molars*

Both teeth are present in *Zinjanthropus* and their crowns are completely intact, as are the exposed roots of the right tooth. The roots of the left M^3 are partly damaged and partly not exposed. Neither tooth is yet in occlusion, but the left tooth is lower down and somewhat nearer the occlusal plane than the right. Both teeth are skewed in shape, so that the mesiobuccal–distolingual axis is very much greater than the mesiolingual–distobuccal axis. This effect is contributed to by the marked reduction of the distobuccal angle of the tooth (pl. 40). In addition, both teeth are in a skewed position. However, the left tooth which is almost in occlusion is not as skewed in position as the less-erupted right M^3; this suggests that, in attaining the position of full eruption, the originally much-skewed tooth settles down to a position which is squarer to the dental arcade.

As seen on the occlusal surface, the reduction of the distobuccal angle is very marked, more so than in either M^1 or M^2 (pl. 20). The occlusal surface presents a highly crenulated appearance, with a great multiplication of cuspules (pl. 40). In this respect, the tooth resembles Robinson's (1956) 'more aberrant molars' of *Paranthropus* (such as SK 31, SK 41 and SK 836) which show a similar tendency, though the M^3's of *Zinjanthropus* are not as markedly triangular as them, being nearer to the less strongly triangular shapes of *Australopithecus*. Despite the crenulation, the primary fissure pattern is clear-cut and the primary cusp outlines are unmistakable.

The protocone is the largest primary cusp and is least subdivided by secondary fissures: it is, nevertheless, beset with some six to seven cusplets which are low and rounded. The protocone makes contact with the metacone over a short distance (1·6 mm. on the left and 2·3 mm. on the right), but the two cusps are contiguous, not continuous. That is, there is no oblique or trigon crest connecting protocone to metacone; instead, their line of contact is marked by a deep fissure, with a tendency to form two cuspules (von Koenigswald's 'double metaconule'—1957, p. 158) on either face of this fissure.

The metacone is the smallest cusp, being much reduced; it is broken into a series of at least eight cusplets by secondary fissures. The metaconal cusplets are smaller and more pointed than those on the protocone.

The hypocone appears large, but its buccal part is more or less separated as an extra cusp in the distal wall of the posterior fovea. The separation is more distinct on the right than on the left M^3; the resulting extra cusp or distoconule is large and is itself subdivided by a transverse secondary fissure into two cusplets. This large extra cusp tallies better with the structure in *Australopithecus* than that in *Paranthropus*, where it is usually smaller. The hypocone proper is broken by secondary fissures into about five cusplets, which are intermediate in form between those of the protocone and those of the metacone.

If the extra distal cusp is not counted as part of the hypocone, the paracone is the second largest cusp. It is subdivided by secondary fissures into five or six low-rounded cusplets.

The posterior fovea is not obvious, because of the complexity of the fissural system, but the central fovea is more easily recognised. A mesial fissure or limb from the central fovea leads to an anterior fovea, more clearly marked on the right M^3; there is no anterior trigon crest between protocone and paracone, but the two cusps are in contact with each other across the shallow mesial limb. The buccal groove is moderately developed on the occlusal surface, but it does not incise the occlusal margin directly; its approach to the margin is blocked by a marginal cusplet which divides the buccal groove Y-fashion about the cusplet. The lingual or oblique groove is better developed and clearly reaches and incises the occlusal margin.

A broad and shallow horizontal groove crosses the buccal face from mesial to distal (pl. 39); there are one or two areas of hypoplastic enamel in the groove and it is itself possibly a manifestation of retarded development or hypoplasia. As mentioned in the discussion on the state of the enamel, the enamel on this face is rugose, irregular, sometimes faintly pitted, with markedly wavy perikymata. There is a possible parallel between the state of the buccal surface and the crenulated state of the occlusal surface: it is conceivable that hypoplasia just at the time when calcification of M^3 was beginning could manifest itself as a highly crenulated occlusal surface of this sort. The whole problem of the 'wrinkles', 'cusplets' and crenulations in primate teeth is too vast to review here, except in the most cursory fashion, but it has been thoroughly discussed by Weidenreich (1937, pp. 62-3, 80, 100-3, 161). Although Weidenreich recognised the widespread occurrence of 'wrinkles' in the Primates, he drew a distinction between the type occurring in New World monkeys and that in the anthropoid apes (p. 100); on the other hand he disagreed with Aichel (1917) in drawing no distinction between the pattern of molar surfaces in the apes and man. To Weidenreich, all crenulations in the latter two groups represent 'a very characteristic feature of the anthropoid dentition, including the hominids'. 'Their general arrangement', he adds, 'is the same within all groups but differs in a specific manner within each individual group' (p. 161). He recognised a definite evolutionary trend in the hominids towards simplification of this general arrangement. Aichel, on the other hand, distinguished between 'cuspidated molars' in man and 'wrinkled molars' in the anthropoid apes, maintaining that there was no genetic connection between the two forms of molar.

When the pattern of crenulation on the *Zinjanthropus* M^3 is compared with that of the orang-utan, distinct differences are seen. In the orang, the crenulations are described as wrinkles or accessory ridges, descending directly from the tips of the cusps, or lying along one or more margins (e.g. Weidenreich, 1937). In *Zinjanthropus*, the crenulations do not take the guise of wrinkles or accessory ridges, but form a multiplicity of cusplets, each more or less delineated, being lined not by parallel fissures but by curved, circumscribing fissures. These cusplets are subequal in height, each reaching the occlusal plane in its own right, and there is no question of their lying on the slopes of the main cusps. The crenulations in *Zinjanthropus* are far more marked than in any other australopithecine teeth; just as it seems that the degree of hypoplasia on the buccal surfaces of the *Zinjanthropus* teeth is most marked. It may be that wrinkles and cusplets result from two different sets of influences playing on the enamel organ during the formative period. Furthermore, wrinkles on the slopes of the major cusps seem to be a species or population characteristic; whereas multiple cusplets seem to be rather an individual variation. It is not impossible, therefore, that wrinkles are genetically determined, whereas multiple cusplets may be a hypoplastic manifestation produced at a stage when the occlusal surface enamel is being laid down.

On the buccal surface of the *Zinjanthropus* M^3, the enamel swells out buccally, close to the cervical line, reaching its maximum bulge 2·0 mm. from the line. This is different from the descriptions of both South African forms of australopithecine, in which the buccal face is said to be smooth and practically vertical (Robinson, 1956); however, a clear buccal bulge is apparent on several of the M^3's illustrated, most notably Sts 52a and Sts 28 (*op. cit.* p. 94, fig. 28 *b, c*). Such a buccal bulge near the cervical line is considered by both Korenhof (1960) and Frisch (1965) to be a trace of the buccal cingulum. The cervical line in the *Zinjanthropus* M^3 is fairly straight as it crosses this face, having no upward peak between the two buccal roots; instead there is a slight shallow peak over the mesiobuccal angle of the tooth (pl. 39).

The buccal groove is faint, broad and shallow. On the right it ends in a very clear pit with puffed, swollen margins, close to the cervical line; there is no pit on the left M³.

On the mesial face, the enamel surface is somewhat flattened near the cervical line; then it tapers gently mesiad, to a maximum mesial projection about 4·0 mm. above the occlusal surface. At this level is a small round or oval area of contact with the most distally projecting portion of M². From the contact area, the mesial face of M³ retreats rather sharply distad to reach the occlusomesial margin. Thus, the mesiodistal diameter of the occlusal surface (12·8, 14·1 mm.) is much less than the maximum mesiodistal diameter of the tooth (15·7, 16·3 mm.). At the maximum mesial bulge of M³, there is a moderately strong buccolingual curve of the mesial surface. Seen from mesial or distal aspects, the buccal face of the tooth is much more nearly vertical, the lingual surface sloping outwards markedly. Hence, the tooth has a very asymmetrical appearance in this view (pl. 40).

The distal face of the Zinjanthropus M³ is highly convex from above downwards. The curve is fairly even over the reduced metacone, but over the prominent hypocone it reaches a maximum distal projection about one-third of the crown height from the cervical line: this area is the most distal part of the whole tooth. From the summit of the distal convexity, the distal face slopes sharply mesiad to the occlusal margin. In buccal or lingual view, therefore, the tooth is markedly asymmetrical, the most projecting parts of the mesial and distal surfaces being nearer the occlusal surface and the cervical line respectively. These are general australopithecine features.

On the lingual face, the enamel just below the cervical line bulges very strongly lingually, even more so than on the premolars (pl. 38). About 3·5 mm. below the line, it reaches its greatest lingual protuberance; below that it turns fairly sharply buccally, and slopes markedly towards the occlusal edge. The slope is much more pronounced than the converse slope on the buccal face. Over the protocone, the enamel is thickened as a downgrowth, which terminates inferiorly as a succession of tubercles separated by short vertical grooves (pl. 38). This distinctive form of Carabelli structure, which may be regarded as a remnant of the cingulum over the protocone (Korenhof, 1960), is comparable with the Carabelli formation in Australopithecus.[1] Its thickness is close on a millimetre, as in Sts 37, but it is not as well developed as in Sts 28 where it reaches a maximum thickness of 2·5 mm. In Zinjanthropus, the formation extends distally as far as the lingual groove and appears to 'overflow' on to the hypocone, for the most distal vertical column of enamel is posterior to the lingual groove. It is better developed than is usual in Paranthropus, except perhaps in SK 831: the latter tooth has a seemingly well-developed cingulum which, however, reaches the occlusal surface and so is difficult to recognise as such.

The lingual groove is better defined on the right M³ than on the left: it is at first deeply incised, but after reaching a pit, continues in a gingival direction as a broad, shallow depression. On the left M³, it is broad and shallow throughout its extent, and there is no pit. The groove disappears over the lingual bulge on this face. Just posterior to the groove is a faint ridge of enamel, with a more distal lingual groove beyond that, on the distal part of the lingual face of the hypocone.

The intact roots of the right M³ are exposed both buccally and in the back wall of the antrum. There are three roots, two buccal and one lingual, well spaced as in Paranthropus. The mesial buccal root is practically vertical; the distal buccal root sweeps distad at a small angle (about 15°) to the vertical. The lingual root flares away from the buccal roots in a lingual (palatal) direction. All three roots are incomplete; they are widely open superiorly and show no indication of convergence of the apical portion. The root length from the cervical line is:

Mesial buccal root (left)	12·7 mm.
Distal buccal root (left)	11·6 mm.
Lingual root (left)	12·4 mm.

The crown heights of the M³'s in the mesial half of the buccal face are 9·7 mm. (left) and 9·4 (right), and 8·9 (left) and 8·7 (right) in the distal half of the buccal face. Relative to their large

---

[1] Frisch's (1965) statement, based on a published photograph, that the M³ of Zinjanthropus shows no trace of cingulum (op. cit. p. 43) is not correct.

crown diameters, these are low teeth. Thus, the lengths of the roots from the cemento-enamel junction are about one-third longer than the heights of the crowns from the cemento-enamel junction to the bucco-occlusal edge. When the state of eruption, the state of the apical portion of the roots, and root length relative to crown height are considered, it is clear that the formation-stage of the *Zinjanthropus* M^3 lies somewhere between molar stages R 2/3 and R 3/4 of Fanning (1961); falls just short of the comparable R 3/4 of Moorrees, Fanning and Hunt (1963); and falls between stage 7 ('alveolar eruption') and stage 8 ('cusp level') of Garn, Lewis and Bonné (1962); while on the four-stage classification of Dalitz (1963), it tallies with his stage 3 ('root formation almost completed but the root canal walls divergent at the apex').

The general pattern of the *Zinjanthropus* M^3's is typical of that of the australopithecines, with some features reminiscent of each of the two South African forms. In shape, cusp pattern and Carabelli structures, the tooth is reminiscent of *Australopithecus*; in crenulations and in root pattern, it is somewhat closer to that of *Paranthropus*. But, according to Robinson (1956, p. 96), 'the most obvious differences between the molars of *Paranthropus* and *Australopithecus* are: (*a*) difference in absolute size, and (*b*) difference in relative sizes of M^2 and M^3'. Tested by these criteria, *Zinjanthropus* differs from both South African forms in respect of (*a*), but in (*b*) is closer to *Australopithecus*.

In its crown diameters, M^3 agrees with M^1 and M^2 in being excessively broad in a buccolingual direction. Although the australopithecine third molars are very variable in size and hence have high standard deviations for their dimensions, the B.L. diameters of the *Zinjanthropus* M^3's, nevertheless, exceed the means for *Paranthropus* by 5·93 S.D.'s (left) and 4·77 S.D.'s (right) and those of *Australopithecus* by 5·73 S.D.'s and 4·85 S.D.'s respectively, and they exceed the maxima for the two South African forms by 2½–3½ mm. In this dimension, *Zinjanthropus* stands alone. The means for the other two forms are much smaller and differ by only 1·3 mm. from each other, but by 4·1 and 5·5 mm. from the average of the two sides in *Zinjanthropus*.

The mesiodistal diameters of *Zinjanthropus* third molars fall within the range of *Paranthropus* M^3's, being only 0·66 and 1·20 S.D.'s above the mean; they exceed the range for *Australopithecus* and surpass the mean by 2·25 S.D.'s (left) and 2·79 S.D.'s (right). Thus, although the B.L. dimension of M^3 maintains the general premolar–molar trend of buccolingual expansion, its M.D. diameter is strikingly shorter than would be expected if the M^1–M^2 trend towards mesiodistal increase were maintained. As a consequence, although the *Zinjanthropus* M^3 remains greater than its homologues in the other australopithecines, it is smaller both in module and in crown area than its own M^2. This is a most striking distinction between the Olduvai specimen and *Paranthropus*; and it is a further point of resemblance to *Australopithecus*. In the individual dimensional trends, too, the same affinity emerges: *Zinjanthropus* and *Australopithecus* both show a diminution in M.D. diameter from M^2 to M^3, but buccolingual diameters which are equal in the two teeth. In contrast, *Paranthropus* shows an increase in both diameters from M^2 to M^3. In no single specimen of *Paranthropus* in which both M^2 and M^3 are present is the M^3 smaller in module or crown area than the M^2. The degree of M.D. and module reduction from M^2 to M^3 in *Zinjanthropus* is even slightly greater than the mean degree of reduction in *Australopithecus*.

4. *Remains of the cingulum*

(*a*) *The lingual cingulum* (*Carabelli structures and protoconal cingulum*)

It has been noted above that Carabelli structures are in evidence on all three maxillary molars of *Zinjanthropus* on each side. Owing to the marked wear on M^1 and M^2, only traces of Carabelli formations can be detected on these teeth, whereas the unworn M^3 has a beautifully preserved pattern of Carabelli tubercles. Thus, it is impossible to detect whether there was a gradient in the intensity of the Carabelli structures from M^1 to M^3 or vice versa. In modern man, the evidence of a number of surveys concurs in showing that a Carabelli cusp may occur on all three upper molars, but is most frequent on the first one (Korenhof, 1960, p. 252). Korenhof has summarised and tabulated the data for the Australopithecinae (Table 50, p. 293), based on the descriptions of Robinson (1956). He classifies the observations by taxon, site, tooth and form of Carabelli manifestation. Thus, he defines three morphological subtypes ('Carabelli structure, types a, b and c'), and, in addition, he recognises a category 'cingulum'. Clearly, however, all of the Carabelli structures are themselves to be regarded as cingular in origin. Both Korenhof (1960) and Frisch (1965) have recently explored the relationship between the Carabelli cusp of hominids and the lingual cingulum, and have arrived at a similar conclusion. According to Korenhof, '...Carabelli's cusp represents a fairly young (Pliocene–Pleistocene) specialisation, which has its origin in a cingulum that, itself, has been reduced' (p. 311), while Frisch sums up as follows: '...we are led to see in the Carabelli cusp of recent man a structure homologous to similar cingulum derivatives found in other Primates, but also typically human in its configuration and mode of occurrence' (p. 50). Hence, for practical purposes, we may lump together all molars said by Korenhof to show Carabelli and cingular features as distinct from those which show neither. In this way, a simplified version of Korenhof's Table 50 has been compiled (Table 46).

From Table 46 it is clear that protoconal cingular remnants or derivatives occur relatively more commonly in M^3 than in M^1 in both taxa. Only a half of the *Paranthropus* M^1's and less than half (37 per cent) of the *Australopithecus* M^1's show these features, whereas most of the M^3's (88 and 82 per cent respectively) possess them. There is a marked difference between the taxa in the frequency of Carabelli/cingular structures on M^2: in *Paranthropus*, M^2 resembles M^1 in having about 50 per cent of teeth thus adorned, whereas every one of twenty-two *Australopithecus* M^2's bears these structures, M^2 being closer to M^3 in this regard. In other words, two out of the three *Paranthropus* upper molars (M^1 and M^2) show a reduced incidence of cingulum remnants, but only one out of the three *Australopithecus* upper molars (M^1) shows such a diminished frequency.

The recent analysis by Frisch (1965) of evolutionary trends in the teeth of hylobatids, pongids and hominids led him to conclude that 'A trend is observed toward complete reduction in the cingulum in all hominoid evolutionary lines' (p. 46). If this conclusion is correct, it might be inferred that *Australopithecus* represents an earlier or more primitive stage of this trend, since only M^1 shows a marked tendency to lose the cingulum; whereas *Paranthropus* represents a more advanced stage, since both M^1 and M^2 show a strong tendency to lose the cingulum. This inference is supported by another line of evidence than the mere frequency: as was mentioned above, the Carabelli/cingular structures, when present in *Australopithecus*, are in general far thicker and better developed than the homologous structures in *Paranthropus*, in which the lingual cingulum is at most poorly developed (Robinson, 1956). As indicated by Robinson and repeated by Frisch, the strongest traces in *Paranthropus* hardly match the weakest cases in *Australopithecus*. Hence, both in frequency and in degree of development, the

Table 46. *The frequency of Carabelli structures and protoconal cingulum in Australopithecinae*

(Simplified after Korenhof, 1960)

	Paranthropus				*Australopithecus*			
	M^1	M^2	M^3	Total	M^1	M^2	M^3	Total
Carabelli structure or cingulum present	15	11	15	41	7	22	9	38
Absent	15	10	2	27	12	—	2	14
Total number	30	21	17	68	19	22	11	52

geologically older *Australopithecus* lies in a more primitive position on this particular evolutionary trend than does the geologically younger *Paranthropus*.

There is one respect in which the observations on the Australopithecinae are apparently at variance with the conclusions of Frisch. He states that not only is there a trend toward complete reduction of the cingulum in *all* hominoid evolutionary lines, but that 'this trend follows in all forms a fixed pattern: The buccal cingulum is lost first, both in the upper and in the lower molars; the lingual cingulum is lost subsequently and apparently much more slowly, first on the premolars, then on the last molar (with the possible exception of *Pan*), and finally on the first two molars' (Frisch, 1965, p. 46). Although the frequencies in the australopithecines are admittedly based on small samples, they do suggest that, in this group at least, the lingual cingulum tended to be lost *first* on the M^1 (and not *last*), while the M^3 (and, in *Australopithecus*, the M^2) were but little affected by the trend to loss of the cingulum. It would not be surprising for variability in the pattern of the trend to be manifest in some hominoids.

Zinjanthropus has Carabelli/cingular structures on all three molars and, where preserved, these are marked in degree. Thus, *Zinjanthropus*, like its near-contemporary *Australopithecus*, is at a primitive stage in the trend towards reduction of the cingulum. Apart from the protoconal manifestations, the distal two molars of *Zinjanthropus* show a distal extension of the Carabelli/cingular structures on to the hypocone—which extension may be regarded as a further indication of the pronounced development of the lingual cingulum in this form. Robinson (1956) does not speak of such an extension in his series of molars, referring always to the cingulum or Carabelli complex extending mesiad from the lingual fissure. However, one of the *Australopithecus* M^3's figured by him (p. 94, fig. 28(*a*)—T.M. 1511 from Sterkfontein) shows an apparent distal extension of Carabelli structures, comparable with, though not as marked as, the corresponding extension in the M^2's and M^3's of *Zinjanthropus*.

In sum, the Carabelli/cingular features on the maxillary molars of *Zinjanthropus* agree with those of *Australopithecus* in showing an early stage in the trend toward reduction of the cingulum.

(b) The buccal cingulum

In each of the maxillary molars of *Zinjanthropus*, features are present which may be identified as vestiges of the buccal cingulum. On M^1 and M^2, they take the form of a small paramolar cusp or buccostyle (to use Korenhof's term), in which respect there is a parallel with *Australopithecus* but not with *Paranthropus*. The traces are slightest on M^3: here the buccal cingulum may perhaps be represented by a cervical swelling of enamel, as well as by the heaped-up margins of a pit at the upper (cervical) end of the buccal groove. A similar cervical swelling seems to characterise some M^3's of *Australopithecus*.

All in all, the buccal cingular remnants on the maxillary molars of *Zinjanthropus* are vestigial and, on Frisch's analysis of trends, would suggest a somewhat more advanced stage in the reduction of the cingulum than that represented by the lingual cingulum: however, even on the buccal surface, the trend toward cingulum reduction is not complete, either in *Zinjanthropus* or *Australopithecus*, whereas scarcely a trace of the buccal cingulum is detectable in *Paranthropus*.

Thus, *Zinjanthropus* agrees with *Australopithecus* in being at a more primitive stage in the reduction of both the buccal and the lingual cingulum.

CHAPTER XVIII

SUMMARY OF CRANIAL AND DENTAL FEATURES OF ZINJANTHROPUS

The cranial vault

The curvature and components of the vault

(1) The parieto-occipital plane of *Zinjanthropus* rises steeply from the external occipital protuberance. The vault of *Australopithecus* is more evenly curved in this area.

(2) The parietotemporal side-walls of the vault also rise steeply, contributing to the full, rounded character of the vault. This feature finds a parallel in the latest cranium of *Australopithecus* from Makapansgat, MLD 37/38.

(3) The vault of *Zinjanthropus*, as seen in norma verticalis, is long-spheroid, whereas that of the reconstructed *Paranthropus* is short-spheroid and that of *Australopithecus* long-ovoid. *Zinjanthropus* is intermediate in this respect between the two South African forms, though approaching more closely to the second reconstruction of *Paranthropus*.

(4) Most dimensions of the parietal bones in *Zinjanthropus* fall within the range of variation of *Australopithecus*, while a few dimensions exceed the range. In anteroposterior extension, the australopithecine parietal is slightly smaller than that of *Homo erectus*, whereas, in mediolateral extension, the australopithecine parietal is far smaller than that of *H. erectus*. In contradistinction, the anteroposterior extent of the parietal bone in pongids is far smaller than in australopithecines.

(5) The chord–arc indices of the edges of the parietal bones show that in *Zinjanthropus*, the parietal bone is slightly less curved along the coronal margin than in *Australopithecus*. The anteroposterior curvature of the parietal bones of *Australopithecus* tends to be greater, and the mediolateral curvature less, than in *H. erectus*.

(6) *Zinjanthropus*, as well as *Australopithecus*, shows parietal predominance over the occipital in sagittal arc dimensions, approaching in this respect sapient man. In all forms of Asian *H. erectus*, as well as in Solo man, the parietal sagittal arc is appreciably smaller than the occipital. Pongidae are in a neutral position of parieto-occipital equality, veering, if anything, slightly towards occipital predominance.

(7) The absolute dimensions of the occipital bone of *Zinjanthropus* agree fairly well with those of three specimens of *Australopithecus*, but all are much smaller than those of the Homininae.

(8) In *Zinjanthropus* and *H. erectus*, the lower scale of the supra-occipital (inion–opisthion) predominates in sagittal length over the upper scale (lambda–inion); whereas in *Australopithecus* and modern man, the upper scale preponderates. It is suggested that this difference is to be understood in terms of the powerful development of the nuchal muscles in the former; such development in primate crania Delattre has shown to be associated with expansion of the lower scale, especially when it occurs in a little-expanded brain-case.

(9) The upper scale of the supra-occipital is much flatter in *Zinjanthropus* than the lower scale; whereas in *Australopithecus* and modern man, the lower scale is flatter than the upper. It is suggested that the difference is due to the upward migration of the nuchal muscles in *Zinjanthropus*, which has carried the inion upwards and backwards, thus tending to flatten out and even to hollow the planum occipitale between lambda and the edge of the nuchal crest. Similar conditions probably occur in *H. erectus*, in whom the hollow between lambda and occipital torus was dubbed by Weidenreich *sulcus supratoralis*.

(10) The sagittal chord–arc index of the entire

occipital bone in *Zinjanthropus* is only 70·5 per cent, which is ten or more points lower than in three specimens of *Australopithecus* (80·0–83·1 per cent) and falls close to the lower extreme of the range of values for the Homininae. This low index is undoubtedly the result of the formidable nuchal crest of *Zinjanthropus* which adds greatly to the disparity between the sagittal occipital arc and chord.

The supra-orbital height index

(11) The supra-orbital height index (which is an expression of the height of the calotte above the upper margin of the orbit compared with the total calvarial height above the F.H.) is lower in *Zinjanthropus* than in *Australopithecus*; it falls at the top of the range in pongids and well outside the range in Homininae. Thus, in *Zinjanthropus*, as in the reconstructed *Paranthropus*, the whole calvaria is hafted on to the facial skeleton at a lower level than in *Australopithecus*.

The sagittal and nuchal crests

(12) *Zinjanthropus* has a sagittal crest, the surviving chord length of which is 52 mm., but the total chord length of which is estimated to have been about 97·5 mm. The surviving part of the crest is highest and the degree of fusion of the two temporal crests most intimate, in a markedly anterior position, some 110 mm. (arc length) in front of the external occipital protuberance. The temporal crests diverge anteriorly where they are powerfully developed, and posteriorly where they are weakly developed. The point of divergence posteriorly is 58·5 mm. from the external occipital protuberance, leaving a substantial 'bare area of the skull', delimited by the temporal crests and the nuchal crest. All this points to the fact that the most strongly developed part of the temporalis muscle was the anterior part and that the area in which the temporal crests first fused to form a sagittal crest was, as Robinson has claimed for *Paranthropus*, far forward on the calotte, and not as Zuckerman has claimed for *Paranthropus*, just in front of the external occipital protuberance.

(13) There is a strong nuchal crest, rising as much as 10 mm. from the surface of the planum nuchale. Although the planum occipitale is somewhat hollowed above the nuchal crest, this is not of such a degree as to create a shelf-like effect along the upper margin of the nuchal crest; the only shelving in this area is on the undersurface of the nuchal crest. For 23 mm. on either side of the external occipital protuberance, this is a simple nuchal crest, to the formation of which no contribution is made by the temporalis muscle. The succeeding 25–28 mm. of the crest on each side is a compound temporal/nuchal crest. Laterally, the temporal and nuchal crests diverge from each other, the latter continuing as a simple nuchal crest adjacent to the occipitomastoid suture. The cranium of *Zinjanthropus* thus demonstrates that the formation of a compound (T/N) nuchal crest, which Robinson had felt was normally precluded in the australopithecines because of the nearly horizontal position of the nuchal plane, has indeed occurred in *Zinjanthropus*. A study of other australopithecine crania makes it highly likely that a compound (T/N) crest must have occurred, not only in the more muscular *Paranthropus*, but also in the less muscular *Australopithecus* (e.g. MLD 1). In the young adult occipital from Makapansgat, for instance, the temporal and nuchal lines are only 1 mm. apart; an older specimen of the same type must certainly have shown the temporal line contributing to the nuchal crest, i.e. forming a compound (T/N) crest.

(14) The arrangement of the components of the nuchal crest in *Zinjanthropus* confirms that a nuchal crest may form, in part at least, from the action of the nuchal muscles alone. The medial and lateral thirds of the nuchal crest in this adolescent are simple nuchal crests; only the intermediate third on each side is a compound (T/N) crest. When this evidence is considered along with the indications of priority in the development of the anterior part of the temporalis muscle, as well as conditions in such specimens as MLD 1 and the Swartkrans crania, it is reasonable to conclude that a simple nuchal crest first came into being as a result of the action of the nuchal muscles alone; then, with the somewhat belated growth backwards of the temporalis muscle, the inferior temporal line eventually came into tangential

SUMMARY OF CRANIAL AND DENTAL FEATURES

contact with the pre-existing nuchal crest, converting part of it into a compound (T/N) crest. This study therefore does not support the view of Ashton and Zuckerman that a nuchal crest 'would not have been present in *Paranthropus* if the posterior fibres of the animal's temporalis muscle had not already reached the superior nuchal line'.

(15) The sequence of events in the formation of both sagittal and nuchal crests differs in *Zinjanthropus* and other australopithecines, from those obtaining in the pongids. Both of these australopithecine distinctions point in the same direction: namely, that the anterior and middle parts of the temporalis muscle must have expanded more rapidly than the posterior parts, so that the temporal lines reached the midline *anteriorly* earlier, while they reached the superior nuchal line (or nuchal crest) later.

The basis cranii externa

The occipital bone

(16) The impressions for all the nuchal muscles in *Zinjanthropus* are especially clearly marked.

(17) There is a small condylar groove, instead of the better-developed condylar fossa of *Australopithecus* and of modern man. The fossa normally receives the posterior edge of the superior articular process of the atlas when the head is extended.

(18) The condyles are small and more similar to those of the Homininae than to those of the Pongidae.

The temporal bone

(19) *Zinjanthropus* has a large and prominent mastoid process, which does not turn inwards nearly as much as that of *H. erectus erectus* IV. The presence of so well-developed a mastoid process in the as yet immature *Zinjanthropus* aligns this creature with the hominids rather than with the pongids. In the australopithecines as a group, the development of a large mastoid process is a constant feature and occurs early in life. Both of these characteristics are hominid rather than pongid (Schultz).

(20) *Zinjanthropus* and other australopithecines resemble the Pekin crania in having a powerful supramastoid crest, separated from the mastoid process below by a deep supramastoid sulcus. This supramastoid crest continues behind into the temporal crest (or inferior temporal line) and in front into the posterior root of the zygoma. As seen from above, it presents an undulating contour, bulging laterally above the mastoid and again as the posterior root of the zygoma, but receding mediad above the external acoustic meatus.

(21) The digastric fossa lies in the same straight line as the foramen processus styloidei and the foramen stylomastoideum, as in modern man; in *H. e. pekinensis*, however, the stylomastoid foramen lies outside the line of the digastric fossa and the foramen processus styloidei. The mastoid notch is limited medially by an occipitomastoid crest, but in *Zinjanthropus* and three other australopithecine crania the occipital contribution to the crest is less than in *H. e. pekinensis*. The crest in *Zinjanthropus* is an ectocranial superstructure which greatly thickens the occipitomastoid suture; whereas in early hominines, such as Swanscombe, the whole thickness of the occipital bone is curved downwards near the occipitomastoid suture; consequently the presence of this kind of 'crest' leads to no marked thickening of the sutures.

(22) The external acoustic meatus opens 5–6 mm. medial to the sagittal plane through auriculare. This slight overhang allies *Zinjanthropus* to the chimpanzee and to modern man; while the orang-utan is closer to Pekin Man in having a more marked overhang of 10–15 mm. The gorilla tympanic is variable.

(23) In *Zinjanthropus*, as in modern man, the tympanic plate lies nearly vertically with upper and lower margins whereas in pongids it tends to lie more transversely with anterior and posterior margins. *Zinjanthropus* resembles *H. e. pekinensis* in being intermediate in this respect between modern man and the pongids, though it approaches somewhat more closely to modern man.

(24) The tympanic plate of *Zinjanthropus* is large both mediolaterally and supero-inferiorly, its proportions resembling more closely those of *H. e. pekinensis* and of modern man than those of the apes.

(25) The surface of the tympanic plate in *Zinjanthropus* is concave from side-to-side and from above downwards, as in modern man, whereas that of Pekin Man resembles those of the pongids in being plane or even convex.

(26) The lateral border of the tympanic plate in *Zinjanthropus* is thinnest in the floor of the porus, most markedly thickened postero-inferiorly and thickened and apparently doubled on itself anterosuperiorly. This pattern occurs, too, in MLD 37/38, save for the absence of doubling. It is suggested that the thinning of the floor in these australopithecines, like the tendency towards splitting or indentation of the floor in three out of six Pekin crania, is a manifestation of incomplete fusion between the two developmental components of the tympanic bone; just as, in crania with generally thin tympanic plates, the same process may result in a foramen of Huschke. In modern man, the thickest part of the tympanic rim is directly inferior and a thinned area lies anterior to that; it is suggested that this relative rotation of the meatal wall is due to the forward growth of the mastoid process, which in modern man points more anteriorly than it does in the australopithecines. This forward growth results in rotation of the thickest portion of the tympanic plate from a postero-inferior to an inferior position, and of the thinner area from an inferior to an anterior position.

(27) The form of the external acoustic porus is elliptical and its long axis is nearly vertical, sloping slightly backwards. This is the commonest form of aperture in modern man, but it occurs, too, in some chimpanzees (Weidenreich).

(28) *Zinjanthropus*, along with other australopithecines, has a large meatal aperture, which clearly aligns them with fossil hominines and distinguishes them from pongids which have smaller openings. The dimensions in modern races of man are very variable.

(29) No styloid process is present in *Zinjanthropus*, nor does it occur in any other australopithecine, *H. e. pekinensis*, nor in pongids. Nevertheless, the foramen stylomastoideum is present and a vaginal process of the tympanic: the latter closely resembles what Weidenreich called the spine of the crista petrosa in *H. e. pekinensis*, and the two structures are undoubtedly closely related to each other.

(30) The petrous portion of the temporal bone lies in almost the same axial line as the tympanic plate, as in modern man, and in contrast with the anthropoid apes in which the two axes are at a marked angle to each other. In *Zinjanthropus* the petrous axis deviates from the tympanic axis by only 30°, whereas in several series of pongids the deviation of the petrous axis from the tympanic axis ranges from 53 to 75°. In this feature, the australopithecines are intermediate between pongids and modern man, but lie nearer to modern man.

(31) The petro-tympanic axial differences reflect a turning of the petrous process from an anteroposterior orientation in pongids to a more transverse one in modern man. This can be demonstrated by measuring the *petro-median angle*, i.e. the angle between the petrous axis and the median plane. The value in *Zinjanthropus* (47·5°) falls well within the hominine range (38–63°) and well outside the pongid range (10–30°). Three specimens of *Australopithecus* have intermediate values (32–33·5°), slightly nearer to those of the pongids. The more prognathous hominoids have the lower angles, the more orthognathous the higher angles. The two features of gnathism and petro-median angle are indirectly related, since each is governed by a common, third factor, the degree of reduction of the cranial base. *Zinjanthropus* agrees with modern man in showing evidence of marked reduction of the cranial base, as reflected in the high petro-median angle and the relatively orthognathous face. *Australopithecus*, on the other hand, shows relatively little reduction of the cranial base compared with *Zinjanthropus*, hence the more marked prognathism and lower values of the petro-median angle. The pongids are the ultimate term in this hominoid morphological series, showing most prognathism and the smallest petro-median angles. Modern orthognathous men are at the other extreme of the series.

(32) Confirmation of the strong degree of reduction of the cranial base in *Zinjanthropus* is provided by a spheno-occipital index (basion–

hormion distance expressed as a percentage of maximum cranial length): the value in *Zinjanthropus* is 14·5 per cent, in recent man 11·8–18·5 per cent, in several series of pongids 22·5–39·9 per cent, while that of *Australopithecus* (Sts 5) is 21·5 per cent. These figures confirm that the shortening of the cranial base in *Zinjanthropus* is well within modern human limits, whereas that of *Australopithecus* is intermediate between those of men and apes, lying slightly closer to the apes.

(33) The shortening of the cranial base in australopithecines is an absolute as well as a relative shortening, the basion–hormion distance ranging from 25 to 31·5 mm., whereas in pongids it ranges from 30 to 50 mm. In modern man the range is from 22 to 31 mm.

The mandibular fossa

(34) The mandibular fossa of *Zinjanthropus* is very large, especially in mediolateral breadth, being exceeded in this dimension only by the means for male orang-utan and gorilla. In its great breadth as well as in its shallow depth, the mandibular fossa of *Zinjanthropus* is aligned with that of pongids rather than with that of hominines. The depth–length and depth–breadth indices of the fossa are thus closer to the pongid than to the hominine values; whereas the length–breadth index overlaps the bottom of the range of hominine means and the top of the range of pongid means. These comments apply as well to the fossae of several specimens of *Paranthropus* and *Australopithecus*. It is suggested that the dimensions of the fossa are governed largely by the growth of the head of the mandible lodged in it and that this, in turn, may be related to the large cheek-tooth size which *Zinjanthropus* and other australopithecines share with pongids.

(35) The anterior wall of the mandibular fossa is well-curved mediolaterally, slopes from anterolateral to posteromedial, and from antero-inferior to posterosuperior. The anteroposterior slope is about 45° to the F.H. In all these features, the fossa of *Zinjanthropus* finds a close parallel in the fossae of both *Australopithecus* and *Paranthropus*, as well as of a number of fossil and modern hominines. On the other hand, this combination of features of the anterior wall has not once been seen by the author in pongid crania.

(36) Although the mandibular fossa of *Zinjanthropus* is relatively shallow, the articular tubercle is rendered salient by two features: the 45° slope of the anterior wall of the fossa and the fact that the preglenoid plane slopes upwards from the articular tubercle towards the superior orbital fissure. These features of the australopithecine tubercle were not taken into consideration in an appraisal by Ashton and Zuckerman of the degree of development of the tubercle in hominoids: the only criterion used was the projection depth of 'the point of maximum convexity of the articular eminence' below the deepest point of the mandibular fossa. On this criterion, it was concluded that there was little reason to separate the condition of the australopithecine articular fossa from that of the great apes. It is submitted that, if these additional criteria for the salience of the eminence had been used, differences might well have emerged, as they have done in the present study.

(37) The entoglenoid process of *Zinjanthropus* is large and robust and is formed wholly by the squamous part of the temporal. In *Australopithecus* (MLD 37/38 and Sts 19), however, the sphenosquamosal suture passes across the entoglenoid process, dividing it into alisphenoid and squamous elements, subequal in size. The arrangement in *Zinjanthropus* thus resembles that in *Paranthropus* and *H. e. pekinensis*, whereas that in *Australopithecus* is closer to that of modern man. In *Zinjanthropus* even foramen spinosum is in the squama, whereas in all other forms it is in the alisphenoid or, at its most lateral, in the line of the sphenosquamous suture.

(38) When *Zinjanthropus* is compared with another big-toothed, big-jawed hominoid, the gorilla, it is seen that whereas structures tend to crowd on to the squama in the former, in the ape there is a contrasting tendency for the same structures, even the entoglenoid process, to crowd on to the alisphenoid. It seems as though gorilla has achieved its very big mandibular fossa by displacing the entoglenoid process medialwards, while *Zinjanthropus* has achieved the same end by

a different means, namely, throwing the outer margin of the fossa outwards.

(39) Confirmation of the two different modes of enlarging the mandibular fossa is afforded by the much greater proportion of the *Zinjanthropus* articular fossa which protrudes laterally beyond the plane of the side-wall of the calvaria, as compared with gorilla. A second line of confirmation is provided by two *ad hoc* indices, an interglenoid–biporial index and an interglenoid–biglenoid index, both of which show that the glenoid fossa in gorilla approaches much nearer to the mid-line than in *Zinjanthropus*.

(40) Since the biglenoid and interglenoid distances are related to the bicondylar and intercondylar breadths of the mandible, it follows that *Zinjanthropus* has achieved an increase in dental, mandibular and condylar size, without narrowing the space between the two halves of the mandible. In gorilla, on the other hand, increase in these features has been accompanied by a narrowing of the intercondylar distance.

(41) Two trends are detectable in the evolution of the hominoid mandibular fossa: first, the tendency to increase or decrease the mandibular fossa, and the direction of any increase, whether medialwards or lateralwards. Secondly, another process has been at work, namely, a hominising tendency towards lateral expansion of the alisphenoid. In *Zinjanthropus*, expansion of the mandibular fossa is evident and it is a lateralising tendency; the lateral expansion of the alisphenoid is just evident, the foramen ovale alone having been engulfed by the alisphenoid. The exact relations of foramen ovale, foramen spinosum, the entoglenoid process and the sphenosquamous suture in any hominoid, it is suggested, will depend on the resultant between these two developmental trends.

(42) The postglenoid process is relatively small and is closely applied to the anterior surface of the tympanic plate. It is thin anteroposteriorly and is not inflated by an extension of temporal pneumatisation. In all these regards, it is nearer to the hominine condition, though an occasional pongid cranium may have so small a process (Ashton and Zuckerman).

(43) Because of the small size of the postglenoid process, a large part of the anterior surface of the tympanic plate is smooth and articular, as in hominines. The arrangement contrasts strongly with the virtual exclusion of the tympanic from the mandibular fossa of pongids, by the large, inflated postglenoid process.

(44) The right postglenoid process of *Zinjanthropus* is backed up by a second process of about the same elevation. A double postglenoid process occurs, too, in *Australopithecus* (Sts 5). However, whereas the second process in *Zinjanthropus* is seemingly a recurved part of the anterior portion of the tympanic, the second process in Sts 5 appears to be a squamosal and not a tympanic derivative.

The sphenoid bone and related structures

(45) The body of the sphenoid has an inflated appearance; the sphenoid air-sinus projects into the sphenoidal rostrum, tending to spread the two plates of the vomer apart from each other.

(46) The pterygoid process is powerfully developed and its base is invaded by a sphenoid air cell. The palatine bone contributes a substantial pyramidal process to the pterygoid fossa between the laminae, as in one specimen of *Australopithecus* (Sts 8).

Certain critical angles and indices of the cranium

The angulation and height of the nuchal area

(47) The planum nuchale of *Zinjanthropus* and other australopithecines faces downwards much more than backwards: in this approach to a horizontal planum nuchale, the australopithecines clearly resemble hominines and contrast strongly with pongids, which possess more steeply tilted nuchal surfaces.

(48) In both *Zinjanthropus* and Sts 5, the nuchal muscles rise only a very short distance above the F.H. As a result the nuchal area height index of the former is 8·2 per cent (or 9·5 per cent if computed without the sagittal crest) and of the latter 8·0 per cent. These values lie well within the range for modern man, in fact close to the modern means; but well outside the lower extreme of the pongid ranges.

SUMMARY OF CRANIAL AND DENTAL FEATURES

(49) Inion in *Zinjanthropus* lies 8·5 mm. below the F.H. when the cranium is orientated in the F.H. In most other australopithecine crania, it seems, the apex of the external occipital protuberance lies almost exactly in the F.H.

(50) The external occipital protuberance of *Zinjanthropus* is powerfully developed and points backwards and downwards. Its apex is hollowed and testifies to strong development of ligamentum nuchae and especially of its dorsal margin, to which the *trapezius* muscle is attached. This evidence suggests that both the ligament and those nuchal muscles which attach to it subserved a postural function greater than in the Homininae. These nuchal features point to the probability that the head of *Zinjanthropus* was not as well balanced on its occipital condyles as in modern man, but tended to incline forward somewhat.

The foramen magnum position index

(51) The horizontal projective distance between opisthion and opisthocranion (i.e. the horizontal occipital length) in *Zinjanthropus* is 37·7 mm. or 21·8 per cent of the maximum cranial length. Although this value is somewhat exaggerated by the large nuchal crest, it nevertheless shows that opisthion is situated farther forward than in pongids, which have a low value for this index.

(52) A second occipital length index may be obtained by determining the projective distance between opisthion and opisthocranion on an extension of the nasion–opisthion line. Because the projection of the opisthocranion falls behind the opisthion, the index is given a minus sign. In *Zinjanthropus* the value is −26·8 per cent (with the nuchal crest) or −18·7 per cent (without the nuchal crest); both values fall well within the hominine range and well outside the pongid range, which clusters about −6·4 per cent. The value for *Zinjanthropus* when allowance is made for the nuchal crest (−18·7 per cent) is similar to that of the crestless *Australopithecus*, Sts 5 (−17·5 per cent).

(53) Since opisthion marks the posterior limit of the foramen magnum, both of these occipital length indices provide good evidence that the foramen magnum of *Zinjanthropus* and other australopithecines is situated well forward on the base of the cranium.

The plane of the foramen magnum

(54) The foramen magnum of *Zinjanthropus* faces almost directly downwards, the foramen lying in a plane set at 7° to the F.H. The value in Sts 5 is 19·5° and in a female gorilla 28·5°. In this respect, *Zinjanthropus* is much nearer to the hominines than to the pongids.

(55) When the plane of the foramen magnum is compared with the plane of the basis cranii, several different anterior termini may be selected. If prosphenion or klition are used, the results are not consistent in hominoids: the opening out of the angle at basion with progressive horizontalising of the foramen magnum is compensated for by closing of the same angle with bending of the basicranial axis and the consequent thrusting of prosphenion or klition into the brain-case. An anterior terminal, the position of which is not directly altered by the bending of the basicranial axis, is nasion. The opisthion–basion–nasion angle of *Zinjanthropus* is 157°, which compares well with figures for the Homininae (145–171°) and departs appreciably from the values in the Pongidae (121–134°). The angle is thus completely hominine and wider than in the anthropoid apes.

(56) From a study of the angles and the sides of the nasion–basion–opisthion triangle, Weidenreich deduced that the nasion–basion line in hominines should amount to 75 per cent of the nasion–opisthion line, whereas the ratio should exceed 85 per cent in pongids. The values in *Zinjanthropus* (82·5 per cent) and *Australopithecus* (Sts 5) (80·0 per cent) lie midway between those of pongids and hominines, the larger-faced Olduvai specimen inclining more towards the pongids.

The porion position indices

(57) The porion position length index of *Zinjanthropus* gives the exaggeratedly high value of 78·3 per cent, since the powerful development of the glabellar region and the consequent anterior displacement of nasion unduly lengthen the line from nasion to the foot-point of the perpendicular

from porion on to the nasion–opisthion line. In any event, the length-position of porion shows no consistent trends during hominoid evolution (Weidenreich).

(58) The porion position height index shows that *Zinjanthropus* manifests the characteristic hominine tendency for porion to rise above the level of the nasion–opisthion line. The markedly elevated position of porion in *Zinjanthropus* contrasts sharply with its depressed position in pongids. The elevation of porion in *Zinjanthropus* is part of the general transformation of the cranial base which accompanies hominisation.

The position of the occipital condyles and the poise of the head

(59) The condylar position index of *Zinjanthropus* is 55·1 per cent, that of Sts 5, 40 (or 39). While the value in the Sterkfontein cranium lies well below the hominine means and ranges and overlaps the top of the gorilla range, the value for *Zinjanthropus* lies in an intermediate position between those of apes and hominines; it is somewhat nearer to that of *H. sapiens*, both in absolute terms and when the differences from the means are reduced to standard deviations. The presence of a strong nuchal crest in *Zinjanthropus* and its absence in Sts 5 suffice to explain the different values between the two australopithecines, for in both of them the condyles lie in the transverse plane of the biporial axis. The presence of a nuchal crest does not, however, explain away all differences: thus, cresting raises the index of a gorilla to about the same value as in the non-crested *Australopithecus*; while cresting in *Zinjanthropus* raises its indicial value to about the bottom of the range for non-crested hominines. Clearly these facts and the indicial values reflect a real difference in the average position of the condyles in australopithecines and gorillas.

(60) The intermediate position of the index in australopithecines, and the intermediate degree of development of the nuchal crest, support the view that the head was not as well balanced as in the Homininae, but was better balanced than in the Pongidae. The cranium of *Zinjanthropus* is characterised by excessive precondylar heaviness, occasioned by the greatly elongated face and the massive teeth. In the face of these challenges to the nuchal muscles, it is likely that the head of *Zinjanthropus* was maintained not quite as horizontally as in the Homininae, but sagged forward slightly, though not nearly to the same degree as in pongids.

The interior of the calvaria
The endocranial surface of the frontal bone

(61) The angle between the frontal squama and the floor of the anterior cranial fossa is much nearer the angle in *H. e. pekinensis* (50°) than that in modern man (about 90°).

(62) There is a strong frontal crest as in hominines; pongids lack this crest or, at best, have a poorly-developed low ridge. In *Zinjanthropus*, as in Pekin Man, the crest is flanked by a deep fossa on either side, leading downwards to a caving of the floor of the anterior cranial fossa as a marked 'olfactory recess'. This conformation aligns *Zinjanthropus* and *H. e. pekinensis* with the pongids, and serves to distinguish them from modern man. It is suggested that the frontal crest is not related to the 'sagittal reinforcing-system' of the cranial vault, as proposed by Weidenreich, but is to be explained rather in terms of variations in the degree and pattern of development of the falx cerebri, which attaches to the frontal crest.

(63) From lateral to medial, the floor of the anterior cranial fossa slopes gently at first, then descends abruptly. Although it cannot be determined whether a foramen caecum is present in *Zinjanthropus*, the position it would occupy is about 16 mm. below the more lateral part of the floor of the fossa as in *H. e. pekinensis*; this depth is greater in a number of pongid crania or as reflected on a number of endocranial casts of pongids. On the other hand, foramen caecum is at the same level as, or just below, the floor in modern man. In this respect, *Zinjanthropus* and Pekin Man are nearer to the pongids than to recent man.

The endocranial surface of the parietal bones

(64) A broad, low Sylvian crest is present near the sphenoid angle of the parietal in *Zinj-*

SUMMARY OF CRANIAL AND DENTAL FEATURES

anthropus: it bears more resemblance to that feature in pongids and in modern man than to the well-developed crest of *H. e. pekinensis*. The Kromdraai *Paranthropus* shows a similar crest moderately well developed.

(65) The mastoid angle of the parietal bone is wide and obtuse, the adjacent lambdoid and squamosal margins being but little notched, in association with the poor expansion of the occipital and temporal squamae. The inner aspect of this angle has a narrow groove, not for the transverse sinus, but for the posterior ramus of the middle meningeal arterial groove. Below and behind the groove is a slight torus angularis: it is, however, not nearly as well developed as in *H. erectus* and produces a bony thickening of only an extra millimetre.

The basis cranii interna: posterior cranial fossa

(66) The cruciate eminence on the inner aspect of the occipital squama is not strongly elevated from the adjacent fossae, especially the sagittal limb. The inferior part of the sagittal limb is a short, broad prominence, as is common in anthropoid apes, rather than a sharp internal occipital crest such as occurs in *Australopithecus* (MLD 1) and modern man.

(67) The part of the cerebral fossae on the occipital bone is smaller than the cerebellar fossae. In fact, however, only the deepest part of the cerebral fossa, accommodating the occipital pole, occurs on the occipital squame; the rest of the cerebral fossa, housing the remainder of the occipital lobe, is to be seen on the lower part of the parietal bones. Taken as a whole, then, the cerebral fossa of *Zinjanthropus* exceeds the size of the cerebellar fossa in the ratio of perhaps 4 to 3; in Pekin Man, the ratio is about 2 to 1, whilst in modern man, the cerebellar fossae are greater longitudinally and transversely than the cerebral fossae.

(68) The cerebellar fossae of *Zinjanthropus* are flattish, their relief being offset by broad shallow grooves for enlarged occipital and marginal sinuses which encroach on the fossae. In MLD 1 (*Australopithecus*), in which there are no unusual occipital or marginal sinus sulci, the cerebellar fossae are deep and rounded. In *Zinjanthropus* the internal occipital protuberance has moved relatively further away from opisthion than in MLD 1, the chord distances being respectively 27·0 and 18·5 mm. On the other hand, the internal occipital protuberance in *Zinjanthropus* is relatively nearer to lambda (26·5 mm.) than is the case in MLD 1 (41·5 mm.). *H. e. pekinensis* is similar to MLD 1 in the small anteroposterior size of the cerebellar fossa and the approximation of the internal occipital protuberance to opisthion.

(69) With the migration *up* the squame of the internal occipital protuberance and the migration *downwards* of the inion, a marked change in the relations of these two landmarks has come about in *Zinjanthropus*. The distance between inion and the internal protuberance is 26·8 mm., the protuberance being on a level about 20 mm. higher than that of inion. In *H. e. pekinensis*, the distance between protuberance and inion is greater (27·5–38·0 mm.), but in all these crania the protuberance is at a lower level than inion.

(70) It seems reasonable to infer that there is a relatively more expanded cerebellum in *Zinjanthropus* than in *Australopithecus*. A comparison of the relevant parts of the endocasts confirms this deduction. In this feature, *Zinjanthropus* seems to have moved in a hominine direction, since, in modern man, the cerebellar fossae are more extensive than the cerebral.

(71) The petrous pyramids make an angle of 85° with each other. The posterior surface is high (21·1 mm. as compared with a maximum of 18 mm. in *H. e. pekinensis*, while it may reach 23 mm. in modern man), nearly vertical and partly hollowed laterally, so that the superior margin overhangs posteriorly. The adjacent anterior and posterior surfaces meet at a sharp angle superiorly. In all these respects, the petrous pyramid of *Zinjanthropus* closely approaches that of modern man; it is even more modern looking than that of *H. e. pekinensis*, which has a blunt rounded superior margin with the adjacent surfaces pressed down, as it were, towards the floor of the fossa, as in apes.

(72) The arrangement of structures on the posterior surface of the petrous, as well as the

vertical slope of this face, are indistinguishable from those in modern man; whereas in none of the pongid crania examined is this combination of features found.

(73) The clivus is long and narrow and, below, is overlapped on each side by a strongly developed tuberculum jugulare which overhangs the inner opening of the hypoglossal canal. This tubercle is altogether absent or only slightly developed in gorilla crania, while in chimpanzee and orang-utan crania, it is slightly better developed, though lacking in some of these, too; but in none of the pongids does it attain the degree of development found in *Zinjanthropus*. In a group of African negroid crania, the tubercle is moderately strong and better developed than in pongids, though not as markedly as in *Zinjanthropus*.

(74) A distinct petro-occipital fissure intervenes in *Zinjanthropus* between the apical part of the petrous and the basilar part of the occipital. There is no tendency for the medial margin of the petrous to overflow as an excrescence or shelf to meet and even articulate with the basi-occipital, such as occurs in a number of chimpanzee crania and in some modern human crania, though not in a small series of gorilla skulls.

(75) Two distinct bony prominences mark the position of the now fused spheno-occipital synchondrosis. This feature should be viewed in the light of a fairly widespread tendency in both hominoid and cercopithecoid Primates for regular or irregular bony prominences to mark the upper surface of the clivus at the approximate level of the synchondrosis.

(76) The dorsum sellae is completely hominine in form, being a solid, imperforate ledge of bone, standing up distinctly from the clivus and the sella turcica. In all pongid crania examined for this feature, the dorsum sellae is either low and squat, consisting largely of a posterior 'overflowing' of bone over the clivus; or completely absent, being represented merely by two small posterior clinoid elevations; or very thin and perforate. The dorsum sellae of *Zinjanthropus* compares well with that of Sts 5 and both are extraordinarily reminiscent of that of modern man. Unfortunately, nothing is known of this area in *H. erectus*.

The foramen magnum

(77) The foramen magnum is heart-shaped, with the 'base' anterior and a blunt-pointed apex posterior. The truncation of the anterior margin of the foramen is not due to excessive pneumatisation of the basi-occipital, since the posterior extension of the sphenoidal air-sinus does not seem to penetrate for more than a short distance behind the probable region of the spheno-occipital synchrondrosis.

(78) As a result of the anteroposterior abbreviation of the foramen magnum, its length of 26·4 mm. is smaller than those of other australopithecines, except Sts 8 which has a generally small foramen. The breadth of 26·1 mm. falls within the range for other australopithecines and at the lower end of the range for modern man.

(79) The foramen magnum breadth/length index of 98·6 per cent is well beyond the range of indices in a number of australopithecines (76·8–85·2 per cent) and, at the same time, exceeds the maxima for all other hominoids recorded in this study.

The basis cranii interna: middle cranial fossa

(80) The anterior surface of the petrous portion is typically hominine. The prominent arcuate eminence projects above the summit of the pyramid; the trigeminal impression is shallow and lies close to the apex of the pyramid.

(81) The tegmen tympani bulges upwards slightly. There are no unusual features in the medial wall of the middle ear cavity, but the mastoid air cells to which the aditus and the mastoid antrum give access are numerous and completely fill the pars mastoidea.

(82) The sella turcica is deep and well defined, in these respects being far more like that of hominines than that of pongids.

(83) The inner opening of the foramen ovale has an anterolateral bay, from which a groove runs laterally; the bay and lateral groove are most probably for the accessory or small meningeal artery, which generally traverses foramen ovale in company with the mandibular division of the trigeminal nerve. The foramen ovale is enclosed by the alisphenoid, very close to the spheno-squamous suture.

SUMMARY OF CRANIAL AND DENTAL FEATURES

(84) The foramen spinosum internally is entirely enclosed by squamous temporal bone, in contrast with the position in modern man, in whom the foramen lies totally within the alisphenoid. Sometimes, however, in modern man, the foramen lies close to the edge of the alisphenoid and may be incompletely walled by the latter bone. The position in *H. e. pekinensis* is somewhat intermediate, though approaching rather more closely to that of modern man.

(85) The groove for the common stem of the middle meningeal artery is short and it bifurcates on the incurved part of the temporal squama (i.e. on the inferior surface of the temporal lobe), as in Sts VII and possibly VIII and in the Taung specimen. On the other hand, the groove for the common stem is longer in Sts I and Sts II and in the Kromdraai cranium, bifurcating only on the lateral surface of the temporal lobe. However, this seems to be a highly variable feature of no real taxonomic significance.

(86) The side wall of the middle cranial fossa, constituted by the temporal squama, is long and low, as in *H. e. pekinensis*. Viewed endocranially, the line of the squamosal suture rises from behind forward to reach its summit in the biporial axis (when the cranium is orientated in the F.H.), then declines at a gentle angle to the pterion. This contour exactly parallels that of modern human crania. In pongids, however, the highest point of the squamosal suture is well behind the biporial axis, at or even just behind the point where the margin of the squama crosses the root of the petrous pyramid. From this point forwards, the suture pursues a long, gently downhill course endocranially. In no pongid cranium examined did the arrangement of the squamosal suture approach that of *Zinjanthropus*.

The pattern of the venous sinuses

(87) The venous sinuses of *Zinjanthropus* exhibit a distinctive variation. The blood from the superior sagittal and straight sinuses apparently drained directly into a large right and a smaller left occipital sinus; thence, as right and left marginal sinuses, directly to the jugular bulb. As it passed over the upper surface of the lateral occipital, each marginal sinus received a sigmoid sinus, which had been formed close to the root of the petrous portion by the confluence of the superior petrosal and petrosquamous sinuses. There is no bony evidence of a transverse sinus having existed. This pattern has been found as well in a *Paranthropus* occipital from Swartkrans, a trace of it is present in another, but it is apparently lacking in a third Swartkrans specimen. It has not been encountered in about 100 pongid crania examined. While common in human infants, it occurs as an exceptionally rare variant in adults of modern man. Only 1 out of 211 Bantu negroid crania showed an almost identical pattern. The choice during ontogeny of the occipital-marginal sinus route, instead of the longer, trans-cerebellar route through the lateral sinus system, may be related to the enlargement of the cerebellum in *Zinjanthropus*.

(88) The groove for the sigmoid sinus in *Zinjanthropus* lies far forward in the posterior cranial fossa, close to, though not under cover of, the petrous part of the temporal. It lies about one-fifth of the way between the anterior and posterior margins of the cerebellar hemispheres, as seen on the plaster endocast. Both *Zinjanthropus* and *Australopithecus* seem to agree more with the hominine trend than with the pongid trend, but there is great variation in both groups.

The thickness of the cranial bones

(89) The calvaria of *Zinjanthropus* is basically thin-walled; it is rendered highly robust in some areas by marked pneumatisation and in others by massive ectocranial superstructures. In these respects, it resembles the crania of pongids and of other australopithecines (especially *Paranthropus*), but differs from the massive, thick-boned cranium of *H. erectus*.

(90) As an illustration of the unusual robusticity of parts of the *Zinjanthropus* cranium, the thickness of the parietomastoid suture just in front of asterion is 22·2 mm. (left) and 24·1 mm. (right). Comparable values in *H. e. pekinensis* range from 15 to 18 mm.; and in modern man from 3·5 to 7·0 mm. The maximum thickness of the occipito-

mastoid suture in *Zinjanthropus* is 26–29 mm., in *H. e. pekinensis* 6·5–8·0 mm., and in modern man 3–6 mm. The great thickness of the mastoid area in *Zinjanthropus*, like that in the pongids, is owing to excessive pneumatisation; on the other hand, in *H. erectus*, it reflects an actual thickening of bone substance, especially of the outer and inner tables.

(91) In parts, the *Zinjanthropus* calvaria is adorned with massive ectocranial embellishments, such as the sagittal and nuchal crests; these resemble comparable structures in pongids and *Paranthropus*, but differ appreciably from the rounded 'sagittal torus' and occipital torus of *H. erectus*.

(92) The parietal bones of *Zinjanthropus* are, in general, very thin, for so robust and massive a cranium; the measurements of thickness drop as low as 3–5 mm. in some areas. In this respect, *Zinjanthropus* and other australopithecines resemble the pongids, since the bones of pongids are extraordinarily thin, except where they are embellished by muscular superstructures or are heavily pneumatised. The thinness of the parietals in australopithecines and pongids contrasts with the great thickness attained by these bones in *H. erectus*: in parts, they are 2–2½ times as thick as in *Zinjanthropus*.

(93) The parietals of *Zinjanthropus* are likely to have thickened somewhat if the individual had lived to an older age, in the light of Getz's demonstration that the thickness of the parietal bone increases with age.

(94) The observations in *Zinjanthropus* do not support the view of Zuckerman that, in the advance of the temporal muscles over the parietal bones, a bony film is laid down over the surface of the bones. The present observations are more in keeping with Scott's view that new bone formation during the migration of the temporal muscles is confined to the 'spreading edge' of the muscle, and that the crest 'moves' by a process of resorption and redeposition of bone.

(95) The reduction in the thickness of the cranial walls during the course of hominid evolution has been supposedly correlated with expansion of the brain-case (Weidenreich). However, while this correlation seems to hold good for the stages from *H. erectus* to modern man, it fails to explain the fact that, between the australopithecines and *H. erectus*, both brain size and cranial bone thickness approximately double. Generalised bony massiveness, as in *H. erectus*, is part of the complex of changes which marked at least one transition from the australopithecines to the early hominine level of organisation.

The endocranial cast of *Zinjanthropus*

Cranial capacity

(96) The volume of the endocranial cast of *Zinjanthropus* is 530 c.c.: this is the first reliable estimate of cranial capacity in one of the robuster australopithecines. It is exactly the same volume as in the biggest adult australopithecine endocast thus far accurately determined (Sts VIII) and is about 100 c.c. larger than the smallest (435 c.c., Sts I). The Taung child, with an endocranial volume of 500–20 c.c., would, however, have had a larger volume as an adult, of the order of 562 c.c., as calculated on the basis of Ashton and Spence's figures for growth of the brain in hominoids. Thus, despite its possessing the most enlarged cheek-teeth of all the Australopithecinae discovered to date, as well as the heaviest jaws and musculature, the cranial capacity of *Zinjanthropus* is no greater than the series of reasonably reliable estimates of the cranial capacity of the small-toothed, small-muscled type of australopithecine.

(97) The sample range of endocranial capacities in seven australopithecines (six calvariae of *Australopithecus* and one of *Zinjanthropus*) is from 435 to 562 c.c. and the mean is 502 c.c. These values overlap with the ranges of pongid capacities, especially those of gorilla, and the means for australopithecines and gorillas are virtually identical. The means for orang-utan and chimpanzee are about 100 c.c. smaller, but the lower part of the australopithecine sample range overlaps the upper part of the ranges for these two smaller-brained pongids. The australopithecine sample range falls far short of the lowest value for *H. erectus* (750).

(98) The range of endocranial capacities in seven australopithecines is thus from 435 to

SUMMARY OF CRANIAL AND DENTAL FEATURES

562 c.c.—a range of 127 c.c. This is smaller than the extent of the range in a sample of 144 chimpanzee crania (160 c.c.) and a sample of 260 orang-utan crania (180 c.c.) and much smaller than the range in a sample of 653 gorilla crania (412 c.c.) and in a sample of nine *H. erectus* specimens (450 c.c.).

(99) If all seven australopithecines here considered were members of a single hominoid population, and if the population variability was of the order of six times the estimated standard deviation, then the most likely estimate of the population range is 361–643 c.c. These figures would obtain if the present sample range is centrally placed in the population range. If, however, through sampling errors, the present australopithecine sample lay either near the bottom or near the top of the population range, estimates of the population range would be 435–717 and 280–562 c.c. respectively.

(100) On this reasoning, it seems highly unlikely that any australopithecine cranium would be smaller in cranial capacity than 280 c.c. or greater than 717 c.c. The latter value is smaller than the largest gorilla capacity (752 c.c.) and the smallest known cranial capacity of *H. erectus* (750 c.c.).

(101) In hominoids with smaller capacities (with means up to about 400 c.c.) variability is moderate, the estimated coefficient of variation ranging from 6·8 to 8·0. In those with larger capacities (with means of over 500 c.c.), including gorilla and *Homo*, variability is high, the estimated coefficient of variation ranging from 12·0 to over 13. Only a part of this enhanced variability is owing to a greater degree of sexual dimorphism in bigger-brained hominoids; thereapart, a real increase of variability seems to accompany increasing brain-size. The estimated population variability of Australopithecinae is not as high as that of other bigger-brained hominoids; we cannot determine at present if this reflects sampling bias, or a real lower variability.

(102) When Jerison's method of estimating the number of cortical neurones is applied to *Zinjanthropus*, *Australopithecus* and other hominoids, it is found that the Australopithecinae show a small but definite advance over the pongids in the number of 'excess neurones' available for improved adaptive capacities. In turn, *H. habilis* shows an improvement on the australopithecines, while *H. erectus* has an increase over *H. habilis*, and *H. sapiens* tops the list. Thus, there is a stepwise progression in the number of 'excess neurones' with succeeding grades of hominisation.

Morphology of the endocranial cast

(103) The sulcal and gyral impressions are well preserved only in the frontal region. The superior, middle and inferior frontal gyri can be identified on both sides, separated by the superior and middle frontal sulci. The inferior frontal gyrus is very prominent as in the endocasts of Sts II and Sts VII, as well as those of earlier Homininae.

(104) The brain-stem is short and squat, and turns sharply downwards well forwards of the hinder extremity of the brain. Behind it are detectable successively, the impressions of the marginal and occipital sinuses, the cerebellar hemispheres extending close to the mid-line behind the brain-stem, and the occipital poles of the cerebral hemispheres. This forward placement of the brain-stem aligns *Zinjanthropus* with other australopithecines and sharply distinguishes it from pongids.

(105) When compared with the endocasts of other australopithecines, that of *Zinjanthropus* is seen to portray a marked parietal expansion, both vertically and transversely.

(106) The cerebellum of *Zinjanthropus* is well underslung, so that it is not seen when the endocast is viewed in norma dorsalis. In this respect it agrees with other Australopithecinae and Homininae, but differs from most Pongidae.

(107) A limited encephalometric analysis confirms the biparietal expansion, and the marked cerebellar development of *Zinjanthropus*.

(108) The low bifurcation of the common trunk of the middle meningeal artery in *Zinjanthropus* resembles that in other australopithecines, but in the distribution of the anterior and posterior trunks *Zinjanthropus* shows several distinctive features. The area of distribution of the anterior trunk is relatively reduced, especially on the left, while that of the posterior trunk is increased. *Zinjanthropus* is the only australopithecine, of those with details of the arterial pattern preserved,

in which the area of supply of the posterior trunk is so extended posteriorly as to include the occipital area. Thus, the incomplete vascularisation of the occiput, which Schepers found a feature of other australopithecine endocasts, does not characterise that of *Zinjanthropus*. Possibly in the Kromdraai specimen, too, the posterior trunk would have supplied the occipital area. The variability of the vascular patterns, even between the two sides of *Zinjanthropus*, shows that they are unlikely to be good indicators of taxonomic affinities.

Metrical characters of the calvaria as a whole

(109) The robust australopithecine crania of *Zinjanthropus* and of *Paranthropus* SK 48 have cranial lengths (173 and c. 170 mm.) much longer than that of *Australopithecus* Sts 5 (146·8 mm.). The difference is owing largely to the strong supraorbital torus and nuchal crest in the former two. The *toro-cristal length and index* give a measure of the degree of development of these structures. The values in *Zinjanthropus* of 44 mm. and 25·43 per cent are higher than in any hominid cranium determined, whereas those of *Australopithecus* Sts 5 (25·8 mm. and 17·57 per cent) are similar to the values in *H. erectus*. Fossil hominines have indices ranging from 13 to 16·76 per cent (*H. erectus* of Asia), 17·5–17·62 per cent (African Rhodesioids) and 10·58–12·95 per cent (Eurasian Neandertalers), while recent human cranial series generally give values of 5–10 per cent.

(110) In *Zinjanthropus* and *Australopithecus*, the greatest breadth of the cranium is the bimastoid, followed by the breadth across the supramastoid crests and the biauricular breadth. The breadths of the brain-case itself—on the parietals and on the temporal squames—are much smaller. In *H. erectus*, the biauricular or supramastoid intercristal breadths are greatest, while the bimastoid breadth is much smaller; the maximum parietal breadth (at torus angularis) and the temporoparietal breadth are, however, greater than the bimastoid. The differences are attributed to greater temporal pneumatisation in the Australopithecinae, inversion of the mastoid process and greater parietal expansion in *H. erectus*.

(111) The australopithecines are characterised by strong lateral development of the mastoid–supramastoid–auriculare complex; whereas only the more anterior parts of this complex, the auriculare–supramastoid components, show excessive lateral development in *H. erectus*.

(112) As determined by the minimum frontal breadth and the transverse frontoparietal index, the australopithecines, *H. erectus* and pongids all show a comparable postero-anterior tapering of the brain-case. Encephalic or endocranial dimensions show that the cranial tapering is based upon a similar encephalic tapering. 'Post-*erectus*' hominines (Neandertaloids and recent man) show much less tapering, consequent upon a lateral expansion of the frontal lobe with increase in the minimum frontal breadth.

(113) In the transverse parieto-occipital index, *Zinjanthropus* and other australopithecines lie near the lower end of the range for the Homininae. The pongid mean lies well within the hominine range.

(114) The basibregmatic height of *Zinjanthropus* shows that this form of australopithecine has a cranial vault of comparable height to that of *Australopithecus*—despite the virtual absence of a forehead in *Zinjanthropus*. Lower hafting of calvaria to face is the factor responsible for the flat forehead, rather than an overall low vault. The values in *H. e. erectus* are only slightly greater than those of the australopithecines, while the top of the pongid range overlaps the australopithecine sample range.

(115) The cranial height–length index aligns *Zinjanthropus* with *H. erectus* in the chamaecranial category, the common factor being marked increase of the cranial length, which measure constitutes the denominator in the index. On the other hand, this index aligns the orthocranial *Australopithecus* Sts 5 with modern man. The cerebral height index confirms this resemblance.

(116) The cranial height–breadth index again shows up the great heightening of Sts 5, its acrocranial value of 106·1 per cent being near the upper reaches of the range for modern man: on the other hand, the Makapansgat specimen, MLD 37/38, which is usually classified as *Australo-*

SUMMARY OF CRANIAL AND DENTAL FEATURES

pithecus, is relatively lower and has an index (89.2 per cent) comparable with that of *Zinjanthropus* (84.1 per cent): both fall into the tapeinocranial category (x–91.9 per cent), as do all early hominines. In the enlargement of the brain-case during the process of hominisation, it seems that the broadening tendency is pre-eminent during the transition from the australopithecine to the pithecanthropine grades of organisation, while the heightening effect predominates during the change from *H. erectus* to *H. sapiens*. To this generalisation Sts 5 is an exception.

(117) The auricular height–breadth index of *Zinjanthropus* is lower than that of *H. e. pekinensis*, although the basibregmatic height–breadth index is higher. It is inferred that the difference is owing to a relatively smaller supra-auricular and larger infra-auricular component of height in *Zinjanthropus*. This confirms that the porion in *Zinjanthropus* has risen higher above the nasion–opisthion baseline than in Pekin Man, as shown by the porion position height index.

(118) The nasion–basion length index in *Zinjanthropus* is the resultant of two seemingly conflicting tendencies—the deflection of the cranial base, which is evident in this creature, and the massiveness of the face. Thus, its value of 82.5 per cent is intermediate between the values for hominines (75 per cent or less) and those for pongids (85 per cent or more). *Australopithecus* (Sts 5), with a smaller face, has a value a little nearer to that of the hominines (80.0 per cent).

The structure of the face (pl. 41)

The supra-orbital torus

(119) The torus in *Zinjanthropus* is far more robust than in any of the australopithecines hitherto discovered. Its vertical thickness of 13.6 mm. and its anteroposterior width of 17.6 mm. exceed the values in a number of other australopithecine crania, though the vertical thickness is exceeded by that in three out of five crania of *H. e. pekinensis*.

(120) There is virtually no forehead, but immediately behind the glabellar region is a broad triangular hollow, *trigonum frontale*, bounded anteriorly by the torus and posterolaterally by the temporal crests. In this respect, *Zinjanthropus* closely resembles *Paranthropus*. From this triangular area, the torus juts forward, but not upward as in pongids.

(121) The torus comprises two distinct moieties: a glabellar part with anterior and superior faces, and a superciliary part with only a single anterosuperior surface and a backing ridge formed by the anterior part of the temporal crests. The different dispositions of the surfaces of the two parts give the torus as a whole a twisted appearance.

(122) The highest part of the torus is at the lateral ends of the glabellar portion, from which point the superciliary components slope steeply downwards laterally to the frontozygomatic suture. The slope and orientation of these lateral parts of the torus differ from the corresponding features in *Paranthropus*.

(123) From a study of pongid as well as australopithecine crania, it is concluded that a single anterosuperior surface of the torus (instead of two surfaces, anterior and superior) results when the anterior part of the abutting temporal crest rises above the level of the torus. In no pongid cranium examined did this arrangement extend further medially than about the lateral half of the superior orbital margin; whereas in *Zinjanthropus* alone, among all the hominoid crania examined, did the crest lie in a supratoral position right across the entire orbit as far as its superomedial angle. The peculiar morphology of the *Zinjanthropus* torus can thus be related to the excessive development of the anterior part of the temporalis muscle. This part of the muscle has utilised a maximum of space anteriorly to gain a purchase, the temporal crest diverging posteriorly only at the superomedial angle of the orbital aperture.

The orbits and the interorbital area

(124) The orbits are fairly high, that on the right being hypsiconch and that on the left mesoconch. The indices (80.6, 86.0 per cent) compare well with those of other Australopithecinae and Homininae, but show an orbit which is not nearly as high as the mean height in pongids (in which mean orbital indices range from 93.3 to 113.5 per

cent), though lower individual values, ranging down to 76 per cent, are encountered in chimpanzees.

(125) The frontal process of the zygomatic is hollowed from above downwards, faces more anteriorly than in other australopithecines, and has a strong downward and forward slope to its junction with the body of the zygomatic. As in Hominidae, but not Pongidae, this process increases in width from above downwards.

(126) A post-marginal process projects immediately behind the abutting temporal crest, just below the zygomaticofrontal suture. It is divided into an anterior and a posterior projection and is clearly related developmentally to several strongly developed fascicles of the anterior part of the temporalis muscle.

(127) The interorbital region is broad and inflated. From the plane of the anterior interorbital breadth (between maxillofrontalia) to that of the posterior interorbital breadth (between lacrimalia), the interorbital width increases by 9·1 mm. from 23·5 to 32·5 mm. This increase, as well as the inflated appearance, is accounted for by the lateral bulging of the fronto-ethmoidal air cells, which are excessively pneumatised.

The nose

(128) The piriform aperture and the lower parts of the nasal bones are set in a central facial hollow, surrounded by the more salient paranasal and subnasal parts of the maxilla, the glabella and the zygomaticomaxillary portions of the face. In this regard, *Zinjanthropus* closely resembles *Paranthropus*, but differs from *Australopithecus*, in which the nasal region is slightly raised.

(129) The upper parts of the nasal bones are, however, not in the central hollow; they are in a high and prominent position, so that nasion is almost coincident with glabella. Once more, this is a feature which *Zinjanthropus* and *Paranthropus* possess in common, whereas in *Australopithecus* nasion is a little lower down, about half-way between glabella and sellion.

(130) The nasal bones of *Zinjanthropus* are of unusual form, being expanded to their maximum width above, but tapering to a constriction below; thus, the inferior breadth is hardly greater than the minimum breadth. In *Paranthropus* (SK 48), although there is a comparable upper expansion, there is an inferior expansion which makes the inferior breadth the greatest breadth. The expansion of the superior parts of the nasals in *Zinjanthropus* may be related to the general inflation of the glabellar region.

(131) The two nasal bones are at a very wide angle to each other, this angle varying between 170 and 180°. Such an arrangement seems to typify the australopithecines. Only in SK 48 is the angle as small as 135°, but it is not impossible that at least some part of this angulation may be owing to distortion.

(132) The lower 4–5 mm. of the nasal bones do not continue the concave curve of the remainder, but turn inwards (posteriorly) towards rhinion, giving a slightly beaked appearance to their lower portion. A similar recurvation seems to have been present in SK 48 (*Paranthropus*) but not in *Australopithecus*. It is a common feature in modern man.

(133) The internasal suture departs from the mid-line above, as the right nasal encroaches on the left and, above that, the left on the right. Such encroachment is frequently encountered in hominine crania.

(134) In keeping with the general facial elongation, the nose and piriform aperture are long. The elongation reflects itself in a low nasal index (45·3 per cent) and aperture index (94·1 per cent). Other australopithecines are shorter-faced and have higher nasal indices (57·7–63·0 per cent) and aperture indices (120·8–145·5 per cent).

(135) The plane of the margin of the piriform aperture looks somewhat upwards, so that a perpendicular dropped from rhinion reaches the nasal floor well behind the margin. In this orientation of the margin, *Zinjanthropus* is somewhat intermediate between anthropoid apes and *H. e. pekinensis*.

(136) *Zinjanthropus* possesses a high anterior nasal spine, in this respect approaching more closely to the Homininae than to the Pongidae. The spine is placed well back in the same plane as the lower part of the nasal bones, as in *Paranthropus*.

SUMMARY OF CRANIAL AND DENTAL FEATURES

In pongids the spine is poorly developed and lies relatively further anteriorly, well in front of the lower parts of the nasals, and practically in the same plane as the sides of the piriform aperture.

(137) There is a tendency for the fossa praenasalis to drop over the floor of the nose to become part of the naso-alveolar clivus, as in modern man. The prenasal fossa is thus almost a subnasal fossa, as, for example, in recent Bushman crania, and in contrast with pongid crania.

The maxilla and zygomatic bones

(138) The infra-orbital foramen is single, as in most other australopithecines (Sts 52a, however, having three infra-orbital foramina on each side) and unlike many pongid crania. It lies well below the orbital margin (29–30 mm.), the distance being greater than in *Paranthropus* (18–23 mm.) and *Australopithecus* (12·2–17·0 mm.).

(139) The measurements and indices of the maxilla confirm the great facial enlargement of *Zinjanthropus* as compared with other australopithecines. Each maxillary dimension is between 20 and 50 per cent greater in *Zinjanthropus* than in the South African forms. The facial indices verify that this enlargement is greater in the direction of heightening or elongation than in the direction of widening. For both superior facial indices, *Zinjanthropus* yields values well in excess of other australopithecines which, in turn, exceed the means for *H. e. pekinensis*. The maxilla itself is responsible for much of the facial enlargement of *Zinjanthropus*, as is confirmed by the high value (72·3 per cent) of the zygomatico-maxillary index.

Prognathism

(140) Although the upper two-thirds of the naso-alveolar clivus of *Zinjanthropus* slopes forward, the lower third is practically vertical and the teeth emerge in a vertical plane. This arrangement contrasts with that in other australopithecines, in which from above downwards hardly any change of contour of the clivus is noticeable when the tooth-roots are reached. This observation is confirmed by the views of the South African and Tanzanian crania in norma verticalis: the subnasal maxilla juts forward from the facial plane to the least extent in *Zinjanthropus* (not even the front teeth being visible in norma verticalis), to the greatest extent in *Australopithecus*, and to an intermediate degree in *Paranthropus*.

(141) Direct measurements of the angles of upper facial, nasal and subnasal prognathism show a marked difference between the orientation of the nasal and subnasal parts of the face in *Zinjanthropus*, as in *Paranthropus*. The nasal profile angle of *Zinjanthropus* places the cranium in the hyperorthognathous category, its angle of 93·5° being close to the mean in Pekin Man (89°), well within the range of individual values in modern man and well outside the range not only of means but of individual values in pongids. The value of 74° in the highly prognathous Sts 5 (*Australopithecus*) lies in the area of the overlapping extremes of pongid and hominine values.

(142) On the other hand, the alveolar profile angle of 62·5° places *Zinjanthropus* in the hyperprognathous category, its value lying well outside the range of means and individual values of the pongids, and at the bottom of the range of modern human *means*.

(143) The facial profile angle, which may be regarded as the resultant of the nasal and alveolar profile angles, is thus intermediate between the hyperorthognathism of the nasal profile and the hyperprognathism of the alveolar profile: this facial angle measures 78° in *Zinjanthropus*, reflecting a somewhat lesser degree of overall prognathism than obtains in *Paranthropus* (75° in SK 46) and far less prognathism than in Sts 5 (50°).

(144) In sum, the direct angle measurements show that *Zinjanthropus* is hyperorthognathous in nasal profile, hyperprognathous in alveolar profile, with an overall prognathous classification. In contrast, Sts 5 is prognathous in nasal profile, ultraprognathous in alveolar profile, with an overall effect which is hyperprognathous.

(145) Flower's gnathic index, the ratio of basion–prosthion to basion–nasion, shows only a small difference in the degree of prognathism between *Zinjanthropus* and Sts 5; the different degrees of development of the glabellar region, the difference in placement of nasion and the

highly different size and dimensions of the jaws, vitiate comparison of the index in the two forms. The terms of the index are simply biological non-comparables.

Infra-orbital and inframalar features

(146) The lateral part of the body of the maxilla is characterised by a distinctive morphology in *Zinjanthropus*, different from that obtaining in the South African australopithecines and in pongids. The latter groups have two powerful ridges on the lateral parts of the body of the maxilla, the anterior of which is the jugum alveolare of the canine and the posterior of which leads downwards from the root of the zygomatic process. In *Zinjanthropus*, there is only a single, broad, rounded ridge, which may be regarded as the conjoined juga alveolaria of the canine and P^3, with the root of the zygomatic process flowing into this single ridge.

(147) The posterior ridge associated with the root of the zygomatic process is further back in the pongids (opposite M^1 or even between the roots of M^1 and M^2) and the South African australopithecines (opposite P^4) than in *Zinjanthropus* (between the roots of P^3 and P^4). That is, the face seems to be tucked in under the zygomatic hafting to a greater extent in australopithecines than in pongids; among the former, this process has been carried further in *Paranthropus* than in *Australopithecus*, but has reached its furthest development in *Zinjanthropus*. This tendency towards tucking in of the lower face seems to have opposed the tendency towards dental enlargement with its implied 'unfolding' of the lower face, as seen in *Zinjanthropus* and *Paranthropus*.

(148) The space between the anterior and posterior maxillary ridges in pongids is occupied by a deep depression, within which a circumscribed area constitutes the canine fossa. In the South African australopithecines, through the more anterior placement of the zygomatic root or the tucking under of the lower face, this depression is reduced to a narrow vertical sulcus directly above P^3; the upper part of the sulcus is in line with the infra-orbital foramen and faces anteriorly, while the lower part in line with P^3 faces laterally. In *Zinjanthropus*, the tucking under of the face and the enlargement and lateral displacement of P^3 and its jugum alveolare have been carried so far that there is no well-developed canine fossa or sulcus, but only a small sulcus leading downwards for 1 cm. below the infra-orbital foramen. This small sulcus corresponds closely with that of *H. e. pekinensis* and Broken Hill, and looks more like a vascular groove than the distinctive morphological entity of pongids and the other australopithecines. The canine fossa, as such, is completely obliterated in *Zinjanthropus*.

(149) The lower border of the zygomatic process in *Zinjanthropus* forms a distinct *incisura malaris* (Weidenreich) or *incurvatio inframalaris frontalis* (S. Sergi), very similar to those of Pekin Man. Furthermore, in both *Zinjanthropus* and *H. e. pekinensis*, the root of the process attaches high above the alveolar margin. In contrast, the lower border of the process in a number of South African australopithecines forms a straight or very slightly curved edge, but no incisure, as in a number of Neandertal crania; it differs from the feature in the latter group in that it inserts some distance above the alveolar margin, whereas in the Neandertals with their shallower alveolar ridges it descends to the alveolar margin.

(150) The australopithecine maxillae correspond neither with the *inflexion type* nor with the *extension type* of maxilla as recognised by Sergi, having rather a combination of features of each type. Thus, *Zinjanthropus* has an inframalar notch but no canine fossa; while *Paranthropus* has a modest canine fossa but no notch.

The attachments of the masseter muscle

(151) The attachment area of the masseter muscle to the zygomatic arch comprises an anterior and a posterior oval area connected by an intermediate constricted impression. The degree and the pattern of excavation betoken a powerfully developed muscle, manifesting the hypertrophy which seems to have characterised all the masticatory muscles of *Zinjanthropus*. The degree of development of the masseteric impressions in *Zinjanthropus* far exceeds that in other australopithecines and is more comparable with conditions in the gorilla.

SUMMARY OF CRANIAL AND DENTAL FEATURES

(152) The masseteric impression of *Zinjanthropus* is placed on a more anterior plane than in pongids. Thus the impression encroaches anteriorly on to the maxilla for almost 18 mm.; in only two out of twenty male gorilla did it encroach beyond the zygomaticomaxillary suture at all and then only for 1–2 mm. Encroachment is commoner in the chimpanzee, the muscular impression often extending 3–5 mm. beyond the suture and in one instance as much as 8 mm. beyond. Posteriorly, the impression stops short in *Zinjanthropus* just in front of the zygomaticotemporal suture and the impression for the most posterior fibres is relatively weak, in comparison with its large size in gorilla. It is likely (following the work of Symons on shifts in the attachments of the masticatory muscles) that the attachment of the masseter would have shifted posteriorly on to the zygomatic process of the temporal bone, with the coming into occlusion of M^3.

(153) Scott's index relating the breadth of the mandibular ramus to the degree of development of the masticatory muscles, as reflected by molar size, makes it possible to estimate the rameal breadth of the *Zinjanthropus* mandible as 66–76 mm. (pl. 42). This compares with rameal breadths of 57, 57·5 and about 69 mm. in three mandibles of *Paranthropus*.

(154) The occlusal plane of *Zinjanthropus* slopes upwards and backwards at $11\frac{1}{2}°$ to the F.H. In *Paranthropus* and *Australopithecus*, on the other hand, the molar part of the occlusal plane is parallel to the F.H., while the anterior part of the arcade turns upwards at an angle of 7·5° in *Australopithecus* and 17° in *Paranthropus*. In all three forms perpendiculars erected on the occlusal plane tangent to the mesial and distal margins of the molar teeth already in occlusion intersect an area along the zygomatic arch which tallies with the attachment of the masseter. It seems that the origin of the masseter is related directly to the molar teeth upon which it is operating. The superficial fibres of masseter must have run very nearly at right angles to the occlusal plane.

(155) It is concluded that, in *Zinjanthropus*, the position of the masseteric impression reflects the upward and backward tilt of the occlusal plane, with the tucking under of the face, and the fact that the M^3's are not yet in occlusion. With the coming into occlusion of the M^3's, not only is the masseter likely to have extended its purchase posteriorly, but it is possible that further adjustments in the occlusal plane might have come about, bringing it more into line with the occlusal plane of other australopithecines.

The palate

(156) The palate is very deep, reaching a depth of 21·5 mm. at the mesial part of M^2, as compared with 15·5, 14·0 and about 9·0 mm. in several specimens of *Paranthropus*. The depth in *Zinjanthropus* is maintained as far forward as the incisive foramen, then shelves steeply downwards to the alveolar margin, as in *Australopithecus* from Sterkfontein and many prognathous recent hominine crania; but in contrast with the arrangement in *Paranthropus*, where the shelving starts as far back as the molar teeth.

(157) There is a distinct torus palatinus medianus, as occurs in *Paranthropus* but not in *Australopithecus*. In *Zinjanthropus* and most specimens of *Paranthropus*, it extends no further forwards than the palatomaxillary suture, whereas, in the Kromdraai specimen, it seems to have extended forwards as a torus maxillaris medianus. The heavy masticatory function in the large-toothed australopithecines might well have elicited the formation of the torus against an appropriate genetic background.

The zygomatic bone

(158) The zygomatic bone of *Zinjanthropus* is large but its metrical predominance over that of other hominoids is not as overwhelming as is the maxillary preponderance. Thus, the distance from the infra-orbital to the masseteric margins is 39·5–44 mm. in *Zinjanthropus*, 30·5–36·8 in *Paranthropus* (SK 48), 31·5 in one specimen of *H. e. pekinensis*, and 27·5 in the Broken Hill cranium. The greatest height of the zygomatic is 71·3 mm. in *Zinjanthropus*, 59·5 in SK 48, 43 in Sts 5, but 65 in *H. e. pekinensis* II and 54 in Broken Hill Man.

(159) The facies malaris of the zygomatic

presents a small anterior surface, bending on itself around a prominent malar tuber, and a large laterally facing surface. These features occur, too, in other australopithecines and, to a degree, in Pekin Man. The frontal process faces mainly anteriorly and so is twisted on the laterally directed body of the zygomatic.

Facial measurements and indices and calvariofacial indices

(160) Measurements confirm the marked elongation of the face of *Zinjanthropus*, its upper facial height of 111·5 mm. exceeding by more than 30 mm. the estimated heights in other australopithecines. Two factors contribute most to this facial elongation: the tucking-under of the lower part of the upper face and the very great depth of the subnasal maxilla, in association with very long dental roots.

(161) The face is very broad and much of this great breadth is made up of the bimaxillary width. Although the zygomatic arches flare markedly to accommodate the temporalis muscles, they do not increase the preponderance of facial width in *Zinjanthropus* as compared with other australopithecines. The overall result is that the bizygomatic breadth of *Zinjanthropus* is not as strikingly in excess of this dimension in other australopithecines, as is the upper facial height of *Zinjanthropus*. The upper facial indices confirm the predominance of facial elongation.

(162) The face of *Zinjanthropus* tapers upwards towards the hafting of the face on to the fore-part of the calvaria. This tapering occurs, too, in *Paranthropus*, and to a lesser degree in *Australopithecus*. The suprafacial breadth indices confirm the trend. Two factors contribute to the tapering effect: the strongly downcurved lateral part of the supra-orbital torus diminishes the superior facial breadth, while the lateral expansion of the maxillae makes for a relative tapering above.

(163) The longitudinal calvariofacial indices place the australopithecines, including *Zinjanthropus*, in an intermediate position between the hominines and the pongids. The transverse calvariofacial indices, on the other hand, relate the australopithecines to the pongids rather than to the hominines: in both the apes and the australopithecines, the relatively unexpanded calvaria and the marked flaring of the zygomatic arches produce these indicial similarities.

The pneumatisation of the cranium

(164) The maxillary sinus of *Zinjanthropus* is voluminous and it extends by means of a variety of recesses into various parts of the maxillary bone (frontal process, nasolacrimal area, the alveolar process and palatine process) and into adjacent bones (zygomatic bones). It does not pneumatise the pterygoid process, which receives rather an extension from the sphenoidal sinus: no evidence could be found of an intrasphenoidal encroachment by the maxillary sinus. In general, the maxillary sinus of *Zinjanthropus* is reminiscent of that of the pongids, especially gorilla and orang, although it falls short in some respects of the pongid degree of 'dominance' of the maxillary sinus; how much of the shortfall is a function of the adolescent age of the Olduvai cranium it is not possible to determine on the present material.

(165) The frontal sinus of *Zinjanthropus* is extensive. Like that of *Gorilla*, it has lateral and posterolateral diverticula, but its relationship to the naso-orbital region and to the ethmoidal sinus system cannot be determined. In its extent, the frontal sinus of *Zinjanthropus* is more comparable with that of pongids than with that of *H. erectus*.

(166) The naso-orbital region of *Zinjanthropus* is appreciably inflated, but it is not possible to determine whether its pneumatisation was owing to encroachment by the frontal and/or maxillary sinuses (as in pongids) or to the proliferation of the initial ethmoidal air cells (as in *Homo*).

(167) The sphenoidal sinus is extensive and spreads right across the floor of the middle cranial fossa. Its recesses and diverticula penetrate virtually to the lateral limit of the greater wing (pterion), the root of the pterygoid formation, the sphenoidal rostrum (thereby widely separating the two alae of the vomer) and the basi-occipital. It would seem to be a slightly more 'dominant' sinus in *Zinjanthropus* than it is in *Gorilla*.

(168) Pneumatisation of the pars mastoidea is

extensive in *Zinjanthropus* and the other australopithecines, though it is not as marked as in *Gorilla*. Temporal pneumatisation in *Zinjanthropus* covers the entire squama and extends into the root of the zygomatic process.

(169) The overall degree of pneumatisation in *Zinjanthropus* is greater than in any previously described australopithecine or hominine; this may be associated with the extreme size of the teeth, jaws and muscular markings in the Olduvai specimen. Further extension of the sinuses in *Zinjanthropus* was to be expected if the individual had lived longer, as pneumatisation has been shown to continue beyond the time of eruption of the third molars in other hominoids.

The shape of the dental arcade and the palate

(170) The maxillary dental arcade of *Zinjanthropus* is an evenly curved parabola, as in the australopithecines and hominines, and unlike the U-shaped arcade of the pongids. In contrast with the pongids, too, *Zinjanthropus* and other australopithecines show no trace of a diastema.

(171) The maxillo-alveolar indices indicate that the Australopithecinae, including *Zinjanthropus*, have a relatively longer and narrower alveolar process than the Homininae, but not so long and narrow as in the Pongidae. On the other hand, the palatal index yields a different picture, the index in *Zinjanthropus* (48·3 per cent) falling well within the pongid range and appreciably below the hominine range. The marked broadening of the molar portion of the alveolar ridge, partly at the expense of the hard palate, accounts for this apparent discrepancy between the 'internal' and 'external' indices. The arcadal index of *Zinjanthropus* (97·6 per cent) lies within the lower part of the hominine range (95–136 per cent).

(172) The maxillo-alveolar, palatal and arcadal indices all point to *Zinjanthropus* having a longer and narrower palato-alveolar component of the maxilla than do the Homininae; the external (maxillo-alveolar) proportions are intermediate between those of Pongidae and Homininae; the internal (palatal) proportions lie well within the pongid range.

(173) Measurements between the buccal faces of corresponding pairs of cheek teeth show that the tooth-rows in *Zinjanthropus* diverge from P^1 to M^3, but that from M^1 backwards the amount of divergence is reduced, thus converting the arcade from a hyperbola to a parabola.

(174) In the canine–incisor region, the posterior margins of the alveoli of *Zinjanthropus* are arranged in a moderate curve, intermediate between the extremely flat curve obtaining in some specimens of *Paranthropus* and the well-arched curve of some maxillae of *Australopithecus*. However, the intermediate form with gentle curve, as manifested by *Zinjanthropus*, occurs in one out of five specimens of *Australopithecus* and two out of seven of *Paranthropus*. In this respect, *Zinjanthropus* falls within the range of variation of both types of South African australopithecine.

The arrangement of teeth in the arcade

(175) There is no trace of either a precanine or a postcanine diastema in *Zinjanthropus*, in contrast with some South African australopithecines in which a minute gap occurs between the lateral incisor and canine.

(176) Both canines and both P^3's show slight labial displacement, while the canines show slight linguodistal rotation. The displacement and the rotation suggest crowding and shortage of space, such as Robinson has detected especially in the large-toothed South African australopithecine, *Paranthropus*. It is suggested that this crowding may be related to an interaction between two opposing morphogenetic tendencies, that leading to flattening of the face including the anterior dental arcade, and that leading to enlargement of the cheek-teeth.

The pattern of attrition and occlusion

(177) Two distinct asymmetrical trends are apparent in the attrition of the maxillary teeth. First, the right teeth are, in general, more heavily worn than the left. Secondly, the gradient of maximum wear passes distad from left to right along the premolar–molar tooth-rows. It is suggested that this occlusal and attritional asymmetry is related to the greater development of the right

temporalis muscle, as reflected in its markings on the calvaria. Whether from the same cause or not, it would seem, too, that the mandibular tooth-rows were not well aligned against the maxillary, i.e. centric occlusion was absent.

(178) The masticatory stresses were so great that every tooth from I^1 to M^2 shows one or more islands of dentine exposure, even though the M^3's had not entered into occlusion; the degree of attrition is far greater than in other australopithecine maxillary dentitions at a comparable eruptive stage. It is probable that the diet was extremely coarse or gritty.

(179) The canines show the typically hominid form of wear from the tip, which has resulted in a nearly flat occlusal surface.

The state of the enamel

(180) Well-defined areas of hypoplastic enamel occur on most of the teeth of *Zinjanthropus*. These correspond in position with what in a modern child would be the 'early childhood ring' ($2\frac{1}{2}$ years), the ring of enamel formed during the latter part of early childhood (4 years) and with the 'later childhood ring' ($4\frac{1}{2}$–5 years). If similar developmental phases applied to *Zinjanthropus*, it is possible that, at three such periods—the actual ages in *Zinjanthropus* probably being a little younger—the Olduvai youth was subject to systemic upsets, which for a time impaired enamel formation.

The size of individual teeth

(181) The most striking feature of the dentition of *Zinjanthropus* is its great size; the cheek-teeth are greater even than those of the *crassidens* (Swartkrans) population of *Paranthropus*, and even approach in size some of the *Gigantopithecus* teeth.

Incisors and canines

(182) As in *Paranthropus*, the incisors are not remarkable for their size. The dimensions of I^1 fall in the upper part of the range for *Paranthropus*, but while the M.D. diameter of this tooth slightly exceeds the range for three *Australopithecus* specimens, the L.L. diameter is slightly smaller than in the latter form. The dimensions of I^2 fall within the ranges for both *Paranthropus* and *Australopithecus*, save only that the L.L. diameter slightly exceeds the range for *Australopithecus*.

(183) The ratio of the modules of I^2 and I^1 suggests that in the Australopithecinae the values are a little lower than those of the Homininae and that, in the degree of reduction of I^2 as compared with I^1, *Zinjanthropus* is slightly nearer to *Paranthropus* than to *Australopithecus*. The range of means for this ratio is 75–82 per cent in the Australopithecinae, 78–91 in the Homininae and 77–85 in the Pongidae.

(184) The ratio of the crown areas of I^2 and I^1 yields values (per cent) for the Australopithecinae of 57–69, for the Homininae of 61–84 and for the Pongidae of 60–72.

(185) Like other australopithecines, *Zinjanthropus* has a small canine. The M.D. diameters (8·7, 8·8 mm.) fall near the mid-value of the range for *Paranthropus* (8·1–9·3 mm.) but at the *bottom* of the range for *Australopithecus* (8·8–9·9 mm.). On the other hand, the L.L. diameters of the *Zinjanthropus* canine (9·7, 9·9 mm.) fall near the *top* of the range for *Australopithecus* (8·7–9·9 mm.), but again close to the mid-value of the range for *Paranthropus* (8·4–11·1 mm.). The module and the crown area fall within the ranges in both South African forms, being a little bigger than the *Paranthropus* mean and a little smaller than the *Australopithecus* mean. With the exception of *H. erectus*, which has very large incisors, especially I^2, all other hominines compared have smaller canine dimensions than the australopithecines; whereas the pongids have appreciably larger canines, only the female chimpanzee having a mean L.L. diameter (9·0 mm.) which is smaller than the australopithecine mean (9·5 mm.).

(186) The canines of *Zinjanthropus* are more reduced relative to the size of the postcanine teeth than those of any of the published specimens of *Paranthropus* and *Australopithecus*. This was determined by comparing each of the crown diameters, the module and the crown area of *Zinjanthropus*, first, with those of P^3 and then with those of P^4. In every comparison, the canine-premolar ratio was smaller in *Zinjanthropus* than

SUMMARY OF CRANIAL AND DENTAL FEATURES

in any other australopithecine maxilla. The contrast between the East and South African forms is greatest in respect of the labiolingual diameter, in which feature the postcanine teeth of *Zinjanthropus* show excessive development; this preponderance tends to be overlooked when comparisons are confined to modules.

(187) In the C/P³ comparisons, *Paranthropus* is nearer to *Zinjanthropus* than to *Australopithecus* in M.D. diameter, in module and in crown area, whilst in B.L. diameter the *Zinjanthropus* P³ is so enlarged that *Paranthropus* stands midway between *Australopithecus* and *Zinjanthropus*. In the C/P⁴ comparisons, for both dimensions and both indices, the *Zinjanthropus* values are closer to the *Paranthropus* mean than the latter is to the *Australopithecus* mean.

(188) The validity of using the C/P³ and C/P⁴ ratios as a measure of the size-ratio between canine and postcanine teeth has been established by the demonstration that, in a number of australopithecines including *Zinjanthropus*, the premolar dimensions bear a relatively constant size relationship to the cheek-tooth row.

The sizes of the premolars

(189) The premolars of *Zinjanthropus* are very enlarged, especially in buccolingual diameter, in which respect both P³ and P⁴ far exceed the range of values for South African australopithecines. The M.D. diameter (mm.) of the P³ of *Zinjanthropus* (10·9) lies just outside the sample range for *Paranthropus* (9·0–10·8) and substantially exceeds that of *Australopithecus* (8·5–9·4). On the other hand, the B.L. diameter (mm.) of the *Zinjanthropus* P³ (17·0) far exceeds the ranges for both South African taxa (13·1–15·3 and 10·7–13·9 respectively). The excess over the mean for *Paranthropus* is significant. Similarly, the M.D. diameters (mm.) of the *Zinjanthropus* P⁴ (11·8, 12·0) lie at the top of the *Paranthropus* range (9·2–11·8) and well above the range for *Australopithecus* (7·8–10·5); whereas the B.L. diameters of P⁴ (18·0, 17·6) far exceed the range for both southern taxa (13·7–16·5 and 12·0–13·9 respectively), the excess over the *Paranthropus* mean being significant. The modules and crown areas of both premolars in *Zinjanthropus* are appreciably greater than those of their South African counterparts.

(190) *Zinjanthropus* shares the buccolingual broadening of P³ and P⁴ with other Australopithecinae, even the smaller-toothed *Australopithecus*. This feature is not present in the Homininae; but it is shared with certain pongids, notably gorilla and orang-utan, but not with chimpanzee.

(191) In *Zinjanthropus* P⁴ is greater than P³ in every respect, as in *Paranthropus*; the two premolars tend on the average towards equality in *Australopithecus*; whilst, in *H. erectus*, P⁴ is smaller than P³. The intermediate mean trend in *Australopithecus* follows because, in two out of five individual specimens, P³ < P⁴, while in the remaining three, P³ > P⁴, as in the Homininae. In every specimen of *Paranthropus*, as well as in *Zinjanthropus*, P³ < P⁴. The general trend in pongids is for P⁴ to be smaller than P³: this reduction is in both M.D. and B.L. diameters in gorilla, but in M.D. only in chimpanzee and orang. In chimpanzee, B.L. diameters are equal in P³ and P⁴; while in orang the B.L. diameter of P⁴ slightly exceeds that of P³.

The sizes of the molars

(192) All the *Zinjanthropus* molars are larger than their counterparts in the two South African australopithecine taxa. Most notably the tendency to B.L. expansion continues beyond the premolars into the molar field; thus the B.L. diameters of all three Olduvai molars on each side exceed those of *Paranthropus* and *Australopithecus*. The same is true for the M.D. diameters of M¹ and M², whereas that of M³ of *Zinjanthropus* falls within the range for *Paranthropus*.

(193) The M.D. diameter (mm.) of M¹ of *Zinjanthropus* (15·2) lies outside the *Paranthropus* range (13·1–14·6) and exceeds the *Paranthropus* mean (13·8) by 2·92 S.D.'s. It likewise exceeds the *Australopithecus* range (11·9–13·2 mm.) and mean (12·6 mm.) by 6·84 S.D.'s. The B.L. diameter (mm.) of the *Zinjanthropus* M¹ (17·7) lies outside the range for *Paranthropus* (12·7–16·6) and exceeds the mean (14·5) by 3·71 S.D.'s; it lies outside the range for *Australopithecus* (12·8–15·1) and exceeds its mean of 13·8 by 6·19 S.D.'s.

(194) The M.D. diameter (mm.) of *Zinjanthropus* M² (17·2) lies far outside the *Paranthropus* range (13·6–15·7) and exceeds its mean of 14·5 by 3·71 S.D.'s. It likewise exceeds the *Australopithecus* range (12·7–15·1 mm.) and mean (13·8 mm.) by 4·30 S.D.'s. The B.L. diameter (mm.) of the *Zinjanthropus* M² (21·0) lies far outside the range for *Paranthropus* (14·2–16·9) and exceeds the mean of 15·9 by no less than 5·97 S.D.'s; the corresponding range for *Australopithecus* is 13·8–17·1 mm., and *Zinjanthropus* here shows an excess of 5·60 S.D.'s over the mean of 15·4 mm.

(195) The M.D. diameters (mm.) of the *Zinjanthropus* M³'s (15·7, 16·3) fall within the *Paranthropus* range (12·8–17·0) and exceed the mean of 15·0 non-significantly. In relation to the *Australopithecus* M³, *Zinjanthropus* shows an excess of 2·25 S.D.'s (left) and 2·79 S.D.'s (right) over the mean of 13·2 mm. In B.L. diameter (mm.), the *Zinjanthropus* M³'s (21·4, 20·5) exceed the *Paranthropus* mean of 16·8 by 5·93 and 4·77 S.D.'s and the *Australopithecus* mean of 15·5 by 5·67 and 4·81 S.D.'s.

(196) In consequence of these large dimensions, the modules and crown areas of all three molars of *Zinjanthropus* greatly exceed both the means and ranges for the other two australopithecine taxa.

(197) The molar dimensions bear a relatively constant size relationship to the cheek-tooth row, in all australopithecine maxillae (including that of *Zinjanthropus*) in which the teeth from P³ to M³, or at least from P⁴ to M³, are preserved.

(198) *Zinjanthropus* resembles *Australopithecus* in that M³ is reduced in size, as compared to M². This reduction affects the M.D. diameter, whereas in both forms the B.L. diameters are virtually the same in M² and M³. In contrast, *Paranthropus* shows an increase in both M.D. and B.L. diameters from M² to M³. The degree of reduction of both the M.D. diameter and the module in *Zinjanthropus* is intermediate between the mean reduction in *Australopithecus* and the mean reduction in *H. e. pekinensis*; however, the reduction in *Zinjanthropus* and *Australopithecus* is not marked, so that M³ remains the second biggest tooth. This sequence (M² > M³ > M¹) is found, too, in male gorilla; whereas in female gorilla, and in chimpanzee and orang-utan of both sexes, M³ is so far reduced that the order M² > M¹ > M³ is found. Among all the hominoid groups considered (including *Dryopithecus*, *Sivapithecus*, *Proconsul*, *Ramapithecus*, *Gigantopithecus*, *Australopithecus*, fossil and living hominines, and living pongids), *Paranthropus* stands alone in possessing the mean size sequence M³ > M² > M¹. This sequence occurs in six out of seven maxillae of *Paranthropus*, but in only one out of six *Australopithecus* maxillae is M³ slightly larger than M². It is suggested that the enlargement of M³ in *Paranthropus* should be regarded as a specialisation, rather than as a primitive hominoid feature.

(199) Flower's dental index in *Zinjanthropus* is 65·8 per cent on the left and 66·2 per cent on the right, placing the fossil in an extremely hypermegadont category, well above the values for Homininae.

(200) The maxillary 'tooth material' (sum of M.D. diameters of left and right I¹–M¹) of *Zinjanthropus* totals 127·7 mm., a far higher value than that of any other member of the Hominidae.

Shape indices of the teeth

(201) While the M.D./B.L. shape indices among hominoids generally decrease from C to P⁴, those of *Zinjanthropus* increase somewhat from P³ to P⁴, as do five out of twelve South African australopithecine specimens, Sangiran IV, Rabat and male gorilla. The absolute values for the shape indices of P³ and P⁴ of *Zinjanthropus* are at the bottom of or below the range for all other hominoids, underlining again the remarkable broadening tendency of the teeth in this creature.

(202) The trend towards a decreasing M.D./B.L. index from M¹ to M³ is general throughout the Hominidae, including *Zinjanthropus*, with the exception of Sangiran IV, in which the indices of M¹ and M² are practically the same, and modern American Whites, in whom the index rises somewhat from M² to M³. The values for the *Zinjanthropus* molars are lower than those of any other hominoids, save for one apparently atypical M³ of *Paranthropus*. They contrast markedly with the much more squarish molars of pongids. The shape

SUMMARY OF CRANIAL AND DENTAL FEATURES

indices confirm that enlargement of the postcanine teeth of *Zinjanthropus* is not equal in all directions: instead there is a buccolingual exaggeration of the crowns, reflecting itself in low crown shape indices.

Morphology of the teeth

(203) The *Zinjanthropus* upper incisors resemble those of the South African australopithecines in possessing a slight to moderate degree of shovelling, and a simple gingival eminence better developed on the lateral than on the central incisors. In metrical features, the incisors are somewhat intermediate between the South African forms, but the degree of reduction from I^1 to I^2 is a little closer to that of *Paranthropus* than to that of *Australopithecus*.

(204) The labial face of the maxillary central incisor is relatively flat, except for a pair of shallow, vertical, distal labial grooves. In both features, *Zinjanthropus* resembles *Australopithecus* rather than *Paranthropus*. The morphological features of the maxillary lateral incisor of *Zinjanthropus* resemble those of Robinson's type B lateral incisor, which is the predominant type in both *Australopithecus* and *Paranthropus*. The labial face of I^2 has no grooves, resembling in this respect *Australopithecus* rather than *Paranthropus*; but, as in *Paranthropus*, the greatest labiolingual diameter is nearer the mesial face. The roots of the incisors are similar in dimensions and morphology to those of *Paranthropus*.

(205) The maxillary canines of *Zinjanthropus* are marked by mesiodistal asymmetry and by a distal cusplet, features which have hitherto been ascribed exclusively to the *mandibular* canines of australopithecines, and especially to those of *Australopithecus*. Since *Zinjanthropus* (and some other hominid specimens) show that maxillary canines may show these features as well, it is suggested that the identification of Sts 3 (a canine from Sterkfontein showing asymmetry and cusplet formation) as a mandibular canine should be reconsidered, as its measurements strongly align it with australopithecine upper canines.

(206) The lingual face of the canines of *Zinjanthropus* is very reminiscent of that of *Australopithecus*, and not at all like the highly distinctive maxillary canines of *Paranthropus*. One feature on the labial face—a marked bulge of the enamel just below the cervical line—is a reminder of *Paranthropus* rather than of *Australopithecus*.

(207) The absolute dimensions of the *Zinjanthropus* canines fall well within the ranges for *Paranthropus*; in comparison with *Australopithecus*, however, the M.D. diameter is small and the L.L. diameter is large. *Zinjanthropus* thus agrees with *Paranthropus* in showing no relative L.L. reduction of the canine, a trend towards which is apparent in the canines of *Australopithecus*. Relative to the size of the postcanine teeth, the canines of *Zinjanthropus* are smaller than in any australopithecine dentition yet described. In most metrical comparisons, *Zinjanthropus* lies closer to *Paranthropus* than the latter does to *Australopithecus*; an exception is the L.L. diameter, in which *Zinjanthropus* is as far removed from *Paranthropus* as the latter is from *Australopithecus*.

(208) The P^3 of *Zinjanthropus* is an exceptionally large tooth, most of its exaggerated size being due to the excessive development of the buccal cusp. This feature appears to throw the anterior premolar out of alignment with the rest of the dental arcade and yields a very high shape index, well beyond the maxima for both *Australopithecus* and *Paranthropus*. Although the morphological differences between the P^3's of *Australopithecus* and *Paranthropus* are not marked, in most respects the P^3 of *Zinjanthropus* resembles that of *Paranthropus*, but in a few respects only is it reminiscent of the premolars of *Australopithecus*. The root system is one of the more distinctive differences between *Australopithecus* and *Paranthropus*: in this feature *Zinjanthropus* is clearly aligned with *Paranthropus*, in having two buccal roots and a single lingual one.

(209) On P^4 of *Zinjanthropus*, the relative development of the two cusps, as reflected in the B.L. diameters of the cusps, is equal: the buccal cusp of P^4 having a slightly smaller B.L. diameter than that of P^3, whereas the lingual cusp of P^4 has a greater diameter than that of P^3. The P^4's in consequence are seemingly displaced lingually. Most

of the crown features of P⁴ in *Zinjanthropus* approximate to those in *Paranthropus*, although the sculpturing of the lingual face suggests an affinity with *Australopithecus*. In measurements, this tooth lies closer to those of *Paranthropus* than of *Australopithecus*, though exceeding in B.L. expansion both of the South African forms. The ratio of P³ to P⁴ in *Zinjanthropus* approximates to that obtaining in *Paranthropus*; the ratio suggests that in *Zinjanthropus* P⁴ has become more molarised than P³, the cheek-teeth in this form having undergone expansion which is graded disto-mesially.

(210) In the morphology of the occlusal, buccal and lingual surfaces and of the roots, the first molar of *Zinjanthropus* has a number of resemblances to that of *Australopithecus* rather than to that of *Paranthropus*. Vestiges of the cingulum have been identified on both lingual and buccal surfaces, aligning the *Zinjanthropus* M¹ with that of *Australopithecus*. The tendency to B.L. expansion in *Zinjanthropus* is continued from the premolar into the molar area, the M¹ being greater in measurements and indices than the largest values for the South African homologues.

(211) Most of the minute morphological features of the M² of *Zinjanthropus* agree more closely with those of *Australopithecus* than with those of *Paranthropus*; the cingular remnants on the lingual surface are not as strongly developed as in some *Australopithecus* M²'s, though they are more definite than those of *Paranthropus*. In dimensions, there is a considerable overlap between the ranges for M² of *Australopithecus* and *Paranthropus*, but that of *Zinjanthropus* far exceeds the values for both of these South African forms. The B.L. predominance is again evident in the M² of *Zinjanthropus*. In crown area, M² of *Zinjanthropus* exceeds the maximum *Paranthropus* M² by 36·1 per cent and the maximum *Australopithecus* M² by 39·9 per cent: in fact, it falls just short of the crown area recorded for a *Gigantopithecus* upper molar (specimen 3), 361·2 mm.² as against 378 mm.²!

(212) The enamel on the occlusal surface of the M³'s of *Zinjanthropus* is highly crenulated: the folds of enamel take the form of multiple minor cusplets replacing each of the four major cusps. The pattern of crenulation is not wrinkled as in the orang-utan. It is suggested that this cuspidate occlusal surface may be another expression of hypoplasia through influences which were operating at the time when M³ was beginning to calcify. The wrinkled form of molar which occurs in many Primates may be genetic in origin, whereas the cuspidate form may be environmental and hypoplastic.

(213) The general pattern of the M³ of *Zinjanthropus* is typical of that of the Australopithecinae, with features reminiscent of both South African forms. In shape, cusp pattern, Carabelli formation and buccal cingular vestige, M³ of *Zinjanthropus* is nearer to that of *Australopithecus*; in crenulations and root pattern it is somewhat closer to that of *Paranthropus*.

(214) When comparisons are based on metrical features, Robinson's two key differences between *Paranthropus* and *Australopithecus* (difference in absolute size between the teeth of the taxa and difference in relative sizes between M² and M³) lead to varying conclusions. In respect of absolute molar size, *Zinjanthropus* significantly exceeds both South African forms; in M²:M³ ratio *Zinjanthropus* is closer to *Australopithecus*, both forms showing a diminution in M.D. diameter from M² to M³, but B.L. diameters which are equal in the two teeth. *Paranthropus*, in contrast, shows an increase in both diameters from M² to M³.

(215) Probable remnants or derivatives of the cingulum may be recognised on all the maxillary molars, weakly developed on the buccal surfaces, pronounced on the lingual. *Zinjanthropus*, like *Australopithecus*, represents an earlier or more primitive stage in the trend towards reduction of the cingulum; *Paranthropus* represents a more advanced stage of reduction, having only weak vestiges on the lingual face and no traces on the buccal surface.

CHAPTER XIX

THE TAXONOMIC STATUS OF ZINJANTHROPUS AND OF THE AUSTRALOPITHECINES IN GENERAL

A. General considerations

1. 'Splitting' and 'lumping': an historical perspective

The well-known trends of 'splitting' and 'lumping' are very much evident in the history of australopithecine taxonomy and nomenclature. Thus, at one time or another, the following generic labels have been applied to supposed members of this group: *Australopithecus, Plesianthropus, Paranthropus, Telanthropus, Meganthropus, Praeanthropus, Zinjanthropus,* and *Hemanthropus* (originally called *Hemianthropus*). We know today that many of these *nomina* did *not* represent valid taxa or good biological units.

By and large, splitting seems inevitable at an early stage in the sequence of discoveries of a group; lumping is equally inevitable at a later stage when many specimens have accumulated, when ranges of variation for the group in question and for related groups have been minutely explored, when the time dimension is better known, and when a deeper understanding has been gained of the functional and cultural capacities associated with a group of fossils showing a particular morphological pattern.

Because we now possess much of the detailed knowledge necessary to detect where affinities lie, the time is ripe for a reconsideration of hominid fossil designations and the combining of previously separated groups of fossils into broader and more valid taxa.

The fact that the author has been led by his studies of *Zinjanthropus* and other australopithecines to regard them all as members of a single genus is based not upon any enhanced biological insight, but upon the irresistible body of evidence that has accumulated over the four decades since *Australopithecus africanus* was named and launched by Dart (1925).

2. Variability: intragroup and intergroup

It is a reasonable proposition that the variability of the australopithecines should be assessed from the variability of living hominoids. In his valuable discussion on primate variability, Schultz (1963) has emphasised that 'All systematic and phylogenetic studies can be merely tentative until they have considered a great many different characters and these with due regard for age, sex and variability...' (p. 85).

One real difficulty in applying this advice to fossils is the fact that the degree of variability of metrical (and even non-metrical) characters varies from taxon to taxon. Thus, for cranial capacity, the coefficient of variation (V) may be as low as 6·8 per cent in chimpanzee and as high as 13·6 per cent in gorilla (male and female), or 12·8 per cent in modern man (Tobias, 1965e). Even if we separate male and female gorilla capacities, the gorilla male V of 10·4 per cent is higher than that of orang-utan or chimpanzee for both sexes combined! Many factors enhance the variance and these may not operate equally in different hominoids.

Not only may a single metrical character vary to different degrees from population to population, but different taxa may show high or low overall variability. Morant (1927) proved that, in contrast with modern man, Neandertal man was remarkably homogeneous in craniometric characters. This was stressed again not long ago by Blanc, by

Vallois and by Caspari (1961), the latter adding that 'in recent man interindividual variability appears to be relatively large as compared with a number of other species' (p. 267). In England, the Biometric School showed wide diversity of variability among different human populations and different great apes.

Remane (1954) pointed out that 'Mindestens seit dem unteren Miozän besassen die Stämme der Pongidae und Hylobatidae eine ungewöhnliche Breite individueller Vielgestaltigkeit'. Schultz (1963) has stressed that: 'The degrees of intraspecific variability in morphological characters can differ widely among recent primates (Schultz, 1947) and most likely have done so among those of the past.' In seeming contrast with Caspari's very general statement about recent man, Schultz goes on: 'Quite generally speaking it appears that recent man, most Old World monkeys and some of the New World, tend to be decidedly less variable than are the great apes, gibbons and certain prosimians.' But he qualifies this by adding, 'Strictly speaking one cannot assess the variability of a species or a population, but merely that of single features, such as skull shape, vertebral formula or hair color, and it is only after one has established the degrees of variability of many different features that one becomes justified to generalise in regard to the uniformity of a given species and the probability that an available sample permits a valid description of typical conditions' (p. 94).

For any trait, the variability in *Australopithecus* may be high, as in some hominoids, low as in others, or different from both groups. We simply do not know if *Australopithecus* (*sensu lato*) was a highly variable taxon in general, nor whether, for example, cranial capacity in this group was of low or high variability.

In the light of such varying degrees of variability, it is scarcely valid to attribute to a fossil taxon, such as *Australopithecus*, the degree of variability established for any one taxon of living higher Primates. At the most, we may hope to obtain no more than a rough guide to the approximate order of magnitude of the variance in fossil hominoids.

3. *The hypodigm[1] and its bearing on the assessment of variability*

Inferences about the variability of a population may differ according to whether the hypodigm[1] for a given australopithecine taxon is derived exclusively *from a single site*, or from two or more sites. This is true even if the sites are no more geographically dispersed than Kromdraai and Swartkrans, which are scarcely two miles apart.

To illustrate this point, we may cite the variability of the teeth of *Australopithecus* (*sensu stricto*) and those of *Paranthropus*. From a study of the teeth, Robinson (1965b) has claimed that 'in many respects *Australopithecus* is considerably more variable than is *Paranthropus*'. As an example, the buccolingual breadth of M_1 in *Paranthropus* ($n=18$) is stated to have a coefficient of variation (V) of 4·05, whereas in *Australopithecus africanus* ($n=$ only 4) V is 8·85. He based his inferences as to the variability of the population solely on the fossils from Sterkfontein (for *Australopithecus*, *sensu stricto*) and from Swartkrans (for *Paranthropus*). Samples based on more than one site present a different picture. For ten *A. africanus* M_1's from three sites, V is 6·4; while for twenty-one *Paranthropus* M_1's from two sites V is 5·8. Thus, only a slight difference in variability remains when teeth from other South African sites are included. This slight difference virtually disappears if, for purposes of this exercise, the Peninj mandibular M_1's from Lake Natron, Tanzania, are included with the *Paranthropus* sample: V then becomes 6·3 ($n=23$) as compared with 6·4 for *A. africanus* ($n=10$). Figures for variability of the buccolingual diameter of mandibular cheek-teeth are given in Table 47 (from Tobias, 1965d).

These figures do not support the conclusion that, in buccolingual crown diameter, *Australopithecus* is 'considerably more variable' than *Paranthropus*. The original samples were, first, too small,

---

[1] The term *hypodigm* was proposed by Simpson (1940) and defined by him as follows: 'The hypodigm of a given taxonomist at a given time and for a given taxon consists of all the specimens personally known to him at that time, considered by him to be unequivocal members of the taxon, and used collectively as the sample on which his inferences as to the population are based' (1961, p. 185).

especially in respect of *Australopithecus*, and secondly, too restricted in origin to permit so broad a generalisation. The more australopithecine sites are added to the pool, the more accurately we may hope to infer the properties of the population.

Table 47. *Coefficients of variation for buccolingual diameters of mandibular teeth*

	Australopithecus (from Makapansgat, Sterkfontein and Taung)	*Paranthropus* (from Kromdraai, Swartkrans and Peninj)
P_3	8·1 ($n = 7$)	7·6 ($n = 16$)
P_4	8·0 ($n = 5$)	7·8 ($n = 17$)
M_1	6·4 ($n = 10$)	6·3 ($n = 23$)
M_2	5·3 ($n = 11$)	5·4 ($n = 15$)

Similarly, the statement that the B.L. diameter of an australopithecine tooth crown is usually more variable than its M.D. diameter (Robinson, 1965b) is not upheld by larger and more representative samples. The relative variability of B.L. and M.D. diameters varies with the tooth and with the taxon (Tobias, 1965d) (*see* Table 48).

Table 48. *Coefficients of variation for* M.D. *and* B.L. *diameters of mandibular teeth*

	Australopithecus (from Makapansgat, Sterkfontein and Taung)		*Paranthropus* (from Kromdraai, Swartkrans and Peninj)	
	M.D.	B.L.	M.D.	B.L.
P_3	8·0	8·1 ($n = 7$)	3·7	7·6 ($n = 16$)
P_4	5·6	8·0 ($n = 5$)	9·5	7·8 ($n = 17$)
M_1	6·3	6·4 ($n = 10$)	5·8	6·3 ($n = 23$)
M_2	6·2	5·3 ($n = 11$)	5·6	5·4 ($n = 15$)

A more variable B.L. than M.D. occurs only in the *Paranthropus* P_3 and the *Australopithecus* P_4. On the other hand, M.D. is more variable than B.L. in the *Paranthropus* P_4 and, to a certain extent, in the *Australopithecus* M_2. In the remaining four teeth, P_3 and M_1 of *Australopithecus* and M_1 and M_2 of *Paranthropus*, the coefficients of variation are virtually the same for M.D. and B.L. diameters. The supposed distinction in dental variability between the B.L. and M.D. diameters certainly cannot be considered proven.

Thus assessments of variability lead to widely differing conclusions, when based on hypodigms of restricted or of broader origin. It follows that valid statements can be made about the comparative variability of taxa only when based on adequate samples and on the total hypodigm for each taxon.

4. *The concept of 'phyletic valence'*

The term 'phyletic valence' has been used to refer in general terms to the ability of a metrical or non-metrical trait to distinguish among a number of different groups (Robinson, 1960, 1965a). Although I have been unable to trace a formal definition of the term in Robinson's writings, nor any metrical expression of it, it would seem to be based upon the variability of a feature within and between groups, irrespective of the phylogenetic, morphological or functional importance of the feature.

Two examples may be given of the use of this term, one applied to a non-metrical trait, and the other to a metrical character.

The non-metrical example is provided by the form of the tympanic plate (Robinson, 1960). Three variants of the feature in *Paranthropus* are illustrated; one of the variants, it is claimed, closely resembles the structure in *Zinjanthropus*. Robinson comments: 'Being thus variable, it is in any event a feature of low phyletic valence.' Here, moderate variability of a feature *within one taxon* apparently permits the attribution of low phyletic valence to it.

The commoner usage is in a situation characterised by high intragroup variability and low intergroup variability, as in the following metrical example. Speaking of the use of the M.D./B.L. crown shape index as a feature differentiating the teeth of *H. habilis* from those of australopithecines, Robinson states, '...a far more trenchant criticism of the use made of the dental length/breadth index by Leakey *et al.* is that analysis of the index shows that it and the features on which it is based have extremely low phyletic valence, so far as hominids are concerned. This is readily apparent from the extremely wide overlap of the ranges for this index for *Australopithecus, Paranthropus, Homo erectus* and *Homo sapiens*' (Robinson, 1965a, p. 122).

These examples and others suggest the implication that a trait which varies markedly in a group, or which fails to distinguish between two or more related groups, should not be used to distinguish a *new* specimen or proposed taxon from previously known groups. Once a trait has been shown to have 'low phyletic valence', it is implied, the use of such a trait in a new taxonomic comparison is of dubious value. Notwithstanding, it would seem to the present writer that the taxonomic usefulness of a character in any specific comparison depends upon its ability to differentiate *between the two taxa being compared*, and *not* upon its ability to differentiate other taxa. The fact that the dental crown shape index may fail to distinguish between certain teeth of *Australopithecus* and *Paranthropus* in no way lessens its usefulness as a mark of distinction between, say, *Australopithecus* and a proposed new taxon, such as *H. habilis*.

The concept of phyletic valence at best suggests a grading of characters into those which are more useful and those less useful in differentiating among a number of subgroups of some higher taxon. But the fact that a feature may in *general* be less useful as a diagnostic criterion does not imply that *in any particular instance* it may not be of some value in differentiating two subgroups. The application of the general statement to the specific instance here is formally invalid. For this reason, those criticisms of the separate taxonomic status of *Zinjanthropus* which depend upon the application of the concept of phyletic valence cannot be considered valid.

5. *The relative usefulness of traits*

A scientifically acceptable measure of 'phyletic valence', or, better, relative usefulness of traits, has been devised for interracial comparisons in modern man by Tildesley (1950) and van Bork-Feltkamp (1950). They gave numerical expression to the usefulness of metrical traits in differentiating among living and cranial populations. First, they computed the 'mean intraracial S.D.', where data for at least thirty populations were available. Then the 'interracial S.D.' was computed—but only for traits for which large numbers of sample means were available. The mean intraracial S.D. was divided by the interracial S.D. to give a single index of 'relative usefulness'. The varying values of this index for different metrical traits permitted Tildesley and van Bork-Feltkamp to arrange the traits in estimated order of usefulness. On such information could be based a recommended list of characters for use in the measurement of 'racial series', living and skeletal—which was the objective originally set by the Second International Congress of Anthropological and Ethnological Sciences (Copenhagen, 1938). There was, however, no implication that a character of estimated low general usefulness might not prove an effective basis of differentiation in any specific comparison.

What van Bork-Feltkamp and Tildesley have done for populations of modern man could theoretically be carried out for prehistoric populations and for higher taxa, living or extinct. The difficulty at higher taxonomic levels, however, is that the number of separate groups per taxon is too small to permit a confident estimate of the intergroup S.D. For instance, for a given metrical trait, one could assemble species means for only four or five hominid species. Clearly, the sample of four or five species means would be insufficient for the computation of the interspecific S.D. This makes it very difficult to compare intraspecific with interspecific variability.

Multivariate analysis, such as has been applied to these problems by Bronowski and Long (1951, 1952, 1953) and by Ashton, Healy and Lipton (1957), is an immensely valuable statistical tool to help assess the overall resemblances among samples: however, it does not set out to answer the same questions as we have discussed above. Its purpose is not to determine the relative usefulness of individual characters, but, in respect of resemblances, 'the discriminant is essentially a comparative test: a device to decide which of two or more stated alternatives is to be preferred' (Bronowski and Long, 1953).

The degree of overlap of ranges of variation provides a crude approximation to Tildesley's index of relative usefulness, when the graphs for different characters are compared. But it can be applied only with adequate samples and with complete hypodigms drawn from all relevant sites. In this way the comparative graphs on dental size and shape were constructed in the present work and in studies related to *H. habilis* (e.g. Tobias, 1965c).

B. *Zinjanthropus* a hominid and an australopithecine

The fossil designated by Leakey (1959a) *Zinjanthropus boisei* and re-allocated by Leakey *et al.* (1964) to *Australopithecus (Zinjanthropus) boisei* has been shown to possess a complex of cranial and dental features which permit it to be assigned to the Australopithecinae. The most important morphological characters leading to this conclusion are:

(1) the relatively small cranial capacity (530 c.c.) which falls within the existing australopithecine sample range of 435–562 c.c., but is much smaller than the existing sample range of 750–1225 c.c. for *H. erectus*;

(2) strongly-developed supra-orbital ridges;

(3) a low sagittal crest in the frontoparietal region of the vault;

(4) a strong nuchal crest which is not continuous with the sagittal crest;

(5) occipital condyles which lie in the coronal plane through the external acoustic apertures (poria);

(6) a low and restricted nuchal area on the occipital bone, with the external occipital protuberance slightly below the level of the F.H.;

(7) a large, pyramidal mastoid process of typical hominid form, despite the immaturity of the individual at death;

(8) the structure of the mandibular fossa, especially the anterior wall, the salient articular tubercle and the small postglenoid process;

(9) massive maxillary bones;

(10) a parabolic dental arcade with no diastema;

(11) spatulate canines wearing down flat from the tip only;

(12) absolutely and relatively large premolars and molars.

All of the above features form a part of Le Gros Clark's formal diagnosis of the genus *Australopithecus* (1964, p. 168) which, it should be stressed, is the only australopithecine genus he recognises. The other features in Le Gros Clark's diagnosis—such as the bicuspid anterior lower premolar and the pronounced molarisation of the deciduous first lower molar—relate to parts which are not available in the type specimen of *Zinjanthropus*.

The wealth of anatomical detail preserved in the *Zinjanthropus* cranium makes it possible to fill out further details in the description of at least this australopithecine and, at the same time, provides additional support for the classification of the australopithecines as hominids rather than pongids. Thus, *Zinjanthropus* is characterised by the following features diagnostic of or common in the family Hominidae:

(1) appreciable flexion and anteroposterior reduction of the cranial base, as reflected in the relatively orthognathous face, the high angle between the petrous axis and the median plane, the low spheno-occipital index and basion–hormion distance, and the markedly elevated position of porion;

(2) relative anterior displacement of the occipital condyles, as shown by the condylar position index, and of foramen magnum, as displayed by the occipital length indices;

(3) the nearly horizontal plane of the foramen magnum, reflected both in the low angle between the plane of the foramen and the F.H., and the wide nasion–basion–opisthion angle;

(4) the vertical slope of the posterior face of the petrous pyramid;

(5) the detailed structure of the external acoustic porus and meatus, tympanic plate and supramastoid crest;

(6) the structure of the dorsum sellae and sella turcica;

(7) the indications of cerebellar enlargement;

(8) the moderate height of the orbits, despite the overall tendency to facial elongation;

(9) the presence of a high anterior nasal spine;

(10) reduction of subnasal prognathism;

(11) the lateral position of the mandibular fossae, showing that, despite condylar, dental and mandibular enlargement, the space between the

left and right halves of the mandible has *not* been appreciably reduced.

It is concluded that, though *Zinjanthropus* shows a number of extreme features as compared with the previously known sample of australopithecines, it nevertheless clearly qualifies for inclusion in the family Hominidae, subfamily Australopithecinae.

Many workers in the field today consider that all the australopithecines can be accommodated within a single genus, *Australopithecus*, rather than two or more genera (*see* next section). If this view came to be accepted universally, and at the same time general agreement were reached that all hominines could be accommodated in *Homo*, the need to maintain two subfamilies, Australopithecinae and Homininae, would fall away (Simpson, 1963, p. 29).

C. The generic and specific status of *Zinjanthropus*

1. *The nomina of* Zinjanthropus

When he published the first account of *Zinjanthropus*, Leakey (1959a) considered it to be generically and specifically distinct from the two South African australopithecine genera which he then supposed existed. What he did has been well stated in formal terms by Simpson (1963, p. 5) as follows:

Much of the complexity and lack of agreement in nomenclature in this field does not...stem from ignorance or flouting of formal procedures but from differences of opinion that cannot be settled by rule or fiat. For example, when Leakey inferred from an Olduvai specimen (which he made a hypodigm) the existence of a taxon that he called *Zinjanthropus boisei* he was using correct taxonomic grammar to express the opinion that the taxon was distinct at both specific and generic categorical levels from any previously named. In equally grammatical expression of other opinions many other nomina, such as *Paranthropus boisei*, *Australopithecus robustus boisei* or *Homo africanus boisei*, might have been proposed and might now be used. Or the specimen might have been and might now be referred to (or added to the hypodigm of) some previously named taxon such as *Paranthropus crassidens*. Any of these alternatives accord equally with the Code[1] and would have equal status before the Commission.[1] Decision among them is a zoological, not a nomenclatural or linguistic question, and it will be made by an eventual consensus of zoologists qualified in this special field.

[1] The International Code of Zoological Nomenclature (Stoll *et al.* 1961).

In 1960, Robinson analysed the diagnostic features adduced in Leakey's original description and concluded that 'separate generic status seems unwarranted and biologically unmeaningful'. He therefore proposed 'that the name of the Olduvai form be *Paranthropus boisei* (Leakey)'.

The position of *Zinjanthropus* was specifically excluded from the classification agreed upon by the Wenner-Gren International Symposium on 'Classification and Human Evolution' held in 1962, because the specimen had not yet been fully published (Campbell, 1962, 1963). Nevertheless, a number of recent books have accepted that *Zinjanthropus* belongs to the same genus and for some even the same species as the robust South African australopithecine (e.g. Hulse, 1963; Mayr, 1963b; Harrison and Weiner, 1964; Oakley, 1964; Le Gros Clark, 1964).

In 1964, Leakey *et al.* proposed to recognise only a single genus, *Australopithecus*, 'with, for the moment, three subgenera (*Australopithecus*, *Paranthropus* and *Zinjanthropus*)' (p. 7). The genus *Zinjanthropus* was thereby formally sunk and the correct nomen for the taxon thus became *Australopithecus* (*Zinjanthropus*) *boisei* or, for short, *A. boisei*. In formal terms, the specimen described in this book has been added to the hypodigm of the genus *Australopithecus* (*sensu lato*). This has been effected without doing violence to the formal diagnosis of the genus *Australopithecus* (*sensu lato*) put forward by Le Gros Clark (1964). The present more detailed assessment of the features of *A. boisei* and of the australopithecines in general has confirmed that the Olduvai specimen should be regarded as a member of the same group as that to which the South African specimens belong—and indeed has confirmed that the two southern forms should not be regarded as generically distinct, but as taxa (?species) within the same genus. This brings us to a consideration of the subdivisions of *Australopithecus*.

[1] The International Commission for Zoological Nomenclature.

TAXONOMIC STATUS

2. *The subdivisions of* Australopithecus

Apart from *A. boisei*, the australopithecines hitherto known from South Africa fall into two broad categories: one of which appears to be larger, with bigger cheek-teeth, jaws and muscle-markings on the skull, than the other. The former is represented by a large sample from Swartkrans and a small sample from Kromdraai; the latter by a large sample from Sterkfontein (Lower Breccia), a moderate sample from Makapansgat and one skull from Taung. While three genera (*Australopithecus*, *Plesianthropus* and *Paranthropus*) were earlier recognised by some as being represented by the fossils from the five South African sites, the accumulation of specimens and their detailed study led to a 'lumping' of *Plesianthropus* and, for most students, of *Paranthropus* into *Australopithecus*. As mentioned in the first chapter, however, Robinson (1963) and a few other workers have continued to regard *Paranthropus* as generically distinct from *Australopithecus* and, as Le Gros Clark (1964, p. 169) points out, '... because of his (Robinson's) intimate acquaintance with most of the original material, his opinion deserves close consideration'.

(a) *The dietary hypothesis*

Robinson's (1962) contribution to the Heberer Festschrift offers formal diagnoses of his proposed australopithecine genera. The first important point is that he considers *Paranthropus* to be 'vegetarian hominids' and *Australopithecus* 'omnivorous hominids'. This suggested ecological difference in diet—which runs through all Robinson's writings on the subject from 1954 to 1965—is itself an inference from morphological characters. Its inclusion as a diagnostic criterion in a formal taxonomic definition of a fossil group is of uncertain validity. More meet for such a definition would be the morphological characters themselves, upon which the inference is based: essentially, these are dental traits. Since these features are basic to Robinson's dietary hypothesis, they will be considered in some detail.

The disparity between front and back teeth. As far as I have been able to trace, the concept of dietary differences was first suggested in any detail in Robinson (1954*d*). There he cited the great disparity in size of the anterior and cheek teeth in *Paranthropus* as an important consideration: *Australopithecus*, on the other hand, shows 'less disparity in size between anterior and posterior elements of the dentition, with appreciably larger canines and smaller premolars and molars than *Paranthropus*' (p. 328). Before considering the implications of the disparity, we should examine the actual data for slightly larger samples. The differences in absolute size of the front teeth are not as great as had been thought (Table 30, p. 147). Thus, on small samples, 2–3 I^1's of *Australopithecus* are slightly larger on both crown dimensions and module than those of *Paranthropus*, but only the L.L. breadth disparity is significant. On the other hand, I^2 of *Paranthropus* is slightly but not significantly larger than that of *Australopithecus* in both crown dimensions and the module. The maxillary canine which Robinson described as 'appreciably larger' in *Australopithecus* than in *Paranthropus* turns out to be larger only in M.D. diameter, there being a significant difference of the means of 1·0 mm. In labiolingual diameter, the means for the two taxa are identical. Thus, the mean module for the *Australopithecus* maxillary canine (9·5 mm.) is only slightly and not significantly greater than that for *Paranthropus* (9·0 mm.).

For the lower teeth,[1] we are again handicapped by very small samples of incisors: for *Australopithecus* these number two of I_1 and two of I_2, although two additional specimens of each lower incisor permit measurements of L.L. but not of M.D. For *Paranthropus*, I_1 numbers six specimens and I_2 four specimens. Comparisons between the two taxa show that both I_1 and I_2 of *Australopithecus* are slightly bigger than those of *Paranthropus* in M.D., L.L. and, hence, module, but none of these differences is significant, except for the M.D. diameter of I_2. Hence, as with the maxillary incisors, it cannot be claimed on the present samples that the mandibular incisors differ significantly in overall

---

[1] The detailed tables of metrical characters and of '*t*' tests for the *mandibular* dentition of *Australopithecus* and *Paranthropus* are not included in the present work, but will be included in volume 5 of the current series on *Olduvai Gorge*.

size between the two taxa. However, the mandibular canines of *Australopithecus* are significantly greater than those of *Paranthropus* in M.D. diameter ($\bar{x} = 9.0$, 8.0 mm.), L.L. diameter ($\bar{x} = 10.2$, 8.3 mm.) and in module ($\bar{x} = 9.6$, 8.1 mm.).

It may be concluded, on the present samples, that the maxillary and mandibular incisors do not differ significantly in size between the two taxa, while the canines are significantly different only in the lower dentition. It is therefore no longer justifiable to adhere to the general statement that the front teeth and, especially, the canines are appreciably larger in *Australopithecus* than in *Paranthropus*. Comparison of the modules for I1, I2 and C reveals only a single significant difference between the two taxa, namely, for the mandibular canines.

On the other hand, six out of ten cheek-teeth of *Paranthropus* are indisputably absolutely larger than those of *Australopithecus*: four of these are upper teeth (P^3, P^4, M^1 and M^3) and only two—lower teeth (P_4 and M_3). In M.D. diameter, all five maxillary cheek-teeth of *Paranthropus* are significantly greater; in B.L. diameter, P^3, P^4 and M^3 are significantly greater, M^1 only just so and M^2 slightly but not significantly larger. Thus, the modules of all the maxillary cheek-teeth except M^2 are greater in *Paranthropus* than in *Australopithecus*.

The distinction between the two taxa is not so clear-cut in the mandibular cheek-teeth. In M.D. diameter, only P_4 and M_3 are significantly greater in *Paranthropus*, M_2 just so, M_1 slightly but not significantly greater, while the mean M.D. of P_3 is slightly but not significantly greater in *Australopithecus*. In B.L. diameter, *Paranthropus* has a slight advantage in P_4, M_1, M_2 and M_3, but only in M_2 is this of borderline significance; while P_3 is slightly but not significantly greater in *Australopithecus*. In module, *Paranthropus* is significantly greater in M_2 and M_3, slightly but not significantly greater in P_4 and M_1, and slightly but not significantly *smaller* in P_3.

In sum, the maxillary cheek-teeth show a more marked preponderance of *Paranthropus* over *Australopithecus* than do the mandibular, whereas the mandibular canine of *Australopithecus* shows a greater size advantage over *Paranthropus* than does the maxillary canine. The differences in absolute dental size between the two taxa are not as great on these newer analyses than appeared to be the case in Robinson's original study (1956).

The disparity between the front and cheek-teeth remains as a feature differentiating the two taxa. To express the disparity, we have compared various parameters of C̱ with those of P^3 and of P^4 (Table 36, p. 156). This may be done in two ways: either by using only specimens in which the teeth in question are in position in the same jaw, or by comparing the means for all teeth, whether *in situ* or isolated. Only four upper jaws (or five sides) of *Paranthropus* and three upper jaws (or four sides) of *Australopithecus* have measurable canines and anterior premolars in position; while two upper jaws of *Paranthropus* and two (or three sides) of *Australopithecus* have measurable canines and distal premolars. For these small samples, the ranges of individual maxillary canine–premolar ratios do not overlap for any of the parameters used (Table 37, p. 157). Likewise, the ratios of the means of all relevant maxillary tooth parameters differ appreciably between the two taxa.

Only two lower jaws (or three sides) of *Paranthropus* and four lower jaws (or five sides) of *Australopithecus* have measurable canines and P_3's in position; while two lower jaws (or three sides) of *Paranthropus* and three lower jaws of *Australopithecus* have measurable canines and P_4's. In these small samples, the ranges of individual mandibular canine–premolar ratios do not overlap for any of the parameters used except for the M.D. diameter in the comparison based upon $\overline{C}/P_3$ ratio (Table 49). The ratios of the means of all relevant mandibular tooth parameters differ appreciably between the two taxa, the differences being more striking in the $\overline{C}/P_4$ ratios than in the C/P_3 ratios.

It may be concluded that there is a real difference between the two taxa in the disparity between the sizes of the canines and the cheek-teeth. In the maxillary teeth it is a disparity caused by differing degrees of enlargement of the cheek-teeth and not significantly added to by differential reduction of the front teeth, whereas in the mandibular teeth, the discrepancy is accounted for mainly by the larger canine of *Australopithecus*.

Table 49. *Mandibular canine–premolar ratios in Paranthropus (P) and Australopithecus (A)**

		$\bar{C}/P_3$		$\bar{C}/P_4$
M.D. crown diameter	P	82 (83–87)	P	73 (71–76)
	A	90 (83–101)	A	94 (83–100)
B.L. crown diameter	P	71 (68–70)	P	64 (55–61)
	A	83 (78–91)	A	83 (68–85)
Module	P	76 (75–77)	P	68 (63–67)
	A	86 (84–92)	A	87 (74–91)

* The figures quoted are the *ratios of the means* for all specimens of $\bar{C}$ and P_3 (or P_4), including isolated teeth; and, in parentheses, the range of ratios for individual mandibles in each of which $\bar{C}$ and P_3 (or P_4) are present. All figures are rounded off to the nearest whole number.

Chipping of the teeth. A further feature adduced as common in the Swartkrans dental material is chipping of enamel from the edge of the occlusal surface (Robinson, 1954*d*). 'That this chipping occurred in life is clearly shown by the fact that most of the roughened areas left after such flaking are smoothed by subsequent use of the tooth' (p. 328). The existence of such chips has been confirmed by me with the aid of a binocular microscope, but they are not confined to, nor even commoner in, the *Paranthropus* material. For example, chips are present on the right P_3 of Sts 52B, M^3 of Sts 28, M^2 of Sts 8, M^1 of Sts 1151, M^1 and M^3 of Sts 53, and on other specimens of *Australopithecus*: these chips are similar in size, character and number per jaw to those found in *Paranthropus* specimens.

Chipping testifies to the teeth having been used for chewing on hard objects. The chips are rather large and it seems unlikely that they would have been produced by chewing on particles of grit, as Robinson has suggested. It seems more likely that chewing on bone fragments would have produced such damage. Grit in the vegetable component of a diet usually acts as an abrasive, accelerating attrition of the teeth, as in modern Kalahari Bushmen, and may well have facilitated the marked attrition—but not the chipping—shown by australopithecine teeth. Hence, the evidence of chipped teeth does not support the suggestion of dietary differences between the two taxa, nor does it provide evidence unequivocally in support of the vegetarian diet. A possibly pertinent observation may be recorded here: several of the chips are in areas of hypoplastic enamel. The hypoplasia probably weakened the enamel in affected areas and rendered it more liable to chip during chewing of strongly resistant material.

Thickened alveolar bone. A third feature is 'the considerably thickened bone around the molar roots' of *Paranthropus* which, according to Robinson, indicates 'that crushing and grinding was the main function involved' (1954*d*, p. 328). It may, however, indicate only the well-known relationship between robusticity of alveolar processes and large size of roots, which, in turn, is commonly though not always related to large dental crowns. Since *Paranthropus* has on the average larger dental crowns and roots than *Australopithecus*, it is to be expected that the alveolar bone would be thicker or more robust. In any event, some of the Makapansgat jaws, with bigger tooth crowns and roots than in many other members of *Australopithecus*, have very robust alveolar processes, comparable with some of those encountered in *Paranthropus*.

Attrition of teeth. Rapid wear of australopithecine teeth is cited as supporting the idea of great masticatory stresses, but this is not a feature differentiating the two taxa. As Robinson states, '*Australopithecus* also has these characters (referring to molar and alveolar massiveness and attritional scratches on the occlusal surface), but to a less marked degree except for the relative rate of wear of the teeth, which is much the same in both groups' (1954*d*, p. 328). Later, however, he cites the relatively rapid rate of dental wear as a feature of *Paranthropus* alone (Robinson, 1963, p. 392), but adds, 'In spite of their small size, the anterior teeth (of *Paranthropus*) do not wear down rapidly'. In adolescent *Zinjanthropus*, the anterior teeth have certainly worn down rapidly. My observations on the original South African material support Robinson's earlier statement that the rate of attrition was much the same in *Paranthropus* as in *Australopithecus*.

The data on which the dietary hypothesis was based have been listed and only the larger size of the cheek-teeth of *Paranthropus* and of the mandibular canine of *Australopithecus* have been shown to be valid differences between the two taxa. Of course, there are other differences between the dentitions of the two taxa (e.g. the different degrees of molarisation in the very small samples of dm_1), but these are not necessarily germane to the ecological interpretation. The inferred ecological differences are crucial to the view that the two taxa are generically distinct. It may well be enquired therefore whether the larger size of the cheek-teeth alone provides adequate evidence upon which to sustain the hypothesis of major ecological differences between the two taxa. Robinson himself does not believe this, for, in his recent analysis, he states, 'The fact that the cheek teeth of *Paranthropus* are larger than those of *Australopithecus* need not be due to anything more than the bodily size difference between them since the former is appreciably more robust than the latter' (1963, p. 392). The evidence of *Zinjanthropus* has shown that there existed a third form of australopithecine, whose cheek-teeth were significantly greater even than those of *Paranthropus*, just as the cheek-teeth of the latter significantly exceeded those of *Australopithecus*. In other words, the australopithecines had differentiated into a series of taxa characterised by differing degrees of enlargement of the cheek-teeth and, naturally, of the supporting structures, muscular prominences, masticatory stress columns, and so on. Such differentiation in the size of cheek-teeth, of itself, provides no evidence of major ecological or adaptive radiation.

It is clear from his latest analysis that the entire case for ecological differences hinges upon the supposed differences of the front teeth. These supposed differences lead Robinson to say, for example, '...obviously the anterior teeth (of *Australopithecus*) were considerably more important than those of *Paranthropus*' and, again, 'The implication of the absolute and relative size difference between the anterior teeth of the two forms clearly is that there was either a difference of behaviour of considerable magnitude between the two, or one of diet, perhaps both' (1963, p. 392).

Even if this study had confirmed that the front teeth of *Australopithecus* were significantly larger than those of *Paranthropus*, the dietary hypothesis would not be the only one capable of explaining such a difference. For instance Washburn, in contesting the dietary hypothesis, has pointed out that 'apes with a purely vegetarian diet nevertheless have large incisors so that small incisors might merely indicate hand feeding' (1963b, p. 565). Robinson has now clearly recognised the possibility that behavioural traits may influence the selective pressures maintaining large front teeth, as an alternative explanation to the dietary hypothesis: this is evident in the last passage quoted above from his 1963 paper, where he speaks of possible behaviour differences 'of considerable magnitude' as an alternative hypothesis to explain the supposed size differences of the anterior teeth.

However, it is now clear that much of the discussion on dietary, ecological and adaptive differences was based upon inadequate factual data. The anterior teeth have been shown *not* to differ significantly between the two taxa, save for the mandibular canine. It is most unlikely that the evidence of this one tooth being significantly bigger in *Australopithecus* than in *Paranthropus* can sustain the entire edifice of the dietary-ecological hypothesis. Clearly the fundamental morphological basis underlying the already-diluted dietary hypothesis has largely fallen away—and so, it seems to me, must the hypothesis itself. This conclusion is based solely upon a reassessment of the very facts which the hypothesis was originally erected to explain.

(b) Other distinctions between Paranthropus and Australopithecus

With the attenuation and, indeed, collapse of the dietary hypothesis, it would seem that the main prop for the generic distinctness of the two taxa falls away too. Some other criteria cited by Robinson in the formal diagnosis of his two genera are given on p. 229 (in each instance, a comment has been added by the present author).

(c) Dental variability

Apart from the characters cited in the formal definitions, various other features have been

TAXONOMIC STATUS

	Australopithecus	*Paranthropus*	Comment
The bony face	'Moderately—not completely—flat'	'Either quite flat or actually dished'	Facial structure seems closely to reflect the size of teeth and supporting structures
Zygomatic arch	'Moderately developed'	'Strongly developed'	Intimately related to the size and development of masseter muscle, which in turn is related to dental size
Temporal fossa	'Of medium size'	'Large'	Closely related to size of temporalis muscle
Lateral pterygoid plate	'Relatively small'	'Strongly developed and large'	Closely related to the size of the medial and lateral pterygoid muscles
Sagittal crest	'Normally absent—may occur in extreme cases'	'Normally present in both sexes'	Related to degree of development of temporalis muscle in a small-brained creature. The Makapansgat *Australopithecus* probably possessed a crest
Palate	'Of more or less even depth'	'Appreciably deeper posteriorly than anteriorly'	Related to height of tooth roots
dm_1	'Incompletely molarised'	'Virtually completely molarised'	The total *A. africanus* sample comprises 4 dm_1 from 2 individuals
Maxillary incisor and canine sockets	'In parabolic curve'	'In almost straight line across front of palate'	Both taxa include intermediate forms, with front tooth sockets in a low curve (1 out of 5 specimens of *Australopithecus*; 2 out of 7 specimens of *Paranthropus*)
Ascending ramus of mandible	'Usually sloping backward and of moderate height'	'Vertical and high'	Probably related to degree of prognathism or orthognathism, hence indirectly to dental and alveolar size

claimed to differentiate the two taxa. One is dental variability: in many respects, it has been claimed, the teeth of *Australopithecus* are considerably more variable than those of *Paranthropus* (Robinson, 1965b). It has been shown above that more representative samples do not bear out this claim.

(d) Brain size and brain quality

Another claim is that the brain quality of *Australopithecus* was somehow superior to that of *Paranthropus*. In my view, this is practically as far-reaching a claim as the suggestion of dietary and ecological differences and it deserves to be examined closely.

Originally, it was contended that *Paranthropus* possessed a larger cranial capacity than *Australopithecus* (see chapter VIII). For instance, a juvenile *Paranthropus* from Swartkrans was said to have a brain-case length 20–25 mm. greater than that of the Taung child and a breadth proportionately greater. 'From this and the adult crania in our possession it seems that the general order of brain size of the adult *P. crassidens* is comparable with that of the smaller specimens of *Pithecanthropus*—roughly 800 cm.^{3}' (Robinson, 1952a, p. 197).

However, this statement was based on crushed and fractured specimens of *Paranthropus*, not one of which allowed anything like a precise estimate of the capacity of this taxon. In 1959, disregarding the earlier estimates, Washburn quoted 450–550 c.c. as 'representative capacities' for the australopithecines. Later that year, *Zinjanthropus* was discovered: its largely intact brain-case revealed that this robust australopithecine, for all its megadont dentition, had a small cranial capacity, subsequently shown to be no bigger than that of *Australopithecus*. Robinson thereafter adopted the range 450–550 c.c. for both taxa, stating, 'The endocranial volume appears to be only about 500 cm.^{3}—I know of no sound evidence at present indicating a brain significantly larger than this' (1961, p. 4). This view, it must be recalled, is based largely on the size of the well-preserved cranium of *Zinjanthropus*, rather than upon precise determinations of any single South African *Paranthropus* cranium—none of which is in a state to permit a precise estimate to be made. In his formal definitions, Robinson (1962) again quotes 450–550 c.c. for both taxa.

At the moment, the sample range for six

Australopithecus crania is 435–562 c.c.; one value for a robust australopithecine, *Zinjanthropus*, is 530 c.c.; no values for *Paranthropus* from South Africa are available. Thus, with the discarding of the earlier high estimates for *Paranthropus*, no evidence exists to suggest that *Paranthropus* had a higher mean cranial capacity than *Australopithecus*. Only the discovery of further, less damaged specimens from the South African sites will finally settle the question.

More recently, it has been suggested that the brain morphology of *Paranthropus* is more pongid-like and that of *Australopithecus* more hominid-like. Speaking of the low supra-orbital height index of *Paranthropus* and *Zinjanthropus*, Robinson (1963) states, 'This feature reflects a significant feature of cranial morphology, which in turn *almost certainly reflects some aspects of brain morphology*. It is significant, therefore, that in these respects *Paranthropus* agrees with the pongids while *Australopithecus* exhibits a condition closely resembling that of the hominines' (p. 405). (Italics mine.) Again, he says elsewhere:

The supra-orbital height index of the latter (*Paranthropus*) falls right in the normal range for pongids whereas that of the former (*Australopithecus*) agrees closely with early hominines. This is not a simple feature as it not only involves considerable alterations in skull architecture, but is doubtless also *a reflection of expansion in the brain*. The latter, it is now well recognised, is evolutionarily a very conservative organ (italics mine).

Rebutting the view of some who regard *Australopithecus* →*Paranthropus* →early hominine as successive members of the same phyletic sequence,[1] Robinson adds:

A very conservative organ thus developed *to essentially the hominine condition*, retrogressed *to a pongid condition* and then rapidly advanced *to essentially the same hominine condition that it previously had*...(italics mine) (1963, p. 395).

The reference to 'a reflection of expansion in the brain' would seem to suggest that the higher forehead of *Australopithecus* was owing, in part at least, to a greater volume of cerebrum in this taxon, as compared with that in *Paranthropus*. Yet, the evidence of the low-browed *Zinjanthropus* with a cranial capacity of 530 c.c. and of the high-browed *Australopithecus* Sts 5 with only 480 c.c.

[1] A view which the present author too does not support.

would suggest that variations of the supra-orbital height index within the australopithecines are not related to differences of 'expansion in the brain'. As was indicated in chapter III, C, the failure of the brain-case of *Zinjanthropus* to rise appreciably above the orbits does not betoken a failure of the brain to expand. Rather, it indicates simply that the whole calvaria is hafted on to the facial skeleton at a lower level than in *Australopithecus*.

Furthermore, such lower hafting provides no evidence of differences in brain morphology, as Robinson (1962, p. 134; 1963, p. 405) has suggested. In fact, the detailed study of the morphology of the *Zinjanthropus* endocast has shown that the brain of this low-browed australopithecine strongly resembled those of the gracile australopithecines and of hominines, and differed markedly in external morphology from those of pongids. The suggestion that the brain of *Paranthropus* was in a 'pongid condition' and that of *Australopithecus* in an essentially 'hominine condition' is not supported by the evidence.

Such evidence as exists suggests that the brains of the australopithecines were essentially similar in size and external form, but differed only in the level at which the brain-case was hafted on to the facial skeleton. In *Paranthropus* and *Zinjanthropus* the hafting is low with a consequent low supra-orbital height index; in *Australopithecus* the hafting is high with a high supra-orbital index.

The difference in level of hafting is a complex question of cranial architectonics. An analogy might be suggested with the difference in cranial form between African pongids and orang-utan: the latter with higher orbits and moderate, non-projecting brow-ridges, has a higher-vaulted calvaria: the former with somewhat lower orbits and marked, anteriorly and superiorly projecting brow-ridges, have lower-vaulted calvariae. Perhaps the difference in supra-orbital height indices or level of hafting of *Australopithecus* and *Paranthropus* is to be explained partly in terms of differences in the brow-ridges, which in turn seem to be related both to the size of the teeth and supporting structures, and to the degree of development of the anterior part of the temporalis muscle. *Australopithecus*, with its rather smaller teeth and relatively weak development of the temporalis muscle, has a modestly developed and vertically thin brow-ridge—and a high forehead. *Paranthropus*, with its bigger teeth and stronger development of the temporalis muscle, has a strongly developed and sometimes vertically thick brow-ridge—and a low forehead (*see* norma facialis, Fig. 21). Possibly, too, a variable growth relationship between the brow-ridge and the orbit may increase the complexity of the analysis, for the supra-orbital height is measured from the *lower* margin of the brow-ridge: this raises the question whether in both australopithecine taxa the ontogenetic thickening of the brow-ridge is entirely upwards, or partly downwards over the orbital aperture, thereby lowering the inferior terminus from which supra-orbital height is measured. A study of the mechanics and growth dynamics of calvarial hafting would throw much light on these crucial morphological variations among the hominoids.

(e) Postcranial skeleton

Robinson (1963, p. 405) has indicated that the ischium of *Paranthropus* is well developed and that the 'bare area' between the acetabular margin and the area of muscular attachment on the ischial tuberosity is 'relatively much longer' than in *Australopithecus*. However, as his detailed study on the os coxae has not yet been published, this aspect will not be considered here. Other postcranial bones are available, but they are relatively few and detailed morphological studies have for the most part not yet been made. At this stage, therefore, the postcranial bones do not assist the definition of the two South African australopithecine taxa.

(f) Conclusion: one australopithecine genus

The foregoing detailed analysis has shown that the supposed dietary–ecological differences between *Australopithecus* and *Paranthropus* rest upon extremely insecure foundations. The dental differences between the two taxa, both those related to the dietary hypothesis of Robinson, and others, are far less marked than has been assumed. Many of the other differences between the crania and mandibles of the two forms depend directly upon the dental differences and the related functional activity of the muscles of mastication (Scott, 1963); this applies to such features as the degree of development of the attachment areas for the masticatory muscles (zygomatic arch, lateral pterygoid plate, sagittal crest, etc.), the robusticity of the alveolar processes, the palate, the brow-ridges, the facial flattening. Dental variability is not a feature which distinguishes the two taxa. There is no evidence that the brains of the two forms differed in size or external form, still less in quality.

One is led to conclude that the two australopithecine taxa are very much more closely related than has commonly been averred hitherto. This study has shown that there is no adequate basis for maintaining that the two taxa are generically distinct. Both groups can more appropriately be considered members of a single genus, *Australopithecus*. This is the view which was proposed by Washburn and Patterson (1951), supported by Oakley (1954, 1956), Dart (1955*a*), Le Gros Clark (1955, 1964) and Leakey *et al.* (1964), and provided a consensus at two recent international discussions on the taxonomy of fossil hominids.

Leakey *et al.* (1964) proposed that, *for the moment*, *Australopithecus* and *Paranthropus* be regarded as subgenera of the genus *Australopithecus*, a view which has been supported by Howell (1965*a*). It is in this sense that these two *nomina* have been used in this book. Since the present analysis has shown that the areas of difference between the two taxa are far smaller than was believed when Leakey and his co-authors wrote the above paper, I now formally propose that these subgenera be sunk and that the two australopithecine taxa from South Africa be considered no more than specifically distinct. The two taxa should thus be designated *Australopithecus africanus* Dart and *Australopithecus robustus* (Broom). Formal definitions will be offered later in this chapter.

3. *The generic status of the Olduvai australopithecine*

The foregoing descriptions and discussion have made clear that (i) there is no justification for separating *Australopithecus* and *Paranthropus* generically; and (ii) *Zinjanthropus* is not generically distinct from either of the South African australopithecines. *Zinjanthropus* is clearly a member of the genus *Australopithecus sensu lato*.

Furthermore, it has been demonstrated that (iii) the differences between *Australopithecus sensu stricto* and *Paranthropus* are not such as would warrant even subgeneric distinction; and (iv) the morphological distance between *Zinjanthropus* and either of the South African australopithecines is roughly comparable with the distance between *Australopithecus sensu stricto* and *Paranthropus*.

It may be inferred that *Zinjanthropus* is not even subgenerically distinct from *Australopithecus sensu stricto* or *Paranthropus*. I therefore propose formally that the subgenus *Zinjanthropus* be sunk and that the australopithecine taxon from Olduvai Bed I be considered no more than specifically distinct. It will be suggested below that it is justified at present to continue regarding the Olduvai australopithecine as a separate species within the genus *Australopithecus*. Its correct designation would be *Australopithecus boisei* (Leakey).

For the remainder of this work, only the unigeneric nomenclature will be used, and the three australopithecine taxa will be referred to as follows:

A. boisei = '*Zinjanthropus*' as used in the preceding part of the book.

A. robustus = '*Paranthropus*' as used above.

A. africanus = '*Australopithecus*' as used above.

4. The specific status of the Olduvai australopithecine

The question now arises: is the Olduvai taxon specifically distinct, or does it fall within *A. robustus* or *A. africanus*? Superficially, it is easy to detect a number of obvious cranial and dental resemblances to *A. robustus*. Closer study, however, shows that in a number of respects the Olduvai specimen is closer to *A. africanus*, while, in still others, it differs from both *A. robustus* and *A. africanus*. Many of these features are metrical, others non-metrical.

An approximate tally of resemblances indicates that the Olduvai specimen has fifty-nine morphological traits in common with *A. robustus*, forty-two with *A. africanus* and sixteen different from both. However, this tally includes a number of features the expression of which is known for only one of the two South African species. For example, in the ratio of the parietal to the occipital sagittal arcs, the Olduvai specimen resembles *A. africanus*, but we do not know this ratio for *A. robustus* because the available specimens are imperfect. If we confine the tally to those features for which we know the expression in both *A. robustus* and *A. africanus* and for which the latter two species are different, the Olduvai specimen has the *A. robustus* pattern in twenty-five traits and the *A. africanus* pattern in eleven traits, while in a further sixteen traits, it differs from both. Needless to say, counting trait resemblances is an extremely superficial method of analysing 'distance': more precise results would be yielded by a multivariate statistical analysis. However, such an analysis has not yet been attempted, since it is considered that it should be undertaken only once further Olduvai australopithecines have been identified and studied.

Even without such a study, it is clear that in some fairly fundamental respects, the Olduvai specimen is different from *A. robustus*. Such differences include the following: the size and especially buccolingual diameters of the cheek-teeth, which significantly exceed those of *A. robustus*; greater disparity between canine and premolar size than in *A. robustus*; marked reduction of the M.D. diameter of M^3 as compared with that of M^2 and equality of B.L. diameters of M^3 and M^2, in both of which respects the Olduvai specimen resembles *A. africanus* and not *A. robustus* (in which both diameters usually increase from M^2 to M^3); the morphology of the labial faces of I^1 and I^2, of the lingual face of the canines, and of the crowns of M^1 and M^2, in all of which respects the Olduvai specimen approximates to those of *A. africanus*; the shape, cusp pattern, Carabelli formation and buccal cingular vestige of M^3, all of which are nearer to those of *A. africanus*; the earlier or more primitive stage in the trend towards reduction of the cingulum, a stage which the Olduvai cranium shares with *A. africanus*, whereas *A. robustus* represents a more advanced stage of reduction, having only weak vestiges of a lingual cingulum and no trace of the buccal cingulum.

These dental traits are basic genetic features probably little affected by short-term functional considerations. They are differences of the same type and order as those which most effectively distinguish *A. robustus* from *A. africanus*.

In addition, the Olduvai specimen differs from *A. robustus* in a number of other respects, some of which are very probably of functional origin and may be related to differences in size of teeth and in vigour and pattern of mastication. These include the curiously foreshortened foramen magnum (which may be only an individual variation); the evidences of marked parietal lobe and cerebellar expansion; the more powerful supra-orbital torus; the elongation of the face as a whole and, especially, of the nose and the maxilla; the flexion of the naso-alveolar clivus; the morphology of the zygomatic buttress and the malar notch, and the absence of even the slightest trace of a canine fossa; the nature and the extent of the masseteric impressions; and the anterior shelving of the palate. Many of these differences simply reflect the still heavier dentition and supporting structures of the Olduvai creature as compared with *A. robustus*, just as differences similar in quality and often in extent distinguish *A. robustus* from *A. africanus*.

It is concluded that the Olduvai australopithecine differs from *A. robustus* in a similar manner to that in which the latter differs from *A. africanus*. These differences, as well as the dental morphological resemblances to *A. africanus* and certain unique features noted, justify us in regarding the Olduvai cranium as the hypodigm of a new species of *Australopithecus*, namely *A. boisei* (Leakey).

The recognition of *A. boisei* as a separate species from *A. robustus* is in keeping with the view of Robinson (1960), who, though he lumps it into *Paranthropus*, retains it as a separate species, *P. boisei*. His attitude is nevertheless one of caution, when he states, 'The validity of separate specific status is not clear on the basis of the single specimen, and it is perhaps wisest to leave it as distinct.' His view was expressed following the preliminary description (Leakey, 1959a) and his (Robinson's) inspection of the original specimen. The more detailed study herein, as well as the newer analyses of the available comparative data, have strengthened the case for regarding *A. boisei* as a separate species. Needless to add, the possibility remains that a larger sample may necessitate a revision of this view.

Howell, too, does not rule out the possible specific distinctness of *A. boisei*. He states, 'The overall resemblance (to the taxon known from Swartkrans and Kromdraai) is so great that "Zinj" may justifiably be attributed to the same taxon, or is at most only specifically distinct' (Howell, 1965a).

Le Gros Clark (pers. comm.) would have no objection to regarding *A. boisei* as a separate species, adding that this interpretation would seem to be supported by the large time and space interval separating Olduvai Bed I, on the one hand, from Swartkrans and Kromdraai on the other. According to faunal dating, Bed I would be upper Villafranchian (Leakey, 1965) and the potassium-argon date is about 1·75 million years (Leakey, Evernden and Curtis, 1961). Opinions differ as to whether this is older than the oldest South African australopithecine sites which, on faunal dating, are Taung and Sterkfontein (Lower Breccia) (Cooke, 1963). On the other hand, it is very likely that the Swartkrans and Kromdraai faunas are younger and 'are close to the Villafranchian-Cromerian boundary, but it is still not certain on which side they lie' (p. 103). Thus, although it may be difficult on present evidence to go all the way with Howell (1965a) in suggesting that the gap is '*at least* 1,000,000 years', it is clear that some considerable time separated the deposits in Olduvai Bed I from those of Swartkrans and Kromdraai, and that this lapse of time was marked by faunal changes. It would not be surprising to find changes in the hominids over the same period.

It is proposed then that *A. boisei* be recognised as a distinct species of *Australopithecus*.

D. Formal definitions of *Australopithecus* and its species

We are now in a position to offer formal definitions of the hominid genus, *Australopithecus*, and of its proposed three species, *A. africanus*, *A.*

robustus and *A. boisei*. These definitions are based solely on morphological considerations; no reference is made to supposed ecological variations. Cultural aspects, to be discussed in the next chapter, are not included in the definitions of the various australopithecine species, since it is not considered that sufficient evidence exists to distinguish among them culturally. Later, when the associated bones and other objects from all the australopithecine sites have been analysed as thoroughly as have those from the Makapansgat limeworks deposit, it is possible that cultural differences may be established between *A. africanus* and one or both of the other species. The definition of the genus *Australopithecus* is a modified version of that proposed by Le Gros Clark (1964); the definitions of *A. africanus* and *A. robustus* are based in part on Robinson's (1962) definitions of *Australopithecus* and *Paranthropus* respectively.

1. Definition of Australopithecus Dart

Australopithecus—a genus of the Hominidae distinguished by the following characters: relatively small cranial capacity, with an average of about 500 c.c. and an estimated population range of about 360 to about 640 c.c.; a relatively thin-walled cranium rendered robust in parts by strong ectocranial superstructures and by marked pneumatisation; strongly-built supra-orbital ridges; moderate to fairly high orbits, with a lower mean height than in pongids; a tendency in individuals with larger cheek-teeth for the formation of a low sagittal crest in the frontoparietal region of the calvaria (but the sagittal crest is not continuous with either the nuchal crest or the occipital torus, whichever is present); occipital condyles well behind the anteroposterior midpoint of the cranial length, but in the same coronal plane as the external acoustic apertures; foramen magnum well forward on the base of the cranium; planum nuchale of occipital bone rising only a short distance above the F.H. and generally facing downwards much more than backwards; inion low and generally close to the Frankfurt plane; a low nuchal crest not continuous with the sagittal crest in heavier-toothed forms, and a slight occipital torus in moderate-toothed forms; consistent development (in immature as well as mature crania) of a pyramidal mastoid process of typical hominine form and relationships; a mandibular fossa which is shallow and mediolaterally broad, but is otherwise constructed on the hominid pattern, especially in the slopes and curvature of the anterior wall and the upward slope of the preglenoid plane, but with a pronounced entoglenoid process and, in some individuals, a moderate development of the postglenoid process; porion elevated in position above the nasion–opisthion line; massive and robust jaws, showing marked individual variation in respect of absolute size; mental eminence absent or slightly indicated; symphysial surface relatively straight and retreating; contour of internal mandibular arch V-shaped or blunt U-shaped; dental arcade parabolic in form with no diastema; moderate-sized, spatulate canines wearing down flat from the tip only; relatively large premolars and molars, the enlargement being more marked in the buccolingual diameter of the crown; lower anterior premolar biscupid with subequal cusps; pronounced molarisation of lower first deciduous molar; progressive increase in size of permanent lower molars from first to third, but M^3 is commonly smaller than M^2; the limb skeleton (so far as it is known) conforming in its main features to the hominid type but differing from *Homo* in a number of details, such as the forward prolongation of the region of the anterior superior iliac spine and a relatively small sacro-iliac surface, the relatively low position (in some individuals) of the ischial tuberosity, and the marked forward prolongation of the intercondylar notch of the femur.

2. Definition of Australopithecus africanus Dart

A species of the genus *Australopithecus* characterised by the following features: more gracile, lighter construction of the cranium; calvaria hafted to facial skeleton at a high level, giving a distinct though not marked forehead and a high supra-orbital height index; ectocranial superstructures and pneumatisation not as marked as in other species; sagittal crest commonly absent though probably present in some individuals; nuchal crest not present, but slight to moderate occipital torus commonly present; bony face of moderate height and varying from moderately flat and orthognathous to markedly prognathous; nasal region slightly elevated from facial plane; ramus of mandible of moderate height and sloping somewhat backward; jaws moderate in size with lesser development of zygomatic arch, lateral pterygoid plate, temporal crest and temporal fossa; palate of more or less even depth, shelving steeply in front of the incisive foramen; premolars and molars of moderate size and not so markedly expanded buccolingually; M^3 smaller than M^2 in mesiodistal diameters, but equal in buccolingual diameters; mandibular canine larger than in other species, and hence more in harmony with the postcanine teeth; degree of molarisation of lower first deciduous molar less complete; cingulum remnants or derivatives present on all maxillary molars, weak on buccal surfaces, pronounced on lingual, representing an earlier or more primitive stage in the trend towards reduction of the cingulum; sockets of anterior teeth arranged in a moderate to marked curve.

3. *Definition of* Australopithecus robustus (*Broom*)

A species of the genus *Australopithecus* characterised by the following features: more robust, heavier construction of the cranium; calvaria hafted to facial skeleton at a low level, giving a low or absent forehead and a low supra-orbital height index; well-developed ectocranial superstructures and degree of pneumatisation (more marked than in *A. africanus*, though not as pronounced as in *A. boisei*); moderate to marked supra-orbital torus with no 'twist' between the medial and lateral components; sagittal crest normally present; small nuchal crest commonly present; bony face of low to moderate height, and flat or orthognathous; nose set in a central facial hollow; ramus of mandible very high and vertical; jaws large and robust with strong development of zygomatic arch, lateral pterygoid plate, temporal crest and temporal fossa; palate deeper posteriorly than anteriorly, shelving gradually from the molar region forwards; premolars and molars of very large size; M^3 commonly larger than M^2 in both buccolingual and mesiodistal diameters; mandibular canine absolutely and relatively small and hence not in harmony with the postcanine teeth; degree of molarisation of lower first deciduous molar more complete; cingulum remnants only weakly represented on lingual face and absent on buccal face of maxillary molars, representing a more advanced stage in reduction of the cingulum; sockets of anterior teeth arranged in a low to moderate curve.

4. *Definition of* Australopithecus boisei (*Leakey*)

A species of the genus *Australopithecus* characterised by the following features: most robust, heaviest construction of the cranium; calvaria hafted to facial skeleton at a low level, giving a virtually absent forehead and a low supra-orbital height index; very pronounced ectocranial superstructures and degree of pneumatisation (more marked than in *A. robustus*); extremely well-developed supra-orbital torus with a 'twist' between the medial and lateral components; well-developed sagittal crest; moderate nuchal crest; plane of foramen magnum nearly horizontal; structure of dorsum sellae and sella turcica typically hominine; cerebellum apparently relatively large; anterior nasal spine high; bony face very high and very flat or orthognathous; nose set in a central facial hollow; ramus of mandible by inference tall and vertical; jaws very large and extremely robust with powerful development of zygomatic arch, lateral pterygoid plate, temporal crest and temporal fossa; palate very deep but shelving steeply only in front of the incisive foramen; premolars and molars extremely large, especially in the buccolingual dimension; M^3 smaller than M^2 in mesiodistal diameters and equal in buccolingual diameter; maxillary canine absolutely and relatively small and hence not in harmony with the postcanine teeth; cingulum remnants or derivatives present on all maxillary molars, weakly developed on buccal surfaces, pronounced on lingual, representing an earlier or more primitive stage in the trend towards reduction of the cingulum; sockets of anterior teeth arranged in a moderate curve.

CHAPTER XX

THE CULTURAL AND PHYLOGENETIC STATUS OF *AUSTRALOPITHECUS BOISEI* AND OF THE AUSTRALOPITHECINES IN GENERAL

A. Cultural status

It has long been known that pebble-tools were made in Africa during the period of the australopithecines. But there have been two schools of thought on the relationship between the fossils and the implements. According to one view, *Australopithecus* was the maker of these stone implements. The notion has been forthrightly expressed by J. D. Clark as recently as 1963: 'To take the example of the australopithecines:—Few would now doubt that they were representative of a hominid form responsible for the Pre-Chelles-Acheul Culture' (p. 356). This view has been espoused by Arambourg (1958), by Oakley (1961, 1964) although he earlier (1954, 1956) took the opposing view, by Washburn (1959), Washburn and Howell (1960), Washburn and De Vore (1961), and others.

Exponents of the second view have doubted whether the australopithecines were stone toolmakers (Robinson and Mason, 1957, 1962; Inskeep, 1959; Mason, 1961; von Koenigswald, 1961, etc.).

In favour of the first view are the contemporaneity of the australopithecines and the pebble-tools, and the discovery of a fragmentary, apparently australopithecine maxilla and a few possible pebble-tools in the same layer near the top of the Makapansgat deposit (Brain, Van Riet Lowe and Dart, 1955; Dart, 1955b). This discovery Dart (1955b) described as 'providing the first concrete evidence that an australopithecine type...was contemporaneous with, and may have been responsible for, the concomitant pebble culture found in this sealed Central Transvaal cavern deposit'. The lithicultural evidence from Makapansgat was, however, not convincing—'An artificial origin for the heavily weathered dolomitic specimens from Makapan was not acceptable to most prehistorians attending the Congress[1] owing to the rapid weathering properties of this rock which obscured the nature of the scars' (J. D. Clark, 1962).

More suggestive was the discovery by Brain (1958) of undoubted stone implements in the breccia of the West Pit of the Sterkfontein excavation (an area which has since come to be known, somewhat misleadingly, as the Sterkfontein Extension Site, though it is not a separate site as some have come to believe). Further implements were recovered by Robinson the following year, along with some hominid fragments identified by him as australopithecine (Robinson and Mason, 1957, 1962). However, Robinson has argued on theoretical grounds against *Australopithecus* having been the author of these Sterkfontein implements, while Tobias (1964b, 1965b) has questioned whether the hominid fragments are all australopithecine and whether two or three of them do not, in fact, represent a more advanced type of hominid such as *Homo habilis*.

What seemed at first to be the most convincing evidence for the association of an australopithecine with early stone tools was provided by the discovery of *Australopithecus boisei* on a living floor along with broken bones and crude artefacts of the Oldowan Culture. *A. boisei* was at once hailed as the maker of the stone implements (Leakey, 1959a, 1960a). This evidence clearly influenced some workers in favour of the australopithecine

[1] The Third Pan-African Congress on Prehistory, Livingstone, 1955.

authorship of the Oldowan Culture. In the rather prophetic words of J. D. Clark, '...should no other more advanced form of man be found in this bed (Bed I), there would be strong reason to accept *Zinjanthropus* as the toolmaker' (Clark, 1961, p. 904). Elsewhere, Clark (1962) summed the position up as follows: 'Since no more advanced hominid than *Australopithecus* is known to have been present at this time anywhere in the Old World there would seem to be little room for continued doubt that the Lower Pleistocene Oldowan industries were made by hominids in the australopithecine pattern' (p. 269).

Subsequently, remains of a more advanced hominid were indeed discovered in Bed I (Leakey, 1961 *a–c*). These remains have been recognised as representing a new and lowly species of *Homo*, namely *H. habilis* (Leakey *et al.* 1964; Tobias, 1964 *a*, *b*, 1965 *a–d*). Furthermore, teeth of *H. habilis* were found on the same living-floor as the type cranium of *A. boisei* and Leakey (1961 *c*) then stated, 'If I am right in believing that the juvenile from FLK NNI is not an australopithecine, but a very remote and truly primitive ancestor of *Homo*, then it is possible (and I stress the fact here that I only use the word "possible" at this stage of the inquiry) that it was this branch of the Hominidae that also made the Oldowan tools at the site FLK I where *Zinjanthropus boisei* was found' (p. 418).

The fact that Leakey changed his mind about the authorship of the Oldowan tools has been levelled at him as a criticism. Yet, Leakey's second thoughts would seem to be not only legitimate and beyond reproach, but scientifically correct. At the time when *A. boisei* was the most advanced creature known alongside the tools, it was reasonable to suggest an association between them. But once a more advanced hominid was identified, no matter what label we apply to it, it is surely more reasonable to attribute the making of the stone tools to the more advanced hominid. When one takes all the evidence bearing on the association between Australopithecinae and stone tools into account, it becomes not only reasonable, but the hypothesis which meets more of the facts than any alternative.

Elsewhere, I have examined critically the evidence of skeletal and cultural occurrences at Olduvai (three levels in Bed I), Taung, Makapansgat, Sterkfontein (Lower and Middle Breccia), Swartkrans, Kromdraai, Garusi and Peninj (Tobias, 1964 *b*, 1965 *b*, *f*). Oakley (1956) has rightly warned that 'it is going to be...difficult to prove or disprove the theory that australopithecines had advanced to the stage of systematic tool-making. The absence of stone implements from layers containing their remains cannot be held to disprove the theory that they were tool-makers, so long as it is possible that the remains represent the food débris of cave-dwelling carnivores' (p. 6). Mason, too, has warned against speculating on the identity of the tool-maker in terms of negative evidence (Robinson and Mason, 1962). Nevertheless, certain correlations seem worthy of consideration:

(1) At every australopithecine site at which stone tools occur in association with hominid remains, there is evidence of the sympatric and synchronic co-existence of a more advanced hominid alongside the australopithecine.

(2) At all sites where *Australopithecus* and a more advanced hominid co-exist, there too we find stone tools.

(3) At every site where early stone tools are found along with associated hominid remains, the skeletal remains include a more advanced hominid, whether or not australopithecine remains are present in addition.

(4) At every early site which has yielded a more advanced hominid, stone tools are present.

Unless we resort to a series of special pleas, it seems that the most reasonable hypothesis to explain these facts is that *Australopithecus* was not the maker of the Oldowan stone implements, but that a more advanced hominid almost certainly was. This hypothesis in no way prejudges the issue of how many kinds of more advanced hominids were involved; present evidence suggests an earlier hominine of the species *H. habilis* and a later hominine of the species *H. erectus*. We are led to conclude that *A. boisei* is unlikely to have been the manufacturer of the Oldowan implements.

This conclusion is not to be construed as denying all implemental activities to *A. boisei*. The australopithecine of Olduvai Bed I may have been a stone tool-maker, as well as *H. habilis*, but the combined presence of the two hominids in the same area and even on the same living-floor makes it almost impossible to decide the question on archaeological grounds. If the stone implements from Olduvai Bed I fell into two distinct groups, clearly distinguishable on typology and technology, and if one assemblage were obviously more rudimentary than the other, it might not be unreasonable tentatively to attribute the more primitive implements to *A. boisei* and the more advanced industry to *H. habilis*. Even if these conditions were fulfilled, such an attribution would have to remain provisional. In any event, there is as yet no evidence for two distinct industries within Bed I.

The fact that Olduvai has not furnished evidence on the implemental activities of *A. boisei* still does not rule out the possibility of a well-developed cultural existence. *A. boisei* is one of the australopithecines and structurally this group is more hominised than the great apes. We have seen that the australopithecines have, on the average, somewhat bigger brains in somewhat smaller bodies, so that their endowment of extra neurones exceeds that of the great apes by about one thousand million on the average. Their front teeth are small and crowded. Furthermore, they possess the general primate structural basis for implemental activities, namely, the ability to *sit upright* (Tobias, 1965*f*) and to manipulate both powerfully and to an extent precisely (Napier, 1959); in addition, they possess the trait of bipedalism, although, as Chopra (1961, 1962) has shown, they were imperfectly adjusted skeletally to upright stance and bipedal gait. Nevertheless, the greater degree of erectness and bipedalism which the anatomical facts permit us to attribute to *Australopithecus* must have enhanced his implemental potential.

It might reasonably be expected that this structural hominisation which characterises all the australopithecines, albeit to differing degrees, would be paralleled by some hominisation of behaviour over the attainments of apes. When Dart first suggested that the australopithecines were capable of violent manual activities (1926, 1929) and of using and fabricating tools of bone, horn and teeth (1957), relatively little was known of the extent to which other higher Primates were capable of tool-using and even, to a degree, of tool-making. Since that time, more and more information has accumulated and today it is clear that the level of implemental activity which the living great apes can attain, in the wild, in captivity and under experimental conditions, far exceeds what had earlier been thought. Proportionately, the resistance to attributing tool-using and tool-making activities to the australopithecines has lessened, for even if the australopithecines showed no greater degree of implemental activity than living great apes, it is clear that a considerable range of cultural activities would have been within their capacity. Nobody, to my knowledge, has suggested that australopithecines were less capable than apes of implemental activity!

On the contrary, much indirect evidence points to the possibility that the australopithecines were culturally advanced as compared with apes. We have cited the evidence of structural hominisation. To this could be added the evidence of ecology, for the australopithecines lived in a habitat providing little natural protection and they had no natural weapons of offence or defence like large canines.

Thus, the indirect evidence points to the need of implemental activities by australopithecines. Direct evidence of association with stone tools, we have seen, is largely lacking or equivocal. The best direct evidence bearing on the cultural status of *Australopithecus* is provided by the osteodontokeratic objects from Makapansgat described by Dart (1957 and many articles between 1955 and 1965—*see* review in Tobias, 1965*f*). The facts and Dart's claims are too well known for me to need to review them here; only the following points need be mentioned:

(1) The analysis of many thousands of bones from the Makapansgat deposit has shown definite evidence of selection of bones, not only from different parts of the body, but of different parts of the same bone. Thus, the ratio of proximal to distal humeral fragments is 33:336, while the ratio of humeri to femora is 518:100.

(2) Tooth-marks of hyaena and porcupine are

absent from all but a handful of the many thousands of bones.

(3) Eighty per cent of over fifty baboon skulls from Taung (21), Sterkfontein (22) and Makapansgat (15) show signs of damage by localised violence.

(4) Large concentrations of ungulate humeri show damage inflicted before fossilisation on the epicondyles, some of which fit the occasional doubly-indented fracture depressions of the baboon crania.

(5) Many of the bone flakes show signs of differential wear and tear along one edge or at one end, but not on the other.

(6) A number of special cases include horn-cores and smaller long bones rammed and lodged up the marrow cavities of broken larger bones.

(7) In several instances, small bones and even stone flakes have been wedged between the condyles of long bones.

(8) Many of the long bones show signs of having been broken by a kind of spiral torsional stress.

(9) Some fragments suggest that stalactites and/or stalagmites were broken off, presumably for use, and some were further fractured.

(10) There is some suggestion of stone-collecting habits: a small number of quartz and quartzite pebbles and fragments have beeen found in the breccia.

(11) So far, the forty-one hominid fragments identified are all of *Australopithecus*. We have no evidence for any other hominid having lived in or near the Makapansgat Limeworks caves during the time that the Limeworks deposit was accumulating.

In sum, Dart's hypothesis on the osteodontokeratic activities of the Australopithecinae is the most reasonable and plausible explanation of the otherwise almost inexplicable mountains of bones, with their selected, fractured and patterned characteristics at Makapansgat. Furthermore, this kind of activity is entirely within the somatic possibilities which might be inferred from behaviour studies on pongids and from anatomical studies of the degree of hominisation of the australopithecines themselves. To accept this general hypothesis is not necessarily to accept everything that Dart has claimed for the osteodontokeratics; nor am I sure that it is justified to ascribe the epithet Osteodontokeratic *Culture* to the implemental activities of the australopithecines: perhaps the phrase 'osteodontokeratic activities' would be more appropriate.

So far, comparable masses of osteodontokeratic objects have not been reported from other australopithecine sites than Makapansgat. However, at Sterkfontein, not more than a fraction of the breccia has been thoroughly searched for broken bone fragments, other than taxonomically identifiable parts. In the Swartkrans cave, where there are large numbers of australopithecine remains, as well as a few pieces possibly representing *H. erectus* ('Telanthropus'), associated faunal remains are conspicuous by their paucity. This could be understood if the australopithecine in that instance were the hunted rather than the hunter, australopithecine flesh having perhaps provided a major item in the diet of the early Transvaal hominine. Perhaps the same applied in Bed I times at Olduvai, as Howell (1965b) has suggested recently.

Hence, the mere absence of hitherto detected osteodontokeratic objects from other australopithecine sites does not weaken Dart's claims based upon Makapansgat. It is suggested that the techniques of *Australopithecus* included toolmaking and that osteodontokeratics provided his major cultural outlet. This does not preclude his having tinkered with stone, but we find no convincing evidence that *Australopithecus* was responsible for the first distinctive stone culture, the Oldowan.

Technologically, it has been suggested (Tobias, 1965f) that *Australopithecus* had not passed beyond the stage of using natural bodily organs—teeth, hands, feet—or the floor, walls and stalactites of caves for making osteodontokeratic objects. It is suggested that they fell short of one intricate conceptual and technological mechanism: the ability to use a tool to make a tool—that is, 'the highest implemental frontier' of Khroustov (1964).

If it is correct that they could not or did not cross the highest implemental frontier of the apes, it may validly be asked: wherein lies the cultural advance of the australopithecines, including *A. boisei*, over the apes? The answer resides in the *frequency of implemental patterns of behaviour*. In apes, tool-using and tool-making are infrequent, sporadic; it cannot be said that the apes' way of life is built around such implemental activities. Survival does not depend on implemental means, but rather on formidable natural defence mechanisms and on the sheltered forest habitat of the living great apes. The australopithecines, on the other hand, lived in a habitat providing little natural protection and they had poor natural bodily defences. Their implemental activities must have come to loom very largely in their pattern of adjustment. Indeed, it would not be too much to claim that their very survival depended on implemental activities. This, it is suggested, is the great step forward of the australopithecines over the apes. They learned to exploit a mental and manipulative capacity, a cultural potentiality, which even apes possess: and they exploited it so effectively that they became dependent on it for survival. As Bartholomew and Birdsell (1953) put it: 'Rather than to say that man is unique in being the "tool-using" animal, it is more accurate to say that man is the only mammal which is continuously dependent on tools for survival.' In this sense, it is suggested that *Australopithecus* had virtually attained the status of manhood. Cultural capacity was the greatest evolutionary asset of the australopithecines: and it was on this aspect of their form and function that selection operated with the greatest vigour.

Robinson (1963) has suggested that cultural capacity was less developed in the robust australopithecines than in the more gracile *A. africanus*. This view seems to be based on the supposed vegetarian habit of *A. robustus* and on the notion that the low brow of the latter betokened a more ape-like brain. Since neither premise is acceptable, it seems that there is no real justification at present for thinking that any of the species of *Australopithecus* was less dependent on cultural activities than any other. The present thesis holds that all of the australopithecines were cultural creatures, all were more proficient than the apes in manipulating and manufacturing, and all had come to depend on their implemental activities, whether with sticks, stones, bones, horns, or stalactites, for survival. It is not impossible that the degree of implemental dependence varied from one australopithecine taxon to another, but the evidence for such variation is as yet lacking.

A. boisei, it is suggested, was subject to the same cultural dependence as the others, even though the concrete evidence may presently be lacking, or may not yet have been recognised as such. If he was the victim of the more skilled hominine hunters of Olduvai, then the presence on the habiline living-floors of his skeletal remains *without bone and horn tools* would be understandable.

B. The place of *Australopithecus boisei* and the other australopithecines in hominid phylogeny

Until the discovery of *A. boisei*, the gracile australopithecine, *A. africanus*, was known from earlier sites tentatively identified as Lower Pleistocene, while the robust australopithecine, *A. robustus*, was known only from later sites, probably of the early Middle Pleistocene. It was possible then to see in *A. robustus* a later hominid which either (*a*) represented an intermediate step in the hominisation process between *A. africanus* and *H. erectus*, or (*b*) represented the end-result of cladistic evolution of the australopithecines, a branch which had *specialised* away from the main line of hominisation. A third possible view, espoused by Robinson (1963), was that the Kromdraai–Swartkrans group represented late survivors of a little-changed, little-hominised, ancestral hominid, from which stock *A. africanus* had risen by further hominisation (Fig. 37).

As for the first alternative, Robinson (1963) has adduced strong arguments against the recognition of a phyletic sequence, *A. africanus* → *A. robustus* → *Homo*, and indeed against the idea that *A. robustus* is on the direct human line at all (1965*a*).

If we discount the first alternative, there remain at least two important views, namely, that *A. robustus* is a surviving 'primitive', or that it has specialised away from the morphology of the presumed ancestral australopithecine (Fig. 37). The discovery of *A. boisei* indicates that already in Lower Pleistocene times at least two different australopithecines were in existence, *A. africanus* and *A. boisei*. In the enlargement of their cheek-teeth, both show a feature which, in any other mammalian line, would tend to be regarded as specialised. The degree of this specialisation differs

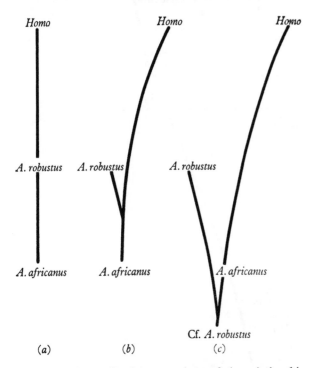

Fig. 37. Three earlier interpretations of the relationship between the Lower Pleistocene *A. africanus* and the Middle Pleistocene *A. robustus* (before the description of *A. boisei* and *H. habilis*).

enormously between the two forms. There seems little doubt that *A. africanus* with its relatively large mandibular canines and moderately enlarged cheek-teeth was the less specialised, and was morphologically closer to the presumed morphology of the ancestral hominines. On the other hand, *A. boisei* with its massive cheek-teeth and supporting structures shows highly specialised features which would tend to place it well off the line of hominisation. Figure 38 attempts to align the known hominids of the Lower and Middle Pleistocene in their position in time and space (after Tobias, 1965c). When the specimens are arranged in this way, it is seen that the large-toothed australopithecines are off the human line, while *A. africanus* is much closer to it. Closer still, it would seem, is the other Lower Pleistocene hominid which has been designated by Leakey et al. (1964) *H. habilis*. These remains of which only cursory descriptions have so far been published (Leakey, 1960c, 1961a–c; Leakey and Leakey, 1964; Napier, 1962; Davis, Day and Napier, 1964; Tobias, 1964a, b, 1965a–d) will be described in full in a later volume of the present series.

The discovery of *A. boisei* thus demonstrates that the large-toothed specialisation was not a late stage in australopithecine evolution. Already, by the beginning of our Pleistocene fossil record, we have at least three kinds of hominid: the least hominised or most specialised, namely *A. boisei*; a moderately hominised, little specialised *A. africanus*; and the most hominised *H. habilis*. All three are roughly synchronic, but only *A. boisei* and *H. habilis* have been shown to be sympatric.

The question now is: upon what sort of ancestral morphology will the Pleistocene phyletic lines converge, if extended *back* in time? If we use the argument of specialisation, it would seem that the ancestral australopithecine would not have shown the extreme specialisations manifested by *A. boisei* and, later, by *A. robustus*. That is, we might have expected the mandibular canines to be somewhat larger and the cheek-teeth to be smaller: but if we strip an *A. boisei* or *A. robustus* of these features and the concomitant modifications in face, jaw and palate structure, we are left with something very like *A. africanus*! Robinson has recently accepted that the ancestral australopithecine would not have shown the 'exaggerated characters seen in the known specimens of *Paranthropus*' (1963, p. 407). He elaborates this statement as follows: 'The canines, for example, will have been larger and therefore more in proportion to the cheek teeth. Body size is likely to have been smaller and therefore probably the skull will have been somewhat more gracile. This early *Paranthropus* will therefore have differed less from the known *Australopithecus* material than does the known, later, *Paranthropus* material' (p. 407). This statement clearly indicates that Robinson now accepts that

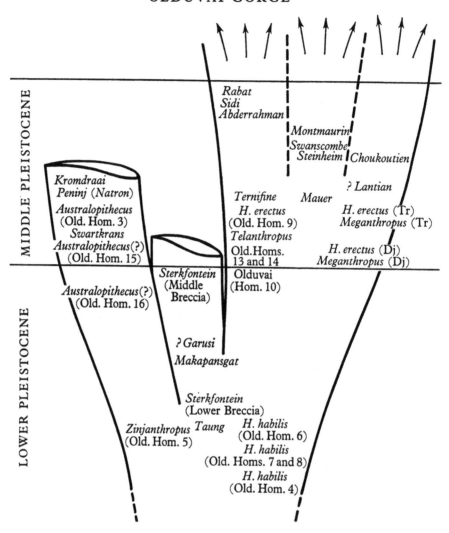

Fig. 38. Schema of Lower and Middle Pleistocene hominids, showing the positions in space and time of the most important specimens discovered to date. Tr—Trinil beds; Dj—Djetis Beds; Old. Hom.—Olduvai hominid.

A. robustus is a creature of specialisation. However, he goes on immediately to say of the ancestral australopithecine, 'but it will nevertheless have been more nearly *Paranthropus* than *Australopithecus* because of diet, absence of forehead, pongid-like ischium, primitive nasal area, and probably many other things of which we are as yet unaware'.

This study has cast serious doubt on the morphological basis of the dietary inference. Furthermore, until we know more about the causal basis of the low forehead, we certainly cannot assume that the ancestral australopithecine with somewhat bigger mandibular canines and smaller cheek-teeth would have possessed no forehead. The 'pongid-like ischium' cannot be discussed here, as the detailed description of the australopithecine pelves has not yet been published. As to the 'primitive nasal area'—which Robinson describes as 'almost ultra pongid (1963, p. 406)—I have shown above that both in *A. boisei* and *A. robustus*, the shape of the nasal margin and floor, while in general reminiscent of that of the chimpanzee, shows a number of clearly hominine departures, such as the posterior placement of the anterior nasal spine and a tendency for the prenasal fossa to 'drop over' from the floor of the nose to become part of the naso-alveolar clivus as in modern man. These features are shared as well with *A. africanus* and even some modern human crania.

Stripped of the paranthropine specialisations and of most of the features just mentioned, our picture of the ancestral australopithecine is virtually indistinguishable from that of *A. africanus*. By indirect inference from morphology, we are led to see in the ancestral australopithecine a creature akin to *A. africanus* and not to *A. boisei* or *A. robustus*.

If we had an adequate Pliocene fossil record, we should not need to extrapolate from the Pleistocene hominids to their presumed ancestor. However, a large gap in the record leaves the middle and upper Pliocene as one of the most tantalising periods in hominid phylogeny. The lower Pliocene and late Miocene have, however, yielded claimants to australopithecine ancestry. Simons has indicated that *Ramapithecus punjabicus* (to which he assigns as well *Kenyapithecus wickeri* Leakey) has dental and facial characters 'so close to *Australopithecus africanus* as to make difficult the drawing of generic distinctions between the two species on the basis of present material' (Simons, 1964*b*, p. 535). He goes on to say, 'Provisionally the two genera, *Ramapithecus* and *Australopithecus*, are retained as distinct because of their considerable time separation. *Ramapithecus punjabicus* is almost certainly man's forerunner of 15 million years ago.' The important point to note is that it is with *A. africanus*, not *A. robustus* or *A. boisei*, that *Ramapithecus* finds its resemblance.

If Simons is correct, both morphological inferences from the Pleistocene fossils, and the evidence of the Mio–Pliocene fossils themselves, would concur in demonstrating that the ancestral australopithecine resembled *A. africanus*, at least dentally and gnathically.

We should thus arrive at a picture of the ancestral australopithecine as unspecialised and relatively small-toothed. At some time not later than the Upper Pliocene, it must have diversified into several lines. A megadontic line (*A. boisei*) emerged with specialised dentition. Another line remained little changed and unspecialised: presently it dichotomised into a progressively more hominised line represented in Africa by *H. habilis* and later in Asia perhaps by *Meganthropus palaeojavanicus*; and a more conservative residual line (*A. africanus*) which, perhaps because of competition, did not long outlast the emergence of this supposed hominine.

The intensive selection pressures which it must be presumed engendered *A. boisei* at the beginning of the Pleistocene must have subsequently relaxed somewhat. Some populations of *A. boisei* then

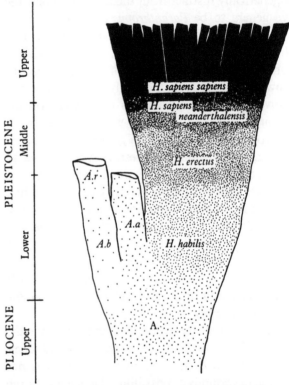

Fig. 39. Provisional schema of hominid phylogeny from Upper Pliocene times to the Upper Pleistocene. Progressive degrees of hominisation are represented by progressively darker shading. A—postulated ancestral australopithecine, cf. *A. africanus*. *A. b*—*A. boisei*. *A. a*—*A. africanus*. *A. r*—*A. robustus*.

moved forward with a moderate reduction in cheek-tooth size, loss of the cingulum probably as part of the same process, shortening of the face and reduction of the jaws—to become the macrodontic *A. robustus* of the Middle Pleistocene. Figure 39 represents a provisional schema of the interpretation proposed here.

We are thus led to recognise two apparent hominid lineages in the Lower and Middle Pleistocene: one line was seemingly specialising away from the main hominising trend and comprised *A. boisei* → *A. robustus*. The other line comprised *A. africanus* → *H. habilis* → later *Homo*, and seems to have been the main line of structural hominisation and of cultural evolution.

It would be easy, at this stage of our knowledge, to exaggerate the distinctness between the two lineages. One recent work has gone so far as to suggest that the members of each line be regarded as generically distinct from those of the other, the two genera to be *Paranthropus* (*A. boisei* and *A. robustus*) and *Homo* (*A. africanus*, *H. erectus* and *H. sapiens*) (Robinson, 1965a). Such a classification *by clade* would be valid if the fossil record showed unequivocally not only that *A. africanus* evolved into *Homo*, but also that fossils on the *A. boisei*–*A. robustus* lineage were completely cut off from contributing to the gene-pool of *A. africanus* and its presumed successors. Yet, the overall resemblances between the australopithecines in the two lineages are so great as to suggest that they belonged to the same evolutionary *grade*, not by parallelism but by homology or real genetic relationship. In fact, the evidence seems to indicate that (1) it is unlikely they were genetically isolated from each other throughout the Lower and Middle Pleistocene; and/or (2) they had a not very remote common ancestry.

As to the first suggestion, we cannot exclude the chance of crossing between *A. africanus* and members of the *A. boisei* → *A. robustus* line. It is not outside the bounds of possibility that such crossing may have led to the 'gracilisation' of *A. boisei* into the later and somewhat toned down *A. robustus*. Even among the australopithecine fossils already known, there is a suggestion of intermediates. Thus, the specimens from Makapansgat, although commonly classed with *A. africanus*, show some features more reminiscent of *A. robustus*: these features include very large, buccolingually expanded molars, very robust jaws and the probable presence of a sagittal crest. In these respects, the Makapansgat specimens seem to show a somewhat nearer approach to *A. robustus* than do the Sterkfontein specimens. This reduces the distinctness of the lineages and renders it less likely that they represented two clades, the members of which should be regarded as generically distinct from each other. The discovery of further specimens and new sites will provide better information on variability and may go further towards closing the gap between the two groups of fossils.

The second line of thinking suggests that, when first encountered in the fossil record, the two hominid lineages had not been distinct for very long. It seems reasonable to infer that the common ancestral australopithecines diverged into the two lineages about the Upper Pliocene or, at the latest, the first part of the Lower Pleistocene. Thus, they might not have been isolated long enough to have attained the distinctness of separate clades and separate genera.

At the present state of our knowledge, therefore, it would seem unjustified to classify the robust australopithecines as generically distinct from the gracile ones. The entire australopithecine grade must for the time being be regarded as providing the substrate from which one or more lines of hominisation emerged.

A. boisei represents one part of this substrate, one extreme in the diverse spectrum of hominids which had appeared upon the African scene by the second half of the Lower Pleistocene. It seems to have been the progenitor of an experimental line of dentally modified and perhaps over-specialised creatures, a line which was to prove unequal to the rigorous challenge of highly competitive hunter-hominines. Its robust but ineffectual descendants survived until the Middle Pleistocene, when the experiment seems to have terminated in genocidal extinction.

REFERENCES

Aichel, O. (1917). Die Beurteilung des recenten und prähistorischen Menschen nach der Zahnform. *Z. Morph. Anthrop.* **20**, 457–550.

Akabori, E. (1933). The non-metric variations in the Japanese skull. *Jap. J. Med. Sci., I. Anat.* **4**, 61–315.

Arambourg, C. (1954). L'hominien fossile de Ternifine (Algérie). *C.R. Acad. Sci., Paris*, **239**, 893–5.

Arambourg, C. (1958). Les stades évolutifs de l'humanité. *The Leech*, **28**, 106–11.

Ariens Kappers, C. U. (1929). The fissures on the frontal lobes of *Pithecanthropus erectus* Dubois compared with those of Neanderthal Man, *Homo recens*, and chimpanzee. *Proc. K. Ned. Akad. Wet.* **32**, 182–95.

Ashton, E. H. (1950). The endocranial capacities of the Australopithecinae. *Proc. Zool. Soc. Lond.* **120**, 715–21.

Ashton, E. H., Healy, M. J. R. and Lipton, S. (1957). The descriptive use of discriminant functions in physical anthropology. *Proc. Roy. Soc.* B, **146**, 552–72.

Ashton, E. H. and Spence, T. F. (1958). Age changes in the cranial capacity and foramen magnum of hominoids. *Proc. Zool. Soc. Lond.* **130**, 169–81.

Ashton, E. H. and Zuckerman, S. (1950). Some quantitative dental characteristics of the chimpanzee, gorilla and orang-utang. *Phil. Trans.* B, **234**, 471–84.

Ashton, E. H. and Zuckerman, S. (1951). Some cranial indices of *Plesianthropus* and other Primates. *Am. J. Phys. Anthrop.* **9**, 283–96.

Ashton, E. H. and Zuckerman, S. (1952). Age changes in the position of the occipital condyles in the chimpanzee and gorilla. *Am. J. Phys. Anthrop.* **10**, 277–88.

Ashton, E. H. and Zuckerman, S. (1954). The anatomy of the articular fossa (fossa mandibularis) in man and apes. *Am. J. Phys. Anthrop.* **12**, 29–61.

Ashton, E. H. and Zuckerman, S. (1956). Cranial crests in the Anthropoidea. *Proc. Zool. Soc. Lond.* **126**, 581–635.

Bartholomew, G. A. and Birdsell, J. B. (1953). Ecology and the protohominids. *Am. Anthrop.* **55**, 481–98.

Baume, L. J. and Becks, H. (1953). The topogenesis of the mandibular permanent molars. *Oral Surg.* **6**, 850–68.

Black, Davidson (1931). On an adolescent skull of *Sinanthropus pekinensis* in comparison with an adult skull of the same species and with other hominid skulls, recent and fossil. *Palaeont. sin.* D, **7**, 1–144.

Black, G. V. (1902). *Descriptive Anatomy of the Human Teeth*, 4th edit., p. 169. Philadelphia: S. S. White.

Bolk, L. (1915). Über Lagerung, Verschiebung und Neigung des Foramen magnum am Schädel der Primaten. *Z. Morph. Anthrop.* **17**, 611–92.

Bolk, L. (1925). On the existence of a dolichocephalic race of gorilla. *Proc. K. Ned. Akad. Wet.* **28**, 204–13.

Boné, E. L. and Dart, R. A. (1955). A catalog of the australopithecine fossils found at the Limeworks, Makapansgat. *Am. J. Phys. Anthrop.* **13**, 621–4.

Bonin, G. von (1963). *The Evolution of the Human Brain*, pp. 1–92. Chicago: University Press.

Bork-Feltkamp, A. J. van (1950). The relative usefulness of various cranial characters for racial comparison. *Man*, **51**, 17–19.

Boule, M. and Vallois, H. V. (1957). *Fossil Men*, pp. 1–535. New York: The Dryden Press.

Brain, C. K. (1958). The Transvaal Ape-man bearing cave deposits. *Transv. Mus. Mem.*, No. 11.

Brain, C. K., Van Riet Lowe, C. and Dart, R. A. (1955). Kafuan stone artefacts in the post-australopithecine breccia at Makapansgat. *Nature, Lond.* **175**, 16.

Breathnach, A. S. (ed.) (1965). *Frazer's Anatomy of the Human Skeleton*, 6th edit., pp. 1–253. London: J. and A. Churchill.

Bronowski, J. and Long, W. M. (1951). Statistical methods in anthropology. *Nature, Lond.* **168**, 794–5.

Bronowski, J. and Long, W. M. (1952). Statistics of discrimination in anthropology. *Am. J. Phys. Anthrop.* **10**, 385–94.

Bronowski, J. and Long, W. M. (1953). The australopithecine milk canines. *Nature, Lond.* **172**, 251.

Broom, R. and Robinson, J. T. (1948). Size of the brain in the ape-man, *Plesianthropus*. *Nature, Lond.* **161**, 438.

Broom, R. and Robinson, J. T. (1950). See Broom, Robinson and Schepers, 1950.

Broom, R. and Robinson, J. T. (1952). Swartkrans ape-man, *Paranthropus crassidens*. *Transv. Mus. Mem.*, No. 6.

Broom, R., Robinson, J. T. and Schepers, G. W. H. (1950). Sterkfontein ape-man, *Plesianthropus*. *Transv. Mus. Mem.*, No. 4.

Broom, R. and Schepers, G. W. H. (1946). The South African fossil ape-men, the Australopithecinae. *Transv. Mus. Mem.*, No. 2.

Browning, H. (1953). The confluence of dural venous sinuses. *Am. J. Anat.* **93**, 307–29.

Butler, P. M. (1939). Studies of the mammalian dentition. Differentiation of the post-canine dentition. *Proc. Zool. Soc. Lond.* B, **109**, 1–36.

Cabibbe, G. (1902). Il processo postglenoideo nei cranii di normali, alienati, criminali in rapporto a quello dei varii ordini di mammiferi. *Anat. Anz.* **20**, 81–95.

Calogero, B. (1959). Ricerche comparative sulla pneumatizzazione della mastoide e dei seni paranasali nell 'uome. *Archo. ital. Lar.* **67**, 1–16.

Campbell, B. (1962). The systematics of man. *Nature, Lond.* **194**, 225–32.

Campbell, B. (1963). Quantitative taxonomy and human evolution. From *Classification and Human Evolution* (ed. S. L. Washburn), pp. 50–74. Chicago: Viking Fund Publications in Anthropology.

Campbell, T. D. (1925). *Dentition and Palate of the Australian Aboriginal*, pp. 1–123. Adelaide: The Hassel Press.

Caspari, E. W. (1961). Some genetic implications of human evolution. From *Social Life of Early Man* (ed. S. L. Washburn), pp. 267–77. Chicago: Aldine Publishing Co.

Cave, A. J. E. (1961). The frontal sinus of the gorilla. *Proc. Zool. Soc. Lond.* **136**, 359–73.

Cave, A. J. E. and Haines, R. W. (1940). The paranasal sinuses of the anthropoid apes. *J. Anat.* **74**, 493–523.

Chopra, S. R. K. (1961). The angle of pelvic torsion in the Primates. *Z. Morph. Anthrop.* **51**, 268–74.

Chopra, S. R. K. (1962). The innominate bone of the Australopithecinae and the problem of erect posture. *Bibl. primat.* **1**, 93–102.

Clark, J. D. (1961). Sites yielding hominid remains in Bed I Olduvai Gorge. *Nature, Lond.* **189**, 903–4.

Clark, J. D. (1962). The problem of the pebble cultures. *VI Cong. Internaz. delle Scienze Preistoriche e Protostoriche*, vol. I. *Relazioni generali*, pp. 265–71. Roma.

Clark, J. D. (1963). Ecology and culture in the African Pleistocene. *S. Afr. J. Sci.* **59**, 353–66.

Clark, W. E. Le Gros (1938*a*). General features of the Swanscombe skull bones. *J. Roy. Anthrop. Inst.* **68**, 58–60.

Clark, W. E. Le Gros (1938*b*). The endocranial cast of the Swanscombe bones. *J. Roy. Anthrop. Inst.* **68**, 61–7.

Clark, W. E. Le Gros (1947). Observations on the anatomy of the fossil Australopithecinae. *J. Anat.* **81**, 300–33.

Clark, W. E. Le Gros (1950*a*). South African fossil hominoids. *Nature, Lond.* **165**, 893–4.

Clark, W. E. Le Gros (1950*b*). New palaeontological evidence bearing on the evolution of the Hominoidea. *Q. J. Geol. Soc. Lond.* **105**, 225–64.

Clark, W. E. Le Gros (1952*a*). A note on certain cranial indices of the Sterkfontein skull No. 5. *Am. J. Phys. Anthrop.* **10**, 119–21.

Clark, W. E. Le Gros (1952*b*). Hominid characters of the australopithecine dentition. *J. Roy. Anthrop. Inst.* **80**, 37–54.

Clark, W. E. Le Gros (1955). *The Fossil Evidence for Human Evolution*, pp. 1–181. Chicago: University Press.

Clark, W. E. Le Gros (1964). *The Fossil Evidence for Human Evolution* (2nd edit.), pp. 1–201. Chicago: University Press.

Clark, W. E. Le Gros and Leakey, L. S. B. (1951). The Miocene Hominoidea of East Africa. *Fossil Mammals of Africa*, pp. 1–117. No. 1. London: British Museum (Natural History).

Colyer, F. (1936). *Variations and Diseases of the Teeth of Animals*, pp. 1–750. London: John Bale, Sons and Danielsson, Ltd.

Connolly, C. J. (1950). *External Morphology of the Primate Brain* (1st edit.), pp. 1–378. Springfield: Charles C. Thomas.

Cooke, H. B. S. (1963). Pleistocene mammal faunas of Africa, with particular reference to Southern Africa. From: *African Ecology and Human Evolution* (eds. F. C. Howell and F. Bourlière), pp. 65–116. Chicago: Aldine Publishing Company.

Coon, C. S. (1963). *The Origin of Races*, pp. 1–724. London: Jonathan Cape.

Coppens, Y. (1964). *Homo habilis* et les nouvelles découvertes d'Oldoway. *Bull. Soc. préhist. fr.* No. 7 (Oct.), pp. 171–6.

Costa Ferreira, A. A. da (1920). Sur le rapprochement et la coalescence des lignes temporales du crâne chez les microcéphales. *C.R. Soc. Biol.* **83**, 1195–6.

Dahlberg, A. A. (1949). The dentition of the American Indian. From *Papers on the Physical Anthropology of the American Indian*, pp. 138–76. New York: Viking Fund.

Dalitz, G. D. (1963). The root development of third molar teeth. *J. Forens. Med.* **10**, 30–5.

Darlington, D. and Lisowski, F. P. (1965). Changes in the sagittal and nuchal crests of the skull of the ferret (*Mustela furo*) after partial removal of the temporal muscles. *VIII Internat. Congr. of Anatomists*, Wiesbaden (August 1965).

Dart, R. A. (1925). *Australopithecus africanus*: the man-ape of South Africa. *Nature, Lond.* **115**, 195–9.

Dart, R. A. (1926). Taungs and its significance. *Nat. Hist. N.Y.* **26**, 315–27.

Dart, R. A. (1929). A note on the Taungs skull. *S. Afr. J. Sci.* **26**, 648–58.

Dart, R. A. (1948*a*). The Makapansgat proto-human *Australopithecus prometheus*. *Am. J. Phys. Anthrop.* **6**, 259–84.

Dart, R. A. (1948*b*). The adolescent mandible of *Australopithecus prometheus*. *Am. J. Phys. Anthrop.* **6**, 391–412.

Dart, R. A. (1949*a*). The cranio-facial fragment of *Australopithecus prometheus*. *Am. J. Phys. Anthrop.* **7**, 187–214.

Dart, R. A. (1949*b*). A second adult palate of *Australopithecus prometheus*. *Am. J. Phys. Anthrop.* **7**, 335–8.

Dart, R. A. (1955*a*). *Australopithecus prometheus* and *Telanthropus capensis*. *Am. J. Phys. Anthrop.* **13**, 67–96.

Dart, R. A. (1955*b*). The first australopithecine fragment from the Makapansgat Pebble Culture stratum. *Nature, Lond.* **176**, 170.

Dart, R. A. (1956). The relationship of brain size and brain pattern to human status. *S. Afr. J. Med. Sci.* **21**, 23–45.

Dart, R. A. (1957). The osteodontokeratic culture of *Australopithecus prometheus*. *Transv. Mus. Mem.*, No. 10.

Dart, R. A. (1962*a*). The most complete *Australopithecus* skull from the pink breccia at Makapansgat. *Actes IV Congr. Panafr. Préhist. l'Etude du Quat.* pp. 337–40.

Dart, R. A. (1962*b*). The Makapansgat pink breccia australopithecine skull. *Am. J. Phys. Anthrop.* **20**, 119–26.

Dart, R. A. (1962*c*). A cleft adult mandible and the nine other lower jaw fragments from Makapansgat. *Am. J. Phys. Anthrop.* **20**, 267–86.

Davis, P. R., Day, M. H. and Napier, J. R. (1964). Hominid fossils from Bed I, Olduvai Gorge, Tanganyika, *Nature, Lond.* **201**, 967–70.

Delattre, A. and Fenart, R. (1960). *L'hominisation du Crâne*, pp. 1–418. Paris: Centre National de la Recherche Scientifique.

De Villiers, H. (1963). A biometrical and morphological study of the skull of the South African Bantu-speaking Negro. Thesis accepted for Ph.D. by Faculty of Science, University of the Witwatersrand, Johannesburg.

Drennan, M. R. (1929). The dentition of a Bushman tribe. *Ann. S. Afr. Mus.* **24**, 61–87.

REFERENCES

Drennan, M. R. (1953). A preliminary note on the Saldanha skull. *S. Afr. J. Sci.* **50**, 7–11.

Dubois, E. (1898). Über die Abhängigkeit des Hirngewichtes von der Körpergrösse. *Arch. Anthrop.* **25**, 1–28, 423–41.

Economo, C. von (1930). Some new methods for studying the brains of exceptional people. *J. Nerv. Ment. Dis.* **72**, 125–34.

Eisler, P. (1912). Die Muskeln des Stammes. From *Handbuch der Anatomie des Menschen* (K. von Bardeleben), Jena (quoted by Weidenreich, 1943, q.v.).

Fanning, E. A. (1961). A longitudinal study of tooth formation and root resorption. *N.Z. Dent. J.* **57**, 202–17.

Flower, W. H. (1879). *Osteological Catalogue. Mus. Roy. Coll. Surg.*, Part 1: Man, pp. 181–95.

Flower, W. H. (1881). On the cranial characters of the natives of the Fiji Islands. *J. Anthrop. Inst.* **10**, 161.

Flower, W. H. (1885). On the size of teeth as a character of race. *J. Anthrop. Inst.* **14**, 183–6.

Frisch, J. E. (1965). *Trends in the Evolution of the Hominoid Dentition. Bibliotheca Primatologica*, Fasc. 3, pp. 1–130. Basle: S. Karger.

Galloway, A. (1941). Palatal measurements of Negro and other crania. *S. Afr. J. Sci.* **37**, 285–92.

Garn, S. M., Lewis, A. B. and Bonné, B. (1962). Third molar formation and its developmental course. *Angle Orthod.* **32**, 270–9.

Getz, B. (1960). Skull thickness in the frontal and parietal regions. *Acta morph. neerl.-scandi.* **3**, 221–8.

Giuffrida-Ruggeri, V. (1912). Über die endocranischen Furchen der *A. meningea media* beim Menschen. *Z. Morph. Anthrop.* **15**, 401–12.

Guggenheim, P. and Cohen, L. B. (1959). External hyperostosis of the mandible angle associated with masseteric hypertrophy. *Archs Otolar.* **70**, 674–80.

Guggenheim, P. and Cohen, L. B. (1961). The nature of masseteric hypertrophy. *Archs Otolar.* **73**, 15–28.

Gyldenstolpe, N. (1928). Zoological results of the Swedish expedition to Central Africa, 1921. Vertebrata 5, *Ark. Zool.* **20A**, 1–76 (quoted by Schultz, 1962, q.v.).

Harris, H. A. (1926). Endocranial form of gorilla skulls with special reference to the existence of dolichocephaly as a normal feature of certain primates. *Am. J. Phys. Anthrop.* **9**, 157–72.

Harrison, G. A. and Weiner, J. S. (1964). Section on 'Human Evolution' in *Human Biology*, by G. A. Harrison, J. S. Weiner, J. M. Tanner and N. A. Barnicot, pp. 63–7. Oxford: Clarendon Press.

Heberer, G. (1960a). '*Zinjanthropus boisei*' und der Status der Praehomininen (Australopithecinae). *Zool. Jb. Syst.* **88** (1), 91–106.

Heberer, G. (1960b). Älteste Menschheit in Afrika. *Natur Volk*, **90**, 309–21.

Heberer, G. (1962). Die Oldoway (Olduvai)-Schlucht (Tanganyika) als Fundert Fossiles hominiden. *Bibl. primat.* **1**, 103–19.

Hoadley, M. F. and Pearson, K. (1929). On measurement of the internal diameters of the skull. *Biometrika*, **21**, 85–123.

Hofer, H. (1960). Studien zum Problem des Gestaltwandels des Schädels der Säugetiere, insbesondere der Primaten. *Z. Morph. Anthrop.* **50**, 299–316.

Hofman, L. (1926–7). L'os temporal des singes et ses cavités pneumatiques. *Archs Anat. Histol. Embriol.* **6**, 141–86.

Holloway, R. L. (1962). A note on sagittal cresting. *Am. J. Phys. Anthrop.* **20**, 527–30.

Howell, F. C. (1965a). Comment on 'New discoveries in Tanganyika: their bearing on hominid evolution' (by P. V. Tobias). *Curr. Anthrop.* **6**, 399–401.

Howell, F. C. (1965b). *Early Man*, pp. 1–200. Life Nature Library. New York: Time Incorporated.

Howes, A. E. (1954). A polygon portrayal of coronal and basal arch dimensions in the horizontal plane. *Am. J. Orthod.* **40**, 811–31.

Hrdlička, A. (1910). Contribution to the anthropology of Central and Smith Sound Eskimo. *Anthrop. Pap. Am. Mus. Nat. Hist.* **5**, Part 2, 177–280.

Hrdlička, A. (1925). Weight of the brain and of the other internal organs in American monkeys. *Am. J. Phys. Anthrop.* **8**, 201–11.

Hrdlička, A. (1930). The skeletal remains of early man. *Smithsonian Misc. Collns*, **83**, 1–379.

Huizinga, J. (1958). Systematic investigations of the position of the greatest breadth of the skull in recent and fossil man. From *Hundert Jahre Neanderthaler* (ed. G. H. R. von Koenigswald), pp. 199–214. Utrecht: Kemink.

Hulse, F. S. (1963). *The Human Species*, pp. 1–504. New York: Random House.

Inskeep, R. R. (1959). Prehistory (Central and West Africa). *Encyclopaedia Britannica*, pp. 328–30.

Jacobson, A. (1966). The Bantu Dentition: a Morphological and Metrical Study of the Teeth, the Jaws and the Bony Palate of several large groups of South African Bantu-speaking Negroids. Thesis submitted for the degree of Doctor of Philosophy in the Department of Anatomy, University of the Witwatersrand, Johannesburg.

Jerison, H. J. (1961). Quantitative analysis of evolution of the brain in mammals. *Science*, **133**, 1012–14.

Jerison, H. J. (1963). Interpreting the evolution of the brain. *Hum. Biol.* **35**, 263–91.

Jovanovic, S. (1958). The anatomy of the frontal sinus in man (trans.). *Bull. Acad. Serb. Sci.* **23**, 145–51 (read in abstract).

Keith, A. (1916). *The Antiquity of Man*. London: Williams and Norgate (quoted by Schepers, 1946, q.v.).

Keith, A. (1931). *New Discoveries Relating to the Antiquity of Man*, pp. 1–512. London: Williams and Norgate.

Keith, A. (1948). *A New Theory of Human Evolution*, pp. 1–451. London: Watts.

Khroustov, H. F. (1964). Formation and highest frontier of the implemental activity of anthropoids. *VII Internat. Congr. Anthrop. Ethnol. Sci., Moscow*, August 1964 (pending).

Knott, J. F. (1882). On the cerebral sinuses and their variations. *J. Anat. Physiol., Lond.* **16**, 27–42.

Koenigswald, G. H. R. von (1955). *Meganthropus* and the Australopithecinae. *Proc. 3rd Pan-Afr. Congr. Prehist.*, 1955, 158–60.

Koenigswald, G. H. R. von (1957). Remarks on *Gigantopithecus* and other hominoid remains from southern China. *Koninkl. Nederl. Akademie van Wetensch.* B, **60**, 153–9.

Koenigswald, G. H. R. von (1958). L'hominisation de l'appareil masticateur et les modifications du régime alimentaire. From *Les Processus de l'Hominisation*, pp. 59–78. Paris: Centre Nat. Rech. Scient.

Koenigswald, G. H. R. von (1960). Remarks on a fossil human molar from Olduvai, East Africa. *Koninkl. Nederl. Akademie van Wetensch.* B, **63**, 20–25.

Koenigswald, G. H. R. von (1961). *Australopithecus und das Problem der Gerö1kulturen. Dt. Gesells. Anthrop.*, Tübingen, April 1961, pp. 139–52.

Köppl, L. (1947). Variabilita splavů v operačním poli zadní jámy lební. *Rozhl. Chir.* **9**, 59 (quoted by Petříková, 1963, q.v.).

Korenhof, C. A. W. (1960). *Morphogenetical Aspects of the Human Upper Molar*, pp. 1–368. Utrecht: Uitgeversmaatschappij Neerlandia.

Korkhaus, G. (1939). Gebiss-, Kiefer-, und Gesichtsorthopädie. From *Handbuch der Zahnheilkunde* (ed. C. Bruhn). Munich: J. F. Bergmann (quoted by Moorrees, 1957, q.v.).

Kraus, B. S. (1964). *The Basis of Human Evolution*, pp. 1–384. New York: Harper and Row.

Krogman, W. M. (1932). The morphological characters of the Australian skull. *J. Anat., Lond.* **66**, 399–413.

Kurth, G. (1960). Progressive Leistungsfähigkeit der humanen Phase bei Praehomininen. *Naturw. Rdsch. Stuttg.* **6**, 215–20.

Laffont, J. and Aaron, Cl. (1961). Etude comparée des différentes techniques de mesure du prognathisme. *Bull. Mém. Soc. Anthrop. Paris*, **2** (XI ser.), 382–9.

Laing, J. D. (1955). The arcadal index. *J. Dent. Ass. S. Afr.* **10**, 376–82.

Leakey, L. S. B. (1959a). A new fossil skull from Olduvai. *Nature, Lond.* **184**, 491–3.

Leakey, L. S. B. (1959b). The newly discovered skull from Olduvai: first photographs of the complete skull. *Illust. Lond. News*, 19 Sept. 1959, pp. 288–9.

Leakey, L. S. B. (1960a). From the Taungs skull to 'Nutcracker Man': Africa as the cradle of mankind and the primates—discoveries of the last thirty-five years. *Illust. Lond. News*, 9 Jan. 1960, p. 44.

Leakey, L. S. B. (1960b). The affinities of the new Olduvai australopithecine (reply to J. T. Robinson). *Nature, Lond.* **186**, 456–8.

Leakey, L. S. B. (1960c). Recent discoveries at Olduvai Gorge. *Nature, Lond.* **188**, 1050–2.

Leakey, L. S. B. (1961a). New links in the chain of human evolution: three major new discoveries from the Olduvai Gorge, Tanganyika. *Illust. Lond. News*, 4 March 1961, pp. 346–8.

Leakey, L. S. B. (1961b). New finds at Olduvai Gorge. *Nature, Lond.* **189**, 649–50.

Leakey, L. S. B. (1961c). The juvenile mandible from Olduvai. *Nature, Lond.* **191**, 417–18.

Leakey, L. S. B. (1965). *Olduvai Gorge 1951–61*, vol. I. *A preliminary report on the geology and fauna*, pp. 1–118. Cambridge University Press.

Leakey, L. S. B., Evernden, J. F. and Curtis, G. H. (1961). Age of Bed I, Olduvai Gorge, Tanganyika. *Nature, Lond.* **191**, 478–9.

Leakey, L. S. B. and Leakey, M. D. (1964). Recent discoveries of fossil hominids in Tanganyika: at Olduvai and near Lake Natron. *Nature, Lond.* **202**, 5–7.

Leakey, L. S. B., Tobias, P. V. and Napier, J. R. (1964). A new species of the genus *Homo* from Olduvai Gorge. *Nature, Lond.* **202**, 7–9.

Lindley, D. V. and Miller, J. C. P. (1953). *Cambridge Elementary Statistical Tables*, pp. 1–35. Cambridge University Press.

Logan, W. H. G. and Kronfeld, R. (1933). Development of the human jaws and surrounding structures from birth to the age of fifteen years. *J. Amer. Dent. Ass.* **20**, 379–427.

Lubosch, W. (1906). Über variationen am Tuberculum articulare des Kiefergelenks des Menschen und ihre morphologische Bedeutung. *Morph. Jb.* **35**, 322–53 (quoted by Weidenreich, 1943, q.v.).

McCown, T. and Keith, A. (1939). *The Stone Age of Mount Carmel. II. The Fossil Human Remains from the Levalloiso-Mousterian*, pp. 1–390. Oxford University Press.

Manouvrier, L. (1893). Mémoire sur les variations normales et les anomalies des os nasaux dans l'espèce humaine. *Bull. Soc. Anthrop., Paris*, **4**, 712–47 (quoted by Weidenreich, 1943, q.v.).

Marchand, F. (1902). Über das Hirngewicht des Menschen. *Abh. Math.-physik. Cl. K. Sachs. Gesellsch. Wiss.* **24**, 391–482 (quoted by Weidenreich, 1941a, q.v.).

Martin, R. (1928). *Lehrbuch der Anthropologie*, 2nd edit. Jena: Gustav Fischer.

Mason, R. J. (1961). The earliest tool-makers in South Africa. *S. Afr. J. Sci.* **57**, 13–16.

Massler, M. and Schour, I. (1946). Growth of the child and the calcification pattern of the teeth. *Am. J. Orthod. Oral Surg.* **32**, 495–517.

Matiegka, J. (1923). Sulci venosi diluviálních lebek z Předmostí. *Anthropologie*, **1**, 31–8.

Maxwell, J. H. and Waggoner, R. W. (1951). Hypertrophy of the masseter muscles. *Ann. Otol. Rhinol. Lar.* **60**, 538–48.

Mayr, E. (1963a). The taxonomic evaluation of fossil hominids. From *Classification and Human Evolution* (ed. S. L. Washburn), pp. 332–46. Chicago: Viking Fund Publications in Anthropology.

Mayr, E. (1963b). *Animal Species and Evolution*, pp. 1–797. Cambridge, Massachusetts: Harvard University Press.

Mednick, L. W. and Washburn, S. L. (1956). The role of the sutures in the growth of the braincase of the infant pig. *Am. J. Phys. Anthrop.* **14**, 175–92.

Montagu, M. F. Ashley (1960). *A Handbook of Anthropometry*, pp. 1–186. Springfield: Charles C. Thomas.

Moorrees, C. F. A. (1957). *The Aleut Dentition*, pp. 1–196. Cambridge, Massachusetts: Harvard University Press.

Moorrees, C. F. A., Fanning, E. A. and Hunt, E. E. (1963). Age variation of formation stages for ten permanent teeth. *J. Dent. Res.* **42**, 1490–502.

Moorrees, C. F. A. and Reed, R. B. (1954). Biometrics of crowding and spacing of the teeth in the mandible. *Am. J. Phys. Anthrop.* **12**, 77–88.

REFERENCES

Morant, G. M. (1927). Studies of palaeolithic man. II. A biometric study of Neanderthaloid skulls and of their relationships to modern racial types. *Ann. Eugen., Lond.* **2**, 318–81.

Morant, G. M. (1930). Studies of palaeolithic man. IV. A biometric study of the upper palaeolithic skulls of Europe and of their relationships to earlier and later types. *Ann. Eugen., Lond.* **4**, 109–214.

Moss, M. L. (1954). Growth of the calvaria in the rat. The determination of osseous morphology. *Am. J. Anat.* **94**, 333–58.

Musiker, R. (1954). *The Australopithecinae: Bibliography*, pp. 1–81. University of Cape Town School of Librarianship.

Napier, J. R. (1959). Fossil metacarpals from Swartkrans. *Fossil Mammals Afr.*, No. 17, pp. 1–18. London: British Museum (Natural History).

Napier, J. R. (1961). Human origins. *Lancet*, Sept. 30, pp. 767–8.

Napier, J. R. (1962). Fossil hand bones from Olduvai Gorge. *Nature, Lond.* **196**, 409–11.

Negus, V. (1965). *The Biology of Respiration*, pp. 1–228. Edinburgh and London: E. and S. Livingstone.

Nikitiuk, B. A. (1964). A study of the temporal and masseter muscles: influence on the skull form of the rhesus monkey (English translation). *Trans. Moscow Soc. Nat.* **14**, 54–64.

Nikitiuk, B. A. (1965). Demonstration at *VIII International Congress of Anatomists*, Wiesbaden (August 1965).

Oakley, K. P. (1954). The dating of the Australopithecinae of Africa. *Am. J. Phys. Anthrop.* **12**, 9–28.

Oakley, K. P. (1956). The earliest tool-makers and the earliest fire-makers. *Antiquity*, **30**, 4–8, 102–7.

Oakley, K. P. (1961). Dating Man's emergence. Presidential address to Section H, British Ass. for the Adv. of Sci., Norwich, 30 Aug.–6 Sept. 1961.

Oakley, K. P. (1964). *Frameworks for Dating Fossil Man*, pp. 1–355. London: Weidenfeld and Nicolson.

Oetteking, B. (1930). Craniology of the North Pacific Coast. *Mem. Amer. Mus. Nat. Hist.* **15**, 1–391.

Olivier, G. (1960). *Pratique Anthropologique*, pp. 1–299. Paris: Vigot Frères.

Oppenheimer, A. (1964). Tool use and crowded teeth in Australopithecinae. *Curr. Anthrop.* **5**, 419–20.

Pedersen, P. O. (1949). The East Greenland Eskimo dentition. *Meddr Grønland*, **142**, 1–256.

Petríková, E. (1963). Confluens sinuum, jeho utváření a variabilita. *Acta Univ. Carol. Medica*, No. 7, 619–37.

Petrovits, L. (1930). Die Übereinstimmung des Kiefergelenkes des neugeborenen Kindes mit dem Kiefergelenk der Anthropoiden. *Anat. Anz.* **69**, 136–44 (quoted by Ashton and Zuckerman, 1954, q.v.).

Pfisster, H. (1903). Die Kapazität des Schädels beim Säugling und älteren Kinde. *Mschr. Psychiat. Neurol.* (quoted by Zuckerman, 1928, q.v.).

Pittard, E. and Kaufmann, H. (1936). Architecture du pariétal chez les crânes des Boschimans. *XVIᵉ Congr. Int. d'Anthrop.*, Bruxelles, 1935, pp. 1–13.

Pycraft, W. P. (1928). Description of the human remains. From *Rhodesian Man and Associated Remains*, pp. 1–51. London: British Museum (Natural History).

Radoievitch, S. and Jovanovic, S. (1959). La pneumatisation du cornet moyen. *Rev. Laryngol.* Mar.–Apr. 1959, parts 3–4, pp. 210–32.

Randall, F. E. (1943–4). The skeletal and dental development and variability of the gorilla. *Hum. Biol.* **15**, 236–54, 307–37; **16**, 23–76.

Raven, H. C. (1950). *The Anatomy of the Gorilla. The Henry Cushier Raven Memorial Volume*, pp. 1–259. New York: Columbia University Press.

Remane, A. (1930). Zur Messtechnik der Primatenzähne. From *Handbuch der biologischen Arbeitsmethoden* (E. Aberhalden), vol. 7 (1), 609–35 (quoted by Moorrees, 1957, q.v.).

Remane, A. (1954). Methodische Probleme der Hominiden-Phylogenie, II. *Z. Morph. Anthrop.* **46**, 225–68.

Remane, A. (1960). Zähne und Gebiss. *Primatologia*, **3** (2), 637–846. Basle: S. Karger.

Remane, A. (1961). Probleme der Systematik der Primaten. *Z. wiss. Zool.* **165**, 1–34 (quoted by Schultz, 1963, q.v.).

Riesenfeld, A. (1955). The variability of the temporal lines, its cause and effects. *Am. J. Phys. Anthrop.* **13**, 599–620.

Robinson, J. T. (1952*a*). The australopithecines and their evolutionary significance. *Proc. Linn. Soc. Lond.* **3**, 196–200.

Robinson, J. T. (1952*b*). Some hominid features of the ape-man dentition. *J. Dent. Ass. S. Afr.* **7**, 102–13.

Robinson, J. T. (1953*a*). *Meganthropus*, Australopithecus and hominids. *Am. J. Phys. Anthrop.* **11**, 1–38.

Robinson, J. T. (1953*b*). *Telanthropus* and its phylogenetic significance. *Am. J. Phys. Anthrop.* **11**, 445–501.

Robinson, J. T. (1954*a*). The genera and species of the Australopithecinae. *Am. J. Phys. Anthrop.* **12**, 181–200.

Robinson, J. T. (1954*b*). The australopithecine occiput. *Nature, Lond.* **174**, 262–3.

Robinson, J. T. (1954*c*). Nuchal crests in australopithecines. *Nature, Lond.* **174**, 1197–8.

Robinson, J. T. (1954*d*). Prehominid dentition and hominid evolution. *Evolution*, **8**, 324–34.

Robinson, J. T. (1954*e*). Phyletic lines in the prehominids. *Z. Morph. Anthrop.* **46**, 269–73.

Robinson, J. T. (1955). Further remarks on the relationship between '*Meganthropus*' and australopithecines. *Am. J. Phys. Anthrop.* **13**, 429–46.

Robinson, J. T. (1956). The dentition of the Australopithecinae. *Transv. Mus. Mem.*, No. 9.

Robinson, J. T. (1958). Cranial cresting patterns and their significance in the Hominoidea. *Am. J. Phys. Anthrop.* **16**, 397–428.

Robinson, J. T. (1960). The affinities of the new Olduvai australopithecine. *Nature, Lond.* **186**, 456–8.

Robinson, J. T. (1961). The australopithecines and their bearing on the origin of man and of stone tool-making. *S. Afr. J. Sci.* **57**, 3–13.

Robinson, J. T. (1962). The origin and adaptive radiation of the australopithecines. From: *Evolution und Hominisation* (ed. G. Kurth), pp. 120–40. Stuttgart: Gustav Fischer.

Robinson, J. T. (1963). Adaptive radiation in the australopithecines and the origin of man. From: *African Ecology and Human Evolution* (ed. F. C. Howell and F. Bourlière), pp. 385–416.

Robinson, J. T. (1965*a*). Homo '*habilis*' and the australopithecines. *Nature, Lond.* **205**, 121–4.
Robinson, J. T. (1965*b*). Comment on 'New discoveries in Tanganyika: their bearing on hominid evolution' (by P. V. Tobias). *Curr. Anthrop.* **6**, 403–6.
Robinson, J. T. and Mason, R. J. (1957). Occurrence of stone artefacts with *Australopithecus* at Sterkfontein. *Nature, Lond.* **180**, 521–4.
Robinson, J. T. and Mason, R. J. (1962). Australopithecines and artefacts at Sterkfontein. *S. Afr. Archaeol. Bull.* **17**, 87–125.
Sauter, M. R. (1941–6). A propos de l'architecture de l'occipital. Comparaisons raciales entre les Boschimans-Hottentots-Griquas et les Suisses brachycéphales. *Anthrop. Paris*, **50**, 469–90.
Schaik, C. van (1958). An improved palatometer. *S. Afr. J. Med. Sci.* **23**, 155–6.
Schepers, G. W. H. (1946). *See* Broom and Schepers, 1946.
Schepers, G. W. H. (1950). *See* Broom, Robinson and Schepers, 1950.
Schultz, A. H. (1936). Characters common to higher primates and characters specific for man. *Qt. Rev. Biol.* **11**, 259–83, 425–55.
Schultz, A. H. (1940). Growth and development of the chimpanzee. *Contr. Embryol. Carneg. Inst.* **28**, 1–63.
Schultz, A. H. (1941*a*). The relative size of the cranial capacity in Primates. *Am. J. Phys. Anthrop.* **28**, 273–87.
Schultz, A. H. (1941*b*). Growth and development of the orang-utan. *Contrib. Embryol. Carneg. Inst.* **29**, 59–110.
Schultz, A. H. (1947). Variability in man and other primates. *Am. J. Phys. Anthrop.* **5** (n.s.), 1–14.
Schultz, A. H. (1950*a*). The physical distinctions of man. *Proc. Amer. Phil. Soc.* **94**, 428–49.
Schultz, A. H. (1950*b*). Morphological observations on gorillas. From *The Henry Cushier Raven Memorial Volume: The Anatomy of the Gorilla*, pp. 227–51. New York: Columbia University Press.
Schultz, A. H. (1957). Past and present views of man's specialisations. *Irish J. Med. Sci.* 6th ser. (380), 341–56.
Schultz, A. H. (1958). Acrocephalo-oligodactylism in a wild chimpanzee. *J. Anat., Lond.* **92**, 568–79.
Schultz, A. H. (1962). Die Schädelkapazität männlicher Gorillas und ihr Höchstwert. *Anthrop. Anz.* **25**, 197–203.
Schultz, A. H. (1963). Age changes, sex differences, and variability as factors in the classification of Primates. From *Classification and Human Evolution* (ed. S. L. Washburn), pp. 85–115. Chicago: Viking Fund Publications in Anthropology.
Schwalbe, G. (1902). Über die Beziehungen zwischen Innenform und Aussenform des Schädels. *Dt. Arch. Klin. Med.* **73**, 359–408 (quoted by Weidenreich, 1943, q.v.).
Scott, J. H. (1954). The growth and function of the muscles of mastication in relation to the development of the facial skeleton and of the dentition. *Am. J. Orthod.* **40**, 429–49.
Scott, J. H. (1957). Muscle growth and function in relation to skeletal morphology. *Am. J. Phys. Anthrop.* **15**, 197–234.

Scott, J. H. (1963). Factors determining skull form in primates. *Symp. Zool. Soc., Lond.* No. 10, pp. 127–34.
Sergi, S. (1947). Sulla morfologia della 'facies anterior corporis maxillae' nei paleantropi di Saccopastore e del monte Circeo. *Riv. Antrop.* **35**, 401–6.
Sergi, S. (1959). *Zinjanthropus boisei* (Leakey) o *Paranthropus boisei*? *Riv. Antrop.* **46**, 254–61.
Sergi, S. (1960*a*). Saggio radiografico di caratteristiche morfologiche del cranio neandertaliano Circeo I. *Acad. Nat. dei Lincei.*, Ser. 8, **28**, 594–9.
Sergi, S. (1960*b*). Röntgenografische darstellung morphologischer Merkmale am Neandertaler Schädel Circeo I. *Anthrop. Anz.* **24**, 160–7.
Shaw, J. C. Middleton (1931). *The Teeth, the Bony Palate and the Mandible in Bantu Races of South Africa*, pp. 1–134. London: John Bale, Sons and Danielsson, Ltd.
Shear, M. (1954). Hereditary hypocalcification of enamel: a report of three cases occurring in one family. *J. Dent. Ass. S. Afr.* **9**, 262–9.
Sicher, H. (1949). *Oral Anatomy*, pp. 1–529. St Louis: C. V. Mosby Co.
Simonetta, A. (1957). Catalogo e sinonimia annotata degli ominoidi fossili ed attuali (1758–1955). *Atti Soc. Tosc. Sci. Nat.*, B, **64**, 53–112.
Simons, E. L. (1961). The phyletic position of *Ramapithecus*. *Postilla*, **57**, 1–9.
Simons, E. L. (1963). Some fallacies in the study of hominid phylogeny. *Science*, **141**, 879–89.
Simons, E. L. (1964*a*). The early relatives of man. *Scientific Am.* **211**, 51–62.
Simons, E. L. (1964*b*). On the mandible of *Ramapithecus*. *Proc. Nat. Acad. Sci.* **51**, 528–35.
Simons, E. L. and Pilbeam, D. R. (1965). Preliminary revision of the *Dryopithecinae* (Pongidae, Anthropoidea). *Folia primat.* **3**, 81–152.
Simpson, G. G. (1940). Types in modern taxonomy. *Am. J. Sci.* **238**, 413–31.
Simpson, G. G. (1961). *Principles of Animal Taxonomy*, pp. 1–247. New York: Columbia University Press.
Simpson, G. G. (1963). The meaning of taxonomic statements. From *Classification and Human Evolution* (ed. S. L. Washburn), pp. 1–31. Chicago: Viking Fund Publications in Anthropology.
Simpson, G. G., Roe, A. and Lewontin, R. C. (1960). *Quantitative Zoology* (revised ed.), pp. 1–440. New York: Harcourt, Brace and Co.
Sollas, W. J. (1926). A sagittal section of the skull of *Australopithecus africanus*. *Q. J. Geol. Soc.* **82**, 1–11.
Sonntag, C. F. (1924). *The Morphology and Evolution of the Apes and Man*, pp. 1–364. London: John Bale, Sons and Danielsson, Ltd.
Stewart, T. D. (1933). The tympanic plate and external auditory meatus in the Eskimo. *Am. J. Phys. Anthrop.* **17**, 481–96.
Stewart, T. D. (1964). A neglected primitive feature of the Swanscombe skull. From *The Swanscombe Skull: a survey of Research on a Pleistocene site* (ed. C. D. Ovey). *Roy. Anthrop. Inst. Occas. Pap.* No. 20, pp. 151–9.
Stoll, N. R. et al. (1961). *International Code of Zoological Nomenclature*, pp. 1–176. London: International Commission on Zoological Nomenclature.

REFERENCES

Straus, W. L. (1953). Contribution to discussion on 'Physical anthropology and the biological basis of human behaviour'. From *An Appraisal of Anthropology Today* (ed. S. Tax), p. 262. Chicago: University Press.

Streeter, G. L. (1915). The development of the venous sinuses of the dura mater in the human embryo. *Amer. J. Anat.* **18**, 145–78.

Streeter, G. L. (1918). The developmental alterations in the vascular system of the brain of the human embryo. *Contrib. Embryol. Carneg. Inst.* **8**, 5–38.

Symons, N. B. B. (1954). The attachment of the muscles of mastication. *Brit. Dent. J.* **96**, 76–81.

Tildesley, M. L. (1921). A first study of the Burmese skull. *Biometrika*, **13**, 176–260.

Tildesley, M. L. (1950). The relative usefulness of various characters on the living for racial comparison. *Man*, **51**, 14–17.

Tobias, P. V. (1958). Studies on the occipital bone in Africa. III. Sex differences and age changes in the occipital curvature and their bearing on the morphogenesis of differences between Bushmen and Negroes. *S. Afr. J. Med. Sci.* **23**, 135–46.

Tobias, P. V. (1959a). Studies on the occipital bone in Africa. I. Pearson's Occipital Index and the Chord-Arc Index in modern African crania: means, minimum values, and variability. *J. Roy. Anthrop. Inst.* **89**, 233–52.

Tobias, P. V. (1959b). Studies on the occipital bone in Africa. II. Resemblances and differences of occipital patterns among modern Africans. *Z. Morph. Anthrop.* **50**, 9–19.

Tobias, P. V. (1959c). Studies on the occipital bone in Africa. IV. Components and correlations of occipital curvature in relation to cranial growth. *Hum. Biol.* **31**, 138–61.

Tobias, P. V. (1959d). Studies on the occipital bone in Africa. V. The occipital curvature in fossil man and the light it throws on the morphogenesis of the Bushman. *Am. J. Phys. Anthrop.* **17**, 1–12.

Tobias, P. V. (1960a). Report on Section II (Human Palaeontology) of 4th Pan-African Congress on Prehistory, Leopoldville. *S. Afr. Archaeol. Bull.* **15**, 6–8.

Tobias, P. V. (1960b). Studies on the occipital bone in Africa. VI. The relative usefulness of Pearson's Occipital Index and the Occipital Chord-Arc Index. *Man*, **60**, 23–5.

Tobias, P. V. (1960c). *Embryos, Fossils, Genes and Anatomy*, pp. 1–25. Johannesburg: Witwatersrand University Press.

Tobias, P. V. (1961). The work of the Gorilla Research Unit in Uganda. *S. Afr. J. Sci.* **57**, 297–8.

Tobias, P. V. (1962). Early members of the genus *Homo* in Africa. From *Evolution und Hominisation* (ed. G. Kurth), pp. 191–204. Stuttgart: Gustav Fischer Verlag.

Tobias, P. V. (1963). Cranial capacity of *Zinjanthropus* and other australopithecines. *Nature, Lond.* **197**, 743–6.

Tobias, P. V. (1964a). The Olduvai Bed I hominine with special reference to its cranial capacity. *Nature, Lond.* **202**, 3–4.

Tobias, P. V. (1964b). The early hominid remains from Tanganyika: *Australopithecus* and *Homo*. *VII Internat. Congr. Anthrop. Ethnol. Sci., Moscow*, August 1964 (pending).

Tobias, P. V. (1965a). *Homo habilis*: last missing link in hominine phylogeny? From: *Festschrift on 65th birthday of Professor Juan Comas* (ed. S. Genoves). Mexico City.

Tobias, P. V. (1965b). New discoveries in Tanganyika: their bearing on hominid evolution. *Curr. Anthrop.* **6**, 391–9.

Tobias, P. V. (1965c). Early man in East Africa. *Science*, **149**, 22–33.

Tobias, P. V. (1965d). Reply to comments of J. T. Robinson and others. *Curr. Anthrop.* **6**, 406–11.

Tobias, P. V. (1965e). Comment on 'Cranial capacity of the hominine from Olduvai Bed I' (by R. L. Holloway). *Nature, Lond.* **208**, 206.

Tobias, P. V. (1965f). *Australopithecus, Homo habilis*, tool-using and tool-making. *S. Afr. Archaeol. Bull.* **20**, 167–192.

Tobias, P. V. and von Koenigswald, G. H. R. (1964). Comparison between the Olduvai hominines and those of Java and some implications for hominid phylogeny. *Nature, Lond.* **204**, 515–18.

Todd, T. W. (1933). Growth and development of the skeleton. From *Growth and Development of the Child*. New York and London: Century Co.

Trevor, J. C. (1950). Anthropometry. From *Chambers' Encyclopaedia*, pp. 458–62. London: George Newnes.

Turner, W. (1884). Report on the human crania and other bones of the skeleton. *Challenger Reports*, **10**, 6.

Twiesselmann, F. (1941). Méthodes pour l'évaluation de l'épaisseur des parois crâniennes. *Bull. Mus. Roy. Hist. Nat. Belg.* **17**, 1–33.

Vallois, H. V. (1954). La capacité crânienne chez les Primates supérieurs et le 'Rubicon cerebral'. *C.R. Acad. Sci. Paris*, **238**, 1349–51.

Vallois, H. V. (1958). La grotte de Fontéchevade: Anthropologie. *Archs. Inst. Paléont. hum.* **29** (2), 5–164.

Van Reenen, J. F. (1961). The use of the concept of tooth material as an indication of tooth size in a group of Kalahari Bushmen. *S. Afr. J. Sci.* **57**, 347–52.

Vogel, C. (1963). Die *spina nasalis anterior* und vergleichbare Strukturen bei rezenten Simiern. *Zool. Anz.* **171**, 273–90.

Wagner, K. (1935). Endocranial diameters and indices. *Biometrika*, **27**, 88–132.

Wagner, K. (1937). *The Craniology of the Oceanic Races*, pp. 1–193. Univ. Anat. Inst. (Anthrop. Section), Oslo.

Waltner, J. G. (1944). Anatomic variations of the lateral and sigmoid sinuses. *Archs Otolar.* **39**, 307–12.

Washburn, S. L. (1942). Technique in primatology. *Anthr. Briefs*, No. 1, pp. 6–12.

Washburn, S. L. (1947). The relation of the temporal muscle to the form of the skull. *Anat. Rec.* **99**, 239–48.

Washburn, S. L. (1959). Speculations on the interrelations of the history of tools and biological evolution. *Hum. Biol.* **31**, 21–31.

Washburn, S. L. (ed.) (1963*a*). *Classification and Human Evolution*, pp. 1–371. Chicago: Viking Fund Publications in Anthropology.

Washburn, S. L. (1963*b*). Comment in *African Ecology and Human Evolution* (ed. F. C. Howell and F. Bourlière), p. 565. Chicago: Aldine Publishing Co.

Washburn, S. L. and De Vore, I. (1961). Social behavior of baboons and early man. From: *Social Life of Early Man* (ed. S. L. Washburn), pp. 91–118. Chicago: Aldine Publishing Co.

Washburn, S. L. and Howell, F. C. (1952). On the identification of the hypophyseal fossa of Solo Man. *Am. J. Phys. Anthrop.* **10**, 13–21.

Washburn, S. L. and Howell, F. C. (1960). Human evolution and culture. From: *Evolution after Darwin*: Vol. II, *The Evolution of Man* (ed. S. Tax), pp. 33–56. Chicago: University Press.

Washburn, S. L. and Patterson, B. (1951). Evolutionary importance of the South African 'man-apes'. *Nature, Lond.* **167**, 650–1.

Wegner, R. N. (1956). Studien über Nebenhöhlen des Schädels. *Wiss. Z. Ernst Moritz Arndt-Univ. Greifswald,* **5**, 1–55.

Weidenreich, F. (1924). Über die pneumatischen Nebenräume des Kopfes. Ein Beitrag zur Kenntnis des Bauprinzips des Knochens, des Schädels und des Körpers. *Z. Anat. Entw.Gesch.* **72**, 55–93.

Weidenreich, F. (1932). Über pithekoide Merkmale bei *Sinanthropus pekinensis* und seine stammesgeschichtliche Beurteilung. *Z. Anat. Entw. Gesch.* **99**, 212–53.

Weidenreich, F. (1936*a*). Observations on the form and proportions of the endocranial casts of *Sinanthropus pekinensis*, other hominids and the great apes: a comparative study of brain size. *Palaeont. sin.* D, **7**, 1–50.

Weidenreich, F. (1936*b*). Über das phylogenetische Wachstum des Hominidengehirns. *Kaibogaku Zasshi,* **9**, 1–14.

Weidenreich, F. (1937). The dentition of *Sinanthropus pekinensis*: a comparative odontography of the hominids. *Palaeont. sin.* **10**, 1–180 and 1–121.

Weidenreich, F. (1941*a*). The brain and its role in the phylogenetic transformation of the human skull. *Trans. Amer. Philos. Soc.* **31** (Part V), 321–442.

Weidenreich, F. (1941*b*). The extremity bones of *Sinanthropus pekinensis*. *Palaeont. sin.* **116**, D, 5, 1–150.

Weidenreich, F. (1943). The skull of *Sinanthropus pekinensis*. *Palaeont. sin.* **127**, 1–486.

Weidenreich, F. (1945). Giant early man from Java and South China. *Anthrop. Pap. Am. Mus. Nat. Hist.* **40**, 1–134.

Weidenreich, F. (1951). Morphology of Solo Man. *Anthrop. Pap. Am. Mus. Nat. Hist.* **43**, 205–90.

Weiner, J. S. and Campbell, B. G. (1964). The taxonomic status of the Swanscombe skull. From *The Swanscombe Skull* (ed. C. D. Ovey), pp. 175–209. London: Roy. Anthrop. Inst.

Weinert, H. (1928). *Pithecanthropus erectus*. *Z. Anat. Entw. Gesch.* **87**, 429–547.

Woo, J. K. (1962). The mandibles and dentition of *Gigantopithecus*. *Palaeont. sin.* D, **11**, No. 146, 1–94.

Woodhall, B. (1936). Variations of the cranial venous sinuses in the region of the torcular Herophili. *Archiv. Surg.* **33**, 297–314.

Woodhall, B. (1939). Anatomy of the cranial blood sinuses with particular reference to the lateral. *Laryngoscope,* **49**, 966–1010.

Woodhall, B. and Seeds, A. E. (1936). Cranial venous sinuses: correlation between skull markings and roentgenograms of the occipital bone. *Archiv. Surg.* **33**, 867–75.

Wood Jones, F. (1931*a*). The non-metrical morphological characters of Hawaiian skulls. *J. Anat.* **65**, 368–78.

Wood Jones, F. (1931*b*). The non-metrical morphological characters of the skulls of prehistoric inhabitants of Guam. *J. Anat.* **65**, 438–45.

Zuckerman, S. (1928). Age-changes in the chimpanzee, with special reference to growth of brain, eruption of teeth, and estimation of age; with a note on the Taungs ape. *Proc. Zool. Soc. Lond.* **1**, 1–42.

Zuckerman, S. (1954). Correlation of change in the evolution of higher primates. From *Evolution as a Process* (ed. J. S. Huxley, A. C. Hardy and E. B. Ford), pp. 301–352. London: George Allen and Unwin.

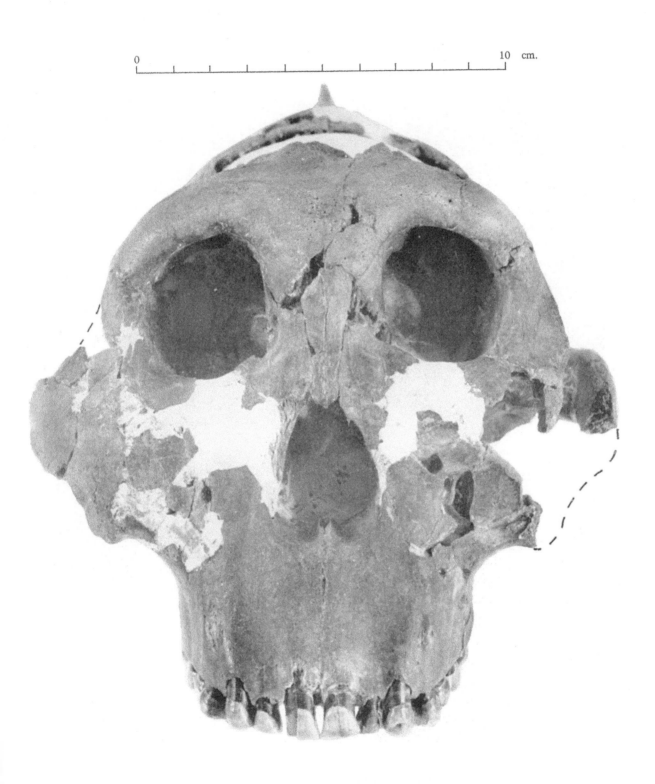

1. Norma facialis of the cranium of *Zinjanthropus*.

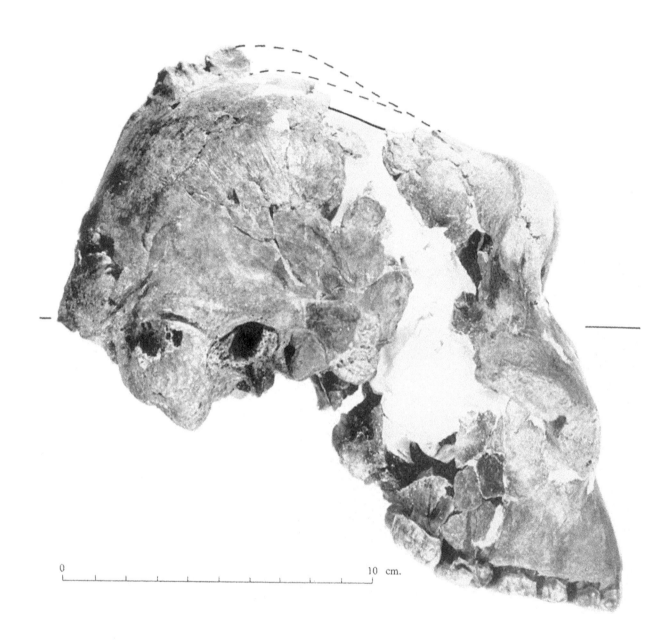

2. Right norma lateralis of the cranium of *Zinjanthropus*.

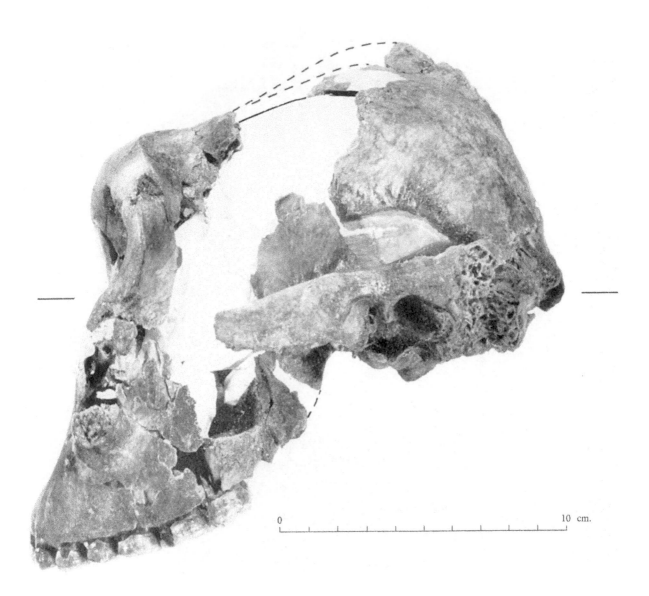

3. Left norma lateralis of the cranium of *Zinjanthropus*.

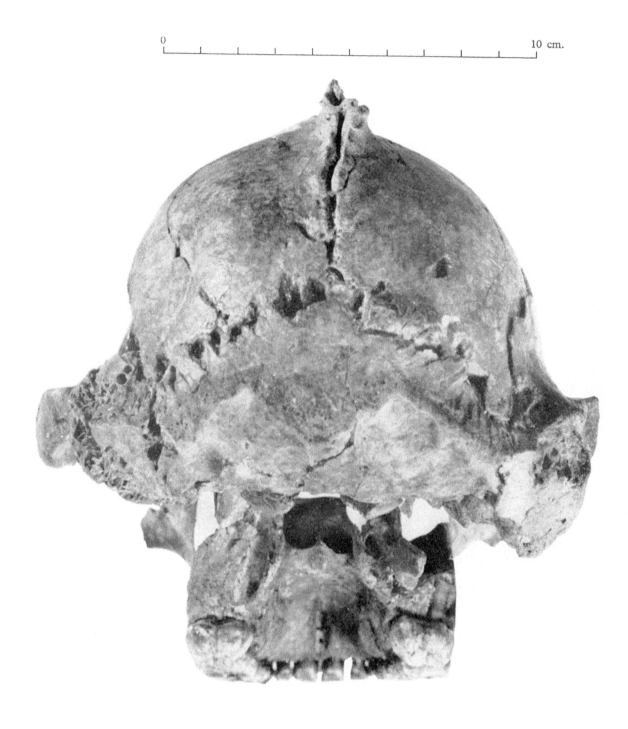

4. Norma occipitalis of the cranium of *Zinjanthropus*.

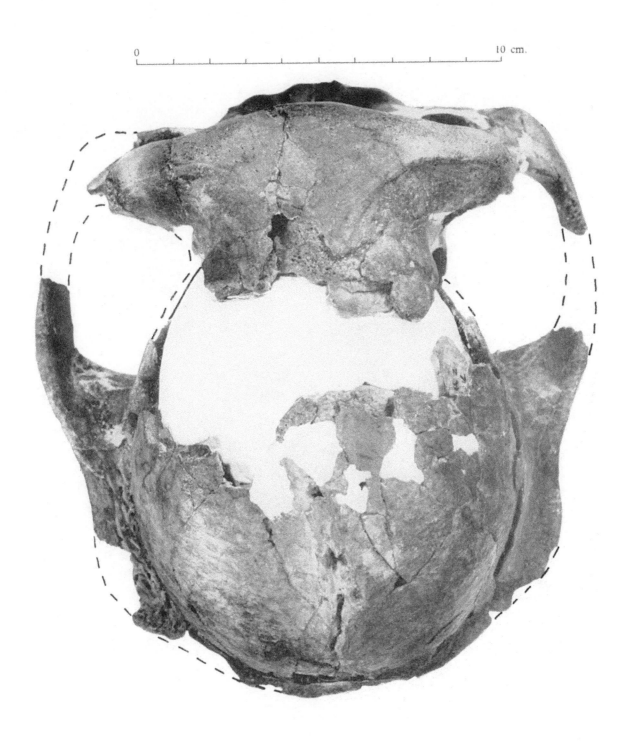

5. Norma verticalis of the cranium of *Zinjanthropus* (in Frankfurt Horizontal).

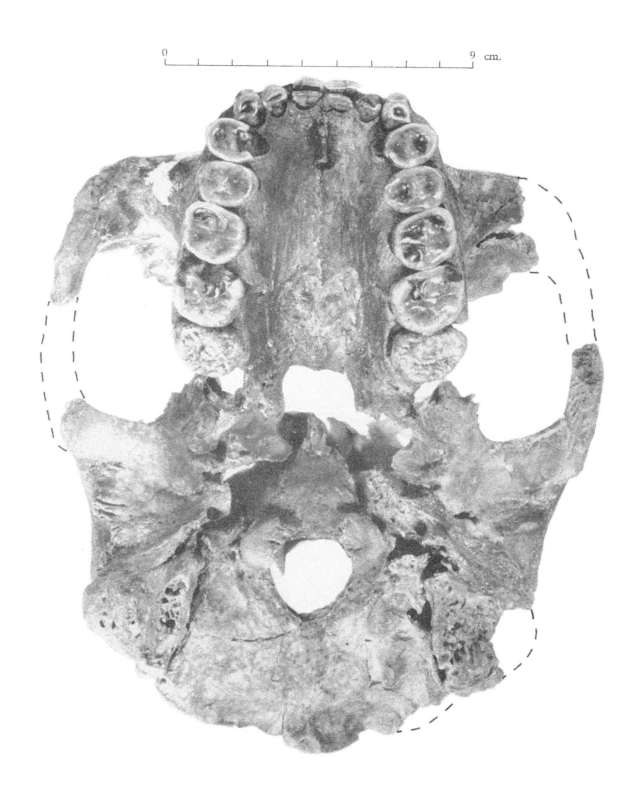

6. Norma basalis of the cranium of *Zinjanthropus* (cranium *not* in Frankfurt Horizontal).

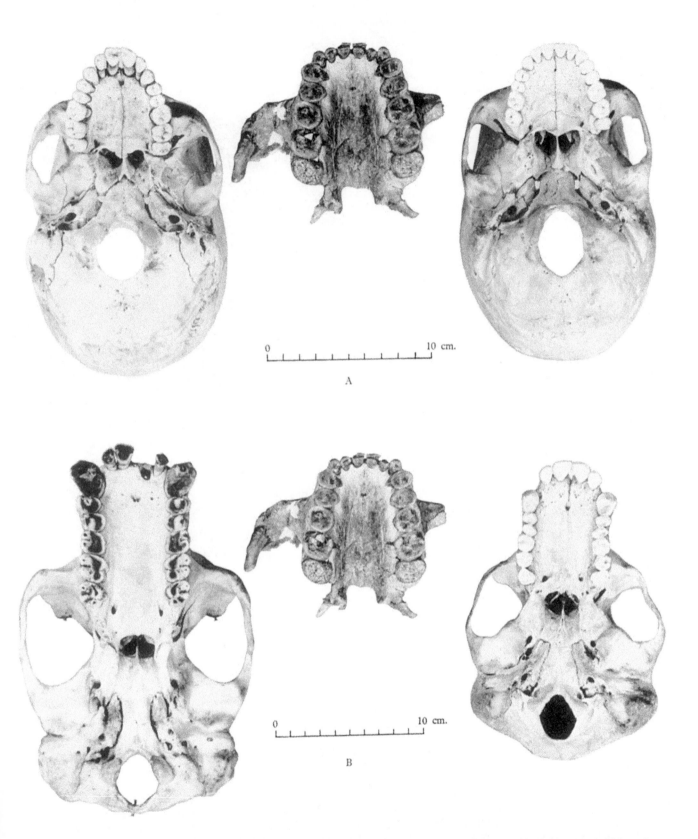

7. The dental arcade and palate of *Zinjanthropus* compared with those of (A) two modern men (Bantu Negroids) and (B) a gorilla (left) and chimpanzee (right).

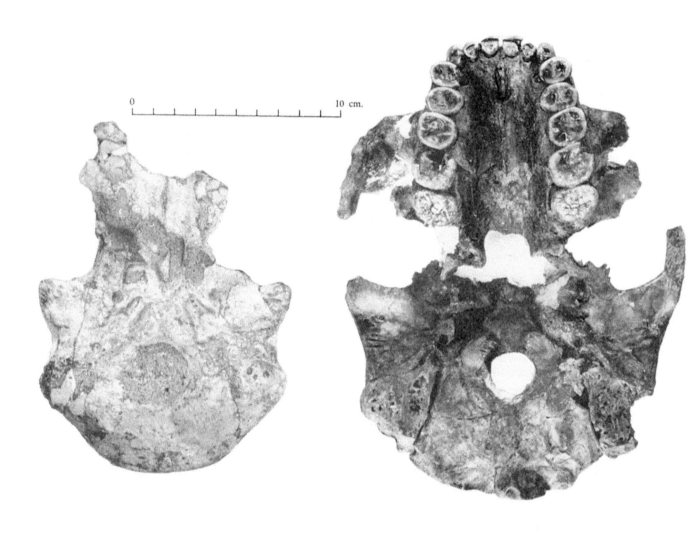

8. Norma basalis of the cranium of *Zinjanthropus* (right) compared with that of the recently discovered cranium of the Makapansgat *Australopithecus*, MLD 37/38 (left). Despite the great differences in dental and palatal size, in pneumatisation, and in muscular markings and crests, there is a fundamental similarity of pattern between the two crania.

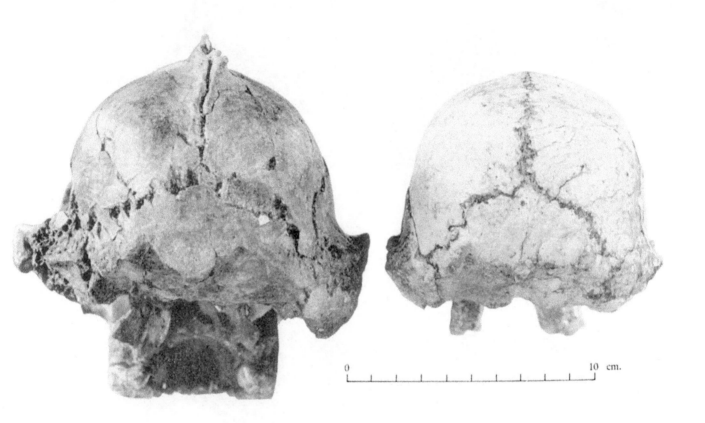

9. Norma occipitalis of the cranium of *Zinjanthropus* (left) compared with that of the recently discovered cranium of the Makapansgat *Australopithecus*, MLD 37/38 (right). Shorn of its crests and considerable pneumatisation, the *Zinjanthropus* cranium bears a strong resemblance to that of MLD 37/38.

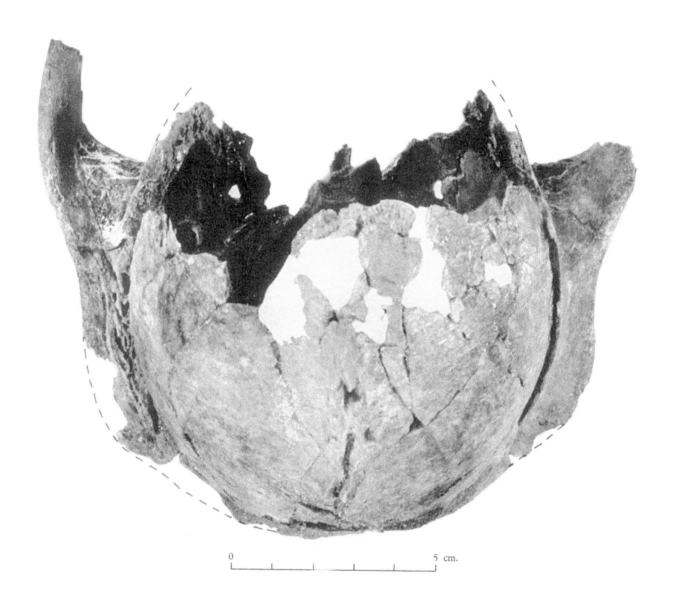

10. The posterior part of the vault of the cranium as seen from above. Note the globular, almost spheroid shape of the vault tapering towards the postorbital constriction; and the supramastoid and zygomatic 'appendages' of the cranial vault.

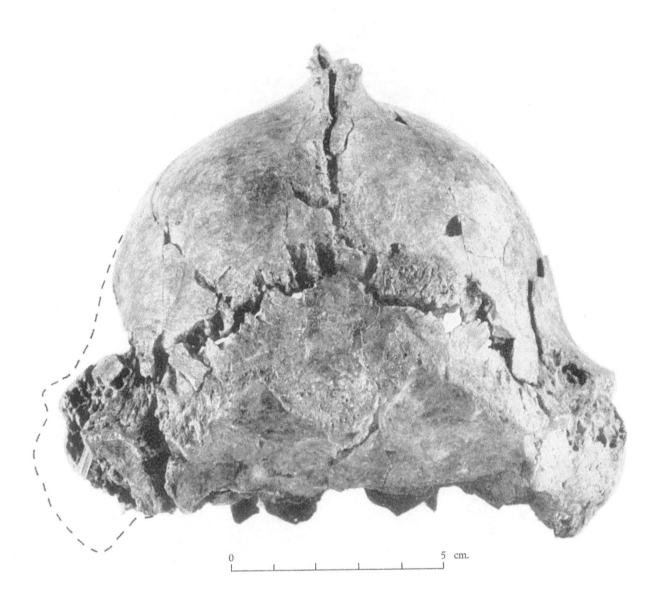

11. The posterior part of the vault of the cranium as seen from the rear. It comprises the articulated parietals, temporals and occipital bone. Note the full rounded curvature of the posterior part of the vault and the great lateral expansion of the pars mastoidea.

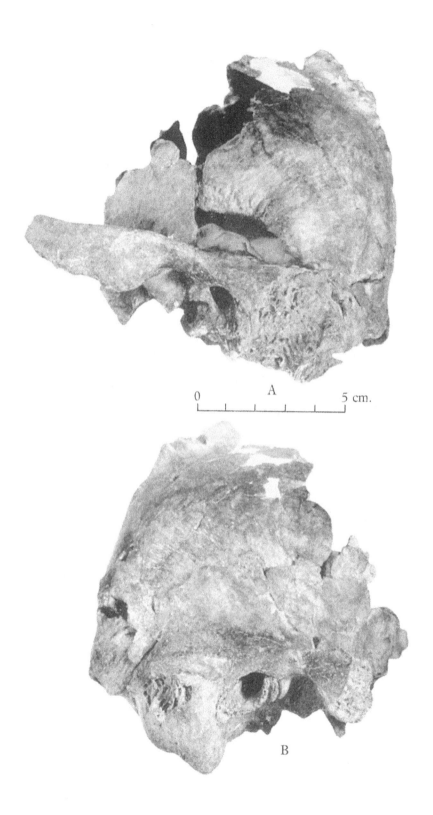

12. The posterior part of the vault of the cranium as seen from (A) the left and (B) the right. The left zygoma is intact, as are the right mastoid process and tympanic plate. Although more than half of the left temporal squama is missing, most of the right is present. Note the relationship between the postglenoid process and the doubled anterior portion of the tympanic plate (on the right).

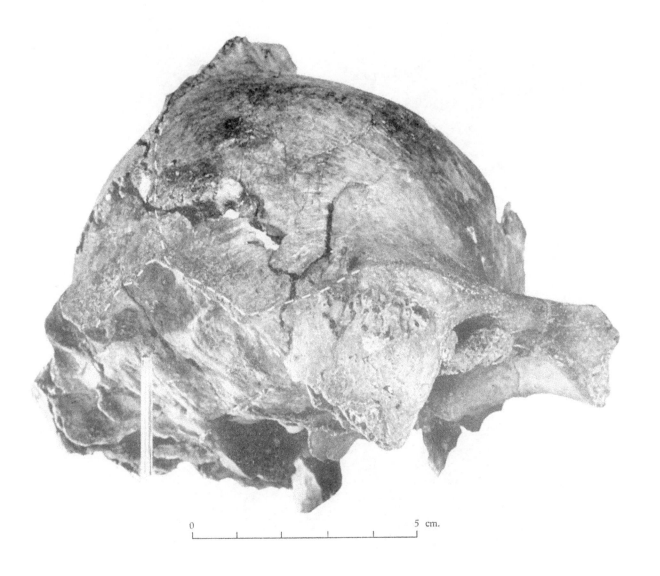

13. The posterior part of the vault of the cranium, seen from below, behind and to the right, to show the formation of the compound (temporal/nuchal) crest and the lateral divergence of the simple temporal and nuchal crests from it.

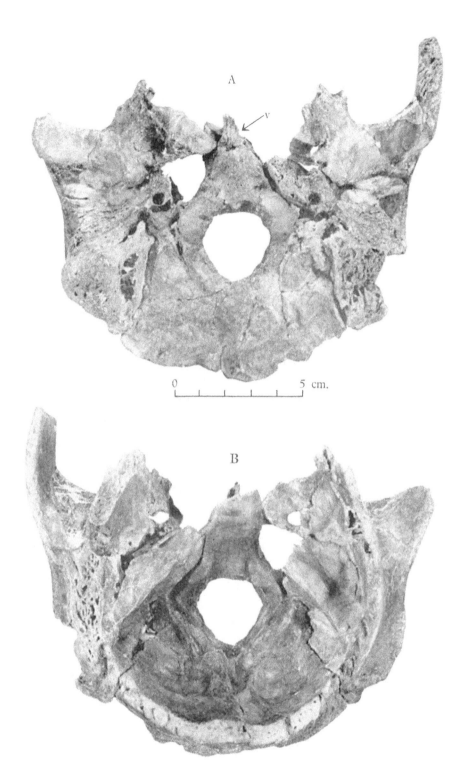

14. The posterior part of the basis cranii externa (A) and interna (B). (A) Note the forward position of the occipital condyles, in relation to the external acoustic meatus. A small part of the vomer (V) is articulated with the sphenoid rostrum, while parts of the alisphenoid may be seen articulating with the squamae of the temporal bones on either side of the body of the sphenoid. The left temporal bone is damaged in three positions: the anterior root of the zygoma, the tympanic plate and the pars mastoidea. (B) The features of the posterior cranial fossa are well shown, including the cerebral and cerebellar fossae and the grooves for the occipital and marginal sinuses. The apex of the right petrous process is missing. The dorsum sellae and the posterior portion of the sella turcica are shown, as well as a slit-like opening in the floor of the hypophyseal fossa.

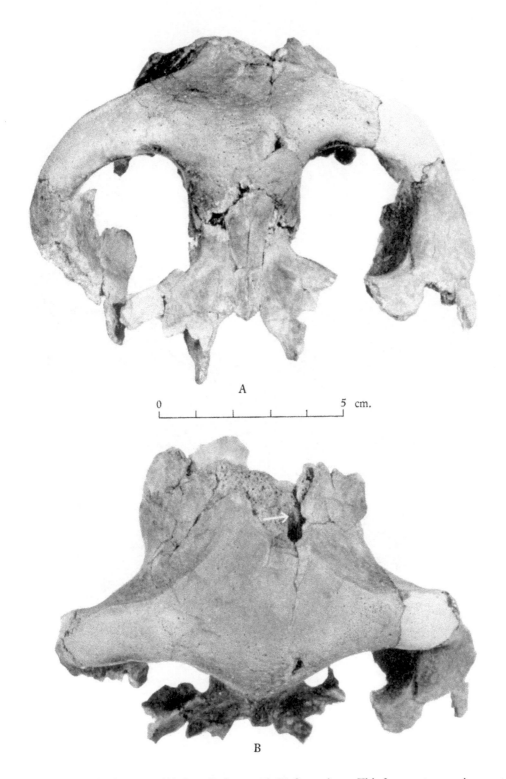

15. The upper calvariofacial fragment (A) from in front and (B) from above. This fragment comprises parts of the frontal, zygomatic and maxillary bones, and the nasal bones. Note the expanded upper part of the nasal bones and their narrow, tapering lower extremity. The twisted nature of the supra-orbital torus, and the powerfully-etched temporal crests converging backwards towards the sagittal crest, are well shown. The arrow on the lower picture marks the highest (most posterior) position reached by the frontal sinus on the left.

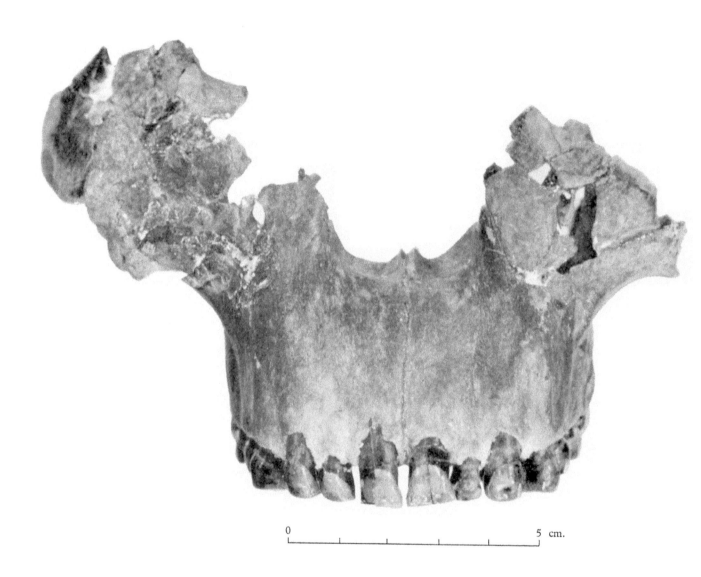

16. Anterior view of the lower facial fragment, comprising most of the maxillae as well as the right zygomatic bone. The anterior nasal spine and the shallow prenasal fossa are shown, as well as the great depth of the naso-alveolar clivus. Note especially the mode of attachment of the root of the zygomatic process of the maxilla to the body of the bone. The absence of a canine fossa is to be noted.

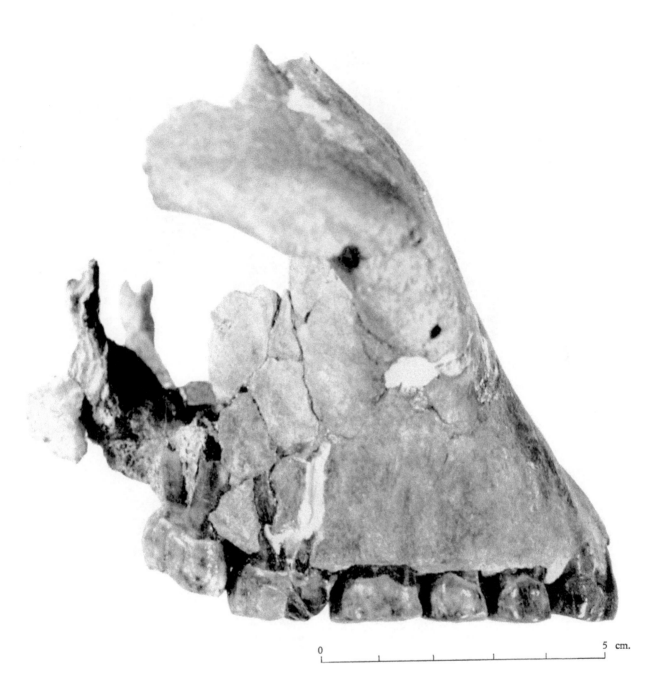

17. The right maxilla seen from the right side. Note the plane of occlusion and attrition of the teeth. The third molar has not yet descended into a position of occlusion; its uncalcified root-tips are visible. The pterygoid process is represented behind the damaged maxillary tuberosity.

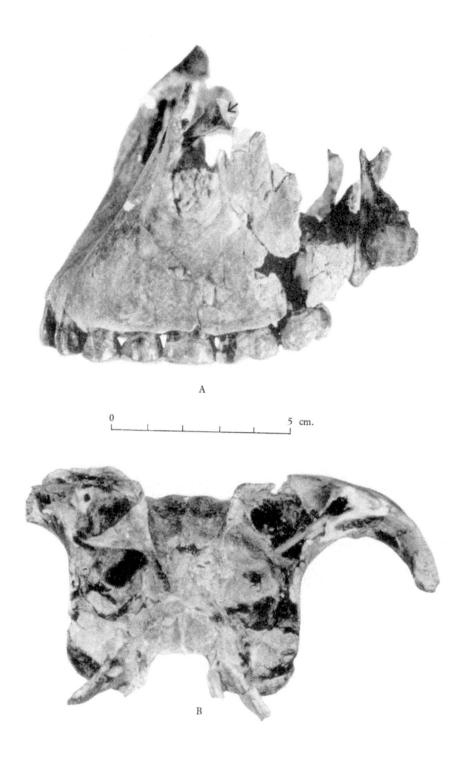

18. The maxilla of *Zinjanthropus*. (A) As seen from the left side. Through a break in the infra-orbital part of the bone, the bony canal by which the infra-orbital nerve traverses the maxillary antrum can be seen (arrow). The abutment of the maxillary tuberosity against the pterygoid process is well shown. (B) As seen from behind, looking into the exposed antra. Note the upper and lower compartments of the antrum separated by a slightly oblique shelf. On the left the upper opening of the bony canal for the infra-orbital nerve is seen near the top of the antrum.

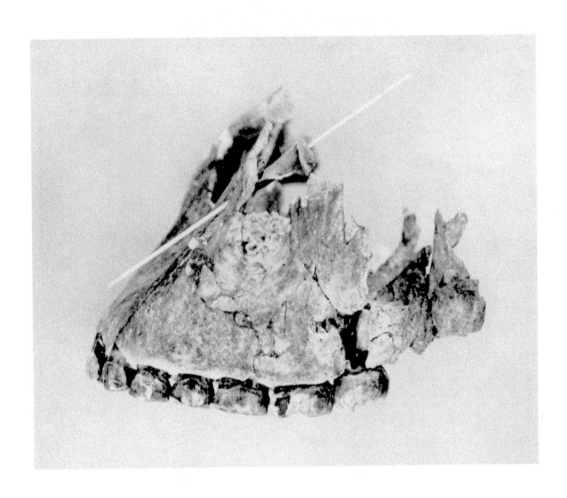

19. Left lateral view of the lower facial fragment: a straw has been passed through the bony canal for the inferior orbital nerve, traversing the maxillary antrum.

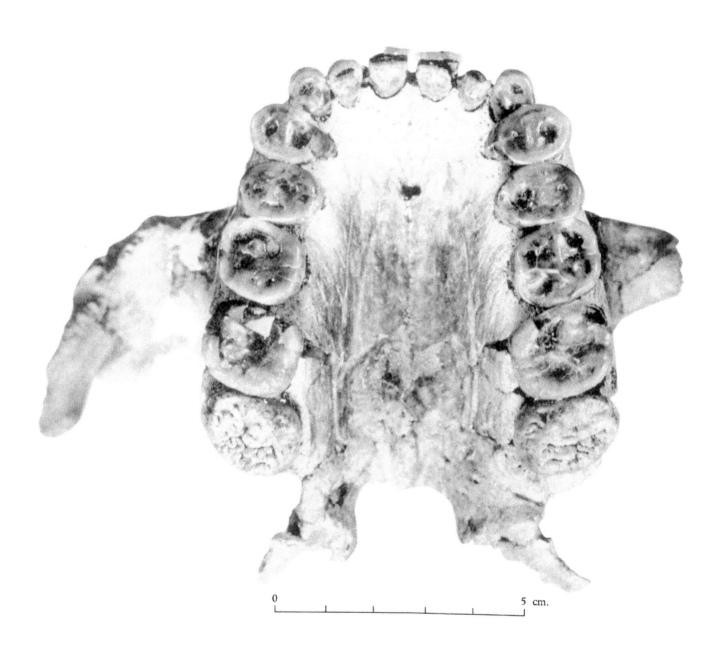

20. Occlusal view of teeth and palate of *Zinjanthropus*. Note the incisive canal, the grooves for the greater palatine nerves and vessels, the pterygoid process, especially the robust lateral pterygoid plate.

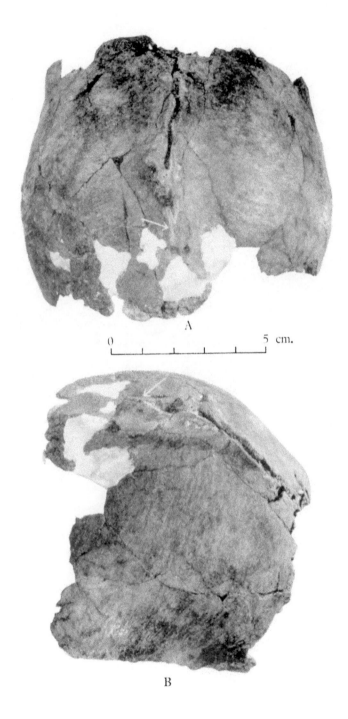

21. The parietal bones of *Zinjanthropus*, (A) from above and somewhat behind; (B) from above and to the left. The double structure of the sagittal crest, and the asymmetry of its components, are apparent. An arrow indicates the point at which the two temporal crests are most intimately fused.

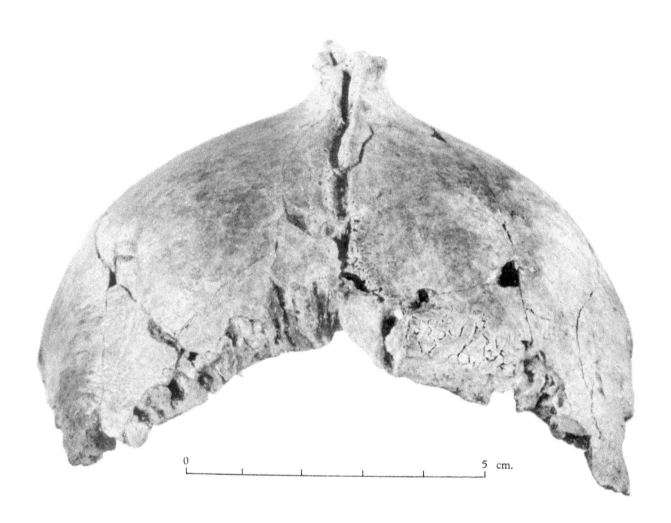

22. The parietal bones seen from below and behind. The two temporal crests comprising the sagittal crest are clearly shown; posteriorly, they diverge near lambda. A small part of the occipital squama is still articulated with the posterior margin of the right parietal bone near lambda.

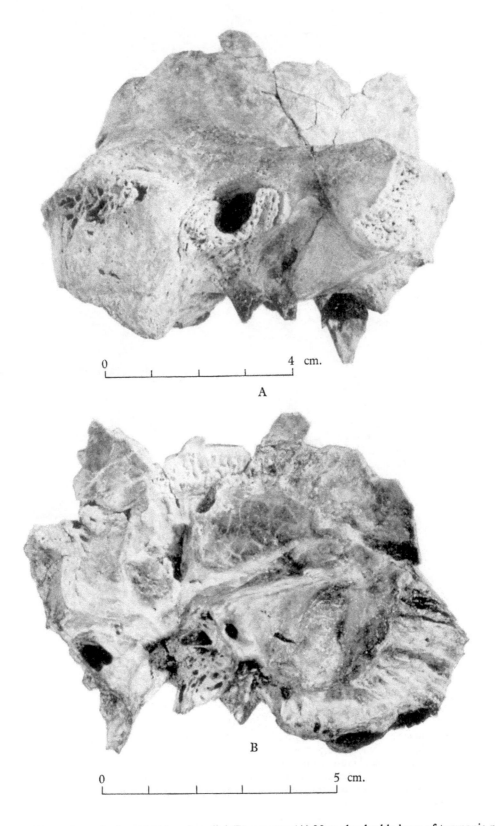

23. The right temporal bone from the lateral (A) and medial (B) aspects. (A) Note the double layer of tympanic plate immediately behind the postglenoid process. The mastoid process is clearly shown. (B) Note the grooves for the middle meningeal artery and its branches and that for the sigmoid sinus. The petrous pyramid is broken off just in front of the internal acoustic meatus. The great thickness of the squamo-occipital suture near asterion is clearly portrayed.

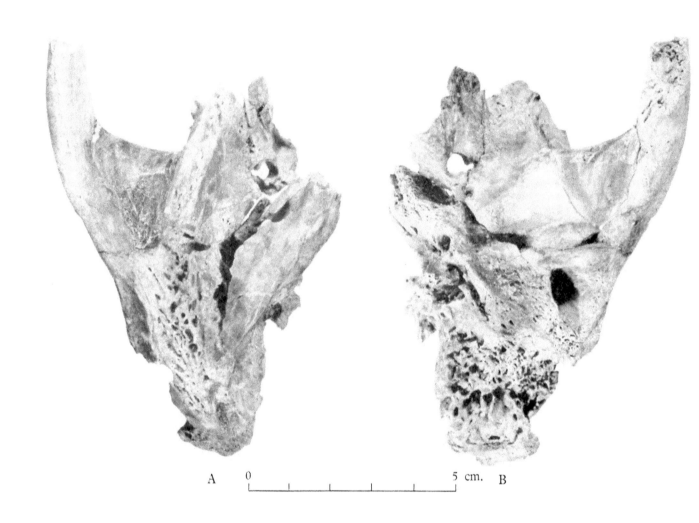

24. The left temporal bone seen (A) from above and (B) from below. Note that part of the anterior root of the zygoma has been reconstructed. (A) Only the anterior part of the squama is present; posterior to it, the pneumatisation of the base of the squama and of the supramastoid crest is apparent. (B) Note the intact sutural surface of the zygoma. Part of the tympanic plate is missing from the lateral margin. The relationship between foramen ovale, foramen spinosum and the squamo-sphenoid suture is clearly shown. Note that the tympanic plate and the petrous process are virtually in the same straight line, as in the Homininae.

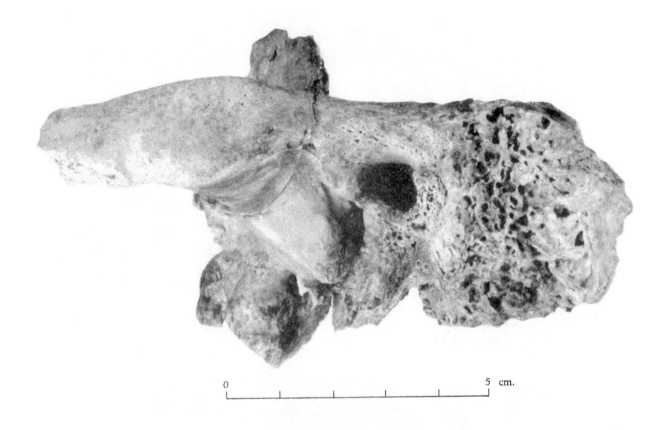

25. Lateral view of the left temporal bone. Note the honeycomb of mastoid air cells exposed by damage to this area. The entoglenoid process is clearly visible just in front of the tympanic plate, part of the margin of which is missing.

26. The occipital bone seen from below and behind. Note the strong nuchal crest with the powerful external occipital protuberance. The temporal crests are indicated by dotted lines, as they converge towards the nuchal crest, to constitute a short length of compound (temporal/nuchal) crest. The 'bare area' between lambda and inion, flanked by the temporal crests, is clearly seen.

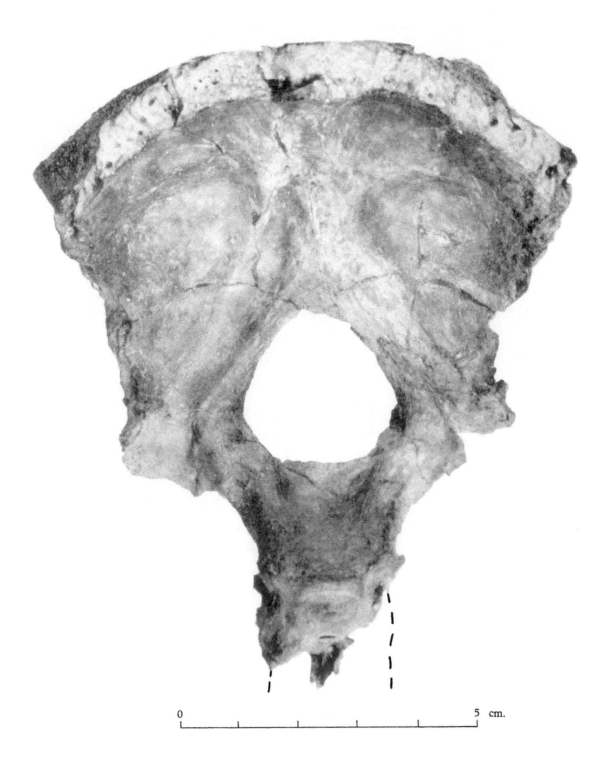

27. The occipital bone from within. Clearly shown are parts of the cerebral and cerebellar fossae; the sulci for the occipital and marginal sinuses leading forwards to the jugular notch; the clivus and part of the body of the sphenoid, including the dorsum sellae and the major part of the hypophyseal fossa.

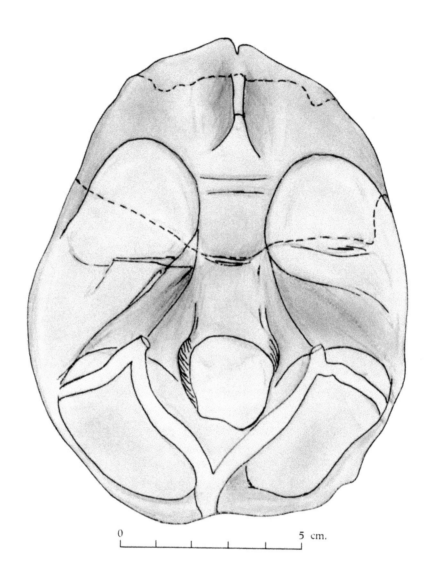

28. The endocranial cast of *Zinjanthropus* as seen in norma ventralis (basalis). Natural size.

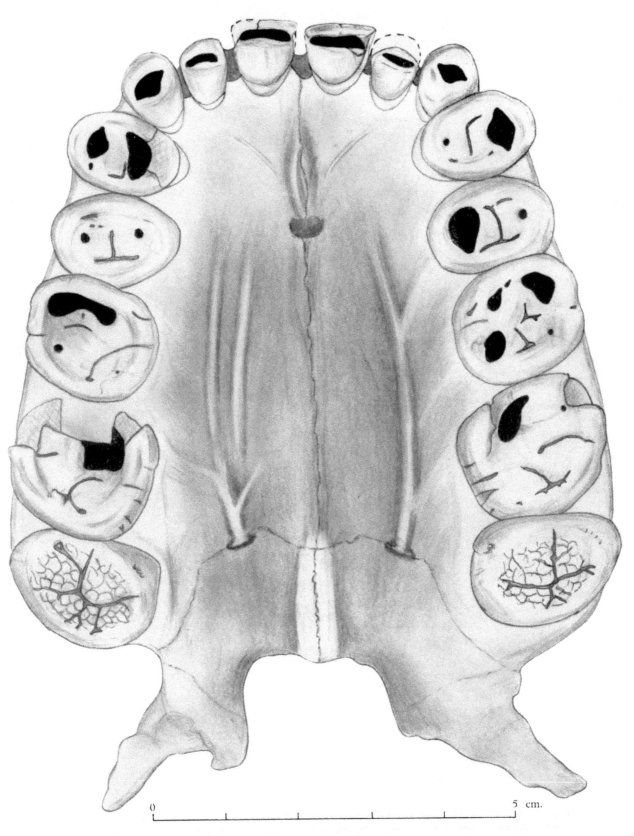

29. The dental arcade and palate of *Zinjanthropus*. The dark areas are areas of dentine exposure. Twice natural size.

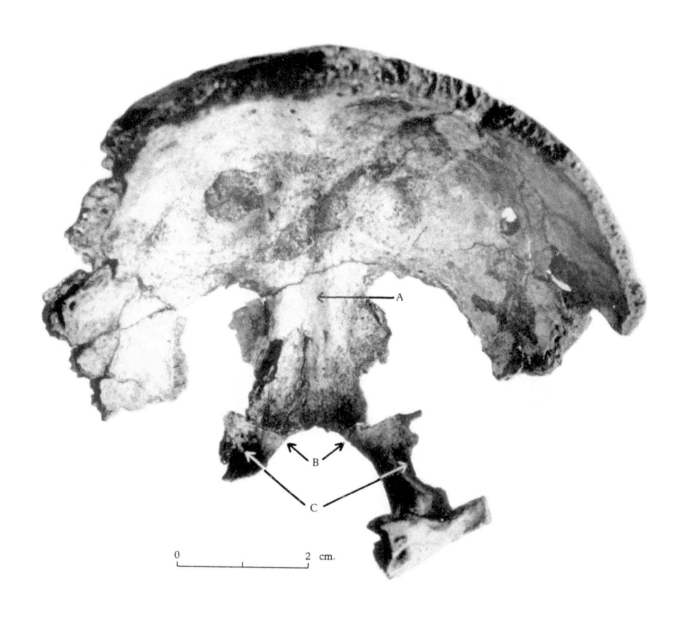

30. The interior of the occipital bone of a juvenile *Paranthropus* from Swartkrans (SK 859), showing clear markings for enlarged occipital (A) and marginal (C) sinuses. (B) the unfused joints between the lateral occipitals and the squamous occipital.

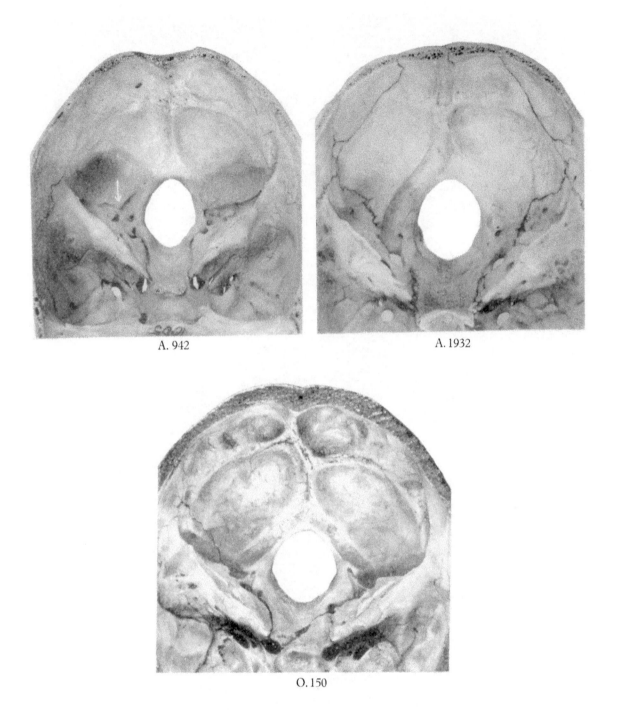

31. The interior of the occipital bones of three Bantu crania with anomalous patterns of the venous sinuses. The arrow on cranium A. 942 indicates the junction of the diminutive right sigmoid sinus with the enlarged right marginal sinus.

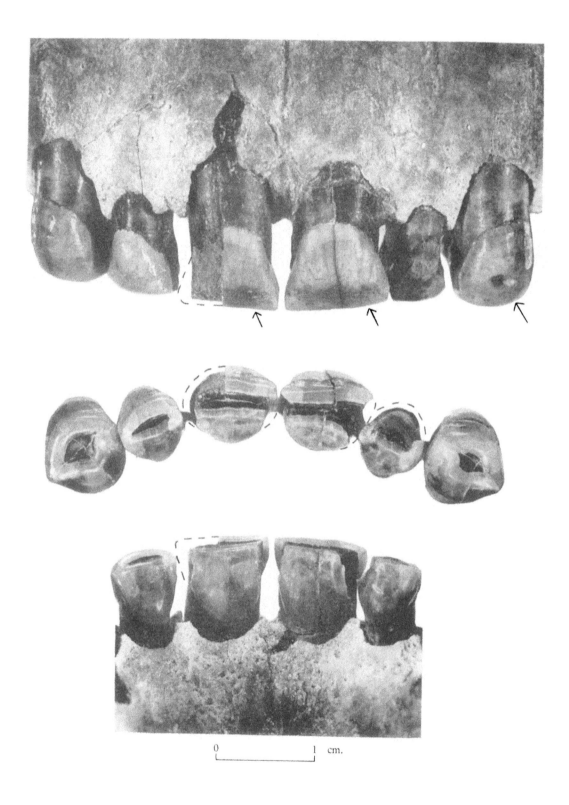

32. Labial, occlusal and lingual views of the front maxillary teeth. Note the distinct hypoplastic area on the labial surface of the I¹ near the occlusal margin, and the hypoplastic pit on the left canine. On the occlusal surface, the black areas represent exposed dentine. Some enamel is missing from three of the four incisors.

33. The tooth-row from lateral incisors to first molars. The marked contrast in tooth-size and position between the anterior teeth and the cheek-teeth is clear. Note the tendency of the P^3's towards buccal displacement and of the P^4's towards lingual displacement.

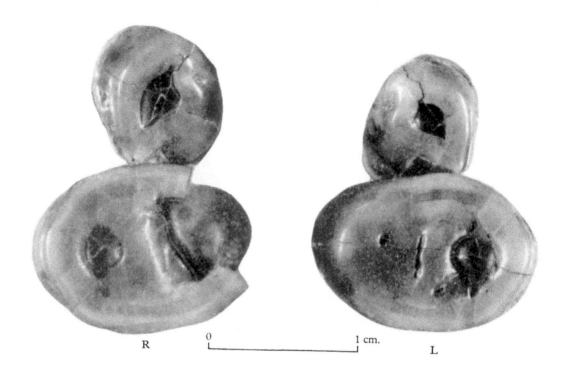

34. The maxillary canines and the first premolars of *Zinjanthropus*. Note the relative sizes and positions of the teeth, the canines being somewhat rotated distally. The fissure between the buccal and lingual cusps of the premolar is well preserved on the left. Note the uneven wear on the left and right premolars.

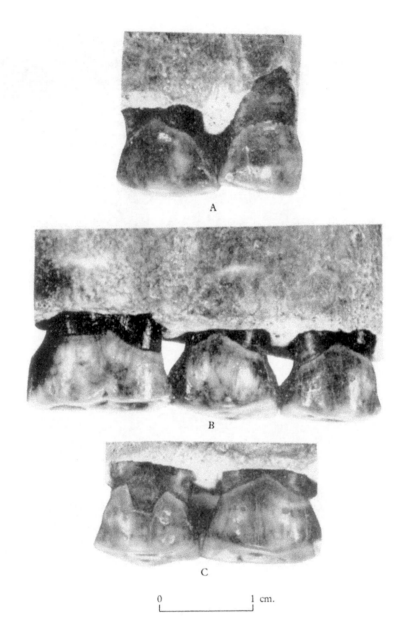

35. (A) Labial aspect of the right canine and P^3, to show the obliquely intersecting occlusal planes. On the canine, note areas of hypoplasia at three distinct levels. (B) Labial aspect of the right premolars and M^1. Hypoplastic areas are visible on the P^3. The cervical line is especially clear. (C) Labial aspect of the left P^3 and P^4. Hypoplastic pits are clearly seen on both teeth. The position of the cervical line is evident, despite the loss of enamel on P^3.

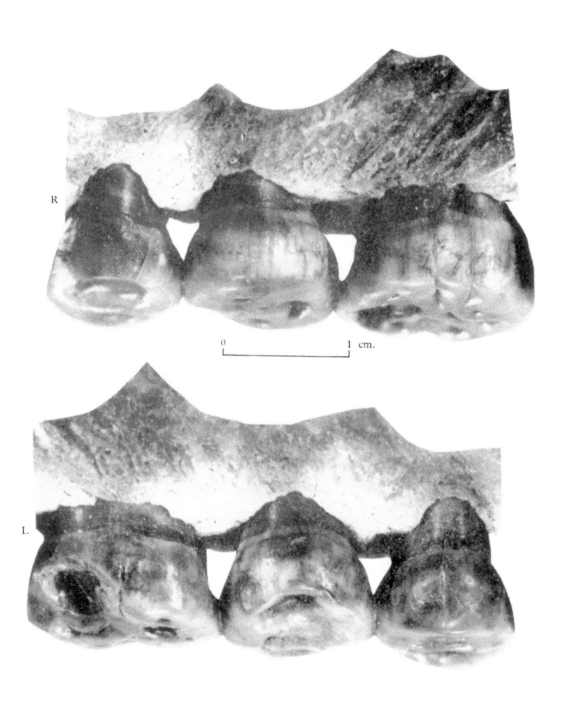

36. The lingual aspect of the premolars and the first molar. Note the deeply indented wear damage on the lingual halves of the left P^4 and M^1, unmatched by comparable attrition on the right. Areas of hypoplasia are well shown on the lingual surface of the left P^3.

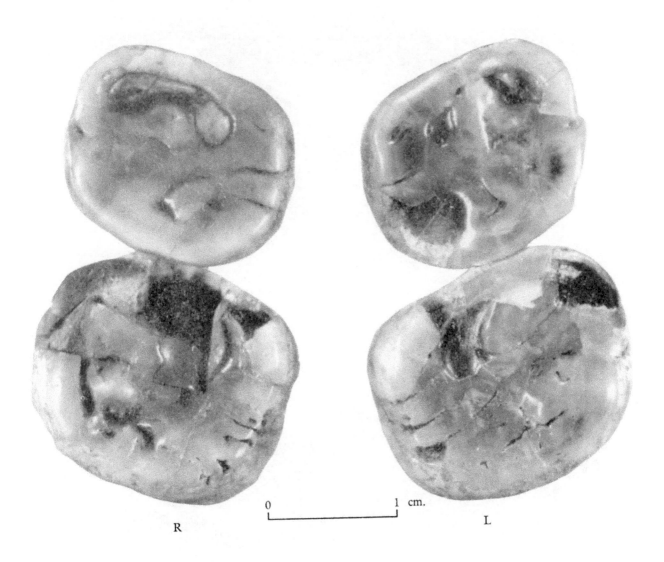

37. Occlusal surfaces of the first and second molars. Note the areas of dentine exposure. Both second molars are damaged in their mesial halves; nevertheless, they are the largest teeth in the maxillary dentition of *Zinjanthropus*. The posterior fovea and the oblique fissure demarcating the hypocone are particularly well preserved on the first molars as well as on the left second molar. The protocone–metacone contact is best seen in the left first molar.

38. The lingual aspect of the molar teeth (M² and M³ on the right; M¹, M² and M³ on the left). The complex Carabelli formations, especially on the third molars, are clearly demonstrated.

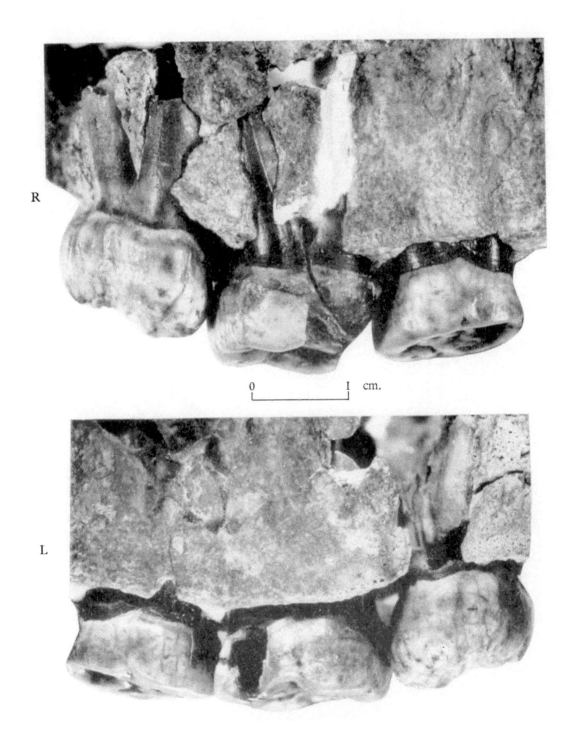

39. Buccal views of the maxillary molars of *Zinjanthropus*. Note the notched attrition, especially on the left second molar. The incompletely calcified roots of the right M³ are revealed because of damage to the alveolar ridge in that area.

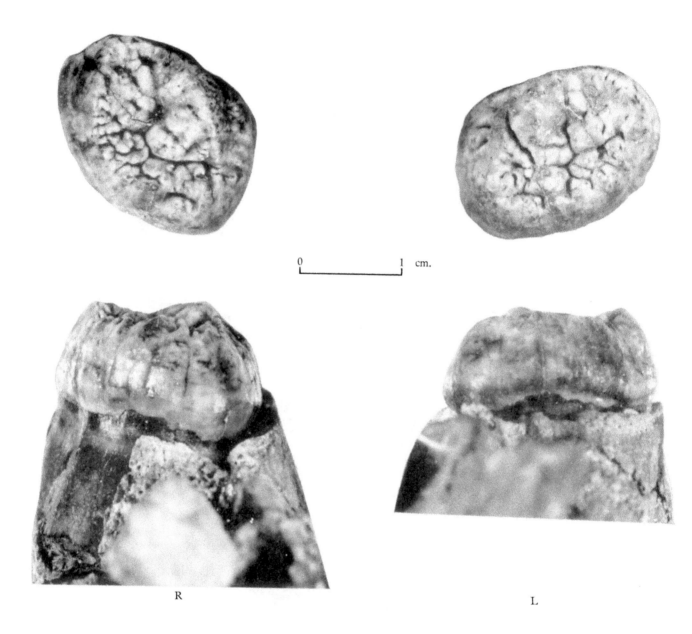

40. Maxillary third molars of *Zinjanthropus*, in occlusal and distal views. The blurred structures in the foreground of the distal views are the pterygoid processes articulating with the maxillary tuberosity.

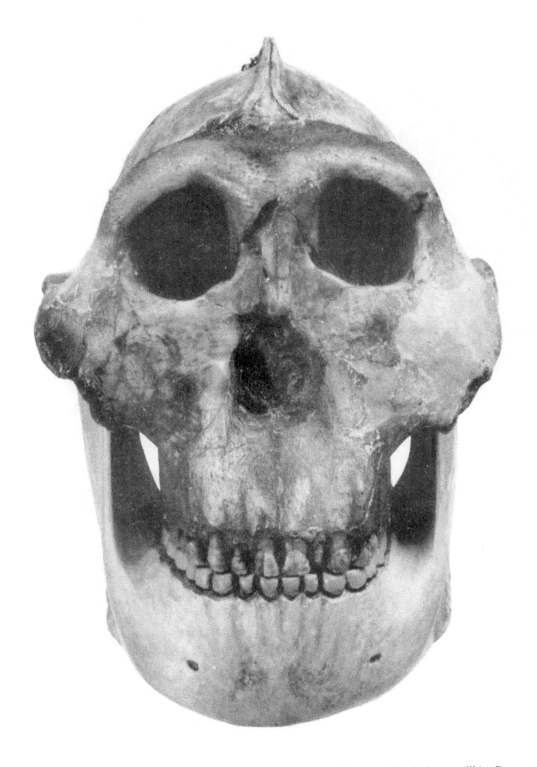

41. Norma facialis of reconstructed skull of *Australopithecus boisei* with a model of the mandible. Reconstruction and model by R. J. Clarke under the author's supervision; photograph by R. Campbell and A. R. Hughes.

Some mandibular dimensions are (mm.): bicondylar width (Martin 65) 149·5; intercondylar width 87·5; intercoronoid breadth (M. 65 (1)) 111; bigonial or angular breadth (M. 66) 126; bimental or anterior mandibular breadth (M.67) 58·4; symphyseal height (M. 69) 55·4; right corpus height at mental foramen (M. 69 (1)) 50·4; right corpus height at M_2 (M. 69 (2)) 44·7; vertical distance of centre of right mental foramen from alveolar margin 26·2 and from basal margin 26·2 (*see also* legend to pl. 42).

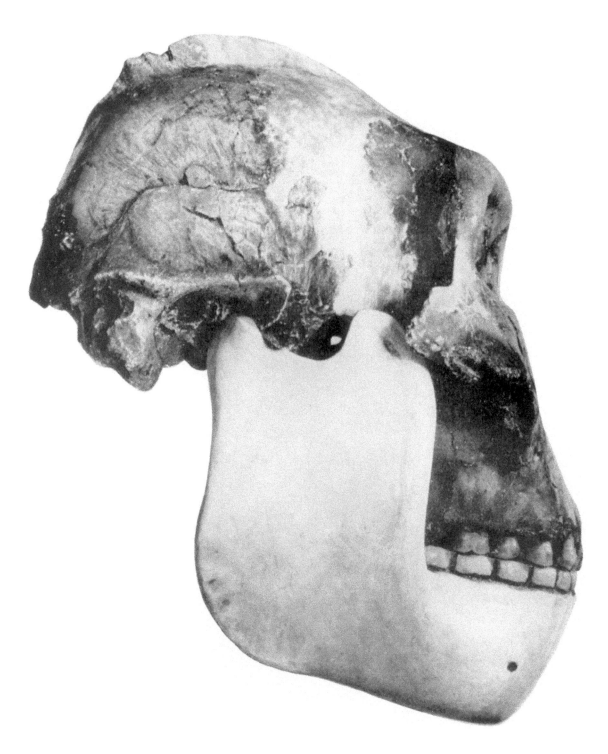

42. Norma lateralis dextralis of reconstructed skull of *A. boisei*. Reconstruction and model of mandible made by R. J. Clarke under the author's supervision; photograph by R. Campbell and A. R. Hughes.

Further mandibular measurements are (mm.): right corpus height at $\bar{C}/P_3$ 53·5, at P_4 48·8, at P_4/M_1 48·4, at M_1 46·4, at M_1/M_2 46·5, at M_2 44·7, at M_2/M_3 44·9; ramus height (M. 70) 106·2; coronoid height (M. 70 (1)) 120; minimum rameal height (M. 70 (2)) 101·9; depth of incisura (M. 70 (3)) 16; minimum rameal breadth (M. 71) 74·5; smallest rameal breadth (M. 71 (*a*)) 74·0; breadth of incisura (M. 71 (1)) 41·5; right premolar–molar chord (P_3–M_3) 75·0.

INDEX OF PERSONS

Aaron, C., 115, 248
Aichel, O., 188, 245
Akabori, E., 32, 245
Anthony, R., 75
Arambourg, C., xvi, 4, 236, 245
Ariëns Kappers, C. U., 91, 245
Ashton, E. H., 16, 18, 23, 36–7, 40, 43–4, 49–51, 77–8, 80–3, 85, 151–2, 161–2, 168–9, 171, 173, 195, 197–8, 204, 222, 245

Baikie, G. G., 142
Bartholomew, G. A., 240, 245
Baume, L. J., 122, 245
Becks, H., 122, 245
Birdsell, J. B., 240, 245
Black, Davidson, 31, 33, 95, 99–100, 130, 245
Black, G. V., 173, 186, 245
Blanc, A. C., 219
Boise, C., 1
Bolk, L., 80, 89, 245
Boné, E. L., 4, 245
Bonin, G. von, 78, 91–3, 245
Bonné, B., 190, 247
Bork-Feltkamp, A. J. van, 222, 245
Borovansky, L., 67
Boule, M., 82, 245
Brain, C. K., 66, 236, 245
Breathnach, A. S., 66, 245
Bronowski, J., 222, 245
Broom, R., 4, 26–7, 32–3, 41–3, 49, 55, 58, 60, 78–9, 84, 92, 95, 98, 102, 109–10, 112, 128–9, 142, 177–8, 231, 235, 245
Brothwell, D., xvi
Browning, H., 66, 68, 70, 245
Buchanan, A. M., 29
Bunak, V. V., 109
Bürkner, K., 32
Butler, P. M., 182, 245

Cabibbe, G., 41, 245
Calogero, B., 130, 245
Campbell, B. G., 1, 4, 59, 73, 82, 224, 245, 252
Campbell, T. D., 122, 132–3, 135, 151, 170, 245
Caspari, E. W., 220, 246
Cave, A. J. E., 126–31, 246
Chopra, S. R. K., 238, 246
Čihak, R., 69
Clark, J. D., 236–7, 246
Clark, W. E. Le Gros, xiii, xvi, 1, 16–18, 22, 43–4, 46–7, 49, 59, 65, 67–8, 78–9, 91, 116, 132, 139, 169, 223–5, 231, 233–4, 246
Cohen, L. B., 119, 247
Colyer, F., 141–2, 246
Connolly, C. J., 78, 246
Cooke, H. B. S., 233, 246

Coon, C. S., 1, 82, 246
Coppens, Y., 1, 246
Costa Ferreira, A. A. da, 24, 246
Curtis, G. H., 233, 248

Dabelow, A., 75
Dahlberg, A. A., 182, 246
Dalitz, G. D., 190, 246
Darlington, D., 75, 246
Dart, R. A., xvi, 1, 3–4, 9, 20–1, 35, 43, 45, 49, 78–9, 84, 95, 117, 137–8, 146, 219, 231, 234, 236, 238–9, 245–6
Davis, P. R., 6, 241, 246
Day, M. H., 6, 241, 246
Delattre, A., 14–15, 89, 193, 246
De Villiers, H., 27, 32, 136, 246
DeVore, I., 236, 252
Drennan, M. R., 98, 151, 246–7
Dubois, E., 86, 94, 247

Economo, C. von, 92, 247
Eisler, P., 108, 247
Evernden, J. F., 233, 248

Fanning, E. A., 143, 190, 247–8
Fenart, R., 14, 89, 246
Fick, L., 75
FitzSimons, V., xvi, 66, 146
Flower, W. H., 103, 116, 132–4, 170, 209, 216, 247
Frisch, J. E., 167, 184–6, 188–9, 191–2, 247

Galloway, A., 136, 247
Gardner, E., 29
Garn, S. M., 190, 247
Getz, B., 74, 204, 247
Giuffrida-Ruggeri, V., 93, 247
Gray, D., 29
Gray, T., 29
Gudden, J. von, 75
Guggenheim, P., 119, 247
Gyldenstolpe, N., 80, 247

Haines, R. W., 126, 246
Harris, H. A., 74, 80, 247
Harrison, G. A., 1, 224, 247
Healy, M. J. R., 222, 245
Heberer, G., 1–2, 225, 247
Henlé, F. G. J., 68
Hoadley, M. F., 97–8, 247
Hofer, H., 130, 247
Hofman, L., 59, 130, 247
Holloway, R. L., 21, 247
Hooijer, D., xvi
Howell, F. C., 60, 231, 233, 236, 239, 247, 252
Howes, A. E., 170–1, 247
Hrdlicka, A., 24, 86, 247
Hughes, A. R., xvi, 6, 31, 77
Hughes, D. R., xvi

Huizinga, J., 99, 247
Hulse, F. S., 1, 224, 247
Hunt, E. E., 190, 248

Inskeep, R. R., 236, 247

Jacob, T., 94
Jacobson, A., 133, 135–6, 170, 247
Jerison, H. J., 86–8, 205, 247
Jovanovic, S., 127–8, 247, 249

Kaufman, T. W., xvi, 77
Kaufmann, H., 11, 249
Kaye, D. B., 69
Keith, A., 3, 16, 64, 67, 78, 84, 247–8
Khroustov, H. F., 239, 247
Klomfass, R., xvi
Knott, J. F., 68, 247
Koenigswald, G. H. R. von, xvi, 1, 4, 94, 159, 183, 187, 236, 247–8, 251
Köppl, L., 68, 248
Korenhof, C. A. W., 145, 184–6, 188–9, 191–2, 248
Korkhaus, G., 133, 248
Kraus, B. S., 248
Krogman, W. M., 32, 248
Kronfeld, R., 143, 248
Kurth, G., 2, 248

Laffont, J., 115, 248
Laing, J. D., 133–4, 136, 248
Leakey, L. S. B., xv, xvi, 1–4, 6, 8–9, 28, 30, 44, 47, 74, 79, 105, 108–10, 116, 119, 126, 145, 155, 157, 167, 169, 182, 221, 223–4, 231–3, 235–7, 241, 243, 246, 248
Leakey, M. D., xv, 1, 6, 8, 79, 182, 241, 248
Le Double, A. F., 29
Le Gros Clark, *see under* Clark
Lessghaft, P., 75
Lewis, A. B., 190, 247
Lewontin, R. C., 84, 147, 250
Lieutaud, J., 68
Lindley, D. V., 84, 147, 248
Lipton, S., 222, 245
Lisowski, F. P., 75, 246
Logan, W. H. G., 143, 248
Long, W. M., 222, 245
Lubosch, W., 37, 248

McKown, T., 16, 248
Manouvrier, L., 111, 248
Marchand, F., 77, 248
Martin, R., 11–14, 16, 60–1, 73, 81, 100–1, 106–8, 110–11, 115, 125, 133, 135, 248
Mason, R. J., 236–7, 248, 250
Massler, M., 143, 248
Matiegka, J., 67–8, 248
Maxwell, J. H., 119, 248
Mayr, E., 1, 224, 248
Mednick, L. W., 89, 116, 248

INDEX OF PERSONS

Miller, J. C. P., 84, 147, 248
Montagu, M. F. A., 133–4, 248
Moorrees, C. F. A., 133, 138, 144–5, 170, 182, 190, 248
Morant, G. M., 16, 219, 249
Morris, H., 29
Moss, M. L., 89, 249
Musiker, R., 3, 249

Napier, J. R., 1–2, 6, 221, 223–4, 231, 237–8, 241, 246, 248–9
Negus, V., 131, 249
Nikitiuk, B. A., 75, 249

Oakley, K. P., xvi, 224, 231, 236–7, 249
Oetteking, B., 32, 249
Olivier, G., 133–4, 144, 249
Oppenheimer, A., 138, 249
O'Rahilly, R., 29

Patterson, B., 231, 252
Pearson, K., 15, 97–9, 247
Pedersen, P. O., 145, 170, 249
Petřiková, E., 68–9, 249
Petrovits, L., 36, 249
Pfisster, H., 77, 249
Pilbeam, D. R., 146, 169, 250
Pittard, E., 11, 249
Pycraft, W. P., 114, 249

Radoievitch, S., 128, 249
Randall, F. E., 80, 83, 249
Raven, H. C., 119, 249
Reed, R. B., 138, 248
Remane, A., 145, 162, 168, 175, 185, 220, 249
Riesenfeld, A., 24, 249
Robinson, J. T., xvi, 1–4, 9–10, 16, 18–19, 21–4, 26–7, 31–2, 35, 41–4, 46, 51, 55, 58, 60, 66, 78–80, 83–4, 92, 95, 98, 102, 104–5, 109–10, 112, 114–16, 120, 128–9, 132, 134–5, 137–8, 140–3, 145–6, 148–9, 151, 153–5, 157–9, 162, 167–8, 173–85, 187–8, 190–2, 194, 213, 217–18, 220–2, 224–31, 233–4, 236–7, 240–2, 244–5, 249–50
Roe, A., 84, 147, 250

Sartono, S., 94
Sauter, M. R., 13, 16, 250
Schaik, C. van, 122, 250
Schepers, G. W. H., 4, 26, 33, 41, 49, 55, 63–5, 67, 71, 78–9, 84, 88, 91–2, 94, 98, 100, 102, 128–9, 177–8, 206, 245, 250
Schour, I., 143, 248
Schultz, A. H., 28, 61, 77–8, 80–1, 83, 86, 112, 195, 219–20, 250
Schwalbe, G., 54, 250
Scott, J. H., 74–5, 120–1, 123, 204, 211, 231, 250
Seeds, A. E., 66, 252
Sergi, S., xvi, 2, 118–19, 210, 250
Shaw, J. C. M., 133, 135, 145, 151, 170, 250
Shdanov, D. A., 75
Shear, M., 142, 250
Sicher, H., 175, 184, 250
Simonetta, A., 4, 250
Simons, E. L., 146, 169, 243, 250
Simpson, G. G., 1, 84, 147, 220, 224, 250
Sollas, W. J., 49, 250
Sonntag, C. F., 119, 250
Spence, T. F., 77–8, 80–1, 204, 245
Stewart, T. D., 29, 33, 250
Stoll, N. R., 224, 250
Straus, W. L., 78, 251
Streeter, G. L., 70, 251
Streit, H., 68
Symons, N. B. B., 121, 211, 251

Tildesley, M. L., 16, 222, 251
Tobias, P. V., 1–2, 4, 6, 15–16, 77, 79, 81, 84, 86–7, 89, 95, 106, 116, 178, 182, 219–21, 223–4, 231, 236, 237–9, 241, 248, 251
Todd, T. W., 78, 251
Trevor, J. C., xvi, 3, 15, 251
Turner, W., 133, 251
Twiesselmann, F., 73, 251

Vallois, H. V., xvi, 73, 78, 80–2, 84, 220, 245, 251
Van Bork-Feltkamp, A. J., *see under* Bork-Feltkamp
Van Reenen, J. F., 170–1, 251
Van Riet Lowe, C., 236, 245
Van Schaik, C., *see under* Schaik
Vernieuwe, J., 68
Vogel, C., 112, 251
Von Bonin, *see under* Bonin
Von Economo, *see under* Economo
Von Koenigswald, *see under* Koenigswald

Waggoner, R. W., 119, 248
Wagner, K., 89, 97–8, 102, 251
Waltner, J. G., 42, 68, 251
Washburn, S. L., 60, 89, 95, 116, 122, 228–9, 231, 236, 248, 251–2
Wegner, R. N., 126, 252
Weidenreich, F., 4, 14–16, 28–35, 37, 39–40, 44–50, 53, 56–7, 59–60, 62, 72–6, 78, 80–3, 91, 94–6, 98–104, 107–8, 110–11, 116–18, 123–5, 127–8, 130–1, 133, 135–6, 145, 154, 158, 163, 170, 178, 187–8, 193, 196, 199–200, 204, 210, 252
Weiner, J. S., 1, 59, 73, 224, 247, 252
Weinert, H., 81, 98, 252
Welcker, H., 77
Wells, L. H., xvi
Wiebes, J. T., 94
Woo, J. K., 94, 168, 252
Woodhall, B., 66, 68, 70, 252
Wood Jones, F., 32, 252

Zoja, L., 29
Zuckerkandl, E., 184
Zuckerman, S., 16, 18, 23–4, 36–7, 40, 43–4, 49–51, 74–5, 77, 122, 151–2, 161–2, 168–9, 171, 173, 194–5, 197–8, 204, 245, 252

INDEX OF SUBJECTS

acrocephaly, 112
adaptation, 228
adaptive capacity, 86–7, 123
aditus ad antrum, 61
African
  great apes, 87, 105; *see also Gorilla, Pan*
  Negroid crania, 58, 115, 202
age changes
  in brain size, 77–9, 204
  in condylar position index, 51
  in parietal thickness, 74, 204
age, individual, of *Zinjanthropus*, 74, 77, 121–2, 129
air cells
  ethmoidal, 109, 128–9, 208, 212
  mastoid, 61, 202
  of orbital plate of frontal, 108–9, 208
akanthion, 114
ala
  major ossis sphenoidei, *see under* sphenoid bone
  vomeris, 7, 129, 212
Aleut, teeth of, 154, 158, 163, 172
alisphenoid, 37–40, 54, 61–2, 197; *see also* sphenoid bone, greater wing
Alouatta, 112
alveolar
  point, *see* prosthion
  process, *see under* process
  profile angle, *see under* angle
  ridge, *see under* process
alveolare, 107, 114, 124
alveolus
  of canine, 137–8, 229
  of incisors, 137–8, 229
  of third molar, 122
American White, teeth of, 152, 154, 158, 163, 167, 171–3, 186, 216
Amerindian crania, 32–3, 35, 43
Anatomy Department
  Cape Town, 3
  Witwatersrand, 1–3, 66, 69, 77, 106, 120
Andamanese, teeth of, 170
ANGLE
  alveolar profile, 115–16, 209
  asterionic, of parietal bone, *see* angle, mastoid
  external orbital, 105–6
  facial profile, 115, 209
  floor/squame, of anterior cranial fossa, 53
  mastoid, of parietal bone, 21, 54, 73, 201
  between nasal bones, 110, 208
  nasal profile, 115–16, 209
  nasion–basion–opisthion, 47–8, 199, 223
  opisthion–basion–prosphenion, 47, 199
  petro-median, 33–4, 57, 196, 223
  petro-tympanic, 33, 196

pterionic, of temporal squame, 54
sphenoid, of parietal bone, 54, 200
angular torus, *see under* torus
anterior cranial fossa, *see under* fossa
anthropoid ape
  crania, 28, 30, 34, 37, 43, 46–8, 57–9, 107, 111, 113, 115, 125–6, 220; *and see* individual species
  teeth, 135, 188
Anthropoidea, 169, 238
antrum
  mastoid, 61, 202
  maxillary, *see under* sinus, air
ape-man, 4
aperture
  for aqueduct of cochlea, 57
  for aqueduct of vestibule, 57
  meatal, 31–3, 49, 196
  piriform, 7, 109–14, 208–9
apex of petrous pyramid, 57–8, 202
apophysis, supernumerary mastoid, 29
ARC
  coronal margin, 10–12
  lambdoid margin, 10–12
  nasomalar, 107
  occipital sagittal, 13, 193
  parietal sagittal, 10–13, 193
  temporal margin, 10–12
  total sagittal, 14
arcadal index (of Laing), *see under* index
arcade, dental, 132–8, 211, 223
arch
  dental, 132–8, 170, 213; dimensions, 132–4
  zygomatic, 8, 119, 121, 124–5, 210, 212, 229
AREA
  infra-orbital, 116–18
  interorbital, 106–9, 208
  naso-alveolar junction, 126
  naso-orbital, 128–9
  pterygomaxillary contact, 127
ARTERY
  anterior tympanic, 41–2
  descending palatine, 126
  greater palatine, 122
  infra-orbital, 117–18, 126
  maxillary, 42
  meningeal, accessory, 62, 202
  meningeal, middle, 54, 60, 62–3, 93–4, 203, 205
  occipital, 26, 29–30
  superficial temporal, 42
articular eminence, *see under* tubercle
articulation, glenoid, *see* joint, temporo-mandibular
asterion, 21–2, 72, 96
ASYMMETRY
  of attrition pattern, 19, 140–1, 213–14
  of canines, 177–9, 217

of jaw function, 19, 140–1
of nuchal crest, 22
of occipitomastoid crest, 29
of temporalis, 19–20, 140–1, 213–14
Atlanthropus mauritanicus, see Homo erectus mauritanicus
atlas, 27, 195
ATTRITION
  diet and, 141, 214, 227
  interproximal (interstitial), 138, 170
  occlusal, 139–42, 213–14, 227
  pattern of, 140–1, 213–14
  rate of, 227
auriculare, 30, 96, 103, 195
Australian
  crania, 17–18, 24, 32, 44, 51, 67, 73–4, 97–8, 170
  teeth, 133, 135–6, 151–4, 158, 163, 170–2
Australopithecinae, 1, 4, 223–4, 237–9
  crania, 35, 40, 43–4, 48, 61, 73–4, 76–9, 81–2, 84–5, 87, 91, 93, 95, 97, 99, 101, 105, 108, 115, 117–18, 128, 205, 207, 212
  teeth, 132, 136, 138, 141, 143, 146, 151–2, 168–9, 172, 184, 191, 213, 215, 221
Australopithecine
  ancestral, 241, 243
  gracile, 1, 79–80, 86, 204, 230, 240, 244
  robust, 1, 64, 70, 79–80, 123, 204, 224, 229, 240, 244
Australopithecus (as subgenus), 2, 219–20, 224, 231
  cranium, xv, 1, 9–16, 18–21, 33–5, 38, 42–50, 54–6, 59, 63, 65–6, 71–3, 78–80, 83, 88–92, 95–9, 101–5, 109, 112–17, 120–2, 130, 132, 193–9, 201, 204–11, 228–30
  teeth, 132, 135–40, 142, 145–9, 151–60, 162–9, 172–87, 189–92, 213–18, 220–1, 225–8
Australopithecus (as genus), 219–20, 222, 224, 231, 233, 236, 242
  definition, 223, 234
A. africanus, 45, 86–7, 146–7, 219, 231–2, 240–1, 243–4
  definition, 234–5
A. boisei, 223–4, 232, 235–7, 240–4
  definition, 235
A. robustus, 146–7, 231–2, 240–4
  boisei, 224, *and see A. boisei*
  crassidens, 24
  definition, 235
AXIS
  basicranial, 46–7, 103
  biporial, 27, 63, 200, 203
  of petrous, 33–4, 57, 196, 223
  of pyramid, 33–4
  of tympanic plate, 33, 196
  vestibular, 14

INDEX OF SUBJECTS

baboon, 75, 141, 239
Bantu
  crania, 32, 34, 37, 67, 69–70, 203
  teeth, 133, 135–6, 151–2, 154, 158, 163, 170–2
bare area
  of cranium, 20–1, 194
  of occipital bone, 21
basal line of face, 103, *and see* length, nasion–basion
basicranial axis, *see under* axis
basi-occipital, *see* occipital bone, pars basilaris
basion, 34, 46–7, 59, 96, 102–3, 129
  –inion length, 45
  –nasion length, 47, 96, 103, 116, 170, 199
  –opisthion length, 48, 199
  –prosthion length, 45, 107, 116
basis cranii, 3, 8, 47, 103
basis cranii externa, 26–42, 195–8
basis cranii interna, 54–63, 202
  shortening of, 34–5, 103, 196–7, 199, 207, 223
  spheno-occipital part, 34–5
bicondylar breadth, *see under* breadth
bidental distances, 134, 137, 213
biglenoid breadth, *see under* breadth
Biometric School, 96, 101, 114, 220
biparietal fragment, 7
bipedalism, 238
biporial breadth, *see under* breadth
Birunga, 105
Boise Fund, xvi
bone tools, 238–40
Boskop calvaria, 95
boss
  frontal, *see* tuber frontale
  parietal, *see* tuber parietale
BRAIN
  –body weight ratio, 78, 86–7
  -case, expansion of, 18, 40, 75, 102, 204, 207; hafting of, 8, 19, 101, 124, 194, 206, 212, 230
  morphology, 18, 230
  quality, 229
  size, 78, 87, 229
  stem, 88–90, 205
  volume, 18, 78
  weight, 78
BREADTH
  arc, of occipital bone, 13
  arcadal, 133–4
  biasterionic, 13, 96, 101
  biauricular, 96, 99–101, 206
  bicondylar, 40, 198
  biglenoid, 40
  biorbital, external, 107; inner, 107
  biparietal (of endocast), 92
  biporial, 40
  bizygomatic, 107, 114, 124–5, 212
  of clivus, 96
  dental arch, 133–4
  ectomolar, 133–4, 136
  endomolar, 133–4, 136
  facial, superior, 107, 124–5, 212
  of foramen magnum, 60–1, 96
  –height index, *see under* index

intercondylar, 40, 198
intercrestal, 99
interglenoid, 40
interorbital, anterior, 107–8, 208; posterior, 107–8, 129, 208
interporial, 96
maxillary, 107, 113–14, 124, 212
maxillo-alveolar, 132–5
maximum, across supramastoid crests, 22, 96, 99, 101, 206; average, of calvaria, 100–1; bimastoid, 22, 96, 99, 206; biparietal, 96, 99–101, 206; cranial, 96, 99–100, 125; intercristal, 96, 99; of posterior cranial fossa, 55; on temporal squames, 96, 99–101, 206; supramastoid intercristal, 96, 100
minimum frontal, 96, 100, 125, 206
nasal, 110
of nasal bones, greatest, 110, 208; inferior, 110, 208; least, 109–10, 208; superior, 110, 208
palatal, 133–5
rameal, 120–1, 211
temporoparietal, 96, 100–1, 206
bregma, 24, 63, 101
  –lambda arc, *see* arc, parietal sagittal
British Council, xvi
British Museum (Natural History), 3, 59, 66, 120, 127
Broken Hill
  cranium, 16, 37, 45, 48, 60, 82, 95, 97–8, 111, 114–15, 117, 120, 123–4, 210–11
  teeth, 135–6
brow ridges, 29, 223, 230–1; *and see* torus, supra-orbital
buccolingual
  diameter, 3, 144–5, 151, 156–64, 166–7, 215–16, 220–1
  expansion, 117, 159, 162, 172–4, 179, 184, 190, 215–17
buccostyle, 183–5, 192
bulb, jugular, 65, 68–70, 203
Burmese crania, 16
Bushman
  crania, 27, 32, 58, 113, 209
  teeth, 136, 151–4, 158, 163, 171–2, 227

calcification of teeth, 141–3, 188
Callicebus, 112
calotte, 3, 8, 16, 24, 102
calvaria, 3, 8–10, 95–103, 206–7
  hafting of, 8, 19, 101, 124, 194, 206, 212, 230
  height of, 96, 230
  shape of, 9–10, 206
  thickness of, 72–6
calvariofacial
  indices, *see under* index
  part of *Zinjanthropus*, 7, 8
CANAL, CANALIS
  carotid, 33, 58, 60
  condylar, 27, 67
  greater palatine, 122
  hypoglossal, 27, 57, 202
  incisive, 112, 126
  infra-orbital, 113, 126
  medullary, 76

pterygopalatine, 127
for superior alveolar nerves, 127
canaliculus
  cochleae, 57
  innominatus, 62
CANINE
  asymmetry, 177, 179, 217
  attrition, 139, 214, 223
  hypoplasia, 141
  metrical characters, 144, 147–57, 179, 214–15, 217, 225–6
  morphology, 177–9, 217
  –premolar ratios, 155–7, 179, 214–15, 217, 226
  root, 179
  shape index, 144, 147–50, 172–4, 179
Canis familiaris, 75
capacity, cranial, *see under* cranial
Carabelli's cusp or tubercle, 189–92, 218
caries, 142
Carnivora, 75, 237
cast, endocranial, *see under* endocranial
Catarrhini, 128
Caucasoid crania, 34, 58
Ceboidea, 49
Cebus, 142
cellulae
  ethmoidales, *see under* air cells
  mastoideae, *see under* air cells
Cercocebus, 142
Cercopithecoidea, 49, 58, 77, 202
Cercopithecus, 58
CEREBELLAR
  arc of occipital bone, 13–15, 193
  chord of occipital bone, 13–15, 193
  index of occipital bone, 13, 15, 193
  indices, 92
cerebellum, 56, 88, 91, 205
  expansion of, 56, 70, 92, 201, 203, 205, 223
  hemispheres, 57, 65, 71, 88–90, 203, 205
CEREBRAL
  arc of occipital bone, 13–15, 193
  chord of occipital bone, 13–15, 193
  height–length index, 92
  hemisphere, 88, 205
  index of occipital bone, 13, 15, 193
  indices, 92
  'Rubicon', 78, 84
cerebrospinal fluid, 78
cerebrum, 88, 230
Chellean man, *see* Olduvai hominid 9
chimpanzee, *see* Pan
  pygmy, *see Pan paniscus*
China, 10
Chinese crania, 32, 97
chipping of teeth, 141, 227
CHORD
  breadth of occipital bone, *see* breadth, biasterionic
  cerebellar, of occipital bone, 13–15, 193
  cerebral, of occipital bone, 13–15, 193
  coronal margin, 7, 10–12, 193
  index of occipital scales, 13–15, 193
  lambdoid margin, 10–13, 193
  occipital sagittal, 13, 15–16, 193–4
  parietal sagittal, 10–13, 193
  premolar–molar, 157–8, 170
  temporal margin, 10–12

INDEX OF SUBJECTS

Choukoutien
 crania, 4, 16, 50, 62, 81, 94, 97, 99, 102, 104, 111, 130, 242
 teeth, 152, 154, 158, 163, 167, 172
 and see Homo erectus pekinensis
cingulum, 178, 184–6, 188–92, 218
 reduction of, 184, 189–92, 218
clade, classification by, 244
cladistic evolution, 240
classification, 1, 4, 219
clivus
 naso-alveolar, 112–14, 126, 209
 occipitosphenoid, 7, 57–8, 96, 202
Colobus, 58
complex
 mastoid–supramastoid–auriculare, 100, 206
 supramastoid–auriculare, 100, 206
components of cranial vault, 9–16
compound temporal–nuchal crest, 21–5, 194–5
concha, middle nasal, 128
condylar position index, 17–18
condyle, condylus, *see under* occipital bone
confluence of sinuses, 67–9
Copenhagen Congress, 222
CORONAL MARGIN of parietal, 10–13, 193
 arc, 11–12
 chord, 10–12
 index, 11, 13, 193
cortical neurones, 86–7, 205
CRANIAL
 base, *see under* basis cranii
 capacity, 77–87, 204–5; of Australopithecinae, 78–87, 204–5; of *A. africanus*, 78–9, 204, 229–30; of *A. boisei*, 77–9, 204, 223, 229–30; of *A. robustus*, 79–80, 229–30; of *Gorilla*, 78, 80–5, 87, 94, 204; of *Homo erectus*, 78, 81–7, 94, 204–5; of *H. habilis*, 84, 205; of *H. sapiens*, 81, 85, 87; of Hylobatidae, 80–1, 84–5; of *Pan*, 81–5, 87, 204; of *Pongo*, 81–3, 85, 87, 204
 crests, *see under* crest
craniogram
 lateral, 9, 17
 median sagittal, xv, 46
 occipital, 20
 vertical, 10
cranium, 3
 reconstruction of, xv
CREST
 frontal, 53, 63, 200
 incisor, 112
 internal frontal, *see* crest, frontal
 mastoid, 28, 72
 nasal, 112, 114
 nuchal, 13–14, 16, 20–6, 43, 56, 73, 95, 193–5, 200, 204, 206, 223
 oblique, of molars, 183, 185, 187
 occipital, external, 26, 45; internal, 55, 69–70, 201
 occipitomastoid, 22, 29–30, 73, 195
 petrous, 30
 sagittal, 1, 7, 17–21, 49, 53, 73–4, 194, 204, 223, 229
 supramastoid, 22–3, 28, 30–1, 195, 223

 temporal, 7, 19–25, 28, 73, 104–6, 194–5, 207
 transverse, of tympanic plate, 30
 trigon, of molars, 183, 187; anterior, of molars, 183
crista
 galli, 128
 nasalis, *see under* crest
 Sylvii, 54, 200–1
Cromerian fauna, 233
crowding of teeth, 138, 213
crown
 area, 144–5, 152, 154–9, 163, 165, 167, 169, 184, 187, 214–16, 218
 height, 145, 189–90
 shape, 144, 146–50, 172–4, 184
cruciate eminence, 55–6, 64, 67, 69, 201
 sagittal limb, 53, 55–6, 59, 65, 201
 transverse limb, 55, 64
C.S.I.R. (S. Africa), xvi
cultural capacity, 219, 236–40
culture
 Oldowan, 236–7
 Osteodontokeratic, 238–40
 pre-Chelles-Acheul, 236
curvature
 of cranial vault, 9–10, 193
 of occipital bone, 13, 15–16, 193–4
 of parietal bone, 10–13, 193
curve of Spee, 141, 170
CUSP
 buccal (of premolars), 139–40, 179–81, 217
 Carabelli, 189–92, 218
 distoconule, 185, 187
 double metaconulus, 183, 187
 hypocone (distolingual), 140, 183, 185–7, 192
 lingual (of premolars), 139–40, 179–81, 217
 metacone (distobuccal), 140, 183, 185, 187
 paracone (mesiobuccal), 140, 181, 183, 187, 189
 paramolar, 183–5, 192
 protocone (mesiolingual), 140, 182–3, 186–7, 189, 191
cusplet, accessory distal (of canine), 177–9, 217

DEFINITIONS
 of australopithecine taxa, 234–5
 of dental measurements, 144–5
 of dental arcade measurements, 132–4
 of maxillo-alveolar measurements, 132–4
 of palatal measurements, 133–4
deformation of cranium, 43
DENTAL
 age, 77, 190, 211
 arcade, 132–8, 213
 eruption, 130, 190
 size, 36, 144–69, 170–1, 214–16; categories, 170–1
 terminology, 3
dentine, exposure of, 139–40
dentition, 3, 132–92, 213–18, 220–1, 225–8

development
 of condyle, 27
 of nuchal crest, 23–5, 51, 195
 of sagittal crest, 21, 24–5, 195
 of tympanic bone, 31, 196
diagnosis, 1, 9, 28, 30, 74, 105, 108–10, 116, 119, 126, 155, 223, 234–5
diameters
 buccolingual, 3, *and see under* buccolingual
 labiolingual, 3, *and see under* labiolingual
 mesiodistal, 3, *and see under* mesiodistal
diastema, absence of, 132, 138, 213, 223
diet of *Zinjanthropus*, 141, 227
dietary hypothesis, 225–8, 231
discovery of *Zinjanthropus*, 1
displacement of teeth, 117, 138, 181, 213, 217
distance
 basion–hormion, 34–5, 96, 196–7, 223
 endobasion–klition, 96
 inion–internal occipital protuberance, 96
 opisthion–klition, 96
diverticula
 of frontal sinus, 127–8, 212
 of maxillary sinus, *see under* recess
Djetis, 242
dorsum sellae, 7, 47, 55, 58–9, 61, 202, 223
Dryopithecus, 146, 169, 216
Duckworth Laboratory, 15
duct, nasolacrimal, 108, 127
dura mater, 53, 58, 78

ecology, 2, 225, 228, 231, 238
ectobasion, 27, 46, 59, 60
Egyptian crania, 97–8
Ehringsdorf cranium, 98, 129
ektoconchion, 107
ektomolare, 133–5
elongation
 of face, 1, 107, 113, 124, 200, 208–9, 212, 223
 of nose, 111, 208
EMINENCE, EMINENTIA
 arcuate, 60, 202
 articular, *see* tubercle, articular
 cruciate, *see under* cruciate
 cruciform, *see under* cruciate
 gingival, 175–7, 217
emissary vein, *see under* vein
enamel, state of, 139, 141–3, 214
encephalometric
 constants, 92–3
 features, 92–3, 205
encephaloscopic features, 87–91, 205–6
endobasion, 48, 96
ENDOCRANIAL
 cast, endocast, 8, 77–94, 205–6; of australopithecines, 3, 64–5, 77, 88–94; of cercopithecoids, 77; of hominines, 3, 67, 77; of pongids, 3, 53, 71, 77, 88–91, 200; of *Zinjanthropus*, 77–94, 203, 205–6
 dimensions, 92, 205
 form, 88–91, 205
 surface, of frontal bone, 7, 53–4, 200; of occipital bone, 55–7, 64–5; of parietal

257

INDEX OF SUBJECTS

ENDOCRANIAL (cont.)
    bone, 54, 200–1; of sphenoid bone, 61–3; of temporal bones, 57, 60
    volume, see cranial capacity
endomolare, 133–5
English crania, 16–18, 44, 51, 136
entoglenoid process, see under process
erect posture, 45, 238
eruption of teeth, 130, 190
Eskimo
    crania, 24, 33
    teeth, 154, 158, 163, 167, 171–2
ethmoid
    air cells, 109
    bone, 7, 109, 128
    orbital plate, 109
European crania, 33–5, 45, 73
euryon, 95, 99
excess (extra) neurones, see under neurones
exoccipital, see occipital bone, pars lateralis
extensors of head, 45, 51–2

FACE, 104–25, 207–11
    breadth of, 107, 124, 209
    elongation of, 107, 124, 200, 208–9, 212, 223
    enlargement of, 113–14, 209
    flatness of, 115, 117, 138, 213
    height of, 107, 212
    verticalising of, 124, 212
facial
    dimensions, 107, 123–5, 212
    indices, upper, 107, 212
    skeleton, 3, 229
FACIES
    anterior partis petrosae, 60
    cerebralis ossis sphenoidei, 61–3
    inferior partis petrosae, 33–4
    infratemporalis ossis sphenoidei, 37, 197
    lateralis ossis zygomatici, 7, 123, 211–12
    malaris ossis zygomatici, see facies lateralis
    orbitalis ossis frontalis, 7, 55, 108
    posterior partis petrosae, 57, 201–2
falx cerebri, 53, 67, 70, 200
Felis domestica, 75
femur, 76
fenestra
    cochleae, 61
    vestibuli, 61
fibula from Bed I, 6
field concept, 182
FISSURE, FISSURA
    Glaserian, see fissure, petrotympanic
    horizontal, of cerebellum, 56, 88
    orbital, superior, 37, 197
    parieto-occipital (of cerebrum), 91
    petro-occipital, 58, 202
    petrosquamous, 60, 62
    petrotympanic, 30, 41–2
    pterygomaxillary, 127
    squamotympanic, see fissure, tympanosquamous
    tympanosquamous, 30, 41–2
Flower's
    dental index, 170, 216
    gnathic index, 103, 116, 209

FORAMEN
    caecum, 53, 200
    carotid, see under canal
    of Huschke, 32, 196
    incisive, 122, 211
    infra-orbital, 113, 116–17, 126–7, 209–10
    jugular, 64–5, 67–8, 70
    lacerum, 58
    magnum, 27, 45–8, 57, 59–61, 67, 89, 199, 202; dimensions, 60, 202; L/B index, 60, 202; plane, 43, 47–8, 199, 223; position index, 45–7, 199; shape, 59–60, 202; tilt, 47, 199
    mastoideum, 28, 68
    occipitale, see foramen magnum
    optic, 55
    ovale, 38–40, 42, 61–2, 129, 198, 202
    palatine, greater, 122
    postcondylar, 27
    postglenoid, 42
    rotundum, 62
    spinosum, 38–40, 61–2, 197–8, 203
    'spurious jugular', 42
    squamosal, 42
    of styloid process, 29, 195
    stylomastoid, 29, 33, 195–6
    of Vesalius, 61
forehead, 101, 104, 206–7, 230–1
formation
    of enamel, 142–3, 188
    pterygoid, see under sphenoid
FOSSA
    canine, 116–19, 210
    cerebellar, 55–6, 67, 201
    cerebral, 55–6, 65, 67, 201
    condylar, 27, 195
    cranii anterior, 53, 55, 128, 200
    cranii media, 54–5, 60–3, 129, 202–3
    cranii posterior, 54–9, 71, 201, 203
    digastric, see incisura mastoidea
    glenoid, see fossa, mandibular
    hypophyseal, 7, 55, 58, 77, 202
    incisive, 112, 114, 122
    infratemporal, 37, 39
    jugular, 33
    lingual, of incisor teeth, 175–6
    mandibular, 35–42, 197–8, 223; anterior wall, 37, 197, 223; dimensions, 35–6, 197; posterior recess, 41; posterior wall, 30, 40–1, 198, 223
    masseteric, 120
    parieto-occipital, 55
    posterior cerebral, see fossa, cerebral
    prenasal, praenasalis, 11–14, 116, 209
    pterygoid, 42, 198
    pterygopalatine, see fossa, pterygoid
    subarcuate, 57
    subnasal, 113, 209
    temporal, 128, 229
    vermian (vermiform), 55, 59
fossula vermiana, see under fossa
fovea
    anterior, of cheek-teeth, 183, 187
    central, of molars, 183, 185, 187
    posterior, of cheek-teeth, 181, 183, 185, 187
fractures of teeth, 140, 227
Frenchman, cranium of, 58, 71

FRONTAL BONE, 7, 19, 53–5, 108
    air cells, 108–9
    endocranial surface, 7, 53–4
    pars orbitalis, 7, 55, 108
    pars nasalis, 7
    squame, 7, 19, 127–8
    temporal surface, 7–8
frontal lobe, of brain, see under lobe
frontomalare
    orbitale, 107
    temporale, 107, 124
frontotemporale, 96
furthest occipital point, see opisthocranion

galea aponeurotica, 20
Galilee cranium, 129
ganglion, semilunar, 62
Garusi, 146, 158, 237, 242
Gasserian ganglion, see ganglion, semilunar
gibbon crania, 18, 21, 58, 81, 84–5, 220
Gibraltar cranium, 60, 98, 124
Gigantopithecus, 79, 116, 144, 168–9, 187, 214, 216, 218
glabella, 8, 19, 45, 95, 103, 106, 109–10, 199, 208
Glaserian fissure, see fissure, petrotympanic
glenoid fossa, see fossa, mandibular
gnathic index (of Flower), see under index
gnathism, 34; and see orthognathism, prognathism
Gorilla
    crania, 12, 14–18, 20–1, 26, 28, 30, 33–6, 38–40, 44, 46–7, 49, 51, 54–5, 57–8, 60–1, 63, 67, 71–5, 77, 79–90, 94, 105–6, 118–21, 126–30, 195, 197–200, 202, 205, 210–12, 219
    teeth, 135–6, 142, 151–2, 154, 158–9, 161–3, 168–74, 215–16
grade, evolutionary, 244
Grimaldi cranium, 136
grooves
    Carabelli, 183, 186
    for sinuses, see under sinus, venous
GROWTH
    of air sinuses, 127–31
    of brain, 77–80
    of brow-ridges, 230–1
    of cranium, 14–15, 51, 74–5
    of facial skeleton, 51, 130
    lateral, of cerebrum, 92, 205
    of mandibular fossa, 36, 39–40
    of mastoid process, 28
Guamanian crania, 32
gyri, 87
gyrus, frontal
    inferior, 54, 87–8, 205
    middle, 54, 87, 205
    superior, 54, 87, 205

hafting
    of calvaria, 8, 19, 101, 124, 194, 206, 212, 230
    of malars, 117, 210
half-men, 4
Hawaiian crania, 32
head-balancing index (of Dart), 45

INDEX OF SUBJECTS

Heidelberg mandible, *see* Mauer
HEIGHT
 alveolar, 107, 113–14, 124
 auricular, 96, 102–3
 basibregmatic, 96, 101–3, 206
 of calvaria, 16, 96, 101–3
 of cerebrum, 92; increase in, 92–3
 facial, superior, 107, 124
 of lambda, 96
 maximum, above basion, 96; above Frankfurt horizontal, 96; of sagittal crest, 96; of temporal squame, 96
 of meatal aperture, 33
 nasal, 110
 of nuchal area, 96
 orbital, 107
 orbito-alveolar, 107, 113–14, 124
 of piriform aperture, 110
Hemanthropus, 4, 183, 219
heteromorphism, *see under* premolars
Hominidae, 4, 78, 149, 169, 184, 191, 223–4
Homininae, 4, 224, 241
 crania, 16, 18, 35–6, 47, 49, 54–5, 63, 73, 76–7, 88, 91, 106, 108–9, 115–16, 120, 122, 124, 194–9, 205, 207–8, 212
 teeth, 132, 141, 143, 151–3, 169, 172, 182, 213, 216
hominisation
 behavioural, 228, 238, 244
 morphological, 40, 47, 49, 87, 101–2, 207, 238, 240–1, 243–4
Hominoidea, 4, 58, 85, 127, 134–5, 152, 202
Homo, 4, 136, 244
 africanus boisei, 224
 erectus, 4, 237, 239–40, 244; crania, 10–16, 32, 36, 47, 50, 59, 73, 76–7, 81–5, 87–8, 93–5, 97, 99–100, 104, 114, 120, 193, 201, 204–7, 223; teeth, 147–9, 152–4, 158, 163, 172–3, 214, 222
 erectus erectus, 4, 242–3; crania, 10–14, 28, 35, 46, 50, 60, 73–5, 81, 84, 91, 94, 96–8, 102–3, 127–9, 195; teeth, 135, 154, 159–60, 164–5, 167, 170
 erectus mauritanicus, 4, 73–4, 147–8
 erectus pekinensis, 4; crania, 10–15, 27–36, 38–9, 46, 48–50, 53–4, 56–7, 60, 62–3, 65–6, 72–6, 81–2, 84, 91, 94, 96–9, 103–4, 106–8, 111, 114–15, 117–18, 123–5, 127–30, 195–7, 200–1, 203–4, 207–11; teeth, 135–6, 147, 151–2, 159–60, 164–5, 167, 170–1, 175, 178
 habilis, 6, 84, 87, 104, 205, 221–3, 236–8, 241–4
 neanderthalensis, 82
 sapiens, 36, 50, 79, 85, 87–8, 97, 200, 205, 207, 222, 244
 sapiens neanderthalensis, 88, 243
 sapiens sapiens, 98, 136, 172, 243; *and see* modern man
 sapiens soloensis, 49, 82
Hopefield cranium, 82, 97–8
horizontal occipital length, *see under* length
hormion, 34, 96
Huschke, foramen of, *see under* foramen
hyaena, 238
Hylobates, 142; *and see* gibbon
Hylobatidae, 80–1, 191, 220
hyperorthognathism, 115–16, 209

hyperprognathism, 34, 115–16, 209
hypodigm, 220–1, 223–4, 233
hypoplasia of enamel, 141–3, 214, 218, 227

implemental
 activities, 236–40
 frontier, 239–40
impression
 masseteric, 119–21, 210–11
 trigeminal, 60, 202
impressiones
 digitatae, 54, 56, 87–8, 205
 gyrorum, *see* impressiones digitatae
INCISORS
 attrition, 139
 central (medial, first), 144, 175, 214, 217
 crown area ratio, 152, 214, 217
 hypoplasia, 141–2, 214
 lateral (second), 144, 175–7, 214, 217
 metrical characters, 144, 147–52, 176–7, 214, 225–6
 module ratio, 151–2, 214, 217
 morphology, 175–7, 217
 roots, 176, 217
 shape index, 144, 147–50
INCISURA
 jugularis, 64–5
 malaris, 117–19, 210
 mastoidea, 29–30, 195
 supra-orbitalis, 105
inclination angle
 of foramen magnum, 47, 199
 of planum nuchale, 23, 43, 198
incurvatio inframalaris frontalis, 118, 210
INDEX
 altitudinal, 101–2, 206
 arcadal (of Laing), 133–4, 136, 213
 auricular height/breadth, 101, 103, 207
 bifrontal/biparietal (of endocast), 100
 breadth/height (of calvaria), 101–2, 206–7
 breadth/length (of nasal bones), 111
 calvariofacial, length, 125, 212; width, 125, 212
 cerebral height/length, 102, 206
 condylar position, 17–18, 45, 49–52, 200, 223
 cranial, 101
 craniofacial, longitudinal, 125; transverse, 125
 depth/breadth, of mandibular fossa, 35–6, 196
 depth/length, of mandibular fossa, 35–6, 196
 Flower's dental, 170, 216; gnathic, 103, 116, 209
 foramen magnum, L/B, 60–1, 202
 frontobiorbital, 125
 frontoparietal, transverse, 100–1, 206
 horizontal occipital length, 45–7, 199
 inion–opisthion curvature, 13–15, 193
 interglenoid/biglenoid, 40, 198
 interglenoid/biporial, 40, 198
 interorbital, 107
 length/breadth, of mandibular fossa, 35–6, 197
 length/breadth, of occipital squama, 13–14

 maxillo-alveolar, 132–6, 213
 molar/rameal, 120–1, 211
 nasal, 111, 208
 of nasal aperture, 111, 208
 nasion–basion length, 101, 207
 nasomalar, 107
 nuchal area height, 17–18, 22, 43–4, 198
 occipital length, 45–7, 101, 199, 223
 occipital sagittal, 13, 15–16, 193–4
 occipitoparietal arc, 14
 orbital, 106–7, 207–8
 palatal, 133–6, 213
 parieto-occipital, transverse, 101, 206
 porion position, height, 48–9, 103, 200; length, 48, 199–200
 of relative usefulness, 222–3
 shape, of teeth, 144, 146–50, 172–4, 216–17, 221–2
 spheno-occipital, 34, 196–7, 223
 supra-orbital height, 17–19, 194, 230
 superior facial, 107, 114, 124, 209, 212
 of tapering (of dental arcade), 134, 213
 torocristal, 97–9, 206
 upper facial, *see* index, superior facial
 vertical, 101–2
 zygomaticofrontal, 125
 zygomaticomaxillary, 107, 114, 124, 209
 zygomatico-suprafacial, 107, 124
Indonesian crania, *see Homo erectus erectus*
infantile pattern of venous sinuses, 70, 203
inion, 8, 14, 20, 56, 95–6, 193, 199
 –internal occipital protuberance distance, 56, 201
 –opisthion dimensions, 13–15, 193
 position of, 43–5, 199
Institut de Paléontologie Humaine (Paris), 3
intercondylar distance, *see* breadth, intercondylar
interglenoid distance, *see* breadth, interglenoid
ischium, 231

Japanese, teeth of, 154, 158, 163, 167
Java man, *see Homo erectus erectus*
joint
 exoccipital/supra-occipital, 66
 intra-occipital, 66
 temporomandibular, 30
juga alveolaria, 117, 210
 of canine, 116–18, 210

Kenyapithecus wickeri, 243
klition, 47, 96, 199
Krapina crania, 68, 111
Kromdraai, 4, 220, 225, 233, 237, 240, 242
 crania, 28, 31, 35, 37, 39–41, 49, 54, 61, 63–5, 72–3, 79, 84, 94, 109, 113–14, 116, 118, 122–3, 130, 139, 201, 203, 206, 211
 teeth, 135, 146, 158, 163, 168, 173, 221

labiolingual diameter, 3, 144–5, 147–51, 153–4, 214–15
La Chapelle-aux-Saints cranium, 37, 48, 98, 118, 124, 135–6
lacrimal bone, 7, 108, 127–8

INDEX OF SUBJECTS

lacrimale, lacrimal point, 107–9, 129, 208
lambda, 7, 21, 55–6, 66, 93, 96
lambda–inion
  dimensions, 13–16, 193
  curvature index, 13, 15–16, 193
lambda–internal occipital protuberance distance, 56, 201
LAMINA
  externa (of calvaria), 54, 74, 128
  horizontalis ossis palatini, 122
  interna (of calvaria), 54, 74, 128
  lateralis processus pterygoidei, 7–9, 42, 229
  medialis processus pterygoidei, 42
  orbitalis ossis ethmoidei, 109
Lantian cranium, 94, 242
Lapp crania, 97–8
lateral occipital, *see under* occipital bone
LENGTH
  basion–opisthion, 48, 199
  cerebellar, 92
  cerebello-temporal, 92
  cerebral, 92, 95, 97
  cranio-basal, *see* length, nasion–basion
  dental arch, 133–4
  facial, lateral, 107; superior, 107
  of foramen magnum, 60–1, 96, 202
  glabella–opisthocranion, 95, 206
  horizontal occipital, 45–6, 199
  maxillo-alveolar, 132–5, 213
  maximum arcadal, 133–4
  maximum cranial, external, 8, 95–8, 206; internal, 95, 97–8
  of nasal bones, 110, 208
  nasion–basion, 47, 96, 103, 116, 170, 199
  nasion–opisthion, 47–8, 96, 199
  palatal, 133–6
  prosthion–inion, 96
  torocristal, 97–9, 206
Léopoldville Congress, 1–2
ligament of apex of dens, 27
ligamentum nuchae, 45, 199
limb-bones, 76
limen coronale, 54
Limnopithecus macinnesi, 169
LINE
  basal, of face, 103
  cervical, 178–81, 183–6, 189
  nuchal, inferior, 26; superior, 21–6, 43, 194–5
  temporal, 19–21, 194; inferior, 21–5, 28, 104, 194–5
lingula, of greater wing, 61
Livingstone Congress, 236
lobe
  frontal, 88, 93, 101
  occipital, 55, 93, 201
  parietal, 91, 93, 205
  temporal, 63, 77, 93, 203
lobuli semilunares, 56
lobulus biventer, 57
lower scale
  of occipital, 13–15, 193
  preponderance, 14–15, 193
lumping (in classification), 219
Luschka, sinus of, *see under* sinus, venous

macrodontic hominids, 243
Makapansgat, 4, 225, 236–9, 242, 244
  crania, 9–11, 13–14, 18, 20–1, 23–4, 27, 29–35, 37–9, 43, 55–6, 61, 65, 79, 84, 96, 100–2, 109–11, 113, 117, 122, 130, 193–4, 196–7, 201, 206, 229
  teeth, 137, 146, 148, 151, 153–4, 156, 158, 163, 177–8, 184–5, 221, 227
mandibular
  head, 36, 197
  ramus, 120–1, 211
Maori crania, 97–8
MARGIN, MARGO
  frontal, of parietal bone, 7, 11, 193
  lambdoid, of parietal bone, 7, 10–13, 54, 99, 193, 201
  occipital, of parietal bone, 10–13, 193
  orbital, 104–6
  of piriform aperture, 111
  sagittal, of parietal bone, 193
  squamosal, of parietal bone, 7, 10–13, 54, 201
  superior, of petrosal part, 57, 201
  supra-orbital, 104–5
  temporal, of parietal bone, *see under* margin, squamosal
masseter, *see under* muscle
masseteric hypertrophy, 119
massiveness
  of cranium, 1, 204
  of face, 1, 103, 207
  of jaws, 130, 204
  of teeth, 130, 144, 204, 214
mastication, 36, 140–1, 211, 214, 227
masticatory apparatus, 123
mastoid
  notch, *see under* incisura
  part of temporal bone, 7, 22, 26, 61, 72, 99, 130, 202
  process, 22, 26, 28–33, 72, 99, 130, 195–6, 206, 223
mastoideale, 96
Mauer mandible, 120–1, 147, 242
MAXILLA, 113–18, 209, 223
  alveolar process, 6, 113, 118, 211
  body of, 113, 116, 210
  dimensions, 107, 113–14, 209
  extension type, 118–19, 210
  frontal process of, 108, 111, 113, 116
  inflexion type, 118–19, 210
  infratemporal surface, 6, 8
  nasal crest, 6
  nasal surface, 7
  orbital surface, 7
  palatine process of, 6, 113, 122
  subnasal part, 6, 209
  tuberosities, 6, 8, 42, 113, 126–7, 132, 134
  zygomatic process of, 6, 113, 116–19, 121, 210
maxillary
  antrum, *see under* sinus
  sinus, *see under* sinus
maxillo
  –alveolar dimensions, 132–5, 213; index, 132, 213
  –facial fragment, 6
  –frontale, 106–8, 208

meatus acusticus
  externus, 28, 40–1, 61, 195, 223
  internus, 57
medulla
  oblongata, 88
  spinalis, 88
megadontic hominids, 243
Meganthropus, 4, 219, 242
  africanus, *see* Garusi
  palaeojavanicus, 159, 243
Melanesian
  crania, 24
  teeth, 170
meningeal vascular markings, 93–4, 205–6
mesiodistal diameter, 3, 144–5, 151, 153–67, 170–1, 214–16, 221
microcephalic cranium, 24
middle ear, 61, 202
migration of muscles, 15, 74–5, 121–2, 211
Miocene hominoids, 169, 243
modelling of cranium, 52
MODERN MAN
  crania, 11–18, 24, 27, 29–32, 35, 37–9, 41, 46–9, 54, 56–64, 66, 68, 70, 72–4, 77–8, 85, 93, 97, 101–3, 106–8, 111, 115, 123, 125, 127, 130, 195–7, 201–3, 206, 219; of Africa, 16; of Europe, 16, 73–4, 111
  teeth, 135–6, 143
module of teeth, 145, 147–59, 163–5, 167–8, 184, 187, 214–16
MOLARS
  lower, 220–1, 226
  upper, 140, 142, 144, 162–9, 172–4, 182–92, 215–16, 218, 220–1; attrition, 140, 186; cusp pattern, 182–3, 185, 187–8, 190, 218; first, 162–5, 182–4, 218; fissures, 182–3, 185–7; hypoplasia, 142, 188, 214; metrical characters, 144, 147–51, 154–5, 157–8, 160–9, 184, 186–7, 190, 215–16, 218, 226; morphology, 182–90, 218; ratio to dental chord, 165–6, 216; reduction of M3, 166–9, 190, 216, 218; relative size, 1, 162, 166–9, 171, 190, 216, 218; roots, 184, 186, 188–90, 218; second, 163–5, 185–7, 218; shape, 144, 147–50, 172–4, 184, 186–7, 216–17; third, 164–6, 187–90, 218
Monaco Congress, 132–3
Monte Circeo cranium, 118
Montmaurin mandible, 147, 242
Moravian crania, 67
morphogenesis of calvaria, 89; *and see under* ontogeny
Moscow Congress, 2
mouth, roof of, 126
MUSCLE, MUSCULUS
  longus capitis, 27, 59
  massetericus, 119–22, 210–11, 229
  obliquus capitis superior, 26, 30
  occipitofrontalis, 20
  pterygoideus, lateralis, 42, 229; medialis, 42, 122, 229
  rectus capitis, anterior, 27; lateralis, 27; posterior major, 26; minor, 26
  semispinalis capitis, 26

INDEX OF SUBJECTS

Muscle, musculus (cont.)
 splenius capitis, 26, 45
 temporalis, 19, 22–5, 74–5, 105–6, 108, 122, 125, 140–1, 194, 204, 207–8, 214, 229–31
 trapezius, 45, 199
Muscles
 extensor, 45, 51–2
 masticatory, 121–2, 130, 210, 231; adjustments, 121–2, 211
 migration of, 15, 74–5, 121–2, 211
 nuchal, 15, 17, 22–3, 25, 43, 45, 51–2, 130, 193, 195, 198
muscularity of *Zinjanthropus*, 23, 44–5
Musée de l'Homme (Paris), 3
Mustela furo, 75

Nasal
 bone, 7, 108–11, 208
 breadth, 109–10, 208
 bridge, 110
 cavity, 112–13
 dimensions, 109–11, 208
 floor, 111–15, 126, 208
 height, 110, 208
 index, 110–11, 208
 region, 115
 roof, 111, 208
 spine, anterior, 112–14, 208, 223; posterior, 122
Nasion, 45–6, 96, 103, 107, 109, 115–16, 124, 127, 208–9
 –alveolare height, 107, 124
 –basion distance, 48, 199
 –nasospinale line, 110, 115
 –opisthion base-line, 45–6, 103, 199–200, 207; distance, 47–8, 96, 199
 –prosthion line, 115
nasospinale, 107, 114–15
 –alveolare line, 107, 114
 –prosthion line, 107, 114–15
National Geographic Society, xvi
National Museum of Kenya (Nairobi), 3, 58, 66–7, 71, 120
Natron, see Peninj
Neandertal calotte, 68, 98
Neandertaloid crania, 29, 46, 48–9, 60, 64, 73–4, 82, 93, 95, 97–8, 101–3, 111, 115, 118, 124, 210, 219
 of Africa, 16, 98, 206
 of Europe, 16, 97–8, 206
near-men, 4
Negro, Negroid crania, 34, 55
Nerve(s)
 cranial, 78
 infra-orbital, 117–18, 126
 palatine, anterior, 126; greater, 122
 superior alveolar, 127
 trigeminal, 62, 202
neurocranium, 101
neurones
 cortical, 86–7, 205
 excess, 86–7, 205, 238
New Caledonian crania, 35–6
New World monkeys, 188, 220
New York Congress, 3
Ngandong crania, 48, 82, 83; see also Solo man

nomenclature
 anatomical, 3
 systematic, 4
Nomina Anatomica, 3
Norma
 basalis, 88–90
 dorsalis (of endocast), 90–1
 facialis, 105, 112
 lateralis (of cranium), 9; (of endocast), 89, 91
 occipitalis, 20
 ventralis (of endocast), 88–90
 verticalis, 10, 115, 193
Norwegian crania, 97–8
nose, 109–13, 208–9
Notch
 frontotemporal, of brain, 91
 jugular, 64–5
 mastoid, 29–30, 195
 supra-orbital, 105
 Sylvian, see notch, frontotemporal
nuchal
 area height index, 17–18, 22, 43–4, 198, 223
 crest, see under crest
 muscles, 15, 17, 22–3, 25, 43, 45, 51–2, 130, 193, 195, 198

Obercassel crania, 68
Occipital
 bone, 14–16, 26–7; pars basilaris, 7, 58–60, 129, 202, 212; pars condylaris, see pars lateralis; pars lateralis, 7, 26–7, 59, 64, 66; pars superior, 66
 cartilaginous, 14–15
 condyles, 27, 36, 45, 49–52, 59, 195, 200, 223
 curvature, 13, 15–16, 193–4
 index, 13, 15–16, 193–4
 membranous, 14–15
 metrical characters, 13–16, 193
 preponderance, 14, 193
 sagittal curvature, 13, 15–16, 193; dimensions, 13–16, 193
 squama, 14–16, 21, 54–6, 195
occipitoparietal
 arc index, 14, 193
 parity, 14
occipitosphenoid fragment, 7
occlusal plane, see under plane
occlusion, 36, 140–1, 213–14
 centric, 140–1, 214
odontofacial adjustment, 117
Olduvai, 1, 2, 6, 146, 231–3, 236–9, 242
 hominid 9, 4, 60, 73, 81, 83, 94, 98
Old World monkeys, 220
omnivorous hominids, 225
ontogeny
 of cranium, 28, 89, 130–1, 231
 of temporal crests, 24–5
 of venous sinuses, 70–1
opisthion, 45–7, 50, 55–6, 64, 96, 199, 201
opisthion–internal occipital protuberance distance, 56, 201
opisthocranion, 8, 44–7, 50, 95
orale, 133–6
orang-utan, see Pongo
orbit, 106–9, 128, 207, 230

orbital
 indices, 107, 207–8
 margin, 105, 108, 113, 126, 209
 measurements, 106–8
orbitale, 17
orthognathism, 34, 138, 196
os coxae, 231
ossicle, sutural (Wormian), 21
ossific centres of tympanic bone, 31–2, 196
ossification of crest, 75

palatal
 dimensions, 132–3, 211, 213
 index, 133, 213
palate, 122–3, 126, 132, 211, 213, 229
 depth, 122, 211
 shelving of, 122, 211
palatine bone, 42, 122, 198
palatometer, 122
Palestinian Neandertal crania, 97
Pan
 crania, 12, 14, 17–21, 26, 28, 30, 33–6, 38–9, 41, 44, 49–51, 54, 58–61, 63, 67, 71, 73, 77, 79, 81–5, 87–90, 94, 106, 112–13, 117–18, 120, 126, 195–6, 202, 205, 208, 211, 219, 242
 teeth, 135–6, 142, 153–4, 158–9, 161–3, 168–70, 172–3, 192, 214–16
 paniscus, 44, 49, 58, 60, 67
Papio, 142
paranasal sinuses, see sinus, air
Paranthropus, 1, 2, 219–20, 224, 231, 241;
 cranium, 9–10, 16, 21, 27–31, 33, 35, 38, 40–1, 43–4, 49, 54, 61, 63–6, 70, 72–3, 78–80, 83, 95–6, 104–18, 120–2, 126, 128, 130, 193–5, 197, 201–3, 206–13, 228–30; teeth, 135–40, 142, 144–6, 162–9, 172–87, 189–92, 213–18, 225–8
 boisei, 2, 224, 233; see also *Australopithecus boisei*, *Zinjanthropus*
 crassidens, 109, 214, 224, 229; see also *A. robustus crassidens*
 robustus, 33, 65, 94; see also *A. robustus*
Parietal
 bone, 10–14, 204; dimensions, 10–13, 193; endocranial surface, 54, 200–1; supra-asterionic part, 66
 lobe expansion, 91, 205–6
 preponderance, 13–14, 193
 sagittal curvature, 10–13, 193; dimensions, 10–13, 193
Parieto-occipital
 area, 20
 contour, 9
 equality, 14
 plane, 24, 193
 ratio, 12–14
parietotemporal walls, 9, 193
Paris Nomina Anatomica, 3
Pars
 basilaris, see under occipital bone
 lateralis, see under occipital bone
 mastoidea, see under mastoid
 nasalis, see under frontal bone
 orbitalis, see under frontal bone
 petrosa, see under temporal bone
 squamosa, see under temporal bone
 tympanica, see under temporal bone

INDEX OF SUBJECTS

pattern of cresting, 21, 23–5, 194–5
Pearson's occipital index, 15
pebble tools, 236
pedomorphism, 32, 70
Pekin man, *see Homo erectus pekinensis*
Peninj, 146, 182, 220–1, 237, 242
perikymata, 141–2, 188
periosteum, 75, 121
Petralona cranium, 98
petro-median angle, *see under* angle
petro-occipital fissure, gap, 58, 202
petrosquamous sinus (of Luschka), *see under* sinus, venous
petrous
    part, 7, 33–5, 54, 57–8, 62, 71
    pyramid, 33–4, 55, 57, 60, 63, 130, 201, 223
    shelf, 58, 202
phyletic
    sequence, 240, 242–3
    valence, 221–2
Piltdown cranium, 67
piriform aperture, 109–14, 208–9
Pithecanthropus, 4; *and see Homo erectus modjokertensis*, 4; *see also H. erectus pekinensis*, 82; *see also H. erectus robustus*, 4; *see also H. erectus soloensis*, 82; *and see* Solo man, Ngandong
PLANE
    biporial transverse, 50, 63–4, 223
    occlusal, 121, 133, 141, 211
    orbito-occipital, 91
    Sylvian vertical, 91
    vestibular, 89
PLANUM
    nuchale, 14, 23, 26–7, 43–5, 194, 198; tilt of, 23, 43, 198
    occipitale, 14–15, 22–3, 193–4
    praeglenoidale, 37, 197
    vermiforme, 59
plate, tympanic, *see under* temporal bone
Platyrrhini, 128
Pleistocene, 241, 243
    Lower, 76, 237, 240–4
    Middle, 76, 240–4
Plesianthropus, 4, 92, 120–1, 219, 225; *and see Australopithecus africanus*
plexus, tentorial, 70
Pliocene, 243
    hominoids, 169
PNEUMATISATION, 126–31, 212–13
    of alisphenoid, 129–30, 212
    of basi-occipital, 59, 129–30
    of basis cranii, 129–30, 212
    of cranium, 6, 41, 72–4, 126–31, 212–13
    of crista galli, 128
    development of, 127–8, 130–1
    of frontal bone, 127–8, 212
    of glabellar region, 110
    interorbital, 108–9
    of lacrimal bone, 127–8
    of lacrimo-ethmoidal region, 128
    of maxilla, 126–8, 212
    of middle nasal concha, 128
    of nasal bone, 128
    of naso-orbital region, 128–9, 212
    of palatine bone, 127

    of perpendicular plate, 128
    of pterygoid process, 127, 129–30, 212
    of sphenoid bone, 129–30
    of sphenoidal rostrum, 129, 212
    of temporal bone, 28, 40, 99–100, 130, 206, 212–13
    of zygomatic bone, 126–7, 212
poise of head, 45, 49–52, 199–200
pole
    frontal, 77
    occipital, 55–6, 67, 88, 93, 201, 205
    temporal, 77, 91
Pongidae, 4, 78, 91, 220, 239–40
    crania, 3, 12, 33, 35–7, 43, 46–9, 53, 59–60, 63, 73–4, 76, 80, 91, 106, 108–9, 112, 116–17, 124, 127, 129, 193, 195–6, 198–200, 202–3, 205, 207–9, 212, 230
    teeth, 132, 136, 152, 161, 168–9, 172, 191, 213–14, 216
Pongo
    crania, 12, 14, 16–18, 30, 33–6, 38–9, 44, 50–1, 54, 58, 60–1, 63, 77–9, 81–3, 85, 88, 94–5, 112–13, 118, 126, 128, 195, 197, 205, 212, 219, 230
    teeth, 135–6, 154, 158–9, 161–3, 168–70, 172–3, 188, 215–16, 218
porcupine, 238
porion, 17, 48–9, 103, 207
position height index, 48–9, 103, 200, 207, 223
position length index, 48, 199
Port Elizabeth Museum, 3
porus acusticus externus, 30–3, 49, 196, 223
postcondylar segment of cranium, 49
postcranial skeleton, 231
postorbital constriction, 7, 10, 39, 53, 100–1, 105, 128
potassium–argon date, 233
Powell-Cotton collection, 3, 59, 66, 120, 126–7
Praeanthropus, 219
precondylar segment of cranium, 49, 51, 200
Předmostí crania, 67–8, 136
preglenoidal plane, *see under* planum
prelacteon, 137
prehominids, 4
PREMOLARS, 179–82, 215, 223, 226–7
    anterior (P^3), 157–9, 172–4, 179–81, 215–17
    attrition, 139
    buccolingual expansion, 117, 159, 162, 172–4, 179–80, 182, 215
    cusps, 179–82, 217–18
    fissures, 179–81
    heteromorphism, 181, 218
    hypoplasia, 142, 214
    metrical characters, 144, 147–51, 154–62, 180–2, 215
    morphology, 179–82, 217–18
    posterior (P^4), 157–9, 172–4, 181–2, 215–18
    ratio P^3/P^4, 159, 162, 182, 218
    ratio to dental chord, 157–8, 215
    roots, 180–2, 217–18
    shape-index, 144, 147–50, 172–4, 216
premolar-molar series, 132, 140, 155, 170, 213

preservation of cranium, 6
Primates, higher, 77, 220
PROCESS, PROCESSUS
    alveolar, of maxilla, 113–15, 118, 137–8, 227
    angular, external, 106
    clinoid, anterior, 58; posterior, 58–9, 202
    condylar, of mandible, 36, 39, 197
    ectobasionic, 27, 59
    entoglenoid, 37–41, 62, 197–8
    frontal, of maxilla, 108, 111, 113, 116
    frontal, of zygomatic bone, 106, 108, 123, 208, 212
    jugular, of occipital bone, 26–7, 64–5, 67
    marginal, of zygomatic bone, 108, 123
    mastoid, 22, 26, 28–33
    palatine, of maxilla, 113
    paramastoid, 27
    petrosal, of sphenoid bone, 58, 62
    postglenoid, 31, 40–2, 198, 223; subdivision of, 41–2, 198
    postmarginal, 108, 123, 208
    pterygoid, *see under* sphenoid bone
    pyramidal, of palatine bone, 42, 198
    spinous, of maxilla, 112
    styloid, 33, 196
    temporal, of zygomatic bone, 8, 119
    vaginal, of styloid process, 33, 196
    zygomatic, of maxilla, 6, 113, 116–19, 121; of temporal bone, 7–8
Proconsul, 169, 216
prodontism, *see* prognathism, dental
PROGNATHISM, 34, 45, 114–16, 138, 196, 209–10
    alveolar, 115, 209
    dental, 114, 138
    maxillary, 114–16, 209
    subnasal, 114–15, 209, 223
promontory of middle ear, 61
Prosimii, 220
prosphenion, 46–7, 55, 199
prosthion, 45, 96, 114, 132, 134, 136
protuberance, occipital
    external, 20, 22, 24, 44–5, 95, 194, 199, 223
    internal, 55–6, 64–5, 70, 96, 201
pterion, 63, 93, 203, 212
pterygoid
    muscles, *see under* muscles
    process, *see under* sphenoid bone
pygmy chimpanzee, *see Pan paniscus*
pyramid, *see* petrous part

Rabat teeth, 148, 152, 154, 158, 163, 172–3, 216, 242
Ramapithecus punjabicus, 169, 216, 243
rameal breadth, *see under* breadth
ramus, mandibular, 120–1, 211, 229
RECESS, RECESSUS
    alveolar, 126
    anterior, 127
    epitympanic, 61
    lacrimal, 127
    olfactory, 53, 200
    palatine, 126
    posterior, of mandibular fossa, 41
    pterygoid, 127
    zygomatic, 126
reconstruction of cranium, xv, 6

INDEX OF SUBJECTS

reinforcing system of cranium
  sagittal, 53, 200
  transverse, 100
Rhesus monkey, 75, 122
rhinion, 111, 208
Rhodesian man, *see* Broken Hill cranium
ridge
  marginal, 175, 178
  median, of endocranial cast, 64
Rijksmuseum (Leiden), 3, 94
Rijksuniversiteit (Utrecht), 3, 94
rim, tympanic, 30–3
ring, childhood
  early, 143, 214
  later, 143, 214
robusticity of cranium, 72
rostrum sphenoidale, 7, 42, 77, 198

Saccopastore cranium, 118
sagittal margin of parietal bone, *see under* margin
Saldanha cranium, *see* Hopefield
Sangiran
  crania, 81, 84, 94
  teeth, 152, 154, 158, 163, 167, 172–4, 216
segment of sagittal crest
  anterior, 19
  intermediate, 19
  posterior, 19–21
sella turcica, 61, 129, 202, 223
sellion, 109, 208
sexual dimorphism
  in body size, 85
  in cranial capacity, 85, 205
siamang, 81, 84, 85
Sidi Abderrahman, 242
simple nuchal crest, 22–5, 194
Sinanthropus, 4; and see *H. erectus pekinensis*
'*Sinanthropus officinalis*', 163, 172
SINUS, AIR
  ethmoidal, 128–9, 212
  frontal, 127–9, 131, 212
  maxillary, 6, 126–7, 129, 131, 212; recesses of, 126–7
  paranasal, 126, 130
  sphenoidal, 42, 59, 60, 127, 129–30, 198, 202, 212
SINUS, VENOUS, 63–71, 203
  cavernous, 62
  development, 70–1
  horizontal, 67
  inferior, 67
  inferior longitudinal, 67
  lateral, 66–70, 203
  marginal, 55–7, 64–70, 88, 201, 203, 205
  occipital, 55–6, 64–70, 88, 201, 203, 205
  pattern, 65, 68–71, 203; foetal, 68
  petrosal, superior, 64–5, 68, 203
  petrosquamous (of Luschka), 42, 64–5, 203
  roentgenologic appearance, 66
  sigmoid, 27, 42, 54, 57, 64–71, 203
  straight, 65, 67, 69–70, 203
  superior sagittal, 63–5, 67–70, 203
  tentorial, 70
  transverse, 42, 54, 64–70, 203

Sivapithecus, see under Dryopithecus
Skhūl crania, 16, 124, 135–6
skull, 3
Solo man, 14, 16, 60, 193
South African Bantu, *see* Bantu
specialisation, 168–9, 182, 216, 241
Spee, curve of, *see under* curve
SPHENOID BONE, 42, 198
  basisphenoid, *see* body
  body, 7, 42, 47, 55, 58–9, 61, 77, 198
  greater wing, 7, 8, 37–40, 42, 55, 61–2
  lesser wing, 7, 54–5
  pterygoid process, 7, 8, 42, 60, 127, 198
  rostrum, 7, 42, 77, 198
sphenoidale, 55
spheno–occipital
  index, *see under* index
  synchondrosis, 27, 42, 58, 96, 129, 202
SPINA, SPINE
  nasal, anterior, 112–14, 208, 223; posterior, 122
  of petrosal crest, 33, 196
  sphenoid, 38–9
  suprameatal, 28
splitting (in classification), 219
Springbok Flats cranium, 120
squama, squame
  occipital, 14–16, 21, 54–6
  temporal, 37–40, 54
standard deviations of mean
  cranial capacity, 83–4
  tooth size, 147–50, 162
staphylion, 133–6
Steinheim cranium, 242
Sterkfontein, 4, 225, 233, 236–7, 239, 242, 244
  crania, 6, 9–14, 16, 18–19, 26–8, 32–4, 37–8, 41–3, 47–50, 55, 58–65, 72, 78–80, 87–9, 92, 94–118, 120–5, 128–30, 132, 197–9, 200, 203–7, 230
  teeth, 132, 135, 139, 142, 146, 148–9, 153–4, 156, 158, 162–3, 165–6, 173, 176–80, 182, 184–5, 188–9, 192, 217, 220–1, 227
SULCUS
  canine, 116–17, 210
  carotid, 61
  cerebral, frontal, middle, 54, 87, 205; superior, 54, 87, 205
  cerebral, lunate, 91
  digastric, 29, and see incisura mastoidea
  interlaminar, 20
  intermaxillary, 122
  maxillary, 117–18
  for middle meningeal artery, 61–2, 201
  for occipital artery, 29
  postglenoid, 41–2
  prenasal, *see under* fossa
  sigmoid, 57, 64–7
  subnasal, *see under* fossa
  superior petrosal, 57, 64–5
  superior sagittal, 54, 63–5, 67
  supramastoid, 22, 28, 195
  supratoral, of frontal bone, 104; of occipital bone, 15, 193
  transverse, 42, 54, 64–7
superstructures, ectocranial, 14, 29, 72–4, 76, 104, 195, 204

supra-occipital, *see under* occipital bone
supra-orbital height index, 16–19, 194
supra-orbital torus, *see under* torus
SUTURA, SUTURE
  coronal, 7
  frontomaxillary, 106
  frontonasal, 111
  frontozygomatic, 105, 108, 123, 128, 207
  intermaxillary, 8
  internasal, 8, 109, 111, 208
  lambdoid, 7, 21–2, 55
  median palatine, 122
  occipitomastoid, 22, 26, 29, 72, 194–5
  occipitotemporal, 21
  palatomaxillary, 113, 122, 211
  parietomastoid, 21, 63, 72
  premaxillary/maxillary, 122
  sagittal, 7, 20, 54, 66, 75, 104
  spheno-ethmoid, 55
  sphenosquamosal, 37–40, 55, 62, 197–8, 202
  squamosal, 54, 63, 130, 203
  zygomaticofrontal, 105, 108, 208
  zygomaticomaxillary, 108, 119–20, 123, 211
  zygomaticotemporal, 6–8, 119, 122, 211
Swanscombe calvaria, 29, 59–60, 67, 73–4, 195, 242
Swartkrans, 4, 220, 225, 233, 237, 239–40, 242
  crania, 10, 18–19, 23–4, 28–33, 35, 37, 40–1, 60–1, 63, 66, 70, 79, 95–6, 104–15, 117–18, 120, 122–6, 130, 194, 203, 206, 229
  teeth, 135, 137–9, 142, 146–7, 152–4, 156, 158, 162–3, 165–6, 168, 173–8, 180–4, 189, 214, 227
Sylvian crest, *see* crista Sylvii
synchondrosis, *see under* spheno-occipital

tabula
  externa, *see under* lamina
  interna, *see under* lamina
Tabūn crania, 16, 98, 124
Talgai cranium, 136
Tasmanian crania, 136, 170
Taung, 3, 4, 6, 45, 49, 62–3, 65, 78–80, 83, 87, 89, 91–2, 106, 109, 111, 146, 163, 184, 203–4, 221, 225, 229, 233, 237, 239, 242
tegmen tympani, 61, 202
Telanthropus capensis, 4, 114, 120, 147, 153, 219, 239, 242
TEMPORAL BONE, 7, 28–42
  mastoid portion, 7, 22, 26, 61, 130, 195, 202
  pars petrosa, 7, 33–5, 54, 60, 71, 196, 201
  pars squamosa, 7, 37–40, 54, 60, 62–3, 130, 197, 203
  pars tympanica, 30–3, 37, 40–1, 195–6, 221, 223
  squame, squama, *see* pars squamosa
  zygomatic process, 7–8, 22, 30, 41, 120–2, 130, 195, 211
tentorium cerebelli, 67, 70
Ternifine
  parietal, 10–11, 242
  teeth, 148, 154, 163, 172, 174

INDEX OF SUBJECTS

THICKNESS (of cranial bones), 72–6, 203–4
  near asterion, 54, 72–3, 203
  near bregma, 74
  near lambda, 73
  of occipitomastoid suture, 72, 203–4
  of parietal bone, 73–6, 204
  of parietomastoid suture, 72, 203
  at tuber parietale, 73–4
tibia from Bed I, 6
tools
  dependence on, 240
  making of, 238–40
  using of, 238–40
tooth
  material, 170–1, 216
  rows, divergence of, 132, 136–7, 213
torcular Herophili, *see* confluence of sinuses
torocristal
  index, 97–9, 206
  length, 97–9, 206
TORUS
  angularis, 54, 73, 99, 201
  frontal, *see* torus, supra-orbital
  mandibular, 123
  maxillary, 122, 211
  median palatine, 122–3, 211
  occipital, 14–15, 95, 193, 204
  parietal, 123
  sagittal, 53, 204
  supra-orbital, 19, 48, 95, 104–6, 124, 206–7, 223, 230–1
    glabellar part, 105, 207
    sagittal length, 104
    superciliary part, 105–6, 207, 212
    thickness, 104, 207
    width, 104
Transvaal Museum (Pretoria), 1–4, 21, 66, 137, 146
transverse frontoparietal index, *see under* index
trigonum frontale, 104–5, 128, 207
Trinil calvaria, 81, 94, 242
trochlea of frontal bone, 104
tuba infra-orbitalis, 126
TUBER, TUBEROSITY
  frontal, 104
  malar, 123
  maxillary, 6, 8, 42, 113, 126–7, 132, 134
  parietal, 73–5
TUBERCLE, TUBERCULUM
  articular, 7, 37, 39, 197, 223
  dentale, 175, 177; *and see* eminence, gingival
  jugular, 57–8, 202
  lingual, 175, 177; *and see* eminence, gingival
  molar (of Zuckerkandl), 184
  paramolar, 183–4
  pharyngeal, 27
  postglenoid, *see under* process
  sellae, 55
  zygomatic, 120
Tuinplaa(t)s cranium, 120
tympanic plate, *see under* temporal bone

Uganda, 105–6
ultraprognathism, 115–16
ungulate humeri, 239
upper scale
  of occipital bone, 13–15, 193
  preponderance, 15, 193
Upper Palaeolithic
  brains, 93
  crania of Europe, 16

VARIABILITY, 219–21
  in cranial capacity, 80, 84–6, 219
  of dental arcade, 138
  latent, 123
  of living hominids, 219–20
  primate, 219–20
  of Tanzanian australopithecines, 2
  of teeth, 220–1, 228–9
  of torocristal index, 99
vascular patterns, 93–4
vascularisation of occipital pole, 94, 206
vault of cranium, 9–25, 193
vegetarian hominids, 225
VEIN
  deep temporal, 42
  emissary, 27, 42; of foramen Vesalii, 62; of postglenoid foramen, 42; of squamosal, 42
  jugular, external, 42; internal, 68, 70; system, 42
vessels, blood, 78
  greater palatine, 122
  infra-orbital, 117–18, 126
vestibular
  axes of cranium, 14
  plane of Delattre, 89
Villafranchian, 233
vomer, 7, 34, 42, 55, 129, 198, 212
  alae, 7, 129, 212

Wenner–Gren Foundation, xvi, 2, 224
West African crania, 16–18, 44, 51
WIDTH
  bicerebellar, of endocast, 92
  biparietal, 92
  bitemporal, 92
  nasal, 110
  orbital, 106–7
Witwatersrand University, xvi, 66, 69, 77, 105, 120
Würmian Neandertalians, 118

Zinjanthropus as subgenus, 1–2, 219, 223–4
  boisei, 1, 223–4; *see also Australopithecus boisei*
Zulu, teeth, 162
zygoma, *see* zygomatic process of temporal bone
ZYGOMATIC
  arch, 8, 119, 121, 124–5, 210, 212, 229
  bone, 6, 108, 119–20, 209, 211–12; body, 7, 108, 121, 123, 211; facies malaris, 7, 123, 211–12; frontal process, 7, 106, 108, 123, 208, 212; marginal process, 108, 123; orbital process, 7; post-marginal process, 108, 123, 208; temporal process, 8, 119, 123; surface, 8
  process, of frontal bone, 7; of maxilla, 6, 113, 116–19, 121, 210; of temporal bone, 6–7, 28, 37, 41, 120, 122, 130, 195
zygomaxillare, 107, 113

For EU product safety concerns, contact us at Calle de José Abascal, 56–1°,
28003 Madrid, Spain or eugpsr@cambridge.org.

www.ingramcontent.com/pod-product-compliance
Ingram Content Group UK Ltd.
Pitfield, Milton Keynes, MK11 3LW, UK
UKHW030904150625
459647UK00025B/2886